Biomes and Ecosystems

Biomes and Ecosystems

EDITOR
Robert Warren Howarth
Cornell University

ASSOCIATE EDITOR
Jacqueline E. Mohan
University of Georgia, Athens

Volume 4
Articles

Qilian Mountains Subalpine Meadows – Zapata Swamp

SALEM PRESS
A Division of EBSCO Publishing
Ipswich, Massachusetts

GREY HOUSE PUBLISHING

Cover photo: Rock Islands, Koror, Palau, Pacific. © Casey Mahaney

Biomes and Ecosystems, 2013, published by Grey House Publishing, Inc., Amenia, NY, under exclusive license from EBSCO Publishing, Inc.

The paper used in these volumes conforms to the American National Standard for Permanence of Paper for Printed Library Materials, X39.48-1992 (R1997).

LIBRARY OF CONGRESS CATALOGING-IN-PUBLICATION DATA

Biomes and ecosystems / Robert Warren Howarth, general editor ; Jacqueline E. Mohan, associate editor.
volumes cm.
Includes bibliographical references and index.
ISBN 978-1-4298-3813-9 (set) -- ISBN 978-1-4298-3814-6 (volume 1) -- ISBN 978-1-4298-3815-3 (volume 2) -- ISBN 978-1-4298-3816-0 (volume 3) -- ISBN 978-1-4298-3817-7 (volume 4) 1. Biotic communities. 2. Ecology. 3. Ecosystem health. I. Howarth, Robert Warren. II. Mohan, Jacqueline Eugenia
QH541.15.B56B64 2013
577.8'2--dc23

2013002800

ebook ISBN: 978-1-4298-3818-4

First Printing

PRINTED IN THE UNITED STATES OF AMERICA

Produced by Golson Media

Contents

Volume 1

Qilian Mountains Subalpine Meadows

Category: Grassland, Tundra, and Human Biomes.
Geographic Location: Asia.
Summary: The alpine meadows of the Qilian Mountains once enjoyed rich biodiversity but now are under increasing pressure from overgrazing and unsustainable use of resources.

The Qilian Mountains are situated at the northeast of the Qinghai-Tibet Plateau, forming the border between Qinghai and Gansu provinces in China. The range is comprised of several parallel ranges that run northwest to southeast, and is bordered by desert areas: the Alashan Plateau to the north and the Qaidam Basin to the south. The range has a temperate continental mountainous climate with cold winters and warm summers. Seasonal rains peak during the summer season.

The mountains are comprised of rocky scree slopes and glaciated peaks. Alpine meadows lie below 10,827 feet (3,300 meters), deciduous shrub is found from 10,827 to 14,764 feet (3,300 to 4,500 meters), and cushion plants grow sparsely above 14,764 feet (4,500 meters). The north-facing slopes are typically wetter and therefore enjoy greater biodiversity, including coniferous forest ecosystems. South-facing slopes are drier, and support alpine meadow and shrub. The Qilian Shan Nature Reserve, on the northern side of the range, includes several habitat types (forest, alpine meadow, and shrub). Inhabitants of the range include Tibetans, Mongolians, Hans, and other ethnic groups.

As natural pastures, the Qilian alpine meadows help store carbon dioxide, but as these pastures become denuded of grass or come under the plow, they are no longer able to store as much carbon, thus degrading their ability to offer the ecological service of slowing greenhouse gas output and global warming. Additionally, alpine rangelands are important because of their high biodiversity and their role in preventing soil erosion.

Seasonal melting of the glaciers in the Qilian Mountains has been a crucial source of water in this dry region. Studies since the 1970s have shown that the glaciers in the Qilian Mountains have been retreating. Natural vegetation once in abundance here helped to conserve the waters and prevented flooding. As the grasslands became denuded, an increased risk of flooding and erosion followed. The streams and rivers that originate in

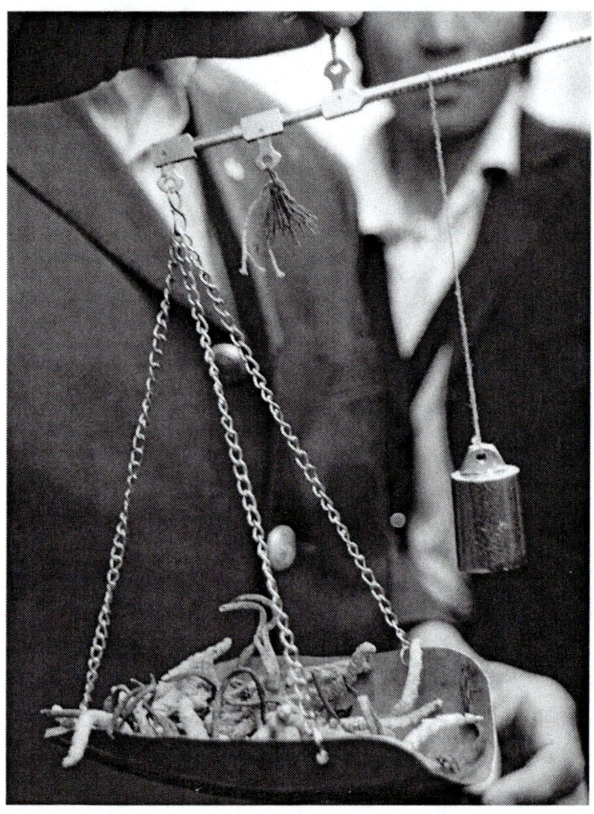

*Weighing caterpillar fungus (*Ophiocordyceps sinensis*) in a Chinese market in 2009. The fungus is highly valued in Chinese medicine; this has led to unsustainable harvesting practices. (Wikimedia/Mario Biondi)*

the Qilian Mountains provide irrigation for agriculture, otherwise impossible in the dry Hexi Corridor in Gansu province.

The meadows of the Qilian Mountains are heavily overgrazed, primarily by yaks and Tibetan sheep, leading to fewer grasses and higher proportions of unpalatable and poisonous plants. Even though sheep and yaks have similar forage demands, sheep herds are typically three times larger than yak herds. Traditionally, herders in this region practiced pastoral nomadism, allowing their animals to move from one area to the next. Herd sizes were limited by what the pastures could sustain. This practice prevented the meadows from becoming overgrazed.

In the 1950s, some of the residents of the Qilian Mountains were organized in brigades, which engaged in animal husbandry as well as agriculture, cultivating oats, barley, and rapeseed or canola. In recent decades, the meadows at lower elevations in the Qilian range have begun to be fenced off and allotted to individual families for grazing and cultivating mustard (for oil production). There have been reports of a decrease in grazing productivity, due to more families in the region owning more animals. With the advent of fenced pastures, animals are concentrated and contained, resulting in overgrazed pastures during the summer months.

Midsummer, many herders move their animals to pastures on the north side of the mountains, even though these pastures belong to other families. One positive side effect of enclosing pastures is a noted improvement in quality of pastures damaged by the zokor, a burrowing rodent similar to the mole. When pastures that were damaged by excessive zokor activity were fenced and reseeded, they recovered productivity within one season. In such enclosed pastures, increased vegetation coverage also leads to improved carbon storage.

Flora and Fauna

Several native plant species here are collected for medicinal uses. Of particular concern is the Chinese caterpillar fungus (*Ophiocordyceps sinensis*), which is found throughout the alpine meadows of the Tibetan plateau, including the Qilian Mountains. This parasitic fungus infects burrowed caterpillars, overtaking the caterpillar body until nothing remains but the caterpillar-shaped fungus. This fungus is highly prized in traditional Chinese medicine, and despite the ability to artificially cultivate this medicinal substance, the quest for financial gain has led to unsustainable harvesting of the caterpillar fungus.

Among the endangered animal species whose natural habitats include the Qilian Mountains are the snow leopard, Tibetan gazelle, ibex, white-lipped deer, argali, and wild yak. These animals are endangered as a result of habitat destruction and poaching. Once common across the Tibetan steppe, the Tibetan gazelle has been poached for its pelt as well as for its horns, which are valued in Chinese traditional medicine; the gazelle is now limited to small enclaves, including the Qilian Mountains.

Conservation Efforts

Many plant species native to the Qilian Mountains are rare or endangered as a result of habitat destruction and overcollection for the traditional medicinal market. Since the 1990s, scientists in China and abroad have been focusing their attention on the genetic and chemical properties of these medicinal herbs with the hope of cultivating them commercially. This would alleviate stress on the wild populations, and some cultivated specimens might be used to repopulate natural environments.

The Qilian Shan Nature Reserve, established in 1987, includes 10,243 square miles (26,530 square kilometers) of land on the northern side of the mountain range located in Gansu province. Though the reserve includes many habitats, in practice it is primarily a no-logging zone protecting the coniferous forests. Livestock graze in the reserve, with detrimental results for the grassland ecosystems.

KATHRYN OTTAWAY

Further Reading

Harris, Richard. *Wildlife Conservation in China: Preserving the Habitat of China's Wild West.* Armonk, NY: M. E. Sharpe, 2007.

Kurschner, Harald, Ulrike Herzschuh, and Dorothea Wagner. "Phytosociological Studies in the North-Eastern Tibetan Plateau (NW China)—A First Contribution to the Subalpine Scrub and Alpine Meadow Vegetation." *Botanische Jahrbucher* 126, no. 3 (2005).

Squires, Victor, et al. *Towards Sustainable Use of Rangelands in North-West China.* Dordrecht, Netherlands: Springer, 2010.

Queensland Tropical Rainforests

Category: Forest Biomes.
Geographic Location: Australia.
Summary: This unique landform contains a World Heritage site with primitive flora of great scientific value—but faces a set of grave environmental threats.

The Queensland Tropical Rainforests biome covers an area of approximately 12,626 square miles (32,700 square kilometers), and is subject to two distinct climatic seasons, commonly known as the wet and the dry. Queensland tropical rainforests occur along narrow coastal plains, foothills, mountain ridges, and tablelands from sea level to 2,953 feet (900 meters), with mountain peaks rising to 5,322 feet (1,622 meters). The now-rare lowland coastal rainforests, such as that at Cape Tribulations, extend to the sea edge, where they may even abut the salt-tolerant mangrove forests.

The rainforest forests are highly varied, with uneven canopies ranging from 98 to 131 feet (30 to 40 meters) in height. There are three sections: the first, in the northern reaches of the biome, features a wet tropics regime; the second section extends from the Whitsunday group to Carmilia; and the third section, which includes the Wargunburra Peninsula, extends inland to include portions of the Normanby Range. The Queensland tropical rainforests are designated one of the World Wildlife Fund's Global 200 ecoregions, and are the largest remnants of Australia's rainforest flora.

Studies have shown that tropical rainforests are remnants of the oldest types of vegetation in Australia. Many species have ancestors dating back to the Cretaceous or early Tertiary period, more than 65 million years ago. Furthermore, the wet tropics microclimate here provides an unparalleled living record of the ecological and evolutionary processes that shaped the flora and fauna of Australia over the past 400 million years. For this reason, Queensland's tropical rainforests have considerable historical and scientific importance.

The largest fragment—about 70 percent—of the tropical rainforest in Queensland (and Australia) occurs as a narrow strip along the east coast; it covers approximately 4.9 million acres (2 million hectares). Such is the biological significance of the region that a large part of this section was inscribed on the World Heritage List in 1988, as the Wet Tropics World Heritage Area. The Kulla (McIlwraith Range) National Park (jointly managed as Cape York

Peninsula Aboriginal Land) is a recent outcome of the Cape York Tenure Resolution Program; it now protects one of the largest pristine tracks of rainforest left in Australia.

Biodiversity

The wet tropics portion of these rainforests have more plant taxa with primitive characteristics than perhaps any other area on Earth. Of the 19 angiosperm families described as the most primitive, 12 occur here. Some 600 to 800 tree species overall occur in Queensland's tropical rainforests. The crowns of the largest forest trees interlock and form a closed canopy that shadows the middle canopy, subcanopy, and forest floor below. The upper canopy supports an aerial garden of climbing vines, as well as a full suite of epiphytes such as giant elkhorns, staghorns, bird's-nest ferns, orchids, mosses, and lichens.

Lowland rainforests are characterized by robust woody lianas, epiphytic ferns, fan palms, and strangler figs. The leaves of the lowland forests generally are larger than those of the uplands, and more of the plants tend to be deciduous. Upland forests characteristically have woody lianas and epiphytic ferns, but tree ferns, climbing vines, and mosses appear to be more abundant. Buff alders, silkwood, crowsfoot elm, milky pine, and black bean trees are all found here, as well as a variety of palms, vines, and orchids.

A butterfly resting on a fern in the Queensland rainforest. The forest contains two-thirds of Australia's butterfly species, half of the birds and one-third of the mammals. (Thinkstock)

The Wet Tropics zone reportedly contains the richest variety of animals and plants in Australia, including two-thirds of the butterfly species, half of the birds, and one-third of the mammals. A high proportion of fauna is endemic (found nowhere else) to the Wet Tropics bioregion, including 70 vertebrate species.

Mammals endemic to the Queensland rainforests include the Atherton antenchinus (a mouselike marsupial), green ring-tailed possum, mahogany glider, yellow-bellied glider, musky-rat kangaroo, and two species of tree kangaroo (Lumholtz's and Bennett's). Iconic rainforest birds include the golden bowerbird and cassowary. There also are many reptiles and amphibians, including the prickly rainforest skink, taipan, Boyd's forest dragon, chameleon gecko, and giant tree frog.

Environmental Threats

It is estimated that about 35 percent or about 4,053 square miles (10,497 square kilometers) of the pre-European extent of Queensland's rainforest has been cleared. Although vegetation-clearing started to decline when the Vegetation Management Act 1999 was introduced, broad-scale clearing of remnant vegetation did not cease until the end of 2006. The main threats to these forests today are the clearing of lowland forests for agricultural (mostly sugar cane) and residential development, leading to further fragmentation and degradation.

Other factors also place incredible pressure on the natural resources of this area and threaten its biodiversity, including stock grazing; mining; feral animals (such as pigs, dogs, cats, and cane toads); invasive plants (such as rubber vine and lantana); the plant pathogen *Phytophthora cinnamomi,* known to cause forest dieback; the chytrid fungus, a temperature-sensitive pathogen that kills frogs; cyclones; altered fire regimes; and global climate change, which is already exacerbating most if not all of these individual threats.

Petina L. Pert

Further Reading

Accad, A., et al. *Remnant Vegetation in Queensland. Analysis of Remnant Vegetation 1997–1999–2000–2001–2003–2005, Including Regional Ecosystem Information.* Brisbane, Australia: Queensland Herbarium, Environmental Protection Agency, 2008.

Keto, A. and K. Scott. *Tropical Rainforests of North Queensland: Their Conservation Significance: A Report to the Australian Heritage Commission by the Rainforest Conservation Society of Queensland.* Canberra: Australian Government Publishing Service, 1986.

Queensland Government. *Building Nature's Resilience. A Draft Biodiversity Strategy for Queensland.* Brisbane, Australia: State of Queensland Department of Environment and Resource Management, 2010.

Sattler, P. S. and R. D. Williams, eds. *The Conservation Status of Queensland Bioregional Ecosystems.* Brisbane: Australia Environmental Protection Agency, 1999.

Stanton, J. P. and D. Stanton. *Vegetation of the Wet Tropics of Queensland Bioregion.* Cairns, Australia: Wet Tropic Management Authority, 2005.

Stork, N. E., et al. "Australian Rainforests in a Global Context." In *Living in a Dynamic Tropical Forest Landscape,* edited by N. E. Stork and S. M. Turton. Oxford, UK: Blackwell Publishing, 2008.

Williams, S. E. and D. W. Hilbert. "Climate Change as a Threat to Biodiversity of Tropical Rainforests in Australia." In *Emerging Threats to Tropical Forests,* edited by W. F. Laurance and C. Peres. Chicago: University of Chicago Press, 2006.

Rajasthan Desert

Category: Desert Biomes.
Geographic Location: Asia.
Summary: A part of the Thar Desert, the Rajasthan hosts a remarkable suite of valuable flora, as well as some vulnerable fauna.

The Rajasthan Desert, a part of the Thar or Great Indian Desert, extends along the borderlands of Pakistan and India between the Indus River to its west, the Sutlej River to its northwest, the Aravali Mountains to its east, and the salty marshland of the Rann of Kutch to the south. The desert covers an area of more than 77,000 square miles (200,000 square kilometers) of mostly barren land, but also features the Luni River, which cuts through the desert on its way to the Arabian Sea.

Level to gently sloping plains are broken by dune fields and low, barren hills; these are interspersed with sandy depressions, and some river terraces and flood plains. A thick series of sedimentary rocks here comprise sandstone, limestone, and shales; the Rajasthan Desert is known as a source of marble and limestone that are widely used in India and exported to other countries as construction material. As a foundation for habitats, the region can be divided into major sand dunes, sandy plains, stony hill areas, generally textured and compact soil lands, and saline terrain.

The arid climate is harsh: Annual rainfall averages a scant 1 inch (25 millimeters), although it can range up to 4–20 inches (100–500 millimeters), because of the unpredictability of the monsoon effects across the region. Most of the rain here falls in the monsoon months, from June to September; winter rains are insignificant. Extremely high temperatures assure very high evaporation rates. Daytime temperatures reach a very hot 120 degrees F (49 degrees C)—and occasionally 124 degrees F (51 degrees C)—but also can dip to a low of 32 degrees F (0 degrees C) at night during December and January. The Rajasthan Desert is sometimes locally called *Marusthali,* meaning "the region of death."

The precious rains play a vital role in the life of all parts of the Rajasthan Desert; the water in many places deposits in tobas, or small ponds, that are the only source of surface water for animals and humans in most parts of the desert. It is for this reason that a major portion of the population lives as nomads. When a toba dries up, these pastoralists gather their tough grazing herds and seek out the next toba. Diversion of water from the Luni

River, along with well water, does support agriculture in some areas of the Rajasthan Desert. Chief crops are barja, or pearl millet; jowar, or coarse millet; moong, or pigeon pea; til, or lentil seeds; moth, a legume; and matira, a type of melon.

Biodiversity

Ecologically, most of the desert's vegetation is considered thorn forest or scrub forest. Although vegetation is sparsely distributed, there is a surprisingly large number of plant species with high economic and ecological value. In fact, important soil-binder species of grasses found in this region have been widely used by ecologists for the eco-restoration of degraded lands around the world. Xerophilous, or drought-adapted, grasses of the biome include *Aristida adscensionis, Lasiurus scindicus,* and *Cenchrus biflorus.* Among the small trees are *Acacia nilotica, Tamarix aphylla,* and *Prosopis cineraria.* Scrub species include *Calligonum polygonoides, Crotalaria* spp., and *Haloxylon recurvum*—all endemic to the Rajasthan Desert, that is, found nowhere else.

Species having commercial importance by yielding fibers for cottage industries or larger textile concerns include khimp (*Leptadenia pyrotechnica*), munja (*Saccharum bengalense*), mudar (*Calotropis procera*), and *Acacia jacquemontii.* Natural dyes are extracted from thumba or bitter apple (*Citrullus colocynthis*). Matira or watermelon (*C. lanatus*) yields non-edible oil for the soap industry.

Famine food plants providing grain are *Cenchrus biflorus, Panicum turgidum,* and *Panicum antidotale.* Plants of medicinal value include isabgol (*Plantago ovate*) and gugul (*Commiphora wightii*).

In the extreme west of the region, there is hardly any tree cover, except for the "king of desert trees," the khejri (*Prosopis cineraria*), which grows only near wells. It is only in this area where also are naturally found the Rajasthan state bird, the great Indian bustard (*Ardeotis nigriceps*); state animal, the chinkara or Indian gazelle (*Gazella bennettii*), and state flower, the rohida (*Tecomella undulata*).

The desert hosts some important reptile species, including the Indian spiny-tailed lizard (*Uromastyx hardwickii*), dwarf gecko (*Tropiocolotes persicus euphorbiacola*), Persian gecko (*Hemidactylus persicus*), desert monitor (*Vara-*

Shrubs growing amid sand dunes in the Rajasthan Desert. Despite the harsh conditions and frequent dry spells, the desert supports a wide range of plant species, including a number of important drought-adapted grasses that are now being used elsewhere in the world to address land degradation. (Wikimedia/Chinmayisk)

nus griseus), and saw-scaled viper (*Echis carinatus sochureki*).

Among the mammals found in the area are desert fox (*Vulpes vulpes*), Indian fox (*Vulpes bengalensis*), desert cat (*Felis silvestris*), hairy-footed gerbil (*Gerbillus gleadowi*), desert hare (*Lepus nigrricollis dayanus*), and long-eared hedgehog (*Hemiechinus auritus*). The human populations of the region have domesticated the camel (*Camelus dromedarius*), upon which they are widely dependent for transportation and other purposes.

Conservation and Threats

The Desert National Park (DNP) covers an area of 1,220 square miles (3,162 square kilometers), split between the Jaisalmer and Barmer Districts of Rajasthan State. More than 100 bird species found in the DNP include a good population of the great Indian bustard, which is locally called godawan. The park also provides vital support for the migratory Houbara bustard (*Chlamydotis undulata*).

In the Rajasthan Desert, persistent environmental concerns include drought, dust storms, searing temperatures, and the viability of the land to support fauna, flora, and human populations. During the extreme dry spells, herd animals are sold off, and soil is challenged to yield enough vegetation to support the food web.

Climate change is already being assigned blame for the western storms that push moisture-laden monsoon winds across the desert without rainfall; the net effect is to delay the full monsoons and the much-needed rain. Extended dry periods such as these would further strain the land's capacity to regenerate seasonal flora, and jeopardize habitats across the region.

Manoj Kumar
Monika Vashistha

Further Reading

Champion, H. G. and S. K. Seth. *A Revised Survey of the Forest Types of India.* Delhi, India: Manager of Publications, 1968.

Gupta, R. and I. Prakash. *Environmental Analysis of the Thar Desert.* Dehradun, India: English Book Depot, 1975.

Mathur, C. M. "Forest Types of Rajasthan." *Indian Forester* 86 (1960).

Tewari, D. N. *Desert Ecosystem.* Dehradun, India: International Book Distributors, 1994.

Ward, D. *The Biology of Deserts.* Oxford, United Kingdom: Oxford University Press, 2009.

Whitford, W. G. *Ecology of Desert Systems.* Waltham, MA: Academic Press, 2002.

Rann of Kutch Seasonal Salt Marsh

Category: Grassland, Tundra, and Human Biomes.

Geographic Location: South Asia.

Summary: An unusual ecosystem, this very large seasonal marshland offers vital refuge to animals as diverse as flamingo and wild ass—and even hosts the world's only inland stand of mangroves.

The Rann of Kutch is a hot, arid, and intermittently inundated biogeographic region extending across the Kutch (or Kachchh) district of northwestern Gujarat, India, and the Sindh province of southern Pakistan. The Rann of Kutch lies mainly on a great peninsula between the Gulf of Kutch and the mouth of the Indus River, and is largely ensconced within the Thar Desert. The phrase *Rann of Kutch* is derived from Hindi *ran,* meaning *desert,* and *Kutch,* referring to an alternately wet and dry place. The Rann of Kutch seasonal salt marsh, geographically split into the Great and Little Ranns, covers an area of some 10,000 square miles (26,000 square kilometers).

Since the Mesozoic era, the Little and Great Ranns were extensions of the shallow Arabian Sea until geological uplift closed off the connection with the sea, creating a vast lake that was still navigable until as recently as the 5th century. Over the centuries since then, silting has created a broad, saline mudflat. With monsoon periods bringing direct rainfall and flooding, and also filling the seasonal upland rivers that drain through the peninsula, parts of the region became a seasonally inundated grassland. While these freshwater inputs are substantial, they are not enough to overcome the underlying saline foundation of the area; therefore, the nature of the inundated grassland is that of a sprawling salt marsh.

During the dry season, roughly October through June, the area resumes its state as a salty mudflat set amidst the Thar Desert. The average summer daytime temperature is extremely high, around 106 degrees F (41 degrees C), and dropping to about 95 degrees F (35 degrees C) at the onset of monsoon season, July through September. Winters are far less humid, with a milder average temperature of 84 degrees F (29 degrees C) by day, and 54 degrees F (12 degrees C) by night.

There are several ranges of low-salt, sandy land islands set 6.5–10 feet (2–3 meters) above the flood level. These *beyts* provide footholds for thorny scrub vegetation and grasses that are not as subject to the halophytic regime present across the great majority of the biome. They offer vital refuge for key animal species of the biome.

Flora and Fauna

One characteristic of salt marshes and salt deserts is that a relatively small number of plant species are capable of tolerating the arid, saline conditions; among such salt-tolerant, or halophytic, plants in the Rann of Kutch are sedge (*Cyperus dwarkensis*), lotebush (*Ziziphus williamsii*), and a regionally unique salt cedar (*Tamarix kutchensis*).

The biome also has stands of halophytic mangroves (*Avicennia* spp.), in particular one that oddly is found some 62 miles (100 kilometers) inland—quite uncharacteristic for this genus. This conglomeration has taken on a spiritual aura, in the sense that its revered 1.7 acres (0.7 hectares) of landlocked greenery is considered a sacred little forest, locally called Shravan Kavadia.

The Rann of Kutch seasonal salt marsh is vital in offering refuge to migratory wading birds, in particular the greater flamingo (*Phoenicopterus ruber*) and lesser flamingo (*P. minor*); cranes, including common crane (*Grus grus*) and Sarus crane (*G. antigone*); and storks such as white stork (*Ciconia ciconia*) and Asian open-bill stork (*Anastomus oscitans*). The region is also a home to the red-wattled lapwing (*Vanellus indicus*), Indian courser (*Cursorius coromandelicus*), more than 12 species of lark, and in all provides habitat to more than 200 bird species.

The region is quite famous for the Indian Wild Ass Sanctuary, which encompasses some 1,930 square miles (5,000 square kilometers) in the Little Rann of Kutch. About 3,000 individuals of the Indian wild ass species (*Equus hemionus khur*) live in this sanctuary, running freely at speeds up to 31 miles per hour (50 kilometers per hour) across the grasslands.

Other larger mammals found here seasonally include blackbuck (*Antilope cervicapra*), the nilgai antelope (*Boselaphus tragocamelus*), chousingha or four-horned antelope (*Tetracerus quadricornis*), and chinkara or Indian gazelle (*Gazella bennettii*), as well as predators such as striped hyena, Indian wolf, jackal, Indian and white-footed foxes, leopard (*Panthera pardus*), and caracal or African lynx (*Caracal caracal*).

Conservation and Threats

In addition to the Wild Ass Sanctuary—which also provides nesting habitat to about 75,000 migratory birds—this biome includes the Chhari Dhand Wetland Conservation Reserve, which protects about 31 square miles (80 square kilometers) on the edge of the Banni Grasslands.

In terms of the human economy, much of this region is relatively unattractive for anything but subsistence farming, grazing, and some fishing; nearly 20 indigenous tribal communities are based here. However, booming industries in logging, farming, local commerce, and international trade in the Indus Delta area to the west, and expansion of the industrial sector in metropolitan India to the east, are driving development around the fringes of the biome. Commercial salt panning, shrimp farming, and poaching also nibble around the edges of the region. Ecotourism here is a mixed blessing, with its benefits of support for sustainable practices, but its drawbacks from an influx of resource-challenging transient population. New projects such as water diversion schemes and a proposed highway to transect the Rann are further threats.

*As many as 3,000 Indian wild asses (*Equus hemionus khur*) range in their sanctuary in the Little Rann of Kutch, seen here. They can reach speeds of up to 31 miles per hour (50 kilometers per hour). (Wikimedia/Sballal)*

Any or all of these activities potentially undermine the long-evolved balances of this ecosystem.

So, too, may the hazards of global warming. Rising sea levels, altered sprouting and breeding cycles due to average temperature changes, and heavier monsoon events are effects that could damage the ability of local flora and fauna to adapt and thrive.

Muraree Lal Meena
Suman Singh

Further Reading

Geevan, C. P. *Ecological-Economic Analysis of Grassland Systems: Resource Dynamics and Management Challenges Kachchh District (Gujarat).* Kachchh, India: Gujarat Institute of Desert Ecology, 2003.

Gujarat Tourism. "Wild Ass Sanctuary—Little Rann." Government of Gujarat. http://www.gujarattourism.com/showpage.aspx?contentid=249.

Puri, G. S., R. K. Gupta, V. M. P. S. Meher-Homji. *Forest Ecology, Vol. 2.* New Delhi, India: Oxford & IBH Publishing Company, 1989.

Singh, Y. D. and V. Vijay Kumar. *Status of Banni Grasslands and Exigency of Restoration.* Kachchh, India: Gujarat Institute of Desert Ecology, 1998.

Wildlife Institute of India (WII). *Ecology of Wild Ass (*Equus Hemionus Khur*) in Little Rann of Kutch.* Dehradun, India: WII, 1993.

Woodruff, C. D. *Coasts: Form, Process and Evolution.* New York: Cambridge University Press, 2002.

Red Butte Creek

Category: Inland Aquatic Biomes.
Geographic Location: North America.
Summary: This pristine, protected watershed was recently compromised by natural and human-made events.

Utah's Red Butte Creek is a perennial stream that flows out of Red Butte Canyon in the Wasatch Mountains and then through Salt Lake City to the Jordan River. Its water quality and surrounding natural habitat have been protected since 1890, first by the United States Army and later by the U.S. Forest Service, which in 1969 declared it a Research Natural Area. Consequently, the creek area has been preserved as one of the most pristine riparian corridors in the intermountain western United States. In recent years, however, both natural and human-caused events have threatened the ecological balance of this biologically rich ecosystem—particularly the 2010 Chevron oil spill.

Red Butte Creek begins its journey at 8,000 feet (2,438 meters) above sea level, at the top of Red Butte Canyon. The creek cascades through a steep ravine to the mouth of the canyon, where it flows through the neighborhoods of east Salt Lake City and collects in a pond in Liberty Park. Eventually, the water reaches the Jordan River. The creek's flow varies seasonally, increasing in the spring months when snow melts at higher altitudes.

Though the flow of water through Red Butte Creek can vary depending on snowmelt at the end of each winter, the spring of 1983 was notable for its runoff. Heavy snows in mid-May, followed by unusually warm weather at the end of that month, caused flooding throughout the area. On May 28, the creek overflowed its banks, altering riparian habitats, which have yet to recover.

Red Butte Canyon has a climate that is semi-arid and characteristic of the eastern boundary of the Great Basin. Long, hot summers and cold, snowy winters are common, while spring and fall are comfortable and relatively wet. Precipitation amounts vary depending on the altitude and the northern sweep of monsoonal moisture during the summer months. Temperatures can exceed 100 degrees F (38 degrees C) in the hottest season, but rarely fall below 0 degrees F (minus 18 degrees C) in the winter.

Biodiversity

Four distinct plant communities thrive in Red Butte Canyon: riparian, grass-forb, oak-maple, and conifers. On the wet banks of the creek, western water birch, mountain alder, and sometimes willow grow. Elevation plays a large part in the distribution of vegetation, with spring grassland communities appearing closer to the mouth of the canyon, while summer-active scrub oak, aspen, and coniferous forests exist mostly in the upper portions of the canyon.

A diverse animal population also can be found living in and around Red Butte Creek. Bobcats, mountain lions, moose, coyotes, mule deer, and elk make the canyon their home. Rodents such as red-backed voles and jumping mice are abundant in the lower grasslands, while red squirrels, Uinta ground squirrels, and chipmunks live at higher altitudes. Native beavers were abundant in the area until the 1980s, when it was feared that they could infect the water supply with the parasite *Giardia lamblia.* Their removal resulted in the loss of marshy environments and many small mammals that lived where the beavers made their dams.

In the water, fish common to the creek include largemouth bass, brown trout, and bluegill, and waterfowl include ducks and Canada geese.

Human Activity and Environmental Threats

Settlers living west of Red Butte Canyon used creek water to irrigate their crops in the 1850s, as water rights were not an issue until the U.S. Army established Fort Douglas at the base of the canyon. Army engineers constructed two reservoirs in 1875, diverting the water from Red Butte Creek to fill them. Worried that nearby quarries would compromise water quality, the engineers declared in territory district court that the waters of the creek belonged to the Army.

A few years later, the U.S. Congress passed a law to protect the watershed. This prevented the sale of land in Red Butte Canyon and stopped further development. When the U.S. Forest Service acquired the canyon in 1969, it gated the property and declared it a Research National Area (RNA). The purpose of the RNA was to maintain natural ecosystems as baseline areas, research their natural processes, and compare the RNA ecosystems with compromised or manipulated areas.

In 2010, Red Butte Canyon and creek suffered two significant ecological disasters when a Chevron pipeline fractured, sending oil gushing into Red Butte Creek. The pipe carried medium crude oil from eastern Utah to the Salt Lake Valley. On the morning of June 12, Salt Lake City police found 50–60 gallons (189–227 liters) pouring into the creek per minute. Downstream from the canyon's mouth, the oil soon ran into the pond at Liberty Park and then into the Jordan River. Birds and other animals were immediately affected, their coats and feathers matted with sticky black oil. In December of the same year, the same pipe fractured in a different place, spewing 21,000 gallons (79,494 liters) into the stream and surrounding areas.

Chevron took full responsibility for the two spills, funding the multimillion-dollar-cleanup that will continue for several years. In 2011, Utah wildlife officials returned 3,000 native cutthroat trout to Red Butte Creek as part of a plan to restore the compromised ecosystem. Unfortunately, it will take many years to fully assess the effect of the oil spills on plant and wildlife habitats here.

Deborah Foss

Further Reading

Arrington, L. J. and T. G. Alexander. "The U.S. Army Overlooks Salt Lake Valley, Fort Douglas 1862–1865." *Utah Historical Quarterly* 33 (1965).

Ehleringer, James R., Lois A. Arnow, Ted Arnow, Irving B. McNulty, and Norman C. Negus. "Red Butte Canyon Research Natural Area: History, Flora, Geology, Climate, and Ecology." *Great Basin Naturalist* 52 (1992).

O'Donoghue, Amy Joi and Josh Smith. "Oil Spill in Red Butte Creek Threatens Waters, Wildlife." *Deseret News,* June 12, 2010.

Red River of the North

Category: Inland Aquatic Biomes.
Geographic Location: North America.
Summary: This important river has had a number of major flooding episodes, to which humans have responded with dam-building, a practice that challenges some species and habitats.

The Red River of the North originates at the confluence of the Bois de Sioux River in North Dakota and the Otter Tail River in Minnesota, and flows north through the Red River Valley, forming part of the boundary between the two states. The river then flows into Canada toward Lake Winnipeg and ultimately into Hudson Bay.

The Red River has its origins in a flatbed lake, Lake Agassiz, and was created from about 9000–8000 B.C.E., toward the end of the Wisconsin Glacial Episode. When the lake finished draining, probably in about 7500 B.C.E., the lake shrank, and the river was all that remained.

Because of the flat plains on either side of it, the Red River of the North regularly floods, creating rich alluvial soils around it. These soils attracted the Métis people; and the rich fauna there led to the arrival of European fur traders, followed by the establishment of towns, the most important being Winnipeg, formerly the center of the Red River Colony.

The river valley has endured major flooding five times since 1826, the most recent being in 2009. The prospect of flooding was a particular concern, and in 1969 the Red River Floodway, a large channel that aims to take water to the east of the city of Winnipeg, was completed. It has been used many times over the years, but the floodway was unable to block the floodwaters in 1997 and 2009, although it reduced the amount of water that otherwise would have inundated Winnipeg.

The river's path from Lake Winnipeg flows through a range of biomes. It begins in a coniferous forest area, then moves through deciduous forest and grassland near where the river water helps irrigate some of the major spring-wheat region.

Biodiversity

The river flows through various northern United States jurisdictions and two Canadian provinces, and there is wide temperature fluctuation, consisting of hot summers when temperatures can rise to 100 degrees F (37 degrees C) and cold winters, often marked by heavy snows, blizzards, and temperatures consistently below freezing. In Fargo, North Dakota, as an example, the average annual temperature is 41 degrees (5 degrees C) with the average summer temperature of 71 degrees F (21 degrees C) and the average winter temperature of 7 degrees F (minus 14 degrees C).

A wide range of flora and fauna have established themselves along and in the Red River of the North. Until the 19th century, bison herds grazed in the grasslands along the river but the herds were destroyed; trappers then decimated many of the communities of squirrels and beavers. These latter animals are once again relatively common, as are prairie dogs, rabbits, and foxes. Also present are pheasants, migrating waterfowl, whitetail deer, mourning doves, partridge, turkeys, squirrels, and rabbits.

The floodplains have provided habitat for migrating shorebirds, and in the river many species of fish have established habitats, although with the runoff from agricultural lands and effluent from cities, these numbers have diminished significantly.

The river water has long been turbid, which has led to the proliferation of mayflies, that in turn have been good for the fish. The best-known fish is the channel catfish, which, along with the carp, continues to attract recreational anglers from

The Red River of the North flooding the Sorlie Bridge in Grand Forks, North Dakota, on April 21, 1997. Major flooding occurred again in 2009. (U.S. Geological Survey)

around the world. The river also is home to northern pike, smallmouth bass, walleye, and sauger.

The lake sturgeon, which was relatively common until the start of the 20th century, has declined because of pollution and the damming of nearby rivers. In 1995, scientists estimated that only a few sturgeon were left. A major restocking program started in 1997, and this species will remain protected until its numbers increase again. Further along the river, in Canada, are abundant trout, pike, and golden eye.

Along the banks of the river, in areas that have not been cleared, are birch trees and many species of birds, including the great gray owl, the official bird of the Canadian province of Manitoba, through which the river flows. Among the waterfowl of the area are various grebes and gulls, loons, gannets, and cormorants.

Environmental Threats

River flooding continues to threaten the river basin area because of a changing climate and the drainage of area prairie wetlands. River dams were constructed to help alleviate the flooding. However, the dams pose other environmental issues to the area by changing or choking off the natural water supply to fish, plants, birds, and mammals.

Justin Corfield

Further Reading

Drache, Hiram M. *The Day of the Bonanza: A History of Bonanza Farming in the Red River Valley of the North.* Danville, IL: Interstate Printers & Publishers, 1981.

Moehlman, Arthur Henry. "The Red River of the North." *Geographical Journal 25,* no. 1 (1935).

Murray, Stanley Norman. "The Valley Comes of Age: A History of Agriculture in the Valley of the Red River of the North 1812–1920." Fargo: North Dakota Institute for Regional Studies, 1967.

Van Dyke, Henry. "The Red River of the North." *Harper's New Monthly* 660, no. 360 (1880).

Red River of the South

Category: Inland Aquatic Biomes.
Geographic Location: North America.
Summary: The Red River of the South is an important watershed in North America, covering critical ecoregions, but it is seriously affected by human activities.

The Red River of the South is a major tributary of the Mississippi and Atchafalaya rivers in the southern United States, and it drains the second largest river basin in the southern Great Plains. This Red River, previously known as the River Rouge, is 1,360 miles (2,189 kilometers) long, flowing generally in a west-to-east direction as it crosses the states of Texas, Oklahoma, Arkansas, and Louisiana. Its watershed covers approximately 65,598 square miles (169,900 square kilometers). The high, middle, and bottom drainages on the basin are located in the first two states, with a small portion of the bottom in the last two.

The climate of the watershed is arid; in the higher parts of the basin, there is a mean annual precipitation of 20–28 inches (508–711 millimeters), with minimum and maximum temperatures in January and July of 28–54 degrees F (minus 2 to 12 degrees C) and 72–95 degrees F (22–35 degrees C), respectively. The moister situation is quite different in bottomland areas, which receive 42–48 inches (1,066–1,219 millimeters) of mean annual precipitation, with minimum and maximum temperatures in January and July of 30–52 degrees F (minus 1 to 11 degrees C) and 72–91 degrees F (22–33 degrees C), respectively. The drainage basin includes different major ecoregions; among these, from west to east, are: Red Prairies, Broken Red Plains, and Red River Bottomlands.

Biodiversity

In the Red Prairies and Broken Red Plains ecoregions, the dominant vegetation includes the prairie midgrasses or shortgrasses, depending upon soil, moisture availability, and grazing pressure. Under less disturbing practices, the typical grasses include the little bluestern (*Schizachyrium scoparium*), Texas wintergrass (*Stipa leuchotricha*), white tridens (*Tridens albescens*), Texas cupgrass (*Eriochloa sericea*), and sideoats gramma (*Boutelowa curtipendula*).

Under heavier grazing conditions, the Buffalograss (*Buchloe dactyloides*), hairy tridens (*Erioneuron pilosum*), and purple threeawn (*Aristida purpurea*) become dominant.

Riparian zones in these ecoregions are usually dominated by pecan (*Carya illinoensis*), mixed with American elm (*Ulmus Americana*), black willow (*Salix nigra*), and little walnut *(Juglans microcarpa).*

The upper parts of the basin include conspicuous wildlife, of types mostly related to grasslands habitats. Across most of the river, the black-tailed prairie dog (*Cynomys ludovicianus*) is one of the key species defining the entire landscape. Associated with this important rodent are other species—including several prairie-dog predators—such as the black-footed ferret (*Mustela nigripes*) and rattlesnakes (*Crotalus* spp.). Some prefer to take up residence in the extensive underground burrow systems the prairie dogs create. Among these are the kit fox (*Vulpes veloxmacrotis*), the burrowing owl (*Athene cunicularia*), and the horned lizard (*Phrynosoma cornutum*).

Even some grassland birds can benefit from the presence of prairie dogs, such as the mountain plover (*Charadrius montanus*), which mainly feeds on the shortgrass around prairie dog towns. With this seemingly one-directional fauna community, the large scale extermination of prairie dogs, historically considered as pests, has had substantial impact on the entire region. The use of poisoning, mechanical reduction, and systematic extermination have transformed most of the area, not only changing the entire aspect of some landscapes, but also reducing significantly other species both related to and dependent on the prairie dogs or their burrows.

In the Red River Bottomlands, the river flows through a series of marshes and swamps, where its flow is dramatically moderated. The plain of the river is composed by fertile deposits with variations in permeability. The original vegetation in the lower region, despite being almost completely depleted, is mainly composed of hardwood forest in which a number of species are common. These include various oaks such as common water oak (*Quercus nigra),* willow oak *(Q. phellos),* blackjack oak (*Q. marilandica),* overcup oak (Q. lyrata), and southern red oak (*Q. falcate).* Other tree species include sweetgum *(Liquidambar styraciflua),* blackgum *(Nyssa sylvatica),* river birch (*Betula nigra),* red maple (*Acer rubrum),* green ash (*Fraxinus pennsylvanica),* and American elm (*Ulmus americana).*

Wildlife in this part of the basin includes deer (*Odocoileus virginianus*), turkeys (*Meleagris* spp.), squirrels (*Sciurus* spp.), rabbits (*Sylvilagus* spp. and *Lepus* spp.), raccoon (*Procyon lotor*), and bobcat (*Lynx rufus*). Perhaps the most remarkable are the endangered bald eagle (*Haliaeetus leucocephalus*) and black bear (*Ursus americanus*).

Environmental Issues

The Red Prairie ecoregion is currently mostly cultivated, in contrast to the Broken Red Plains ecoregion that is predominately use as grazing land. In the western stretches of the watershed, grasslands are common, but so are dry lands and irrigated ranges for farming of cotton, wheat, and grain sorghum. Additionally, the area is used for cattle, sheep and goat grazing. Also, most of the bottomland natural woodland has been cleared for cropland and improved pastures, although some woodland still occurs in very poorly drained and flooded areas.

The economy of the region is largely influenced by water, both directly and indirectly, for water processing, drinking water, and irrigation related to farming. Because of past problems, this resource is monitored for pollution and environmental damage. The potential effects of climate change related to rising temperatures and lower precipitation regimes in the southern Great Plains are complex; their impact on the species of this biome, however, is sure to be deleterious.

José F. González-Maya
Marylin Bejarano

Further Reading

Benke, A. and C. Cushing. *Rivers of North America.* San Diego, CA: Academic Press, 2005.

Griffith G. E., S. A. Bryce, J. M. Omernik, and A. C. Rogers. "Ecoregions of Texas." Austin: Texas Commission on Environmental Quality, 2004.

Pair, John C. "Stress-Tolerant Trees for the Southern Great Plains." *Journal of Arboriculture* 20, no. 2 (1994).

Red Sea

Category: Marine and Oceanic Biomes.
Geographic Location: Middle East.
Summary: This predominantly shallow inlet of warm saline waters is known for its high biodiversity and unique coral reefs.

Formed by Arabia splitting from Africa, the Red Sea is a long, narrow inlet of the Indian Ocean between Africa and Asia. Informal usage sometimes includes as part of the sea the two gulfs to its north, Aqaba and Suez. The International Hydrographic Organization defined the precise limits of the Red Sea, most recently in 1953: In the north, it is bound by the southern limits of the Gulfs of Suez to the south point of Shadwan Island, and west to the coast of Africa and Aqaba through Tiran Island, and westward to the coast of the Sinai Peninsula.

The world's northernmost tropical sea, it is home to numerous species of coral, 1,200 species of fish, and more than 1,000 species of invertebrates. The name Red Sea, used since ancient times, may refer to blooms of sea sawdust or to the arcane use of colors to refer to cardinal directions (in this case, south).

The sea's coastline is about 1,400 miles (2,253 kilometers) long and it is an average of 174 miles (280 kilometers) wide. The average depth is 1,600 feet (488 meters), but the water may exceed 8,000 feet (2,438 meters) at its center point. Most of the sea is shallow, and a quarter of it is less than 200 feet (61 meters) deep.

The Red Sea is surrounded by arid land masses, and due to low rainfall and the high rate of evaporation in the north, it has become one of the most saline bodies of water in the world. Salinity ranges from 36 parts per thousand (ppt) in the south, where water is exchanged with the Gulf of Aden, to 41 ppt in the north, with an average salinity of 40 ppt. By contrast, worldwide average seawater salinity is 35 ppt. Some of the salinity is contributed by highly salty brines emanating from sub-seafloor hydrothermal vents at temperatures approaching 140 degrees F (60 degrees C), a result of the continued widening of the sea due to the tectonic movement of the Red Sea Rift.

Rainfall over the sea and its coasts is low, at about 2.3 inches (58.4 millimeters) per year, usually in the form of brief thunderstorms. Wind is dominated by persistent northwest winds in the north. The rest of the Red Sea is subject to winds that are seasonally variable in both speed and direction, with speed generally increasing northward. Wind-induced currents shift much of the sediment in the sea, and affect the erosion and accretion of coastal rock exposure. The climate results from the northeasterly and southwesterly monsoons, and the surface water temperature is among the hottest in the world year round, averaging 72 degrees F (22 degrees C), with summer highs around 90 degrees F (32 degrees C).

The health of the coral reefs here is maintained in part by the great depths of the sea and the efficiency of its water circulation. The water mass of the sea exchanges its water with the Arabian

Sea via the Gulf of Aden to the south. The Red Sea is rich in minerals with many sediment constituents. These include: nanofossils, foraminifera, pteropods, siliceous fossils, tuffites, volcanic ash, montmorillionite, cristobalite, zeolites, quartz, feldspar, mica, clay minerals, sulfide minerals, aragonite, calcite, dolomite, chalcedony, magnesite, gypsum, anhydrite, halite, hematite, siderite, pyrite, and rhodochrosite. Despite this diverse sediment, the water is exceptionally clear because of the lack of river discharge and the low levels of precipitation.

Biodiversity

The harsh conditions of the hydrothermal vents are inhospitable to most forms of life. The water is hot and highly saline, with high concentrations of heavy metals, and there is little light or oxygen at the bottom of the rift. Until recently, the vents were believed to be lifeless, but recent studies have discovered that like many other inhospitable parts of the world, they are home to extremophile bacteria and archaea—microbes adapted to extreme conditions. Some such species in the Red Sea include *Halorhabdus tiamatea,* which is related to *H. utahensis* of the Great Salt Lake in Utah; *H. contractile,* which forms tentaclelike protrusions of unknown purpose; and *Salisphaera shabanensis,* which lives at the interface between the sludgy brine and the seawater.

The sea is surrounded by salt flats characterized by evaporite-carbonate deposits, which form shoals similar to those of the Persian Gulf; long flat salt pans covered in fine-grained alkali salt sediment; and salt marshes that are home to halophytic, or salt-tolerant, plant life and migratory birds.

The coastal zones include 28 stands of mangrove (*Avicennia marina* and *Rhyzophora mucronata*), trees that are specially adapted to low oxygen and high salinity. *A. marina* has developed salt glands to rid itself of excess salt by excreting it through the leaves. *R. mucronata* is found only in the mangrove stands of the Farasan Archipelago, in the southern part of the sea, which is richer in nutrients due to the proximity of the Indian Ocean.

Nearly 100 species of algae are found in mangrove stands, including the sea's characteristic red algaes, the most common species of which are *Bostrychia tenella, Spyridia filamentosa,* and *Laurencia papillosa.* These types dominate in shallow mangrove pools. Blue-green algae is the least diverse algae group, but also the most widespread, growing in both benthic and epiphytic forms throughout mangrove ecosystems.

Local wildlife in the mangrove stands includes mangrove crab species such as the very abundant *Uca inversa inversa, Metapograpsus messor, Portunus pelagicus, Ocypode saraten,* and *Macrophthalamus telescopius,* as well as a species of mudskipper fish (*Oxudercinae*).

The Red Sea is the most northern coral-reef ecosystem. The most common reef in the Red Sea is fringing reef, which (unlike atolls and barrier reefs) grows close to and often directly from the shoreline. The back-reef area has the least species diversity and largely consists of seagrass meadows and the fish that feed from them.

Seagrasses are the only flowering plants that can live completely submerged, and seagrass ecosystems are home to numerous rays and echinoderms: sea cucumbers, sea urchins, and starfish. The reefs are home to reef sharks, dolphins, and numerous species of mollusks. About 10 percent of the fish found around the coral reefs—about 120 species—are endemic, or found only here.

The Red Sea coast also is home to four species of seaturtles: green, hawksbill, loggerhead, and leatherback. Greens and hawksbills usually forage around the reefs; the greens eat algae and seagrass, while the hawksbills feed on sponges and invertebrates. Leatherbacks, which feed on jellyfish, are usually found farther from the reefs.

Migratory birds in the Red Sea coastal zones include swallows, songbirds, stocks, cranes, and birds of prey.

Fish, phytoplankton, and zooplankton populations in the Red Sea closely resemble those of the Indian Ocean. Common species include the shagreen ray (*Raja fullonica*) and the black-striped pipefish (*Syngnathus algeriensis*), which anchors itself to seagrass by wrapping its tail around it. Endemic species include the tiger bass (*Terapon jarbua*), which feeds on insects and plant matter; the leopard torpedo (*Torpedo panthera*), a ray

Hammerhead Sharks

Hammerheads are solitary, nomadic predators that patrol the reefs and are studiously avoided by other reef sharks, though bull sharks may prey on the juveniles. Pilot fish often accompany hammerheads, which are used as parasitic hosts by numerous crustaceans called copepods. These include *Alebion carchariae, A. elegans, Nesippus orientalis, N. crypturus, Eudactylina pollex, Kroyeria gemursa,* and *Nemesis atlantica.*

The hammerhead has an especially diverse diet, including invertebrates such as crabs, lobster, and squid; smaller sharks; and such fish as sardine, sea catfish, boxfish, toadfish, porgy, croaker, grouper, and porcupine fish. They are particularly fond of skates and rays, and their predation may have been one of the factors leading the leopard torpedo (*Torpedo panthera*) to develop its electrical defense. The great hammerhead is apparently immune to the venom of the stingray, and the ray's spines are often found lodged in the shark's mouth—to no effect. The hammerhead's distinctive head, called an expanded cephalofoil, is used to head-butt a ray to stun it and force it down to the sea bottom, where the hammerhead can pin the ray down with its head while biting the ray's pectoral fins to cripple it.

*The Red Sea is home to the great hammerhead shark (*Sphyrna mokarran*), the largest hammerhead species in the world, which can reach a maximum length of 20 feet (6 meters). (Thinkstock)*

capable of generating electricity as a defense and feeding mechanism; and the great hammerhead shark (*Sphyrna mokarran*), the largest hammerhead species, with a maximum length of 20 feet (6 meters).

Human and Environmental Threats

The tidal amplitude of the sea is very low, and in the mangrove stands, it is often only about 1 foot (0.3 meter), which helps preserve the stability of the mangrove ecosystems. The roots of the mangroves provide oyster habitats and increase sediment deposits, resulting in sinks for heavy metals. Disrupting the mangrove stands could recontaminate the sea with heavy metals and other pollutants. Mangroves on the Egyptian coast are protected for this reason, and because they help protect the coast from erosion.

Because the land surrounding the Red Sea is arid, there is elevated demand for freshwater to meet both drinking and industrial demands. Currently, 18 desalination plants along the Saudi Arabian coast work to harvest seawater and strip away the salts. This results in discharges of warm, highly saline water and treatment chemicals, which puts local fish stocks in jeopardy.

The use of the Red Sea water as a coolant by factories and refineries has a similar effect. At present, these effects seem to be localized to the coastal zone immediately surrounding the individual plant, refinery, or factory. Commercial fishing in the sea focuses on lobster, tuna, bonito, herring, sardine, and anchovy.

Although warm water temperatures in other seas are a major cause of coral bleaching, that is apparently not the case here. The corals of the southern Red Sea have been under study to determine how they are able to dodge this threat.

The greatest danger to the Red Sea's coral seems to be contaminants introduced into the water, particularly because most of the coral is so close to the coast, where there is not only industrial runoff, but also waste from hotels and offshore oil spillage. Hotel waste has been a considerable problem since at least the 1990s. Egyptian governments have recently banned the use of plastic bags in supermarkets in an attempt to reduce the prevalence of the bags in the sea's waters and reef systems.

Bill Kte'pi

Further Reading

Benayahu, Y. and Y. Loya. "Space Partitioning by Stony Corals, Soft Corals, and Benthic Algae on the Coral Reefs of the Northern Gulf of Eilat." *Helgoland Marine Research* 30, nos. 1–4 (1977).

Botros, G. A. "Fishes of the Red Sea." *Oceanography and Marine Biology Annual Review* 9 (1971).

Fishelson, L. "Ecology of Coral Reefs in the Gulf of Aqaba Influenced by Pollution." *Oecologia* 12, no. 1 (1973).

International Hydrographic Organization. *Limits of Oceans and Seas, Special Publication No. 28*. Monte Carlo: International Hydrographic Organization, 1953.

Saifullah, S. M. "Mangrove Ecosystem of Saudi Arabian Red Sea Coast—An Overview." *Journal of the Faculty of Marine Science* 7 (1994).

Sherman, Kenneth, Ezekiel N. Okemwa, and Micheni J. Ntiba, eds. *Large Marine Ecosystems of the Indian Ocean: Assessment, Sustainability, and Management*. New York: Wiley-Blackwell, 1998.

Zakai, David and Nanette E. Chadwick-Furman. "Impacts of Intensive Recreational Diving on Reef Corals at Eilat, Northern Red Sea." *Biological Conservation* 105, no. 2 (2002).

Red Sea Coral Reefs

Category: Marine and Oceanic Biomes.
Geographic Location: Middle East.
Summary: The Red Sea coral reefs embody unique structural formations, and are home to a high diversity of marine fauna and flora.

The coral reefs of the Red Sea have been evolving for 40 million years, ever since the breakup of the Arabian and African continental plates. The Red Sea now possesses some of the most diverse and abundant coral reef ecosystems on Earth, important to the fishery and tourism industries of multiple Middle Eastern countries, and to the biota of the nearby Indian Ocean, for which these reefs often act as a nursery. According to the United Nations Environment Programme (UNEP), about 30 percent of the world's reefs occur within the area covered by the Indian Ocean, Red Sea, and Gulf of Suez.

The Red Sea is a long, narrow body of water extending about 1,305 miles (2,100 kilometers), with a central trough reaching depths of more than 6,562 feet (2,000 meters) and a continental shelf that is 984–1,312 feet (300–400 meters) deep or less. The Red Sea proper is generally characterized by well-developed fringing reefs found almost continuously along both coasts, but less in the southern portion. In this area, some inlets and partly enclosed bays have incomplete fringing reefs.

The northern and central portions of the Red Sea have the best-developed fringing reefs, but other reef structures—such as the atoll-like ones bordered by deep water, barrier reef structures, and offshore patch and bank reefs—are also common.

The Red Sea rift system splits into the gulfs of Suez and Aqaba, both of which are morphologically different. The Gulf of Suez is a wide, fairly shallow—maximum depth of 279 feet (85 meters)—basin dominated by sand and sediment, with few corals or mangroves. The reefs of the Gulf of Suez are discontinuous fringing reefs along the western side, whereas the eastern side has much smaller, fragmented fringing and patch reefs. The Gulf of Eilat (Aqaba) is fairly deep, reaching a depth of 6,562 feet (2,000 meters), with varied sublittoral ecosystems, and is characterized by narrow fringing reefs and vertical dropoffs.

South of the gulfs of Suez and Aqaba lie extensive and continuous fringing reefs extending all the way down to Halaib, at the border of Sudan. Reef complexes in the central Red Sea are found along the coast at about 2–6 miles (3–10 kilometers) offshore, on a series of narrow underwater banks of tectonic origin. In the southern Red Sea, these banks are much wider and give rise to the Suakin, Dahlak, and Farasan archipelagos, which at times resemble atolls. Beyond the banks, the seabed drops rapidly to more than 3,281 feet (1,000 meters). In the southern Red Sea, the quality, complexity, and extent of reefs decrease because of shallower bathymetry, higher turbidity, and greater freshwater input.

Sediment trapped by the immense depth and a little freshwater input from rivers or rainfall contribute to exceptionally clear water in the Red Sea. The exchange of water with the Indian Ocean diminishes the effect of extreme temperatures and salinity, thus creating a suitable environment for coral-reef development. The comparative calmness and lack of severe storms mean that coral growth is less restricted by exposure to wave action than oceanic reefs. Water pollution was long restricted to industrial areas where desalination plants were located, but recent observations indicate that healthy reef areas are beginning to show signs of deterioration due to poor water quality.

Biodiversity

The Red Sea has rich coral fauna, most of which are found in the central Red Sea and in the Gulf of Aqaba. There are approximately 233 species in 56 genera of stony corals, and up to 125 soft coral species in the Red Sea. The number of stony coral species decline significantly toward the south. On average, coral diversity is greater in the Gulf of Aqaba, northern Red Sea, and central Red Sea than in the south, with nearly double the number of coral species.

The dominant genera of stony corals are *Acropora, Montipora, Pocillopora, Stylophora, Pavona, Leptoseris, Fungia, Porites, Favia,* and *Leptastrea.* Also, the pipe organ *Tubipora musica* has been reported in the Red Sea. Geographically, coral diversity varies quite considerably in the Red Sea due to changes in water temperatures, salinity, sediment loading, light, and anthropogenic effects.

Other fauna and flora in the Red Sea include green (*Chelonia mydas*), loggerhead (*Caretta caretta*), leatherback (*Dermochelys coriacea*), and hawksbill (*Eretmochelys imbricata*) seaturtles, which reportedly use the area to breed on the Sinai coast. Giant clams (*Tridacna maxima*) are common; sharks (*Selachimorpha*), dolphins (*Delphinidae*), and dugons are seen frequently.

The Red Sea also is an important ornithological site during spring and autumn migrations for many species of storks, ospreys, terns, herons, and gulls. Reports document 13 species of marine mammals, 325 endemic (found nowhere else) species of fish, and two species of mangroves here.

Environmental Threats

Destruction of wide reef areas in the northern and central Red Sea has been related to abnormally high numbers of the large black-spined urchin (*Diadema setosum*), a species that feeds on algal films and may damage coral and its spat; and also due to widespread white band disease and bleaching, as well as intensive predation by the corallivorous gastropod (*Drupella conzus).* The two latter causes are major contributors to coral mortality of the important and dominant coral branching species *Acropora hemprichi.*

The once-pristine coral reefs of the Red Sea are now seen to be in decline; a decrease of 20–30 percent in coral cover in the Red Sea from 1987 to 1996 was largely attributed to an expanding tourism industry. Other anthropogenic effects that have contributed to coral reduction include fin damage, swimming and diving, ships dropping anchor, coastal development, sedimentation, dredging, and construction of artificial beaches and desalination plants.

Among the other negative factors are sewage and nutrient-loading from hotels and resorts, high commercial boat and tanker traffic, oil spills, chemical and thermal effluents, disposal of sewage and septic tanks from dive boats, dynamiting, lobster catching, and spearfishing.

Overfishing damages reefs as it upsets the natural balance of the ecosystem by removing large predators, thus allowing population explosions of smaller species, or by removing important herbivores that graze on algal patches, allowing overgrowth of algae that outcompetes coral species for space on the substrate.

Lucia M. Gutierrez

Further Reading

Cesar, Herman. "Economic Valuation of the Egyptian Red Sea Coral Reefs." *Monitoring, Verification, and Evaluation (MVE) Unit of the Egyptian*

Environmental Policy Program. Washington, DC: Chemonics International, 2003.

Fishelson, L. "Ecology of Coral Reefs in the Gulf of Aqaba Influenced by Pollution." *Oecologia* 12, no. 1 (1973).

The Regional Organization for the Conservation of the Environment of the Red Sea and Gulf of Aden (PERSGA). *The Status of Coral Reefs in the Red Sea and Gulf of Aden: 2009.* Jeddah, Saudi Arabia: PERSGA, February 2010.

Spalding, Mark D., Corinna Ravilious, and Edmund P. Green. *World Atlas of Coral Reefs.* World Conservation Monitoring Centre. Berkeley: University of California Press, 2001.

Reelfoot Lake

Category: Inland Aquatic Biomes.
Geographic Location: North America.
Summary: An important migratory bird base, Reelfoot Lake faces a range of natural and anthropogenic threats.

Reelfoot Lake is a shallow natural lake near the Mississippi River in northwest Tennessee. The lake and its wetlands are 15 miles (24 kilometers) long, 5 miles (8 kilometers) wide, and encompass 25,000 acres (10,117 hectares), of which 15,000 acres (6,070 hectares) is water. The lake is swampy in some parts, with bayou-like ditches or basins, and is located in Reelfoot Lake State Park, known for its pairs of nesting eagles and cyprus trees. In 1910, the state of Tennessee designated the lake a fish and game preserve, and it has remained a popular spot for recreational activities and commercial fishing.

Climate and Habitats

The climate at Reelfoot Lake is moderate, with average temperatures of 25–50 degrees F (4–10 degrees C) in January and 68–89 (20–32 degrees C) in July. Rainfall averages around 4 inches (101 millimeters) per month, and ranges from a low of 3 inches (68 millimeters) in August to a high of 6 inches (146 millimeters) in May.

Reelfoot Lake today operates under an interim water management plan developed as part of an environmental assessment that considers private, commercial, and agricultural interests, as well as wildlife resources.

The park includes approximately 800 acres (323 hectares) of refuge lands managed under cooperative farming agreements; some 290 acres (117 hectares) managed to provide habitat for a variety of neotropical migrants, migrating shorebirds, and wintering waterfowl; and 6,000 acres (2,428 hectares) of forested habitats, including cypress swamps and bottomland hardwoods.

The park monitors waterfowl and eagle populations during the winter months, scrutinizing the nesting activities of bald eagles, and also monitors artificial nesting structures that are maintained for eastern bluebirds and wood ducks.

Biodiversity

The northern edge of the refuge is about 3 miles from the Mississippi River, and has been used as a major stopover and wintering area for waterfowl within the Mississippi Flyway. It is also a designated Important Bird Area by the American Bird Conservancy. Wintering mallards may exceed 400,000 birds; during very cold winters, the numbers of Canada geese may top 100,000 individuals. Other birds commonly seen here include egret, herons, nuthatches, and wild turkeys. In total, more than 230 bird species have been documented on the refuge, in addition to 52 species of mammals, and 75 species of reptiles and amphibians.

Fish species include crappie (a kind of sunfish); largemouth, smallmouth, spotted, and white bass; striped bass and striped bass-white bass hybrids; walleye; bluegill; catfish; sauger; saugeye; musky; northern pike; and rainbow, brook, brown, and lake trout.

Wildlife of the Reelfoot Lake biome also includes white-tailed deer, turtles, squirrels, beaver, and mink. Local vegetation features bald cyprus, cottonwoods, and walnut trees. Also located at the lake are populations of *Anopheles walkeri,* a species of mosquito that transmits human malaria.

Two bald eagles perched at the top of a tree in the Reelfoot National Wildlife Refuge in Tennessee. Over 230 species of birds can be found in the Reelfoot Lake ecoregion. (U.S. Fish and Wildlife Service/David Haggard)

Environmental Issues

Water quality continues to be a problem for Reelfoot Lake, caused by excess sediment and nutrients from nearby cropland. If sedimentation continues to build, Reelfoot Lake will fill in over the next 60–200 years. Cropland on the eastern shore of the lake, largely featuring cotton and soybeans, is vulnerable to runoff from eroding steep hillsides, which in turn can enter the lake. Gullies formed by cropped, forested, and grassland areas deposit a significant amount of sediment into the lake, which is augmented by man-made stream channelization, lake eutrophication from nutrient runoff, and alleged contamination from pesticides.

The bigger issue surrounding the Reelfoot Lake has been the loss of some 59 percent of the original 2 million acres (810,000 hectares) of wetlands in Tennessee, as recorded through 1990. This has had a dramatic effect on bird life throughout the region, especially wetland-dependent birds such as the least bitterns. In another loss-of-balance threat, nonnative carp in the lake pose a problem to many other fish species as they aggressively compete for phytoplankton and zooplankton.

Justin Corfield

Further Reading

Baker, C. L. "Reelfoot Lake Biological Station." *American Institute of Biological Sciences Bulletin* 6, no. 1 (1956).

Eddy, Samuel. "The Plankton of Reelfoot Lake, Tennessee." *Transactions of the American Microscopical Society* 49, no. 3 (1930).

Garnier, A. F. "November Bird-Life at Reelfoot Lake, Tennessee." *Wilson Bulletin* 28, no. 1 (1916).

Mizelle, John D. and John P. Cronin. "Studies on Monogenetic Trematodes X. Gill Parasites from Reelfoot Lake Fishes." *American Midland Naturalist* 30, no. 1 (1943).

Nelson, Wilbur A. "Reelfoot—An Earthquake Lake." *National Geographic Magazine* 45 (1924).

Winstead, Nicholas A. and Sammy L. King, "Least Bittern Nesting Sites at Reelfoot Lake, Tennessee." *Southeastern Naturalist* 5, no. 2 (2006).

Rennell Island

Category: Marine and Oceanic Biomes.
Geographic Location: Pacific Ocean.
Summary: Home to one of the world's largest coral atolls, and incorporating the largest Pacific-island lake, Rennell Island supports a variety of distinct species of flora and fauna.

Rennell Island is the main island of the two islands that comprise the Rennell and Bellona Province of the Solomon Islands. Known locally as Mugaba, Rennell Island has a land area of some 250 square miles (647 square kilometers). Considered the world's second-largest raised coral atoll, Rennell is approximately 50 miles (80 kilometers) long and

9 miles (14 kilometers) wide, with a population of about 3,000 people, most of whom are of Polynesian descent. Renbelian and English are spoken, as well as a dialect known as Pigen.

The island is home to the largest insular lake in the Pacific region, Lake Tegano, which is listed as a World Heritage Site by the United Nations Educational, Scientific, and Cultural Organization (UNESCO).

Rennell has a tropical climate; temperature averages 73–84 degrees F (23–28 C) year round, with rainfall totaling approximately 157 inches (400 centimeters) annually. The island lies within the cyclone path, and the last major one was Nina, in 1993.

Human Influence

Rennell Island was first settled sometime before 1400 C.E., when clansmen from Wallis Island (Uvea) crossed the Pacific. The United Kingdom established a protectorate over the Solomon Islands, including Rennell, in 1893, which existed until self-governance was declared in 1979. Today, subsistence farming takes place on Rennell Island, with the major crops including sweet potatoes, slippery cabbage, and taro.

In 1933, explorer and amateur botanist Charles Templeton Crocker visited Rennell, in a trek known as the Templeton Crocker Expedition that resulted in a series of short reports on seaweed, fish, reptiles, and amphibians endemic (found only here) to the island, including a new species of sea snake. Other expeditions to survey the island and chart fauna and flora species occurred over the years, resulting in reports that concluded that the coral reefs of the Solomon Islands, including Rennell Island, were less spectacular than those in other locations. The reports added that these islands lacked the luxuriance of the Great Barrier Reef in the Coral Sea off Australia.

The poverty of Rennell Island's coral reef was viewed as relative to the area's geologically recent elevation, and its steep and sometimes vertical gradients that discouraged the growth of coral and vegetation. The coral collections of Rennell Island have all, or virtually all, Indo-Pacific hermatypic scleractinians (zooxanthellate algae that support hard corals) present. Despite Crocker's pronouncements, Rennell Island is now considered one of the best global examples of reef-coral generic diversity.

Biodiversity

Rennell Island is covered in dense forest with a canopy averaging 65 feet (20 meters) tall. Local flora includes low scrub forest on the karst ridge, tall forest in the interior, and beach flora along Lake Tegano. The lake's vegetation features 312 species of diatoms and algae, including some endemic species.

Throughout the Solomon Islands, more than 700 species of algae are present, but no comprehensive collection has been made of marine algae near Rennell Island. Seaweed diversity near Rennell is low compared to other places. Seagrass beds, which help reduce surface erosion and bind sediments, are typically inshore of coral reefs. Although no comprehensive list of seagrasses near Rennell has been compiled, at least seven species have been noted, including *Cymdocea, Enhalus, Halodule, Halophila,* and *Syringodium.*

More than 26 species of mangrove are found near Rennell Island, representing more than 40 percent of the world's mangrove species. Mangroves protect coral reefs by removing and binding sediment that comes down from rivers. They also help recycle nutrients within lagoons, and shelter some bait-fish species that migrate between lagoons and coral reefs. Bait fish, which are captured for use in tuna fishing, are an important part of the Rennell Island economy. More than 200 species of marine fish have been identified near Rennell Island, many of which are eaten by local residents.

Rennell and Bellona are home to several endemic bird species, including the Rennell starling (*Aplonis insularis*), bare-eyed white-eye (*Woodfordia superciliosa*), Rennell shrikebill (*Clytorhynchus hamlini*), and the Rennell fantail (*Rhipidura rennelliana*). The Solomons white ibis (*Threskiornis molucca pygmaeus*) is a dwarf subspecies of the Australian white ibis that is endemic to the islands.

Rennell Island also is home to 25 species of ants and 11 species of bats, including the Rennell flying fox, which is unique to the island.

Lake Tegano, at 18 miles (29 kilometers) long and 6 miles (10 kilometers) wide, comprises nearly 18 percent of the total area of Rennell Island. With a maximum depth of about 145 feet (44 meters), Lake Tegano is a mixture of brackish, fresh, and salt water. Home to eels, water snakes, and birds, Lake Tegano sits in a central basin on the island that once served as the lagoon.

Environmental Concerns

The sparse populated here has helped keep the natural vegetation mostly intact, and there are no serious invasive species of animals or plants. Climate change has begun to impact Lake Tegano via rising water levels and more persistent salinity. Climate change also has begun to adversely affect plant growth in low-lying areas, which has meant reduced taro and coconut harvests, both staple foods for the islanders.

Stephen T. Schroth
Jason A. Helfer

Further Reading

Gillespie, A. and W. C. G. Burns, eds. *Climate Change in the South Pacific: Impacts and Responses in Australia, New Zealand, and Small Island States.* Norwell, MA: Kluwer, 2010.

McKibben, B. *Deep Economy: Economics As If the World Mattered.* London: Oneworld Publications, 2007.

Wolff, T. "The Fauna of Rennell and Bellona, Solomon Islands." *Philosophical Transactions of the Royal Society of London* B255 (1969).

Rhine River

Category: Inland Aquatic Biomes.
Geographic Location: Europe.
Summary: A mid-size river with vast economic and political significance, the Rhine has rebounded somewhat from long degradation resulting from river training and industrial development.

The Rhine is a Western European river with great historical, economic, and ecological significance, as well as the source of drinking water for 20 million people. Flowing through the countries of Germany, Austria, Switzerland, France, Netherlands, Luxembourg, Belgium, and even a small portion of Italy, the Rhine today is 766 miles (1,233 kilometers) long—65 miles (105 kilometers) shorter than its natural length due to river realignment projects called *river training*—which has been ongoing since the 19th century.

People along the shores of the river have modified the Rhine to fit population growth and industrialization. Thus, a notoriously wild river has become little more than a canal to further industry and trade. A string of environmental disasters in the late 20th century spurred efforts to increase biological diversity, as well as reduce flood risk.

The history of artificial modifications must be considered when describing the Rhine's geography. Projects to straighten the river led to the removal of thousands of islands, as well as oxbows, meadows, and braided beds. River training also standardized the Rhine's width to 656 feet (200 meters) at its base in Switzerland, eventually widening to 3,281 feet (1,000 meters) in Holland. Moreover, it set the Rhine's depth to a minimum of 5.6–8.2 feet (1.7–2.5 meters).

Because the river flows through eight European countries, the seasonal weather fluctuations vary, but temperatures can dip to 30 degrees F (minus 1 degrees C) during the winter and climb to 80 degrees F (26 degrees C) in summer.

River Flow

The Rhine is part of a sprawling canal network, giving it access to the Baltic Sea (via the Herne River), into France (via the Marne), the Mediterranean (via Rhone), and the Black Sea (via Main-Danube). The Rhine has become a flowing highway crucial for regional commerce; it is the second-busiest waterway (after the Mississippi River) internationally.

Geographers divide the Rhine into High Rhine, Upper Rhine, Middle Rhine, Lower Rhine, and Rhine Delta. The Rhine's headwaters—the

Aare, Vorderrhein, Hinterrhein, and Alpenrhein streams—flow from the Swiss Alps to Lake Constance. Lake Constance to Basel, Switzerland, is the High Rhine, where a staircase of locks and dams provides energy for local industry. The Upper Rhine is from Basel to Bingen, Germany, which lies in the Rift Valley, a natural floodplain at the foot of the Alps. This is where North Sea-bound marine traffic begins—a distance of 540 miles (869 kilometers)—making Basel Switzerland's only port.

Between Bingen and Bonn, Germany, is the Middle Rhine, where the Rhenish Slate Mountains, vineyards, and castles captivate tourists. The steepness of the surrounding canyon has spared its natural beauty from intensive development. This contrasts with the Lower Rhine, a region from Bonn to the Pannerden Canal (near the German-Dutch border) that is heavily developed for industrial and energy production.

The Rhine Delta is entirely in the Netherlands, and divides into three branches—Waal, Lek, and Ijssel. None of the branches has reached the North Sea since the 1986 Delta Project severed all of them, and diverted the Rhine's waters into a sluice network.

Since 1872, the New Waterway Canal has connected Rotterdam, Netherlands, to the Rhine, an artery for Rotterdam's Europoort (the world's second-busiest port after Shanghai).

Biodiversity

The river's geographic distinctions are based on fish population variations that no longer exist, as river training and pollution have radically changed wildlife patterns. In 1800, the Rhine boasted 47 fish species, but by the 1970s, half were extinct, and the other half were largely sustained by hatcheries. Similarly, from 1915 to 1971, the number of indigenous invertebrate macrofauna species fell from 80 to 27. These species could not adapt to the changing river.

Straightening the Rhine more than doubled its flow rate, increased its oxygen content, and reduced its temperature, but later industrial and energy production warmed, salted, blocked, and polluted the Rhine. Indigenous species had to compete with invaders brought along by long-distance shipping and canal connections, as well as intentional introductions, such as the zander (*Sander lucioperca)*—known as pike-perch—unleashed to compensate for extirpated salmon.

Since the 18th century, 45 macroinvertebrate species—such as beetles, worms, and snails—have invaded the Rhine and now compose up to 11 percent of the river's total species. Many indigenous species, which depended on fish or flood patterns that no longer existed, simultaneously competed against invading crustaceans and mollusks better suited to the new conditions.

Some river segments host relatively healthy fish communities. In the Lower Rhine, carp, minnows, chubs, and tench still thrive, while the much-less altered Upper Rhine hosts at least 36 species of fish, including sunfish, rainbow trout, roach, carp, silver bream, burbot, and stone loach.

The Rhine's shores have similarly changed in the past two centuries. Only 15 percent of the original floodplain remains, as the local population of 50 million humans has built farms, cities, and factories. The floodplain forests (mainly oak, elm, and lime) of the Upper Rhine are Europe's equivalent of rainforests, due to their superlative diversity of plant and bird species. These forests now only occupy 0.75 percent of their former range in the Upper Rhine, of which 70 percent consists of tree plantations. Moose, bears, and wolves went extinct here during the Middle Ages, and insects like mayfly and caddisfly (sedge-flies) are highly sensitive to water pollution; now their populations fluctuate with the Rhine's variable cleanliness. Floodplain forests regenerate when annual floods return, though they require 200 years to reach maturity.

The 40-mile (65-kilometer) Middle Rhine section, listed as a United Nations Educational, Scientific, and Cultural Organization (UNESCO) World Heritage Site since June 2002, is home to rare and endemic—found nowhere else—species such as the white-backed woodpecker, red kite, long-eared owl, and icterine warbler. Other birds flying over, nesting, and breeding along the Rhine include peregrines, black redstart, white stork, parakeet, eagle, goose, and lapwing.

Environmental Issues

The heart of German industry (especially energy, mining, textiles, agriculture, and chemicals) lies along the Lower Rhine, an area that bore heavy environmental costs until efforts to reduce pollution began in the 1970s. These industrial clusters are closely linked. For instance, dams and power plants both depend on and supply coal firms, while chemical factories use coal for energy, and its byproducts to make dyes and fertilizer for textiles and agriculture. Hundreds of power plants require water for cooling and return warmed water that is detrimental to sensitive organisms. Numerous dams dot the Rhine to generate hydroelectric power.

Like heat, chloride pollution may garner less attention than chemical spills, but it still unsettles the Rhine's natural balance. Until the signing of the Convention on Chlorides in 1976, salinization posed a threat to farmers, city dwellers, and other organisms dependent on the Rhine for fresh water.

Every ton of mined coal requires thousands of gallons (liters) of water for cleaning and pumping, and Alsatian potash mines similarly dumped salty refuse. The Netherlands advocated controls on salt dumping after World War II to mitigate the impact on low-lying areas from both the North Sea and the Rhine.

Textile and, later, chemical factories dumped numerous byproducts into the river, a problem exacerbated by plant growth along the river's banks. Some 6,000 chemicals have been identified in the Rhine, the results of dumping from the chemical industry, and agricultural runoff from artificial fertilizers and pesticides from the farms that border the river.

In 1969, Hoechst Chemicals accidentally dumped the insecticide Endosulfan into the Main River (a Rhine tributary), and in 1986, millions of gallons (liters) of water used to extinguish the fire at Basel's Sandoz chemical factory seeped into the Rhine, killing almost all aquatic life downstream until Koblenz, Germany. These two events galvanized public opinion that pressured governments to mitigate the ill effects of industrial development on the Rhine's ecosystem.

Worries about pollution were not new, as many had noted the Rhine's deterioration early on. In 1901, German parliamentarians called the Rhine a "sewer," and their commission reported the Rhine was red (because of sewage) from Ludwigshafen to Worms and yellowish near Mannheim (home to a large paper-pulp industry). Various complaints surfaced over the decades, and treaties were drawn to protect the river and surrounding areas. There were several treaties to protect salmon in the 19th century, though the fish still disappeared.

International Efforts

The World Wars and Great Depression stalled international negotiations, and it was not until 1950 that Switzerland, France, Luxembourg, Germany, and the Netherlands jointly founded the International Commission for Protection of the Rhine (ICPR). It was the Rhine's first antipollution institution. Although the ICPR systemically measured the river's water quality, the organization lacked funding, legal power, and even a functioning secretariat (until 1953), as fluvial states ensured a rapid economic recovery took precedence.

But by 1995, the ICPR exceeded its goals of reducing major pollutant emissions by more than 80 percent. The reintroduction of salmon became symbolic of the organization's success, and since 1990, some 5,000 salmon have migrated between the Rhine and North Sea, a feat made feasible by the construction of fish channels to circumvent dams.

Four "once in a century" floods in the 1980s and 1990s prompted a radical reformulation of river management. A shorter and faster Rhine was more prone to flooding, because the spring melt of Alpine glaciers coincided more often with the flooding of Rhine tributaries. Instead of constricting the Rhine further with dikes and levees, the ICPR set out to combine flood reduction with re-naturalization. In 2001, it presented Rhine 2020, a plan to increase biodiversity and relieve pressure on flood infrastructure through a patchwork of revitalized fluvial ecosystems.

Rhine 2020 also includes more stringent pollution regulations, especially against difficult-to-detect micropollutants, which can accumulate in organisms. Acres of cropland have been converted to pasture to limit erosion and nutrient runoff from farms.

The ICPR's efforts have largely succeeded in cleaning up the waters of the Rhine. Currently, treatment plants refine more than 96 percent of wastewater. Oxygen levels are suitable for indigenous species. Once-threatened beavers now maintain a stable population. Reduced pollution, runoff, and siltation allows Rotterdam to dredge only 35 million cubic feet (1 million cubic meters) from its harbor—only one-tenth of what was previously required.

Climate change poses another set of challenges for the management of the Rhine, from water flow and flood control to species preservation. Warming air and water temperatures have led already to faster evapotranspiration, while lower rainfall during summers has stressed riparian ecosystems along the Rhine—even as earlier spring snowmelts are disrupting some habitats.

Troy Vettese

Further Reading

Blackbourn, David. *The Conquest of Nature: Water, Landscape, and the Making of Modern Germany.* New York: Norton, 2006.

Cioc, Mark. *The Rhine: An Eco-biography, 1815–2000.* Seattle: University of Washington Press, 2002.

Disco, Cornelis. "Accepting Father Rhine? Technological Fixes, Vigilance, and Transnational Lobbies as 'European' Strategies of Dutch Municipal Water Supplies 1900-1975." *Environment and History* 13, no. 4 (2007).

Frijters, Ine D. and Jan Leentvaar. *Rhine Case Study.* New York: United Nations Educational, Scientific, and Cultural Organization (UNESCO), 2003.

International Commission for Protection of the Rhine (ICPR). "ICPR Brochures." http://www.iksr.org/index.php?id=254&L=3.

Leuven, Rob S. E. W., H. J. Rob Lenders, and Abraham Vaate. "The River Rhine: A Global Highway for Dispersal of Aquatic Invasive Species." *Biological Invasions* 11, no. 9 (2009).

Rhône River

Category: Inland Aquatic Biomes.
Geographic Location: Europe.
Summary: This major European river supports fertile lands where wildlife and plants flourish.

Descending precipitously out of the western Alps from an altitude of nearly 1.1 miles (1,800 meters) high, winding below the Jungfrau, Matterhorn, and Mont Blanc peaks, the Rhône River's relentless flow originates from the melting glacier bearing its name. The river then flows through a granite-based valley of braided streams and wooded flood plains, meandering 504 miles (812 kilometers) to its silt-strewn mouth at the Mediterranean Sea. From the soaring Alps in Switzerland, through France, and into the Mediterranean, the Rhône plummets sharply as it claims over 37,838 square miles (98,000 square kilometers) of watershed fed

A passageway through the ice of the Rhône Glacier in the Alps in Switzerland. The glacier is the origin point of the waters of the Rhône River. (Thinkstock)

by incessant glacial meltwater enhanced by periodic rain and snowfall.

The river's floods can be prodigious, rising to 10 or more times its expected annual flow in just one season. The Rhône has long been the gift of the Alps to the awaiting Gulf of Lion coastal region in southern France, nourishing along the way the forests, fisheries and flyways of the seasonally drier Provence region. More recently, the river is one of the few major freshwater sources for this sea—since the Nile River was heavily dammed in the last century at Aswan.

The Rhône-Alps climate is continental and mediterranean, with temperatures averaging 38 degrees F (3 degrees C) in winter and 67 degrees F (20 degrees C) in summer. Rainfall averages 3 inches (71 millimeters) monthly in summer and 2 inches (56 millimeters) monthly in winter, with an annual total of 32 inches (825 millimeters).

Biodiversity

Agriculture in the Rhône valley largely covers the low areas, plains, and islands. The Rhône's extensive flood plain provides wildlife with varied habitats: First, along the rapid descent through its alpine forested and glaciated valley, before abruptly bending southward at the city of Lyon, then watering mixed hardwood forests on its way into the wide brackish delta, the Camargue, south of Arles and west of Marseilles. Dry sclerophyll forests dominate the Mediterranean coastline around the Camargue; these areas are characterized by vegetation that is fire-adapted, if not fire-dependent, for reproduction.

The greater Rhône valley is a conduit for both native and invasive vegetation, wildlife, bird-flyways, and fisheries—from the warm, arid, southern Mediterranean climate to the cooler, more moist northwestern European maritime zone. The river valley is the only north-to-south watercourse in western Europe, suturing the landscapes that blend together the different flora and fauna from the North and Mediterranean Sea bioregions.

Fifteen types of forests have been identified in the upper Rhône, revealing the diversity of tree species that protect the soils of the watershed. Alpine species, such as grey alder, German tamarisk, and sea buckthorn, are replaced by ash, oak, hornbeam, and common apple as the river moves south between the Alps and the Massif Central through the warmer sections of Provence. Here are stands of fire-tolerant Aleppo oak, as well as stone pine, which requires moderate fire for its cones to release seed and re-forest the landscape. Also prevalent are wild orchids, which thrive in the lime-rich soil.

Thus, the Rhône's watershed provides habitat for both alpine and boreal forest vegetation, as well as ecotone, or overlap, areas. Oak, maple, ash, and alder are common in the rich uplands, while semiarid scrub forests and salt marshes are spread widely about the river's convergence with the Mediterranean.

However, agriculture, irrigated croplands, and vineyards have replaced much of the riparian woodlands, extensive wetlands, and original alpine forests, beginning gradually in the late Middle Ages. By the 14th century, forested areas had been significantly reduced for cultivation purposes. Such activity from the lowlands to the alpine meadows has driven wildlife to the edges of the watershed. Wolves, lynx, ibex, chamois, beavers, and mountain goats are among the more than 38 larger animal species still extant in the mountainous reaches of the watershed, which provide the most sheltered habitats.

The Rhône River basin is home to many protected species, nationally and locally, including 16 types of amphibians, 14 reptiles, 133 birds including the golden eagle, and the mammals. The biome is frequented by such birds as the crested falcon, Egyptian vulture, Lammergeier or bearded vulture, and many species of falcon, buzzard, and owl. Black and royal kites, and ospreys have also been spotted among the ranks of raptors.

Fish found in the Rhône include trout, pike, pike-perch, salmon, bream, catfish, and carp. While sport fishing is common, the sport is strictly policed for ecological reasons, particularly for the presence of bioaccumulated toxic chemicals.

Human Influences

Approximately 20 million people crowd the basin, with the river's natural flow now controlled by 19 dams that generate up to one-quarter of French hydroelectric power. Below its glacier source in Switzerland, the river flows into and out of Lake Geneva, where nearly the entire lakeshore has been affected by agriculture. Four nuclear power plants use huge amounts of the river's ample and mostly regulated flow of water to cool the fission reactors. After agriculture and navigation, the production of energy has had the most significant impacts on the river and the valley's ecological surroundings.

Large and small industries have been established throughout the region, most notably aluminum and chemical plants in Valais, oil refineries at Lyon, and the refineries and steel mills at Fos.

The Rhône has long attracted tourists, and this industry plays an increasingly important role in the regional economy. Recreational activities, from skiing and climbing in the Alps to horseback riding in the Camargue, help underpin the economy, but deliver various stresses to the biome.

Environmental Issues

Wetland loss exacerbates flooding and pollution from agriculture, industry, and nuclear power plants. Reporters noted in 2005 that sufficient polychlorinated biphenyls (PCBs) had been dumped in the river near Lyon to contaminate the river perch and other food fish to an extent such that fishing in the river had to be abandoned.

Deforestation, pollution, and erosion accompanying settlement, grazing, forestry, and agriculture have generated groundwater loss, which has triggered the retreat of native forest cover and altered the river's course from braided streambeds to a dammed series of controlled reservoirs, decreasing the biotic diversity of its fisheries. Dams block shad, lampreys, and eels moving up and down the river. With the extensive dam development, sturgeon became extirpated here in the 1970s, leading to a sturgeon recovery plan that is still in its incipient stage.

Climate change has the potential to upset the predictable amounts and timing of snowmelt that feeds the river and waters the valleys that nourish local wildlife and pasturelands. Rising temperatures are already driving many species to higher ground, where the flora base of the food web is under greater pressure to provide sustenance. Rainfall levels are more unpredictable; there have been a high number of major floods of the Rhône and other Alps-sourced rivers in the past two decades—but some longer-term projections suggest that drought could be the more severe threat.

Joseph V. Siry

Further Reading

Braudel, Fernand. *The Identity of France, Vol. 1.* New York: Harper & Row, 1986.

Palmer, Margaret A. Dennis P. Lettenmaier, N. LeRoy Poff, Sandra L. Postel, Brian Richter, and Richard Warner. "Climate Change and River Ecosystems: Protection and Adaptation Options." *Environmental Management* 44, no. 6 (2009).

Pautou, Guy, Jacky Girel, and Jean-Luc Borel. "Initial Repercussions and Hydroelectric Developments in the French Upper Rhone Valley: A Lesson for Predictive Scenarios Propositions." *Environmental Management* 16, no. 2 (1992).

Troussellier, H., H. Schäfer, N. Batailler, L. Bernard, C. Courties, P. Lebaron, G. Muyzer, P. Servais, and J. Vives-Rego. "Bacterial Activity and Genetic Richness Along an Estuarine Gradient (Rhone River Plume, France)." *Aquatic Microbial Ecology* 28, (May 16, 2002).

Riau Plantation Forests

Category: Forest Biomes.
Geographic Location: Indonesia.
Summary: These forest plantations of *Acacia* spp. provide raw materials for sustainable pulp, paper, and wood processing industries.

The government of Indonesia began aggressive support of forest plantation development in 1985, specifically targeting about 22.2 million acres (9 million hectares) of unproductive or degraded land around the archipelago. Forest plantation aims under this program are to provide raw materials for the timber and fiber (pulp and paper) industries, to improve the quality of local and regional environments, to provide job opportunities, and otherwise enhance human welfare in the localities.

Riau, a province in east-central Sumatra Island, has the largest forest plantation in Indonesia, with *Acacia* as the dominant genus. In 2010, total forest area in Riau province was about 21.3 million acres (8.6 million hectares), including critical lands of some 3.7 million acres (1.5 million hectares). Within the main forest region, Riau now hosts a total of 1.6 million acres (658,000 hectares) of *Acacia* forest plantation.

Acacia was selected as the featured tree genus in forest plantations here because of its ability to grow rapidly on marginal lands, its ample production of seeds, relatively simple silviculture techniques, and its multi-use wood that is suitable for pulp and paper manufacture. Two species of *Acacia* are commonly planted on different soil types, e.g. *Acacia mangium* is commonly planted on mineral soil of lowland forest, while *A. crassicarpa* can be planted on peat swamp area.

The wood of *A. mangium* or *A. crassicarpa* has suitable physical characteristics for pulp and paper industries. Both species produce long fibers, with favorable balances of cellulose and lignin, along with other characteristics that serve the end product well. For construction timber, *A. mangium* wood is better suited, with end-uses comprising plywood, furniture, light construction, handicrafts, utensils, and flooring.

Acacia is commonly planted as an even-aged monoculture system with planting space of 43–97 square feet (4–9 square meters) under intensive, if straightforward, silviculture practices. In the early development of forest plantation, *Acacia mangium* was planted on degraded land and *Imperata* grassland, so named for the genus of tough, opportunistic grasses that colonize fire-denuded or otherwise cleared areas. Seedlings of *A. mangium* are produced from generative and vegetative propagation. The trees can reach a height of 115 feet (35 meters), and a diameter at breast height of 20 inches (50 centimeters).

Acacia mangium leaves and seed pods. Acacia trees extend over 1.6 million acres (658,000 hectares) of the Riau Plantation Forests. (Wikimedia/ J. M. Garg)

The life cycle of *Acacia* for pulp and paper is about five to six years, and more than eight years for construction timber. The biomass volume output, depending on the quality of seedlings, runs to about 8–17 cubic yards per acre (20-40 cubic meters per hectare) per year. This inherently strong rate of growth is accentuated by the tropical climate these forests enjoy.

Environmental Issues

The sustainability of these forest plantations depends on site conditions, management intensity, and various impacts on the soil of the local ecosystem. The productivity of the Riau forest plantations were recorded as declining by the third generation. Improving seedling quality and returning mulched, unused biomass such as bark, twigs, and leaves to the ground along with phosphate fertilizer, have been the main responses of the industry. This is not necessarily positive for other species in the biome.

However, there are some indications of *Acacia* forest plantations making positive impacts on biodiversity—when compared with the previous land-use prior to and during its manifestation as *Imperata* grassland. More than one dozen species of *Acacia* seedlings actually get planted in the average plantation, due to inexact nursery practices. Among other biota, micro-flora, particularly mushrooms, are strongly represented here. One study showed 55 species of fungi from four families in an uneven but pervasive density across the research area.

Invertebrate soil fauna in the *Acacia* plantations tend to consist of reasonably dense populations of such featured types as the classes *Diplopoda* (millipedes) and *Arachnida* (spiders and mites), orders

Hymenoptera (ants, wasps, bees, and sawflies) and *Coleoptera* (beetles), and families *Blattidae* (cockroaches) and *Lumbricidae* (earthworms).

Important in considering the net value of the forest plantation program in Riau is the fact that launching and maintaining these forests has occurred during an historical period of extremely fast deforestation of this region. At the time of inception in 1985, Riau forest cover had declined to a still-robust 75 percent. However, while these mainly monoculture *Acacia* plantations have been established, the province has seen its forest cover overall diminished to less than 35 percent—in a span of roughly 20 years.

Logging, mining, and clearance for hydrocarbon extraction sites have been major factors. Other types of plantation forests—for palm oil and rubber commodities—have expanded alongside the *Acacia* projects. Clearly, the biota of the Riau province of Sumatra has embarked on a path of radical alteration. For a sustainable, ecologically sound role to be played by these plantation forests in the future of this biome, the stewards of the program must work on ways to integrate the stressed native species into surrounding habitats, provide natural migration corridors, and help to assemble more robust habitat areas from a fragmented landscape.

Hesti Lestari Tata
Nina Mindawati

Further Reading

Arisman, H. and E. Haridyanto. "*Acacia Mangium*: A Historical Perspective on its Cultivation." In K. Potter, A. Rimbawanto, and C. Beadle. Canberra, eds., *Heart Rot and Root Rot in Tropical Acacia Plantations*: Australian Centre for International Agricultural Research, 2006.

Hardiyanto, E.B., S. Anshori, and D. Sulistyono. "Early Result of Site Management in Acacia Mangium Plantation at PT Musi Hutan Persada, South Sumatra, Indonesia." In E. K. S. Nambiar, J. Ranger, A. Tiarks, and T. Toma, eds., *Site Management and Productivity in Tropical Plantation Forests.* Bogor, Indonesia: Center for International Forestry Research, 2004.

Sudarmalik, Mindawati N. *Site Characteristics of First and Second Rotation Acacia Crassicarpa Stands in Riau. (Proceedings National Seminar of Indonesian Wood Research Society XI.)* Palangka Raya, Indonesia: Faculty of Agriculture of Palangka Raya University, 2008.

Sudirno. "Plantation Forest Development Program in Riau Province." The Forest Dialogue. http://environment.yale.edu/tfd/uploads/Sudirno_HTI_English_ABSTRACT.pdf.

Rietvlei Wetland

Category: Inland Aquatic Biomes.
Geographic Location: Africa.
Summary: A hot spot for biodiversity and a birder's paradise, the Rietvlei Wetland Reserve offers a unique haven for nature and recreation in an otherwise developed region.

The Rietvlei Wetland Reserve is located in South Africa near the city of Cape Town and is part of the Table Bay Nature Reserve, spanning 2,175 acres (880 hectares). The wetland received its first official recognition in 1984, when the South African government proclaimed it a Nature Area. In 1993, the government officially declared the wetlands the Rietvlei Wetland Reserve, and now the area is identified as a Protected Natural Environment within South Africa. Future plans focus on the government declaring the wetlands area a Provincial Natural Reserve. The name *Rietvlei* originates in the Afrikaans language, with *riet* and *vlei* meaning *reed* and *marsh.*

Rietvlei is a wetland of approximately 1,638 acres (663 hectares) located on the floodplain of the Diep River. The wetland is characterized as a coastal freshwater wetland, but at times the salinity can rise as high as 13 parts per thousand as a result of evapotranspiration. (Seawater is generally 35 parts per thousand.)

The climate is moderate, ranging from a low of 47 degrees F (9 degrees C) during the region's two coldest winter months of July and August, rising to as much as 78 degrees F (25 degrees C) during the summer months of January and February; there is a rainy season of May to September. Total annual rainfall is approximately 27 inches (686.2 millimeters).

The Wetlands

The wetland occupies the low-lying area between two barrier dune systems. The landward edge was formed during the last interglacial period, while the seaward edge is part of an older dune system. The reserve is elongated parallel to the coast for about 3 miles (5 kilometers) and is 328 feet (100 meters) wide. It drains into Milnerton Lagoon and finally into Table Bay.

The wetland is characterized by a permanent freshwater lake, shallow seasonally inundated pans, extensive reedbeds, true riverine habitat, and a tidal lagoon with salt marshes open to the sea. The dominant vegetation is the reed *Phragmites australis.* The site receives about 17 inches (432 millimeters) of rainfall per year, and the climate is defined as semiarid.

There are five distinct wetland plant communities: perennial wetland, reed-marsh, sedge-marsh, open pans, and sedge pans. The perennial wetland has little aquatic vegetation, but does include *Ruppia*, *Potamogeton* and *Enteromorpha.* Dominant in the reed-marsh is *Phragmites*; in the sedge-marsh, *Bolboschoenus* and *Juncus* prevail. In the open pans are macrophytes, mainly *Limosella* and *Salicornia*, and in the sedge pans are *Bolboschoenus* in summer, and *Aponogeton* and *Spiloxene* in winter.

Anecdotal reports state that the wetland was considerably deeper in the past. Increased siltation since the late 1800s has markedly reduced the depth of the Diep estuary. Salt intrusion due to sea-level rise, tied at least in part to global warming, also has affected the Rietvlei wetland. It has been predicted that with sea-level rise of 3 feet (1 meter) or more, the Rietvlei will effectively become salt water and connected directly to the Milnerton Lagoon.

Wetlands provide the essential ecosystem service of filtering water for human use. Even though little direct water use in the area depends on the Rietvlei wetland, the health of the nearby lagoon is very much intertwined with surrounding wetlands. Invasive species are a problem for the functionality of the reserve. Specifically, Port Jackson willow (*Acacia saligna*), rooikrans (*Acacia cyclops*), and kikuyu grass (*Pennisetum clandestinum*) are a problem.

Biodiversity

The Rietvlei wetland is recognized as an Important Bird Area by BirdLife International, with up to 180 identified species, including pelicans, flamingos, ducks, coots, herons, plovers, weavers, and swallows. The wetland provides habitat for both residential and migratory birds.

The lagoon offers a nursery to several coastal fish, such as the harder and mullet. The Southern African Foundation for the Conservation of Coastal Birds (SANCCOB) manages a rehabilitation facility at Rietvlei. Although much of the surrounding region is becoming developed, the Rietvlei Wetland Reserve offers a sanctuary for such wildlife as the lesser flamingo (*Phoenicopterus minor*), and recreational activities.

Environmental and Conservation Issues

The most dramatic human modification to the wetlands occurred in 1974–76 when the entire northwest region was dredged to provide material for the construction of docks used for recreation purposes. To facilitate the dredging process, salt water was pumped into the marsh, creating the present-day permanent lake.

Conserving the Rietvlei wetland presents several challenges. Residential development of the municipality of Milnerton is increasing on the periphery of the wetland ecosystem, which has resulted in siltation caused by erosion, making the substratum very muddy. Development-induced erosion also has caused occasional eutrophication problems, leading in 2006–07 to fish death and blue-green algae formation. At this time, the reed vegetation is filtering the incoming canal water adequately, preventing water-quality problems in the lake or lagoon. It is unclear, however, whether water quality will continue to improve with increased human influence, or to what degree climate-change-induced soil degradation will make stabilization and recovery of the biome a tougher task.

PETER BAAS

Further Reading

Carr, Andrew S., et al. "Molecular Fingerprinting of Wetland Organic Matter Using Pyrolysis-GC/MS: An Example From the Southern Cape Coastline of South Africa." *Journal of Paleolimnology* 44, no. 4 (2010).

Grindley, J. R. and S. Dudley. "Estuaries of the Cape: Part II: Synopses of Available Information on Individual Systems." *CSIR Research Report* 427 (1988).

Hughes, P., et al. "The Possible Impacts of Sea-Level Rise on the Diep River/Rietvlei System, Cape Town." *South African Journal of Science* 89, no. 10 (1993).

Rio de La Plata

Category: Marine and Oceanic Biomes.
Geographic Location: South America.
Summary: This river is the second-largest estuary on the continent and represents an important area for the coastal fisheries of Argentina and Uruguay.

Rio de la Plata is an immense, funnel-shaped river estuary that runs from the juncture of two major rivers—the Uruguay and Paraná—to the South Atlantic Ocean. With Argentina on its south shore and Uruguay to the north, Rio de la Plata extends 186 miles (300 kilometers), broadening from 25 miles (40 kilometers) wide at the juncture to 124 miles (200 kilometers) wide near the sea. The immediate drainage basin covers 11,583 square miles (30,000 square kilometers). The combined hydrographic system of all three rivers extends from the subequatorial zone through the tropics, and includes part of Bolivia, Paraguay, Uruguay, Brazil, and Argentina.

As with any estuary, seawater meets freshwater to form an overall brackish mix. However, here a submerged shoal, the Barra del Indio, represents a geomorphological barrier in the Rio de la Plata, and defines a salinity front characterized by strong vertical salinity stratification. Marine waters (saltier and denser) penetrate further into the estuary along the bottom from this point, while freshwater advances oceanward on the surface, forming a salt wedge.

Hydrology

The high turbidity front in the inner part of the estuary constrains photosynthesis, and food chains are probably detritus-based more than founded upon plankton or algae. Immediately offshore from the turbidity front, however, water is less turbid, and phytoplankton increases. All the valuable species for the coastal fisheries concentrate near this area.

The physical characteristics of this estuary play a key role in the biodiversity of the benthos or bottom-dwellers, the nekton or free-swimmers, and the marine mammals. The reproductive processes of fish species occur in the outer part of the estuary.

The riverine zone of the Rio de la Plata, located in the inner part of the estuary, is characterized by shallowness with an average depth of less than 23 feet (7 meters) and a vertically homogeneous water column occupied by freshwater (very low average salinity of 0.225 parts per thousand). Water temperature averages 73 degrees F (23 degrees C) at the surface and 74 degrees F (23 degrees C) at the bottom.

In the estuarine section of the river, the average depth increases up to 23–82 feet (7–25 meters). Water temperatures here average 72 degrees F (22 degrees C) at the surface and 68 degrees F (20 degrees C) near the bottom. Beyond the estuary, in the marine zone, the water column is vertically homogeneous and characterized by high salinity (more than 30 parts per thousand). The temperature mean value in this offshore area is 68 degrees F (20 degrees C) along the entire column.

The shallow, highly turbid tidal river and the outer estuary open to the shelf are separated by a turbidity front, closely related to the salinity one. Its extent and location are highly variable, however, depending on the river discharge and wind forcing. Winds are extreme in the Río de la Plata region, and water column stratification is disrupted and the salt wedge becomes well mixed after several hours of strong onshore winds. Upstream of the turbidity front, within the tidal river, primary production is strongly light-limited. Downstream of this maximum, the concentrations of dissolved inorganic nutrients generally decrease rapidly as phytoplankton biomass increases along the salinity gradient. Upstream and downstream of the turbidity-salinity-temperature front, the species composition of fishes is completely different.

Biodiversity

There are several hundred species of fish in the Rio de la Plata biome, with around 100 of them ranging in both the estuary and in upstream habitats of the Uruguay and Paraná Rivers. Among the most abundant in the estuary proper, particularly toward the marine end, are whitemouth croaker (*Micropogonias furnieri*), stripped weakfish (*Cynoscion guatucupa*), king weakfish (*Macrodon ancylodon*), black drum (*Pogonias cromis*), Patagonian smoothhound (*Mustelus schmitti*), flounder (*Paralichthys orbignyanus*), and eagle ray (*Myliobatis goodei*). Hake (*Merluccius hubbsi*) is a migratory species that appears along the ocean front from May to October.

Additional fish species recorded here include red porgy (*Sparus pagrus*), hawkfish (*Cheilodactylus bergi*), and Parona leatherjack (*Parona signata*). Many tropical fish, such as the Buenos Aires tetra (*Hyphessobrycon anisitsi*), inhabit the Rio de la Plata; they are prey for the larger fauna, as well as being prized for the aquarium industry. There is also an abundance here of Argentine squid (*Illex argentinus*), quite valuable in the food web and as a commercial fishery take.

Many types of ray and shark patrol the estuary, and among their preferred prey is La Plata dolphin (*Pontoporia blainvillei*), a hallmark marine mammal species here. The La Plata dolphin is one of the smallest cetaceans, and among the river dolphins of the world, it is probably the one species that spends most of its life in saltwater and brackish environments. Green, leatherback, and loggerhead sea turtles depend on the marine and beach habitats of the Rio de la Plata at different points of their life cycles.

Along the beaches, lagoons, and marshes that line the estuary, and in some of the quite-dense forest that has crept up to the very edge of the Rio de la Plata, many resident and migratory bird populations are in evidence. Plovers, albatross, herons, and southern lapwings—occasionally joined by the flightless greater rhea—feast upon mollusks, small fish, seaweeds, and crustaceans. Hummingbirds—at least nine species—buzz around for nectar. A plethora of parrots haunt the forest, along with species like green-barred woodpecker, red-rumped warbling finch, and scarlet-headed blackbird.

Human Interaction

Human effect on the estuary through agriculture, cattle raising, and industrial and port activities have seriously damaged the environment and threaten the sustainability of various habitats around the Rio de la Plata. On the two shores of the estuary lie the capitals of each country: Buenos Aires, Argentina, and Montevideo, Uruguay—together with more than 13 million inhabitants. Ports, dredging, and commercial fisheries cause different ecological pressures. Upstream, several cities and other industrial centers located near the banks of the main tributary rivers also discharge wastes into the waters. The Rio de la Plata system is highly sensitive to changes in nutrient loading and freshwater input, which may modify the ecosystem structure by the development of harmful algal blooms and consequent eutrophication (nutrient-overload and oxygen depletion).

Land-use patterns, especially logging, agricultural clearance for food crops and biofuels, and general human infrastructure construction, all combine to accelerate forest loss and resultant erosion, and greatly increased sediment supply to the estuary. Drastic changes in the quantity and quality of organic inflows also exert a strong influence on the distribution, composition, and metabolism of the aquatic communities here.

Two important and relatively novel consequences of human impact should be considered in this environment: climate change and biological invasions. Annual mean temperature and rainfall significantly increased over the last 100 years—as has the mean sea level. These climatic tendencies, coupled with the direct effects of the increase in the human population, have generated an increase in algal bloom frequencies in the estuary. Such changes often favor invasion by exotic species; two important cases of this were *Limnoperna fortunei* and *Rapana venosa,* two mollusk species that first were recorded here in the 1990s. Both had extensive economic and ecological impact.

There are some efforts to protect native species and habitats. Near the confluence with the Paraná River, for instance, is the Reserva Ribera Norte, a sanctuary that hosts a wonderful range of habitats—willow forest, aliso forest, ceibo forest, riverine brushland, freshwater marsh, and others—and gives respite to a very broad range of flora and fauna. Since human population will tend to increase in the region, and climate change seems certain to intensify, such conservation initiatives in and around this important and unique biome are highly desirable.

María Gabriela Palomo

Further Reading

Acha, M. E., et al. "An Overview of Physical and Ecological Processes in the Rio de la Plata Estuary." *Continental Shelf Research* 28 (2008).

Mianzan, H. W., et al. "The Río de la Plata Estuary, Argentina-Uruguay." In U. Seeliger, ed., *Ecological Studies: Coastal Marine Ecosystems of Latin America*. Berlin, Germany: Springer-Verlag, 2001.

Quiros, R., J. A. Bechara, and E. K. de Resende. "Fish Diversity and Ecology, Habitats and Fisheries for the Un-Dammed Riverine Axis Paraguay-Parana-Rio de la Plata (Southern South America)." *Aquatic Ecosystem Health & Management* 10, no. 2 (2007).

Rio Grande

Category: Inland Aquatic Biomes.
Geographic Location: North America.

Summary: This river supports both humans and a diversity of biota from its headwaters in the southern Colorado Rockies to its mouth at the Gulf of Mexico, along the border of Texas and Mexico.

Originating as snowmelt east of the Continental Divide in the San Juan Mountains at Stony Pass, Colorado, the Rio Grande River travels approximately 1,900 miles (3,057 kilometers) to the Gulf of Mexico. The riparian environment of the Rio Grande biome encompasses close to 356,000 square miles (922,000 square kilometers), and this transnational basin covers portions of Colorado, New Mexico, and Texas in the United States; and Chihuahua, Coahuila, Nuevo Leon, and Tamaulipas in Mexico. The Rio Grande itself is also called Rio Bravo, mainly within Mexico.

Eight notable tributaries contribute to the flows of the Rio Grande: the Conejos River, Red River, Rio Chama, Jemez River, Rio Puerco, Rio Conchos, Pecos River, and the Rio de San Juan. Only the Colorado River exceeds the Rio Grande in size in the American southwest. However, much of the once-navigable, long river of alpine, desert, and coastal plains landscapes has been over-allocated for human purposes.

Geology, Geography, and Ecology

Within the high peaks of the San Juan Mountains in southwestern Colorado, the Rio Grande begins as meltwater near 12,500 feet (3,810 meters) in elevation. The primarily glacial topography here is marked by moraines, U-shaped valleys, and rugged montane spires. The subalpine to alpine vegetation shifts from aspen, pine, and willow trees to a mix of fir, pine, and spruce; and above the timberline, only tundra grasses pervade.

The majority of the Rio Grande High Country near the river's source falls under the supervision of the U.S. Forest Service within the Rio Grande National Forest—a territory equivalent to the size of Rhode Island. Rainbow and brown trout flourish in the crystal headwaters of the river.

As the river flows east by southeast, it drains off of the uplift dome of the San Juan Mountains into the intermountain depression of the San Luis

The Rio Grande at Santa Elena Canyon in Big Bend National Park on the Texas and Mexico border. The Canyon Section of the Rio Grande passes through the Chihuahuan Desert, where the heat and aridity limits vegetation to desert varieties, except for the mesquite and salt cedars (tamarisk) covering the floodplain. (Thinkstock)

Valley—between the San Juan range and the Sangre de Cristo Mountains—at an altitude of 9,000 feet (2,743 meters). From central Colorado to northern Mexico, tectonic plate extension, or divergent force, resulted in the Rio Grande rift. The deepest part of the northern Rio Grande rift is located in the San Luis Valley. Both western yellow and piñon pine are prominent across the valley here. Shifting course from southeast to south, the Rio Grande approaches the Taos Plateau north of the Colorado-New Mexico border. Millennia of fluvial erosion formed the Rio Grande Gorge in the basalt-rich landscape of the Taos Plateau region.

Flanked by the Colorado Plateau to the west, and to the east a series of mountain chains—from the southern Sangre de Cristo Mountains to the Franklin Mountains—the river transitions from the Taos Plateau to the Bolson Section. Beginning north of Santa Fe in the Española Valley, the Bolson Section extends south into the Trans-Pecos region of Texas. Arroyos, sand dunes, short grasses, and scrub vegetation such as sagebrush, piñon, juniper, and cedar signal the aridity associated with this stretch of Rio Grande environs that reaches the northern Chihuahuan Desert. Cottonwood galleries line the banks of the river and compete with invasive salt cedars.

Ponderosa pine, aspen, spruce, and fir trees intermingle on forested mountain slopes of the northern midstream basin. The Bolson Section provides sanctuary for migratory waterfowl such as geese, ducks, and sandhill cranes. Additionally, the endangered silvery minnow resides in the waters of the Middle Rio Grande basin.

Extending from the Quitman Mountains to Redford, Texas, the Presidio Section of the Rio Grande is dominated by the very harsh conditions of the Chihuahuan Desert. The Canyon Section stretches from Redford to Del Rio, Texas, and likewise exists within the Chihuahuan Desert. Big Bend National Park is located in the Canyon Section, a series of three canyons—the Santa Elena, Mariscal, and Boquillas Canyons—that are carved by the great river. Persistent heat and aridity limit vegetation to desert varieties, except for the mesquite and salt cedars (tamarisk) that blanket the floodplain. Common desert flora includes: sotol, ocotillo, Spanish dagger (yucca), candelilla, ceniza (silverleaf), chino grass, cacti, and lechuguilla (agave).

At higher plain elevations, Mexican buckeye, Mexican walnut, desert willow, Fresno (ash), and Mexican persimmon thrive. In the Chisos Mountains, piñon pine, juniper, mountain mahogany, evergreen sumac, and bluestem and sideoats grama grasses coexist with desert flora.

The climate shifts from arid to semiarid near Del Rio, then transitions to subtropical as the river approaches the Gulf of Mexico within the Coastal Plains Section of the Rio Grande. Texas live oaks rise above mesquite, chaparral, prickly pear cactus, and wildflowers along the river in this stretch. Swamp cypress trees emerge in the marshlands. In the delta region of the Coastal Plains, native ebony trees accompany prickly pear.

The Lower Rio Grande Valley is home to the only wild Muscovy duck population in the United States. Local fauna also include white-tailed deer, but the ocelot and jaguarundi are diminishing in number.

Human Interaction

Humans have long been inter-related with Rio Grande ecosystems. In Sandia Cave, near Albuquerque, New Mexico, within the Cibola National Forest, archaeologists from the University of New Mexico's department of anthropology discovered evidence of one of the earliest human settlements on the North American continent. Prior to the Sandia site, the Folsom culture of approximately 10,000 years ago was generally accepted as the earliest known civilization in this part of the world. However, geological evidence dates the Sandia remnants as slightly older than the Folsom artifacts. The faunal remains in the cave include sloth, wolf, horse, bison, camel, mastodon, and mammoth, and illustrate the diversity of the archaic Rio Grande biota.

The river has played a central role in geopolitical shifts in this region of North America. After more than two centuries under the colonial authority of New Spain, the Rio Grande basin transferred to Mexican control following the 1821 Mexican Revolution. However, the basin quickly became a contested zone and, in 1836, the Texas Republic claimed the Rio Grande as incorporated territory. Subsequent to the annexation of Texas to the United States and the U.S.-Mexican War, the Rio Grande—from El Paso to the Gulf of Mexico—was negotiated as the international boundary between the two countries as a condition of the 1948 Treaty of Guadalupe Hidalgo.

Agriculture and domestic livestock ranching involving cattle, sheep, goats, horses, hogs, and chickens command the greatest amount of Rio Grande basin resources, and have prompted significant ecological shifts. Considered a Mesoamerican transplant, corn has been cultivated for many centuries by Native Americans settled along the Great River, or Rio Grande. Likewise, for centuries humans have diverted water from the river to irrigate crops. Although the origin of *acequias,* or ditch irrigation networks, is debated by indigenous and Spanish descendants, the gravity-fed, traditional water-harvesting method supports the cultivation of native agricultural products and foreign cultigens including chiles, onions, potatoes, wheat, barley, oats, lettuce, watermelon, fruit trees, pecans, tomatoes, cotton, and alfalfa.

Environmental Issues

Since the mid-19th century, increased settlement in Rio Grande valleys has exhausted the natural flows of the river, and the competition among jurisdictions on both sides of the international border has exacerbated the limited availability of Rio Grande water in a predominately arid region. Interstate and international pressure for an equitable distribution of the river's volume resulted in the Mexican Treaty of 1906 and the Rio Grande Compact of 1938. The 1906 negotiation guaranteed Mexico 60,000 acre-feet (74 million cubic meters) of annual water to its canal near El Paso, and the 1938 agreement apportioned water among the states.

The construction of Elephant Butte Dam north of Truth or Consequences, New Mexico, was intended to facilitate the agreements and provide for continued development in the agricultural valleys above and below the dam. The U.S. government lifted a ban on new Rio Grande dams in 1907, which resulted in the construction of the Rio Grande and Abiquiu reservoirs and the Cochiti, Caballo, Amistad, and Falcon Dams. As a result, the river intermittently stops flowing at various points between El Paso and the delta, and the flora and fauna of the Rio Grande biome are constantly stressed.

Twentieth- and 21st-century environmental activism has alleviated some of the ecological impacts caused by human modification of Rio Grande ecosystems. Near the river's source, the Rio Grande Headwaters Restoration Project devised a plan to mitigate the impacts of streambank erosion, flooding, endangered species, and invasive plants. Elk have returned to riparian areas where grasses and willows have rebounded. Midstream improvements include flow meters to promote water conservation, reintroduction of native flora to compete with invasive salt cedars, nurturing endangered species such as the silvery minnow back to healthy numbers, and controlled over-bank flooding to support native flora and fauna.

The U.S. Fish and Wildlife service funds many of the local restoration projects. In the Lower Rio Grande Valley, the National Park Service hires workers to eliminate salt cedars in Big Bend National Park; the U.S. Department of Agriculture, in cooperation with the World Wildlife Fund, also provides resources to eradicate the destructive salt cedars in the Presidio Section of the river and to hem in the giant cane (*Arundo donax*) downstream from Big Bend. The University of Texas at El Paso developed and supervised a 372-acre (150-hectare) recovery of the Rio Grande bosque, or riparian gallery forest, a wetlands preserve for native vegetation, Mexican free-tailed bats, and 221 bird species.

Rising temperatures due to climate change could impact snowmelt-driven water levels and seasonal flow patterns, which would affect wildlife habitats. Increased incidence of drought is a symptom of stresses yet to come in this environment.

Matthew Alexander

Further Reading

Baxter, David. *Big River, Rio Grande.* Austin: University of Texas Press, 2009.

Belcher, Robert C. *The Geomorphic Evolution of the Rio Grande.* Waco, TX: Baylor University Press, 1975.

Finch, Deborah M., and Joseph A. Tainter. *Ecology, Diversity, and Sustainability of the Middle Rio Grande Basin.* Fort Collins, CO: U.S. Department of Agriculture, Forest Service, Rocky Mountain Forest and Range Experiment Station, 1995.

Gilpin, Laura. *The Rio Grande: River of Destiny.* New York: Duell, Sloan, and Pearce, 1949.

Maxwell, Ross A. *The Big Bend of the Rio Grande: A Guide to the Rocks, Landscape, Geologic History, and Settlers of the Area of Big Bend National Park.* Austin: University of Texas, 1968.

Rio Plátano

Category: Inland Aquatic Biomes.
Geographic Location: Central America.
Summary: This biome and Biosphere Reserve includes 35 ecosystems that represent habitat for more than half of Honduras's biodiversity.

The Plátano River biome is located on the Caribbean slopes of Honduras, within the Mesoamerican Biological Corridor. It includes beaches, dunes, lagoons, wetlands, lowlands, and mountains, containing 35 ecosystems (11 aquatic and 24 terrestrial) at altitudes ranging up to 4,350 feet (1,326 meters). The wetlands dominate the biome, and the aquatic regimes throughout the region model the species compositions. The fauna have wide seasonal variations. Along the river region, annual precipitation varies from 63 to 142 inches (1,600 to 3,600 millimeters) and temperatures average 80 degrees F (27 degrees C). The region is hit by four tropical storms and two hurricanes in an average decade.

The Plátano River biome comprises 2,027 square miles (5,250 square kilometers) within the La Mosquitia region; it was declared a Biosphere Reserve in 2000 by the United Nations Educational, Scientific, and Cultural Organization (UNESCO).

Biodiversity

Broadleaf evergreen forest is the dominant terrestrial ecosystem surrounding the Plátano River. Ten of the local ecosystems are not represented in the protected areas of the country—but all the ecosystems here represent an important part of the nation's biodiversity, which includes 10 percent of the flora, 27 percent of the amphibians, 36 percent of the reptiles, 57 percent of the birds, 68 percent of the mammals, and 70 percent of the freshwater fish, all totaling more than 700 vertebrate species. Endemism (species found only here) covers 30 flora, five amphibian, two reptile, and four freshwater fish species in the La Mosquitia lowlands of the biome.

Documented fauna include 39 species of mammals, such as Baird's tapir, white-headed capuchin, mantled howler and spider monkeys, brown-throated sloth, paca, kinkajou, white-nosed coati, Central American otter, puma, collared peccary, and white-lipped peccary. Rare or endangered species include the giant anteater, jaguar, ocelot, margay, Caribbean West Indian manatee, and Central American tapir.

More than 370 bird species have been recorded, including the king vulture; harpy eagle; great curassow; crested guan; scarlet, green and military macaws; and well over 125 species of reptiles and amphibian, including at least seven poisonous snakes; American crocodile; green iguana; and green, loggerhead, and leatherback seaturtles.

Broadleaf riparian forests along the river and its tributaries have tall canopies and are usually dominated by tree species such as *Albizia* spp., *Calophyllum* spp., *Inga* spp., *Cecropia* spp., *Ficus* spp., *Lonchocarpus* spp., *Ochroma* spp., and *Luehea* spp. Although most of the forests associated with the Plátano watershed are poorly known, it is clear that among the most important trees are the economically important *Swietenia macrophylla* and the ecologically important *Apeiba membranacea*, *Bursera simaruba*, and *Carapa guianensis*.

Ecological Zones

There are six ecological zones (EZ) within the Plátano River biome.

The marine EZ includes sea and coastal land where the rocky-coral-reef-clump habitats are present, with variations from the subtidal zone to immediate offshore areas down to about 656 feet (200 meters).

The beach EZ is characterized by homogeneous sandy substrates extending along 40 miles (65 kilometers) of the coastline, where the main river flows to the sea. The intertidal zone is used by four marine turtles species for nesting, mainly *Caretta caretta* and *Dermochelys coriacea*, and occasionally *Chelonia mydas* and *Eretmochelys imbricata*. In this EZ, the sand dune ecosystem is found, including herbaceous and shrub flora, sand dunes, and swaths of semideciduous forest.

The coastal wetland EZ has different ecosystems, depending on the saline influence. The salty lagoons with the highest saline input include marine species; the temporary lagoon populations depend on the season's regimes; and the canal ecosystem here is represented by the Plaplaya canal, which is used for transportation, and the Amatigni canal, where a dam was installed. Apart from the lagoon ecosystem, other habitat types here are mangrove, swamp forest, swamp semideciduous forest, and herbaceous swamps with palm ecosystems.

The river EZ historically has been used as a corridor for humans and wildlife, connecting all the ecological zones of the biome. In this zone are the main and tributary rivers and their estuaries, the meanders, the riparian forests, the seasonally flooded alluvial forests, and the agro-ecological gallery forest plant and animal communities.

The savanna EZ is an important influence on the water runoff and flood patterns, and is considered a seasonal local migratory site for many vertebrates. Eight ecosystems are located here: flooded savanna, floodplain, floodplain with pine, islets of thicket, savanna saturated with pine, semideciduous gallery forest, broadleaf forest, and submontane pine savanna.

In the broadleaf forest EZ, there are four types of broadleaf evergreen forests, each with different species compositions depending on the terrain and water availability.

Threats and Conservation

Some of the main threats to this biome include agricultural expansion by small farmers and cattle ranchers, which generally reduce the size of the remaining forests. Intensive extraction of both precious woods and wildlife seriously threatens the area, especially with the heavy logging of caoba (*Swietenia macrophylla*). Uncontrolled commercial hunting of wild animals is another problem, with the related challenge of introduced invasive species of many kinds.

As with most river systems in the tropics, only small portions are protected. The Rio Plátano Biosphere Reserve represents one of the few governmental protection initiatives for the Plátano watershed, but still has serious shortcomings from low active management practices and scant operative protection.

Climate change is of increasing concern to this tropical area, where the dry season has become more pronounced both from global warming and the continuously expanding deforestation over the last 20 years.

José F. González-Maya
Amancay A. Cepeda
Jan Schipper

Further Reading

Fraser, Elizabeth Ann. "Conservation Versus Survival: A Cultural Ecological Study of Changing Settlement Patterns, Cultures, and Land Use in the Rio Platano Biosphere Reserve of Northeast Honduras." Louisiana State University and Agricultural and Mechanical College. http://etd.lsu.edu/docs/available/etd-0707103-124904/unrestricted/Fraser_dis.pdf.

House, P., L. A. Padilla, O. E. Munguía, and C. Molinero. *Reserva de Hombre y la Biósfera del Río Plátano.* Tegucigalpa, Honduras: Diagnóstico Ambiental, 2002.

Stewart, Douglas Ian. *After the Trees—Living on the Transamazon Highway.* Austin: University of Texas Press, 1994.

United Nations Educational, Scientific, and Cultural Organization (UNESCO). "World Heritage Reports—Leveraging Conservation at the Landscape Level." http://unesdoc.unesco.org/images/0015/001508/150878e.pdf.

Rocas Atoll

Category: Marine and Oceanic Biomes.
Geographic Location: South America.
Summary: The only atoll in the southwestern Atlantic, Rocas has enormous ecological importance as a zone of shelter and reproduction for several species of seabirds and marine turtles.

The first biological marine reserve to be created in Brazil, the Rocas Atoll is a reef formation located 162 miles (260 kilometers) off the coast of the state of Rio Grande do Norte in northeastern Brazil, and 91 miles (146 kilometers) west of the Fernando de Noronha Archipelago. The atoll was discovered by Gonçalo Coelho, who left Lisbon in May 1503 with a fleet of six ships and Florentine navigator Americo Vespucci among the crew members. The expedition, which aimed to recognize and explore the territory that belonged to the Portuguese crown according to the Treaty of Tordesillas, was financed by Portuguese traders who were interested in raw materials from the new continent. The atoll was first documented on the 16th-century map of Cantino, while the first detailed chart, dating to 1852, was provided by Captain Lieutenant Phillip Lee.

One of the smallest atolls on Earth, and the only example in the southwestern Atlantic, Rocas (from the Spanish word for *rocks*) was formed primarily by vermetids, or worm snails, and coralline algae—rather than corals. The elliptical atoll is 2.3 miles (3.7 kilometers) long and 1.6 miles (2.5 kilometers) wide, with a surface of 79,074 acres (32,000 hectares), of which 17 acres (6.7 hectares) is land. Its reef ring, which is visible during low tide, consists of a natural wall 5 feet (1.5 meters) high bordered by sand banks, interrupted by a 656-foot (200-meter) passage on the northern side and by a smaller channel on the western side. Inside the atoll is a large lagoon of 2.7 square miles (7.1 square kilometers), in addition to several shallow pools 3–16 feet (1–5 meters) deep.

*The Rocas Atoll is the second-largest breeding area for green turtles (*Chelonia mydas*) in Brazil. In 1993, a permanent research station dedicated to the study of seaturtles was established in the area. (Wikimedia/Brocken Inaglory)*

The two small islands, Farol (meaning *lighthouse*) and Cemitério (*cemetery*), were formed by submarine mountain rock substrate and white "false" sand, which is composed of limestone, broken coral, and bones of birds and fish. At low tide, both islands become connected to Rocas, and the water in the pools is renewed. During high tide, only the islands remain exposed. Farol, the larger of the two, has a stretched "S" form; it is approximately 3,281 feet (1,000 meters) long and 656 feet (200 meters) wide, and reaches a maximum height of 10 feet (3 meters) above sea level.

The climate is tropical, and the rainy season lasts from January to August. Average annual rainfall is 29 inches (731.5 millimeters), but from March to July, this volume can mount to 7 inches (190 millimeters) in 24 hours. The average temperature varies from 74 to 89 degrees F (23.5–31.5 degrees C), and the relative humidity remains around 81.5 percent throughout the year.

Biodiversity

The Rocas Atoll has dense herbaceous vegetation that is resistant to high levels of salinity, sunlight, and variation of the tides. Species of amaranths anchored by dense tangles of rhizomes are found closer to the sea, whereas species of *Portulacaceae, Cyperaceae, Graminae,* and *Amaryllidaceae* grow farther inland. The area also has scattered coconut trees introduced by fishers.

A total about 150,000 individuals of 30 different species of birds have been identified in the atoll area. Five of them are endemic, or found nowhere else on the planet: the masked and the brow booby (*Sula dactylatra* and *S. leucogaster,* respectively), the brown and the black noddy (*Anous stolidus* and *A. minutus*), and the sooty tern (*Sterna fuscata*). Rocas shelters the largest breeding colonies of *Sula dactylatra* and *Anous stolidus* in Brazil, and of *Sterna fuscata* in the South Atlantic.

Migratory species include the white tern (*Gygis alba*), red-footed booby (*Sula sula*), magnificent frigate (*Fregata magnificans*), and red-billed tropic bird (*Phaethon aethereus*), characterized by its long tailfeathers.

Two species of lizard occur on the island: the endemic *Mabuya maculata* and *Tupinambis teguxim.* Although the latter was introduced to control rat populations, it preys on both the eggs and live young of birds and turtles. There are no mammals on the island.

Rocas Atoll is the second-largest reproductive area for green turtles (*Chelonia mydas*) in Brazil, after the volcanic island of Trindade. Since 1987, a program of the combined Projecto Tartarugas Marinhas (TAMAR) and the Brazilian Institute of Environment and Renewable Natural Resources (IBAMA) has been monitoring major populations of marine turtles in the archipelago. In 1990, the TAMAR-IBAMA project launched its first actions aimed at preserving these species. By the end of 1993, a permanent research station was established in the area.

Marine biodiversity is rich in the atoll, which is used by many fish species for spawning and as a refuge for juvenile fish. The shallow, warm waters also provide habitat for benthic, or bottom-dwelling, organisms such as algae (100 species), sponges (44 species), and coral (seven species). Of the 150 species of fish identified in the Rocas reserve, only two are exclusive to the area.

Conservation Efforts

Rocas is the first Brazilian biological marine reserve, created June 5, 1979. Together with Fernando de Noronha, the Rocas Atoll bears important habitats for migratory seabird populations and for marine-turtle nesting, and is a key site for the protection of biodiversity and endangered species in the Southern Atlantic. In 2001, the Biological Marine Reserve of Rocas Atoll was designated a World Heritage Site.

The waters surrounding the Rocas Atoll harbor important stocks of commercial fishes and crustaceans, which were one draw for heavy fishing activity around the atoll in the past. Visits to this natural reserve now are limited to scientific expeditions; the trip to Rocas takes about 26 hours, departing from Natal in Rio Grande do Norte. In general, teams consist of two IBAMA agents responsible for watching over the area, and additional members including scientists, students, and volunteers.

Warming temperatures could disturb the critical habitats here, disturb the intertidal zones and cause inundation above the shoreline, potentially weaken the immune response of the corals, and generally stress both flora and fauna. The same pressures could also allow new invasive species to enter the atoll.

Tatiana Coutto

Further Reading

Kenji Papa de Kikuchi, Ruy. "Rocas Atoll, Southwestern Equatorial Atlantic, Brazil." In C. Schobbenhaus, D. A. Campos, E. T. Queiroz, M. Winge, and M. Berbert-Born, eds., *Sítios Geológicos e Paleontológicos do Brasil*. Brasilia: Brazilian Commission of Geological and Palaeobiological Sites, 2002.

Schulz Neto, A. "Aspects of Seabird Biology at Atol das Rocas Biological Reserve, Rio Grande do Norte, Brazil." *El Hornero* 15, no. 1 (1998).

United Nations Educational, Scientific, and Cultural Organization (UNESCO). "Brazilian Atlantic Islands: Fernando de Noronha and Atol das Rocas Reserves." UNESCO World Heritage Centre. http://whc.unesco.org/en/list/1000.

United Nations Environment Programme (UNEP). "Brazilian Atlantic Islands: Fernando de Noronha Archipelago and Atol Das Rocas Reserves." http://www.unep-wcmc.org/medialibrary/2011/06/29/4f077436/Brazilian%20Atlantic%20Islands.pdf.

Rogerstown Estuary

Category: Marine and Oceanic Biomes.
Geographic Location: Europe.
Summary: Ireland's Rogerstown Estuary biome is a small marine habitat that has become a very important location for migratory birds.

Located on the east coast of Ireland some 16 miles (25 kilometers) north of Dublin, just south of the village of Rush, the Rogerstown Estuary covers a total area of 1.4 square miles (3.6 square kilometers) and is one of the major sites on the east coast of Ireland for wintering wildfowl and waders, and for birds traveling to and from the Arctic.

The estuary river meanders to the Irish Sea, with most of it having silted up over the years. This swampy and marshy area has resisted construction, and also has defied attempts over many centuries to claim much of it for agricultural use. This happens even though the mouth of the estuary is narrow, separating the Donabate Beach to the south and the Rush Beach to the north. A bridge once spanned the estuary, but it has long ago been dismantled.

The embankment, which in large part shelters the estuary from the bitter winds of the Irish Sea, has now been developed for housing along what is now Burrow Road, Portrane. Long before the houses arrived and the land was farmed, this bluff had one of the lowest variations between high and low tide of the various spots measured along Ireland's east coast. This largely perpetual swamp led to migratory birds favoring Rogerstown Estuary, which includes 484 acres (196 hectares) registered as a Nature Reserve and 588 acres (238 hectares) identified as a sanctuary for wildfowl. Historically, the Rogerstown Estuary was within the Pale—the area administered by the English from Dublin.

The climate of the Rogerstown Estuary and vicinity is cool, averaging 49 degrees F (9 degrees C) year round. There is little seasonal variation, from a mean 40–41 degrees F (4–5 degrees C) in January and February to a mean 58 degrees F (14 degrees C) in July and August. Rain falls approximately 200 days a year, averaging 37 inches (940 millimeters) annually, with wide fluctuations.

Biodiversity

Rogerstown Estuary attracts many species of birds, and the pale-bellied Brent (or Brant) geese are the most common species. The geese feed on eelgrass, seaweed and sea lettuce. As many as 8,722 of that species of goose were counted during 1980–86, and its worldwide population has been on the increase, which has put pressure on estuaries like Rogerstown to support this thriving fauna.

Numerous other bird species use the estuary throughout the year. From fall to early spring, Rogerstown visitors include shelduck, wigeon, teal, shoveler, goldeneye, and red-breasted merganser, as well as little egret, buzzard, sparrowhawk, peregrine, and kestrel. Also attracted to the area are such waders as golden and grey plover, lapwing, knot, dunlin, black-tailed godwit, curlew, redshank, and greenshank. In fall, little stint, curlew sandpiper, and ruff use the estuary.

Because there is a garbage dump nearby, large numbers of gulls are found in the area. In the fields and hedgerows are yellowhammers and finches in winter, and common warblers including sedge, willow, chiffchaff, and blackcap visit in summer. The lesser yellowleg visits the estuary for a short time—usually for just six days each July.

Vegetation in the estuary is comprised of saltwater marshes, raised saltmarsh, wet meadows, and riverine shallows and creeks; with silver birch, alder, and larch tree stands nearby.

Effects of Human Activity

The town of Rush, on the northern fringes of the estuary, has been inhabited since Neolithic times, and was the site of a Roman fort. The swampy estuary allowed smugglers including the legendary Jack Connor to hide here, and the infamous pirate Luke Ryan was born here.

The first major threat to the estuary in modern times was the construction of the railway bridge across it in the 1840s for the main railway line from Dublin to Belfast. Known as the Great Northern Railway Bridge, it cut the estuary into two parts, but it had less effect on the wildlife than it might have had in other ecosystems, as both sides are still saltwater marshes, with raised saltmarshes and wet meadows around most of them.

An environment that attracts so much bird life is also a place where many people fish. The Rush Golf Club at the mouth of the estuary is one such hub of activity. More intrusively, the Rush Sailing Club operates from the sheltered Rogerstown Harbor, located along a pier in the estuary. The biggest problem that faces the estuary in the long term, however, is a waste dump along the site. Although birds have been able to find food among the refuse, the long-term effect of pollution on the estuary could be far more serious, with a major long-term effect on the bird life.

The lesser yellowleg (Tringa flavipes), shown here, visits the Rogerstown Estuary for about six days each July. It is one of thousands of avian visitors to the estuary. (Wikimedia/Nigel)

Further, global climate change, and the warmer air and water temperatures that coincide with it, could alter roosting patterns within the estuary, and expand the number of species that frequent the site—attracting still more avians to compete over a rich but still limited food web here. On the ground, grasses and trees could be stressed by coastal erosion from rising seas, and higher siltation levels from the runoff caused by harsher storms inland.

Justin Corfield

Further Reading

Colgan, Nathaniel. *Flora of the County Dublin: Flowering Plants, Higher Cryptogams, and Characeae.* Dublin, Ireland: Hodges, Figgis & Co., 1904.

Doodson, A. T. and R. H. Corkan. "The Principal Constituent of the Tides in the English and Irish Channels." *Philosophical Transactions of the Royal Society of London. Series A, Containing Papers of a Mathematical or Physical Character* 231 (1933).

Dudley, Steve, Tim Benton, and Peter Fraser. *Rare Birds—Day by Day.* London: T. & A. D. Poyser, 1996.

Hutchinson, Clive. *Birds in Ireland.* Dublin, Ireland: Country House, 1986.

Hutchinson, Clive. *Ireland's Wetlands and Their Birds.* Dublin, Ireland: Irish Wildbird Conservancy, 1979.

Rub' al Khali Desert

Category: Desert Biomes.
Geographic Location: Middle East.
Summary: Known as the Empty Quarter, this desert nevertheless supports an array of drought-tolerant plants and some opportunistic fauna.

The Rub' al Khali Desert is the southernmost of the Arabian Peninsula major deserts, and is the largest sand desert on Earth. Located primarily in southeastern Saudi Arabia, and also spreading to parts of Yemen, Oman, and the United Arab Emirates (UAE), this desert is often called the Empty Quarter. The Rub' al Khali Desert is relatively uninhabited and infrequently explored, although some nomadic Bedouins, such as the Al-Murrah, range its perimeter.

Rub' al Khali comprises roughly 200,000 square miles (520,000 square kilometers); it is 620 miles (1,000 kilometers) across, east to west; and 310 miles (500 kilometers) wide, north to south. Surface elevation varies from 2,625 feet (800 meters) in the southwest to sea level not far from the Persian Gulf in the northeast. Feldspar in the mineral base produces a reddish-orange tint to the sand dunes rising to 820 feet (250 meters). Gravel and gypsum stretches cover much of its area.

Across the core of the desert are features of hardened calcium carbonate, gypsum, marl, and clay; these are thought to have been shallow lakes 5,000–37,000 years ago. Along with opal deposits, the calcium carbonate in particular indicates bygone plants and algae. Bits of flint and petrified wood found in quartzite mounds also speak to bygone vegetation and wetlands that likely hosted reeds, cattails, and salt cedar trees.

The Rub' al Khali evidences extreme temperature swings, with summer temperatures in some places shifting from below 32 degrees F (0 degrees C) at night to over 140 degrees F (60 degrees C) at noon. On average, the daily maximum temperature is 117 degrees F (47 degrees C) in July and August, and the daily minimum is 54 degrees F (12 degrees C) in January and February. The biome has very low humidity with scarcely any precipitation; in winter, exceedingly light rains may fall in the northern area, or mists may blow in off the Arabian Sea, stimulating vegetation for the following years. Annual rainfall of less than 1.4 inches (35 millimeters) indicates a hyperarid climate; there is no regular seasonal precipitation.

Shamal winds, generally from the northwest, shape the dunes and keep them shifting. The *shamal* also kicks up sandstorms. The southern monsoon is felt here in February and March, not as a rainfall event, but as strong persistent winds from the south.

Flora

The sea of shifting sand dunes, extremely high temperatures, and scant rainfall produce one of the driest places on Earth—only very hardy plants are supported here. Floral biodiversity is limited, with some sources reporting a total of 37 species of flora: 20 in the sandy desert and 17 on its edges. The ground-hugging woody shrub *Calligonum crinitum* is found on dune slopes. In the pans between dunes and scattered drainage channels are scrub flora such as *Dipterygium glaucum, Limeum arabicum,* and the fine-leafed succulent *Zygophyllum mandavillei;* sedge (*Cyperus conglomeratus*); and the trees *Acacia ehrenbergiana* and ghaf (*Prosopis cineraria*).

Other flora include the fire bush, or abal (*Calligonum comosum*), found generally in the UAE portions of the desert; and annual herbs such as *Danthonia forskohlii.* Oman's part of the desert has the Wahiba sands, with ghaf woodlands that extend approximately 50 miles (85 kilometers) in length by 12 miles (20 kilometers) wide. Other widely established flora in the Rub' al Khali Desert biome include saltbush (*Cornulaca arabica*), flowering parasitic plants called dhanun, and desert candle (*Caulanthus inflatus*).

Fauna

Fauna found more generally around the fringe scrubland and scattered stands of woodland include birds such as warblers, pied wheatears, ravens, falcons, and long-legged buzzards. The greater flamingo (*Phoenicopterus ruber*) took up breeding grounds at the Persian Gulf verge of the desert in the 1990s, following a 70-year hiatus. Houbara bustards (*Chlamydotis undulata*) have also been sighted.

Among mammals, the sand and mountain gazelle (*Gazelle subgutturosa*) and white or Arabian oryx (*Oryx leucoryx*) are protected in the Uruq Bani Ma'arid preserve on the western edge of the desert. The Nubian ibex (*Capra nubiana*) and Arabian oryx (*Oryx leucoryx*) are two endangered species that find some refuge here. Regardless of such human mandates, the Arabian wolf (*Canis lupus arabs*), striped hyaena (*Hyaena hyaena*), sand cat (*Felis margarita*), and Ruppell's fox (*Vulpes rueppellii*) prowl for such ungulates, as well as for smaller prey such as the brown hare (*Lepus capensis*).

The Rub' al Khali clearly had a more forgiving climate in prehistoric times. Fossilized bones of oryx, gazelle, camels, wild asses, and other large mammals have been found in the petrified lake mud, pointing to robust spreads of vegetation capable of supporting herds of herbivores. Some shellfish fossils are also reported. There is even fossil evidence of hippopotamus and water buffalo.

Tribes and Outsiders

Bedouin tribes such as the Al-Murrah and Al-Dawasir typically range the northern side of the Rub' al Khali Desert. To the east are seen their nomadic brethren the Al-Manasir and Al-Duru; while to the south and west are the Al-Kathir, Al-Rawashid, Al-Manahil, and Sa'ar tribes. These peoples have adapted and evolved their traditional ways of living in pastoral balance within this biome since time immemorial. Subsistence comes to them in the form of herding, small-scale farming, and hunting.

At the other end of the human activity scale are modern industrial enterprises from around the world that come here to explore, develop, and exploit such hidden riches of the Rub' al Khali region as oil, natural gas, sulfur, and phosphates. Shaybah, Saudi Arabia, near the eastern edge of the desert, has been a crude oil production site for more than 50 years. Infrastructure construction and pollution are threats to the environment of the Rub' al Khali, as are wildlife poaching, off-road driving, larger-scale agriculture—and even the incidence of camel and goat overgrazing by the nomadic herds.

CONNIE S. EIGENMANN

Further Reading

Clark, Arthur. "Lakes of the Rub' al-Khali." *Saudi Aramco World* 40, no. 3 (1989).

Dunham, K. M. "Population Growth of Mountain Gazelles (*Gazella Gazella*) Reintroduced to Central Arabia." *Biological Conservation* 81 (1997).

Thesiger, Wilfred. *Arabian Sands*. London: Penguin Books, 1959.

Rural Areas

Category: Grassland, Tundra, and Human Biomes.
Geographic Location: Global.
Summary: Rural areas, in addition to farms and farmers, contain forests, minerals, water, wildlife, and landscapes, and are the scenes of various transformation and development schemes.

Rural areas, regardless of where they are, consist of farms, forests, villages, ranches, small towns, and absence of busy roads or other large infrastructure. These are frequently food producing areas that contain such vital resources as water, minerals, fossil fuels, forests, and wildlife. Rural areas are valued for their scenic landscapes and natural environment, which people also use for various recreational activities.

Many indigenous groups with unique cultures, beliefs, and lifestyles tend to inhabit rural areas. Currently, pastoral communities are considered critical to ensuring global food demands and energy security. Rural areas are adaptable to climate change, can help preserve biodiversity, and conserve indigenous cultures. Therefore, many countries are working to preserve them.

Definition

The word *rural* is complex and its strict definition contested, making it difficult to draw a line between urban and rural areas. Derived from the Latin word *rus*, which means *open space*, it is synonymous with the words *country* and *countryside* in English-speaking countries. *Rural* also simply means places that have few people or cities. The term is a subjective state of mind for some and an objective quantitative measure for others. The most generalized definition of rural areas, one that's been used for hundreds of years, is that they are not urbanized.

The U.S. Census Bureau's classification of *rural* includes all territory with a population density of less than 500 people per square mile (2.6 square kilometers). Other countries use different measurements to differentiate rural areas from urban areas.

Transformation

Rural areas have been transforming rapidly, especially since humans settled and adopted agriculture. It was around settled areas, mostly surrounding river basins, that several human civilizations evolved. Over time, as civilizations evolved and cities were established as centers of commerce and administration, rural areas became thought of simply as sources of food.

The Industrial Revolution of the 19th century accelerated the creation of cities throughout the world. Since then, cities have grown, attracting millions of migrants from rural areas. Even though the world has become increasingly urbanized with more people living in metropolitan areas, rural communities still play a significant role in sustaining the urban population. The rate of urbanization varies by country, as developed countries are more

urbanized than developing nations. In the United States, for example, about 79 percent of Americans live in cities—but in Nepal, only 15 percent do. The type of urbanization also varies, but rural areas today, except in a few countries, generally have access to technology, transportation, commerce, and communication.

Hybrid Spaces

The rapid transformation of both rural areas and urbanization is blurring the distinction between the two, resulting in hybrid spaces, which have characteristics of both types of communities. Hybrid spaces include networks of agricultural production that combine human, nonhuman, and technological components, and they have processes of development for social or economic transformation purposes.

Some of these areas include recreational activities that depend on complex technologies and on particular landscapes and climates, transportation patterns comprising human-machine relations, and hybrid networks of local and extralocal agencies. The pervasive influence of urbanization and adoption of nonagricultural economic activities, urban cultural practices, and consumption patterns have made *rural* a hybrid concept.

In North America, some such areas are known as *suburban* or *exurban,* and they are characterized by rural landscapes and settlements that are tightly linked to urban labor markets and service centers. In recent years, real estate developers have attracted urban residents to suburbs by invoking images of an idyllic rural lifestyle, proximity to nature, solidarity and community spirit, high-quality public services, and cultural diversity. Such development also has raised concerns about environmental sustainability and inefficient use of resources.

Characteristics of Rural Areas

Despite rural-urban overlap, rural areas have several distinct characteristics, chiefly the closer attachment of people to nature. Often, rural areas are regarded as the last frontier, full of wilderness and open spaces. People who live in rural areas are viewed as being simple, hard-working, innocent, virtuous, peaceful, and living in harmony with nature. Also, rural societies are considered to have tightly connected structures of families and communities. Over time, these characteristics have been idealized and mythologized through art, literature, media, and movies.

The expansion of urban areas in the United States and Europe led to the growth of this romanticized version of rural areas. During the colonial period, Europeans exported their version of rural areas to colonies worldwide. They even tried to convert the existing landscape to reflect their ideas of countryside. British rulers, for example, established rural estates to re-create English country gardens in India, South Africa, and the West Indies. The transportation of rural ideas was not all one-way, but also flowed back from the colonies to Europe. Several crops were introduced to the Old World from the New World, most notably potatoes, beans, maize, and spices.

Economic Factors

Economically, rural areas are attached to natural resources, mostly agriculture and extractive industries such as mining, fishing, and forestry. In the past, these activities were the chief source of food and income for rural families. Even today, millions of people in developing countries depend on subsistence agriculture to support their families. This dependence on traditional agriculture is considered the primary reason for nondevelopment of rural areas. However, such theories are contested. As cities grew and became associated with advanced economy, rural areas became synonymous with backward economies plagued by poverty, traditionalism, and lack of opportunities.

Several factors contributed to the underdevelopment of rural areas. Historically, the abundance of resources in rural areas resulted in the exploitation of those resources through farming, forestry, mining, quarrying, fishing, hunting, and energy production. The growth of cities, industries, and expansion of a capitalist global economy facilitated the extraction of resources from rural areas. In the process, rural people abandoned traditional agriculture and became part of the global economic system by working for companies and

People working in a rice paddy in a rural area of Nepal. Rates of urbanization vary a great deal between developed and developing countries: while about 79 percent of Americans live in urban areas, by contrast, in Nepal, a developing country, only 15 percent are city dwellers. (World Bank/Curt Carnemark)

corporations. This transformation brought drastic changes to rural economies.

When the natural resources were exhausted, rural dwellers, instead of going back to traditional occupations, looked for other means of income. Many bucolic communities attracted industries at the cost of environmental and cultural degradation. In the United States, some rural communities now depend on such enterprises as waste facilities and penitentiaries for employment and revenue.

In pursuit of a better life and economic opportunities, thousands of people migrated to urban areas. Many of the families remaining in the rural areas, especially areas that have transitioned away from fully agriculture-based communities, now depend on government subsidies to fulfill their basic needs.

As the cities grew in size and population, the press of urbanization weighed heavily on urban dwellers, causing some to move to the countryside. In recent years, the urban-to-rural shift has in a sense become a major industry: rural tourism.

Rural communities all over the United States have adopted tourism as their major source of income. Rural tourism is flourishing in places that contain unique culture, biodiversity, and quaint, natural settings. Increasingly, however, rural areas are becoming economically vulnerable as they become dependent on a single industry or activity. The dependent economy, along with other factors, has made rural areas lag urban or suburban areas in development scale. As a result, rural development is a major objective for governments in both developed and developing countries.

Rural Development

Rural development strategies differ from country to country. The strategies depend on the nature

and availability of natural and social capital that can be mobilized for development.

Conventional approaches to rural development were based on the concept of modernization, which assumes the underdevelopment of rural areas. It assumed that underdevelopment could be overcome with sustainable economic growth that eventually would improve the living conditions of rural people and raise those areas to national standards of development.

Over the past several decades, many governments have been trying to update and improve living conditions in rural communities through enhanced agricultural services, economic conditions, infrastructure, and social values. Agricultural modernization involves mechanized farming practices, application of agrochemicals, and adoption of biotechnology. Economic modernization often entails diversifying rural economies by establishing small industries. Infrastructure updates include expansion of roads, electrical lines, telecommunications, and increased housing options. Social modernization promotes modern education, challenges superstitions and traditional beliefs, and raises awareness about civic rights.

After 1950, this four-pillar process became the dominant rural development paradigm in many parts of the world. Despite decades of investment, however, many rural areas have not been successfully developed along these lines. Since the early 1980s, criticism of this model has focused on environmental degradation versus sustainability, and the drawbacks of dependency on transnational banks and corporations.

A number of alternative paradigms have been proposed and applied. These approaches center on sustainable development, community participation, involvement of nongovernmental organizations, and environmentally friendly approaches. However, there is no universal model to achieve these goals, which have become vague and broad.

Future of Rural Areas

The future of rural areas depends on sociopolitical decisions and environmental changes associated with climate change. Policies of local, national, and international agencies on rural areas will influence the appearance of rural landscapes, the structure of their economies, the pattern of their settlement, and the dimension of the populations. The future also depends on the status of biodiversity, construction of infrastructure, commercialization of rural resources, and the quality of life of rural people.

Policies can have positive or negative effects. The greatest threat to rural areas, however, is climate change. On the positive side, rising global temperatures could increase the yield of a few food crops such as wheat, rice, corn, and sugarcane. The negative effects of rising temperatures include possible water shortages, reduced soil fertility, pest mobility, and increased frequency of flooding and wildfires. Such changes will disrupt the human-nature relationship in rural areas.

Rural tourism, which depends on the natural environment, will be hit hard by climate change as nonagricultural economic activities are forced to relocate, putting millions of people at risk. The response will be migration. Thousands of people have already moved out of rural areas and are living in urban slums or in neighboring countries as climate refugees. Rural areas, despite their crucial role in the urban world, have never been so vulnerable in recent history.

History also suggests that rural areas have the capacity to fight back through resiliency and adoption of new techniques. The fight against climate change has begun, for example, as rural areas from Africa to Asia are responding to the challenges facing them by adopting new techniques to conserve water, growing drought-resistant crops, and planting trees.

Bucolic communities also are diversifying their economies and moving away from dependence on a single activity. There are programs being implemented to raise awareness about the impact of climate change and ways to reduce carbon footprints. These efforts will take time to yield results. In the meantime, rural areas depend more than ever on urbanites to reduce emission of hydrocarbon gases.

Krishna Roka

Further Reading

Halseth, Greg, Sean Markey, and David Bruce. "The Next Rural Economies: Constructing Landesa." Rural Development Institute. http://www.landesa.org.

Shucksmith, Mark, David L. Brown, Sally Shortall, and Jo Vergunst. *Rural Transformations and Rural Policies in the U.S. and UK.* New York: Routledge, 2012.

Woods, Michael. *Rural.* New York: Routledge, 2011.

Ryukyu Islands

Category: Marine and Oceanic Biomes.
Geographic Location: Pacific Ocean.
Summary: The Ryukyu Archipelago is a biodiversity hot spot, rich in threatened endemic species and abounding in coral reefs.

The Ryukyu Archipelago is a chain of subtropical islands, also known as Nansei Shoto, that runs along the oceanic trench of the same name, and extends for approximately 621 miles (1,000 kilometers) from Kyushu in Japan to the eastern coast of Taiwan. This region, belonging to the Japanese Kagoshima prefecture, is divided into the Satsunan Islands in the north and the Ryukyu Shoto in the south. Amami-oshima and Okinawa-jima are the largest islands of these respective areas.

The climate of the Ryukyu Archipelago is considerably warmer than that of mainland Japan. However, due to the latitudinal gradient separating the islands, differences can be observed. The Satsunan Islands traverse a climatic cutoff, with the northern islands exposed to a temperate climate and the Amami islands subjected to a clear subtropical climate. Similarly, most of the Ryukyu Shoto exhibit subtropical flora, but the southwestern part, encompassing the Yaeyama Islands, is located in a tropical climatic zone.

Yakushima Island here, one of the wettest places on Earth with rainfall totalling as much as 394 inches (10,000 millimeters) per year, exhibits a variety of the climate and flora evident in the Ryukyus. While typical subtropical forests populate the lower slopes of the island, deciduous and coniferous forests such as the millennial yakusugi are located at higher altitudes, and in the surroundings of the Miyanoura-dake peak, reaching 6,348 feet (1,935 meters), alpine habitats can be covered by snow in winter. Overall, the general weather in the Ryukyus is marked by mild winters and hot summers, with high average precipitation. Extreme meteorological conditions can be recorded during the rainy season and the subsequent arrival of typhoons from July to September.

Biodiversity

Most of the 200 islands forming the Ryukyu archipelago are made from coral, and some of them are volcanic, such as Iwo-tori-shima and Suwanose-jima. Their geography is variable, but most of them are hilly or mountainous and covered with dense subtropical vegetation, including mangrove forests in some coastal areas. This allows abundant and diverse wildlife to flourish. The Ryukyus have become an evolutionary shelter for many species. Nowadays, the shelters are home to a large number of endemic (found only here) animals, including the venomous Habu snake, the threatened Okinawa rail, Amami jay, Ryukyu tip-nosed frog, and Ryukyu flying fox.

Moreover, the islands are a natural haven for numerous migratory species. Of the birds found here, 80 percent are non-resident varieties. Ryukyu's birds include the Amami woodcock, Izu thrush, Japanese paradise flycatcher, Narcissus flycatcher, Okinawa rail, ruddy turnstone, Ryukyu kingfisher, minivet, robin, and Scops owl.

Endangered loggerhead and green turtles use the islands, from which their migration starts, as nesting sites. About half of the island chain's amphibians are endemic to the islands, including the sword-tail and Anderson's crocodile newt; and such frogs as Holst's, Otton, Ishikawa's, Ryukyu tip-nosed, Namiye's, and the Kampira Falls frogs. Among the local lizards are Kishinoue's giant skink and Kuroiwa's ground gecko.

Living Fossils

Certain animals, such as the Okinawa woodpecker and the Ryukyu long-tailed giant rat, are emblem-

The Iriomote Cat

At the western tip of the Ryukyu Archipelago, on a mountainous island called Iriomote, an ancient animal has remained in its primitive form. The Iriomote cat is a solitary wildcat, about the size of a normal domestic cat but phenotypically close to the leopard from which it is thought to have separated in evolution millions of years ago. Its grayish fur exhibits dark spots forming rows, while its tail is short, thick, and bushy.

The Iriomote cat is able to sheathe its claws and swim, so it is able to cross rivers and hunt fish. While this cat has inhabited the Ryukyus since far before the human era, its population today has been reduced to about 100 individuals, as a result of habitat destruction and overhunting in recent decades, and it is still declining.

This Iriomote cat, now preserved by taxidermy, was rescued by the Iriomote Wildlife Conservation Center in the Ryukyu Islands. (Wikimedia/Purplepumpkins)

atic of the Ryukyu Islands. The most fascinating ones are probably the so-called living fossils. This oxymoronic term, coined by Charles Darwin, designates a living species that appears similar to one that otherwise is known only from fossils and has no close living relatives. Living fossils typically have endured through the millennia because of their confinement in an area with reduced competition for resources, and they represent unique remnants of the past.

The Amami rabbit is the most renowned illustration of a living fossil in this part of the world, and has become the symbol of the eponymous region. This species is notable for its small ears and sounds it makes to call its extended family members, features that are the result of 10–20 million years of genetic isolation. Although extremely furtive, this dark-furred, short-legged rabbit can be spotted at night on rare occasions. During the day, it sleeps in burrows or caves, making it particularly vulnerable to human activities such as deforestation.

Human activity and other factors, such as road kills, habitat fragmentation, and predation by feral cats and introduced mongooses, have contributed to population declines—leading to the classification of this species as endangered by the International Union for Conservation of Nature.

Environmental Threats

The biota of the Ryukyu Islands is already deeply endangered as much in the coastal waters as on land. The abundant coral reefs, which are part of the World Wildlife Fund's Global 200 ecoregions, feature the largest colony of blue corals in the world, and are prey to sedimentation and eutrophication, or nutrient-overload and oxygen depletion. While the islands have a low human population and are poorly developed, local agriculture and fisheries, as well as increasing urbanization, have been putting the ecosystems under stress and causing devastating effects on wildlife.

As an island chain, climate change with its associated warming seawater and heavier storms has started delivering detrimental effects on the coral reefs. The prospect for continually rising sea levels, meanwhile, is leading to great concern over flooding and erosion of shoreline habitats.

Christian Vincenot

Further Reading

Kerr, George H. *Okinawa: The History of an Island People.* North Clarendon, VT: Tuttle Publishing, 2000.

Pearson, Richard J. *Archaeology of the Ryukyu Islands: A Regional Chronology from 3000 B.C. to the Historic Period.* Honolulu: University of Hawaii Press, 1969.

Røkkum, Arne. *Nature, Ritual, and Society in Japan's Ryukyu Islands (Japan Anthropology Workshop Series).* London: Routledge, 2006.

S

Sahara Desert

Category: Desert Biomes.
Geographic Location: Northern Africa.
Summary: The largest hot desert in the world, the Sahara is home to diverse hardy species of plants and animals that have evolved and adapted to its rigors.

The Sahara Desert is the largest hot desert in the world, second in scope only to the vast arid wastes of frozen Antarctica. The Sahara stretches across 11 mostly large countries in northern Africa, from the Atlantic Ocean to the Red Sea, blanketing more than 3.5 million square miles (9.0 million square kilometers), more than one-fourth of the entire continent. The core desert biome encompasses about half that area, while the greater Sahara includes a full range of semiarid and related ecosystems.

The main desert consists of gravel plains, sand dunes, sand seas, salt flats, harshly weathered bare rock, a sprinkling of precious oases, and a topography ranging from the Atlas mountain range in its northwest to vast depressions in its interior. Sand dunes stand up to 590 feet (180 meters) in some areas; these are eclipsed in height by the mountains, among which volcanic Emi Koussi, in Chad, is tallest at 11,204 feet (3,415 meters). The lowest point is Qattara Depression, 436 feet (133 meters) below sea level, found amidst salt pans in northwestern Egypt. The world's longest river, the Nile, flows south-to-north through the entire eastern side of the Sahara region; the mighty Niger defines its border along the southwest.

In this land of stark contrasts, the temperature range is no less extreme. A single day typically can sustain a high of over 100 degrees F (38 degrees C) and a low of 32 degrees F (0 degrees C). But the heat can climb much higher: one of the highest temperatures ever recorded on Earth occurred in the offshoot Libyan Desert in 1922, 136 degrees F (58 degrees C). Yet, there has been snowfall here, as well, although it is rare. One such celebrated event occurred in the sand dunes of southern Algeria in 1979.

Rainfall amounts to about 1 inch (2.5 centimeters) per year, averaged across the vast arid zone. Many thousands of years ago, due to low-pressure weather systems in the Mediterranean that were driven by the glaciation upon Europe, this region of northern Africa was treated to a fairly high precipitation regime, and the landscape was verdant. Upon glacial retreat, the northern expanse of the Sahara became steadily more arid, even while monsoon climate prevailed across its southern reaches. Once the monsoon pattern shifted away to the south and east—roughly 5,400 years ago—the desertification of today's Sahara was complete.

Biota

Approximately 500 plant species have been identified across the Sahara, a rather low count for so vast an area. Arid-adapted plants, or xerophytes, predominate. Xerophytes tend to have very deep and well-developed root systems to tap groundwater; specialized internal tissues to store moisture; and reduced-size leaves often modified into spines and thorns, sometimes covered by a waxy layer—all to resist transpiration and promote water retention. Other prevalent plant types are the ephemeral varieties—those that sprout, grow, and regenerate only when the relatively rare moist periods occur in the desert. A third, fairly abundant, type is the halophyte, or salt-tolerant group.

In the Sahara, frequently seen trees, shrubs, and herbs include acacia, salt cedar, cypress, olive, fig, magaria tree, date palm, doum palm, oleander, thyme, African peyote cactus, and African welwitchsia. Grasses include lovegrass, or eragrostis; desert bunchgrass; and three-awn, or aristida. Those species with higher moisture requirements tend to cluster in transition areas around the fringes of the core desert, or at oases, where aquifer water has managed to percolate up fairly close to the surface. Among the fauna, large mammals are represented by dorcas gazelle, dama deer, oryx, dromedary camel, spotted hyena, and the endangered Sarahan cheetah. Smaller mammals include gerbil, kangaroo rat, Lybian striped weasel, slender mongoose, and cape hare.

The reptiles of the Sahara range from skinks, chameleons, and toads to horned vipers, desert crocodiles, and monitor lizards. Guinea fowl, Nubian bustard, pale crag martin, fan-tailed raven, black-throated fire finch, and African silverbill are among the distinctive avian species of the biome.

Human Interaction

The human population in the Sahara desert has been estimated to be between 2.5 million and 4 million, yielding one of the lowest population densities in the world. It is believed that the desert has been inhabited for the last 6,000 years. The dominant ethnic groups and tribes include Berbers, Tuaregs, Toubou, Beja (Sahrawi), Zinghawa, Songhai, Kanuri, Fulani, Hausa, and Nubians. The main language spoken in the desert among the tribes is Arabic. Indeed, the name *Sahara* itself derives from the Arabic word for desert, which is *sahra.* This, in turn, is a form of the term *asharu,* meaning yellowish-red.

Today, the majority of the people who dwell here are nomadic pastoralists, moving from place to place with their herd animals, which are mainly sheep and camels. Except scattered oasis villages, a few crossroads settlements, mining towns, and

Grasses, palm trees, and shrubs surround a small body of water at the Ubari Oasis in the portion of the Sahara found in southwestern Libya. Oases like this, where aquifer water reaches the surface, support vegetation that requires more water than is available in the rest of the desert. (Wikimedia/Sfivat)

cities around the fringes where extraction industries are based, the majority of the people are of this transient nature. Some people around the edges of the Sahara, however, practice agriculture grounded in irrigation from the fertile Nile River Valley, for instance. The potential also exists for drilling wells into some of the newly discovered aquifers that underlie various parts of the desert.

Deposits of iron ore in Algeria and Mauritania, copper in Mauritania, oil and gas in Algeria and Libya, and phosphates in Morocco and Western Sahara have fueled economies and led to population increases in those areas. Each development, of course, has also raised concerns about ecological damage and habitat disruption. Several trans-Saharan highway plans have been proposed, and some partly built, but these largely remain uncompleted or abandoned.

Global warming projections for the Sahara Desert biome are extremely divergent. Some records show, for instance, that climate change has already caused the Sahara to spread beyond some of its historical boundaries. In Sudan, rainfall has declined steadily for decades, as desert has encroached upon farmlands and grazing areas; increased scarcity of fertile soil and available water has fed into such conflicts as the warfare in the Darfur region of Sudan.

On the other hand, more recent studies point to increased rainfall and the spread of vegetation in previously barren areas of Chad, southwestern Egypt, and western Sudan. Specifically, new stands of acacia trees have been recorded in these areas. Researchers have posited this is due to the capacity of hotter air to hold more moisture, and thus to release more rain. A similar scenario may be playing out in parts of Moroccan-controlled Western Sahara, where pastoralists have reported an expansion of grazing lands fed by heavier rainfall in recent years.

TEMITOPE ISRAEL BOROKINI

Further Reading

Gearon, Eamonn. *The Sahara: A Cultural History.* New York: Oxford University Press, 2011.

Owen, James. "Sahara Desert Greening Due to Climate Change?" *National Geographic News.* http://news.nationalgeographic.com/news/2009/07/090731-green-sahara.html.

Strahler, Arthur N. and Alan H. Strahler. *Modern Physical Geography, 3rd Ed.* New York: John Wiley & Sons, 1987.

Wickens, Gerald E. *Ecophysiology of Economic Plants in Arid and Semi-Arid Lands.* Berlin, Germany: Springer, 1998.

Sahelian Acacia Savanna

Category: Grassland, Tundra, and Human Biomes.

Geographic Location: Africa.

Summary: This enormous band of semiarid habitat in subtropical Africa abuts the Sahara Desert; it is known for the periodic occurrence of severe droughts and famine.

An enormous band of arid and semiarid habitat, composed of three characteristic ecoregions, crosses the width of Africa. The Sahelian Acacia Savanna biome occupies the middle third, spanning the continent from the Atlantic coast of Senegal in the east to the Red Sea coast of Sudan. This biome has pronounced seasonality in precipitation, with the rains falling from May to September, followed by a six- to eight-month period of drought. Precipitation varies from 79 inches (2,000 millimeters) per year in the north of the Sahelian acacia savanna, where it transitions into the Sahara Desert, to about 236 inches (6,000 millimeters) per year in the south, where it borders the wetter Sudanian savanna. In fact, the name of the Sahelian region as a whole derives from an Arabic word for *shore* or *coast*, referring to how this *sahel* is a transition in type from the sandier, less vegetated Sahara Desert.

Historical and current human activity in the region is strongly tied to the availability of water, with rain-fed agriculture uncommon in the more arid northern regions, but more viable in the wetter southern regions. Typical crops include millet and sorghum, which may be abandoned in drought years. Under these conditions, abandoned land is degraded as drought intensifies the effects of the loss of native vegetation. Transhumance, the seasonal migration of nomadic people and their livestock from the drier north to the wetter south, has been by far the most important traditional pastoral use of the Sahel.

Vegetation

Native vegetation here varies with latitude and topography and includes both wooded grassland and deciduous bushland. Among the most common woody species are *Acacia tortilis* and *A. laeta*. Most trees and shrubs are deciduous, dropping their leaves for much of the yearly dry period. In addition to providing forage for goats and cattle, many native savanna trees provide for other human uses. Gum arabic, a derivative of the sap of two acacia species, especially *A. senegal*, has a variety of traditional and commercial uses that make this species of great human importance.

Other tree species serve as valuable sources of fuel, fence wood, and traditional medicines. By their very presence, savanna trees also provide services such as soil stabilization and the creation of fertile sites for small-scale crop production. Annual grasses such as *Cenchrus biflorus*, *Schoenefeldia gracilis*, and others are also common.

Fauna

The Sahelian acacia savanna is sparsely populated, with somewhat higher human population densities in its wetter, southern portions. This vast ecoregion has been greatly altered by a combination of drought and human-caused alterations in the landscape, particularly grazing and browsing by livestock. Its once large and diverse herds of native ungulates and their associated predators have been significantly diminished by these changes, and also by over-hunting for sport and food, made possible by modern vehicles and firearms. However, many of these species continue to persist within scattered protected areas established in the region. Some of the countries in this region are among the world's poorest, and resources for conservation are limited.

Compared with other tropical parts of Africa and other continents, the Sahel ecoregion today does not support a high diversity of species. However, it does contain a few dozen endemic (found only here) vertebrate species, including 10 reptile and four gerbil species (genus *Gerbillus*). Diverse and abundant communities of native ungulates, including gazelle, oryx, and hartebeest species, were once widespread but now are mostly confined to a few protected areas. Associated large predators including the African wild dog (*Lycaon pictus*), cheetah (*Acinonyx jubatus*), and lion (*Panthera leo*) have suffered similar fates, as has the West African giraffe (*Giraffa camelopardus*).

Two species of ungulate, the scimitar-horned oryx (*Oryx dammah*) and the bubal hartebeest (*Alcelaphus busephalus busephalus*), a subspecies of the common hartebeest, are now thought to be extinct in the wild. The region's wetlands, both permanent ones like those in the inner Niger delta and ephemeral ones that form elsewhere here in the wet season, provide a critical link for annual bird migrations along the Afrotropical-Palearctic flyway.

Across this region, vertebrate biomass has clearly shifted from native ungulates to cattle and goats. In the east, camels are important livestock. The effects of livestock and grazing on native vegetation have been most pronounced near the permanent waterholes.

Environmental Threats

The future prospects of the Sahelian acacia savanna are strongly tied both to the actions of its human population and to changes linked to the global climate. Though population densities are comparatively low, from one to five people per 0.4 square mile (1 square kilometer) in the north to 50 to 100 people per 0.4 square mile (1 square kilometer) in the south, the human population here continues to grow at around 3 percent per year, a rate that will double its size in about 20 years.

In a part of the world where poverty levels and rates of land degradation are high, a warming and drying climate is likely to mean that the chance of famine in the Sahel, a region already known for this problem, will remain high. Further degradation of the land and disruption of various habitats will go hand in hand with this crisis.

The history of other parts of the developing world show that programs to improve education—especially of women—and to alleviate poverty are effective long-term solutions to slowing population growth. Continued technical and financial support for agriculture from the developed world will also be needed as this growing population moves further into the 21st century. Restoration of its degraded lands and protection for its native megafauna—now mostly restricted to several national parks scattered throughout the region—will also require more effort. Increased attention to all three areas would pay dividends to its burgeoning human population and help increase the political stability of the region.

JOHN MULL

Further Reading

Bourliere, F., ed. *Ecosystems of the World, Tropical Savannas.* Paris: United Nations Educational, Scientific, and Cultural Organization (UNESCO), 1983.

Raynaut, C. and E. Gregoire et al. *Societies and Nature in the Sahel.* London: Routledge, 1997 .

Salgado, S. *Sahel: The End of the Road.* Berkeley: University of California Press, 2004.

Stewart, R. "Desertification in the Sahel." http://oceanworld.tamu.edu/resources/environment-book/desertificationinsahel.html.

Salcombe-Kingsbridge Estuary

Category: Marine and Oceanic Biomes.
Geographic Location: Europe.
Summary: This unusual estuary in southwestern Britain has diverse flora and invertebrate fauna, but must contend with invasive species.

The Salcombe-Kingsbridge estuary is in the South Hams district of the county of Devon in England. Salcombe and Kingsbridge are small towns lying near the mouth and head of the estuary. An estuary, the coastal part of a river valley that is influenced by the tide as it approaches the sea, typically extends from the upper limit of brackish water to the increasingly saline marine conditions of the open sea. The Salcombe-Kingsbridge is no different in this regard.

Hydrology

Sediment washed down by the river and an environment sheltered from wave action have led to the development here of extensive intertidal sandflats, mudflats, and sediment-filled subtidal channels. These wetlands at the margins of the land and the sea form a link between marine subtidal, marine intertidal, freshwater, and terrestrial ecosystems.

The Salcombe-Kingsbridge Estuary biome is a specific type of estuary called a ria. The word *ria* (from the Galician language of northwestern Spain) refers to an unglaciated river valley partially flooded by rising sea levels. As typical of rias, this estuary has a treelike form, with an irregular coastline and numerous side channels. Also typical of rias is the absence of a large river. The Salcombe-Kingsbridge Estuary instead is fed by a few small streams that rise from nearby springs. Consequently, the estuary is primarily a marine system, with limited freshwater inflow, and it covers a large area relative to the size of the feeder streams.

This estuary is macrotidal, with a tidal range of 15 feet (5 meters). The tidal influence runs 5 miles (8 kilometers) inland to Kingsbridge. The shoreline is 30 miles (49 kilometers) long, and the estuary covers an area of 1,665 acres (674 hectares), with an intertidal area of 1,102 acres (446 hectares).

The estuary has been designated as being within the South Devon Area of Outstanding Natural Beauty; it also forms part of the South Devon Heritage Coast. It supports some rare habitats and species, leading to its designation as a site of special scientific interest and a local nature reserve.

The lower estuary consists of rocky shoreline and sandy bays, and is partially cut off from the open sea by a sandbar. The bar is exposed during low spring tides; when strong southerly winds blow, entrance to the estuary is more difficult regardless of tides.

Biodiversity

In the lower estuary, the seabed consists of sand and mud, supporting numerous species of burrowing fauna, and colonized by important seagrass beds and especially eelgrass (*Zostera marina*) beds. Eelgrass, also called seawrack, is unusual because it is a flowering plant that lives in fully marine conditions. The good water quality and the eelgrass beds in the Salcombe-Kingsbridge Estuary provide a home for one of Britain's rarities: the spiny seahorse (*Hippocampus histrix*). Eelgrass also helps stabilize loose estuarine sediment and provides an important nursery habitat for juvenile fish.

In the upper estuary, large areas of intertidal mudflats are exposed at low tide. These flats contain abundant tube-living and burrowing worms, bivalves, anemones, snails, and small crustaceans. These in turn provide important food sources for wading birds and fish within the estuary.

The estuary also provides conditions suitable for the fan mussel (*Atrina fragilis*). This mussel is large, often growing to more than 16 inches (400 millimeters). It is a long-lived, bivalve mollusk, and occurs sparsely in a few locations around Britain. It is one of the United Kingdom's rarest and most threatened species. Four types have been found in the Salcombe-Kingsbridge estuary, one of which is particularly unusual because it lives intertidally rather than subtidally.

Numerous species of resident waders and wildfowl feed, nest, and overwinter in the sheltered waters of the creeks, and seabirds forage on the outer shores of the estuary or dive for fish beyond the sandbar. Migrant waders also feed on marine invertebrates from the mudflats before or during their long journeys. Larger animals including otters, dolphins, seals, and basking sharks also come into the estuary to feed.

The estuary is now home to several invasive species, including a seaweed commonly known as japweed (*Sargassum muticum*) and the gastropod known as slipper limpet (*Crepidula fornicata*). The numbers of slipper limpets, a kind of marine snail, have increased drastically over recent years. They can occur in vast numbers, often completely smothering the seabed. They are a serious pest of oyster beds and have detrimental effects on scallop fisheries in the estuary. Most scallops here now have slipper limpets on their shells, which may affect their ability to move and their vulnerability to predation. This development also has commercial implications. Scallops covered with slipper limpets require more time for sorting and cleaning, and are less acceptable at market. The buildup of pseudofeces from the slipper limpets is also changing the chemical and nutrient nature of the seabed, with potential effects on biodiversity.

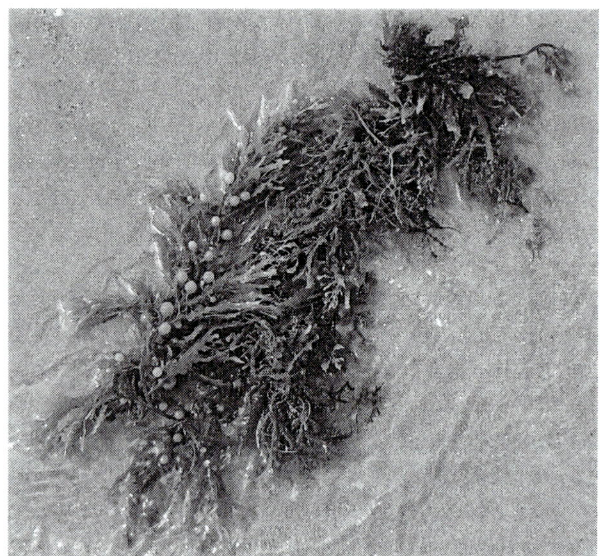

*Japweed, or Japanese wireweed (*Sargassum muticum*), shown here, is one of the invasive species now affecting other organisms in the Salcombe-Kingsbridge estuary. (Wikimedia/Graça Gaspar)*

Effects of Human Activity

The naturally sheltered harbor formed by the estuary means that the waterway has been used and valued by people for hundreds of years. There are records from the 16th century of seine fishers and herring fishers based at Salcombe. The estuary then developed as a successful center for shipbuilding, and as a port for fishing and trade. During the 1800s, Salcombe was a major center for the fruit trade, with oranges, lemons, and pineapples shipped here from the Mediterranean, the Azores, and the Caribbean. The estuary is still used as a minor fishing port, primarily for shellfish such as the scallops and edible crabs (*Cancer pagurus*).

In 1987, the Batson Creek channel in the estuary was dredged to improve access for local fishing boats. A total approaching 21,000 cubic feet (16,000 cubic meters) of mud was excavated and subsequently dumped on the seabed, smothering the biodiverse habitats on the underwater cliff and boulder slope off Scoble Point. Despite strong tidal currents, the dredge spoil did not get washed away. Between 1999 and 2002, a project to create an artificial reef on top of the dredge spoil was completed. Hundreds of large granite blocks were dropped from a barge. Recruitment of marine life to the artificial reef occurred rapidly. Within days, the blocks were covered by sea firs (hydroids), and after several years, the blocks were so covered with life that they were scarcely visible and now form a good replacement for the original habitat.

Climate change may impact the area by creating rising sea levels, which in turn could foster greater coastal erosion. The big question is whether to prepare for such an event by creating bulwarks against erosion and rising sea levels, or to allow nature to change the face of the ria, as it has over the eons.

Angus C. Jackson

Further Reading

Born, Anne. *The History of Kingsbridge and Salcombe.* Fremont, CA: Orchard Publications, 2002.

South Devon Area of Outstanding Natural Beauty (AONB). "The Salcombe and Kingsbridge Estuary." 2010. http://www.southdevonaonb.org.uk/downloads.asp?PageId=294.

Waterhouse, Gordon. *The Wildlife of the Salcombe and Kingsbridge Estuary.* Fremont, CA: Orchard Publications, 1999.

Salween Estuary

Category: Marine and Oceanic Biomes.
Geographic Location: Southeast Asia.

Summary: Where one of southeast Asia's major rivers meets the Andaman Sea, a rich wetland ecosystem is threatened by dam construction and other human activity.

The Salween Estuary is the lowest segment of the longest river flowing through Burma (Myanmar). The Salween descends through narrow gorges from the Tang-ku-la Mountains of eastern Tibet, China, to the Andaman Sea, entering the Gulf of Martaban at Moulmein, where it joins the Ataran and Gyaing Rivers to form an alluvial delta. The Salween is known by many names, depending upon the location: in China, the Nu Jiang or Nu; in Burma, Thanlwin; in Tibet, Gyalmo Ngulchu, to name a few. The river is traversable by boats only for a comparatively small distance, and its principal economic role has been as a conduit for teak logs to be floated downstream for manufacturing and export.

The regular flow of the river, and its contribution to balancing the saltwater in the estuary, has been threatened in recent years by climate change impacts on the Tibetan and Himalayan seasonal glacier melt and on monsoon rains—predictable patterns of each have been altered—as well as the construction of hydroelectric dams upstream. Continued conflict between the central government of Burma and ethnic minority groups seeking autonomy, particularly the Karenni, has hampered the construction of some infrastructure projects, while environmental concerns have become entangled in the political struggles. However, the estuary itself was spared the worst of the damage inflicted by the 2008 cyclone Nargis.

Political secrecy and cultural chaos in the region mean that accurate figures for water flow and sediment discharge through the Salween Estuary are not entirely satisfactory. However, various historic studies had placed the Salween as among the 20 largest in the world in each category. Some contemporary research indicates the estuary carries and deposits a highly organic yield; this is consistent with the monsoonal climate, the river's relatively narrow floodplain and limited catchment area, few tributaries, and the presence upstream of extensive forested areas on steep slopes.

This heavy organic content has contributed to the absence of seagrasses in the Gulf of Martaban, as is also the case with the delta of the Irrawaddy, Burma's other major river delivering a heavy sediment load. The Salween Estuary area has daily tides and seasonal flows marking the meeting of freshwater with the saltwater of the Andaman Sea.

Biodiversity

The Salween Estuary is home to approximately 140 species of fish and a diverse range of turtles, including the giant Asian pond terrapin and the bigheaded turtle. Mammals found in and around the wetlands here include the fishing cat, Asian small-clawed otter, and Siamese crocodile. The golden eye monkey, small panda, wild ox, and wild donkey are also found in the Salween floodplain. Plants of the biome include *Dendrobium fytchianum,* a rare white orchid discovered in the late 1800s and prized among collectors. On the whole, the estuary supports considerable biodiversity, and offers nutrition and income-generating opportunities for numerous local inhabitants.

The estuary receives much of the sediment carried by the Salween prior to its ultimate deposition in the sea. This contributes to the fertility of the area. Beaches and mudflats in the estuary are important feeding and spawning grounds for birds and a variety of fish, crabs, squid, and prawns, which are fundamental to the food web, as well as vital to local fisheries both for domestic consumption and, potentially, future export. Approximately one-third of the fish species in the Salween as a whole are endemic, meaning found nowhere else; many of these range into the estuary.

Effects of Human Activity

The Salween Estuary joins the Irrawaddy Estuary, their deltas, and the region's coastal rainforests in forming the Burmese coastal ecozone, which also extends to parts of the Bangladesh coastline. Most of this area has already been transformed by making it more practical for agricultural and aquaculture uses. As a result, much of the original lowland swamps and evergreen forest have been cleared, as have the deciduous dipeterocarp, montane evergreen, and mangrove forests around the mouth of the Salween.

Inevitably, this has already affected biodiversity in the estuary, which historically supported one of the most diverse systems in the country. Rice plantations, for example, have spread monoculture-dominated vegetation communities across much of the area. Some nonnative species have taken advantage of the invasive opportunities this presents. Global warming and its concomitant sea-level rise means the estuary may become more highly saline further upstream—a trend that damming will only accelerate. By blocking sediment flows, the dams will also make more difficult the restoration of wetlands that will be eroded by sea-level rise and damaged by saltwater intrusion.

These human actions have also affected the long-term flow of sediment into the comparatively shallow Gulf of Martaban, especially since upland forests have also been cleared and replaced by bamboo plantation forests. The extent to which these changes will magnify if several planned major dams are built upstream is not fully clear, but it is evident that because much of the river passes through Chinese territory and in part forms the border with Thailand, the challenges and solutions will require transboundary participation.

John Walsh

Further Reading

American Orchid Society. "Collector's Item: *Dendrobium fytchianum* Bateman ex Rchb.f." 2012. http://www.aos.org/Default.aspx?id=167.

Bird, M. I., R. A. J. Robinson, N. Win Oo, M. Maung Aye, X. X. Lu, D. L. Higgitt et al. "A Preliminary Estimate of Organic Carbon Transport by the Ayeyarwady (Irrawaddy) and Thanlwin (Salween) Rivers of Myanmar." *Quaternary International* 186, no. 1 (2008).

Clarke, J. E. "Biodiversity and Protected Areas: Regional Report." *Regional Environmental Technical Assistance* 5771 (2011).

Meade, Robert H. "River-Sediment Inputs to Major River Deltas." In John D. Milliman and Bilal U. Haq, eds., *Sea-Level Rise and Coastal Subsidence:*

Causes, Consequences, and Strategies. Dordrecht, Netherlands: Kluwer Academic Publishers, 1996.

Mon Youth Progressive Organization (MYPO). *In the Balance—Salween Dams Threaten Downstream Communities in Burma.* Yangon, Burma: Salween Watch, 2007.

Salween River

Category: Inland Aquatic Biomes.
Geographic Location: Southeast Asia.
Summary: Flowing through rugged, mountainous terrain, the Salween is relatively unspoiled but in jeopardy from mining, logging, and dam construction.

The Salween River of Burma (Myanmar) runs through a rugged, mountainous, north-south series of narrow valleys and gorges; it is approximately 1,500 miles (2,400 kilometers) in length. Glaciers in the eastern highlands of the Tibetan Plateau form the sources of the Salween River. Its path takes it southward through Yunnan Province in China, then along parts of the Burma-Thailand border, and finally into the Andaman Sea.

The Salween begins approximately 13,100 feet (4,000 meters) above sea level. Only one southeast Asian river, the Mekong, is longer than the Salween River. Throughout its history, the Salween has been a free-flowing river, but recent pushes from various governments and commercial interests to add as many as 16 major dams along the Salween could change the river and its ecosystem quickly. Dams remain one of the greatest threats to the wetlands and river.

This great river is known by many names as it courses through China, Thailand, and Burma; the Chinese call it the *Nu Jiang,* or *angry river,* perhaps reflecting its rapid flow as it streams through sheer canyons in its upstream segments. Fishing is a primary source of income and food for many human inhabitants of the 105,000-square-mile (272,000-square-kilometer) Salween watershed.

View of a dramatic bend in the Salween River (known as the Nujiang or "angry river" in Chinese) as it flows through China's northern Yunnan Province. The Salween River area is home to approximately 140 different species of fish and 7,000 species of plants, and as many as 80 of these are considered endangered. (Thinkstock)

From its beginning in the Qinghai Mountains on the Tibetan Plateau through its terminus at the Andaman Sea in Burma, the river undergoes dramatic changes. Portions of it are known as the Grand Canyon of the East, while other segments water woodland- and agricultural-dominated plateaus and valleys. The Salween is recognized as being in relatively pristine condition, with much of its biome intact.

Recent plans to dam the river have met with fierce resistance in some quarters, especially among indigenous peoples, as such measures would dramatically change both the nature of the river and its biota, endangering many species of flora and fauna that rely upon the Salween's free-flowing waters, and jeopardizing the livelihood of many humans. Fish spawning grounds and migratory pathways are of particular concern

Biodiversity

Among a diverse flora, montane evergreen forests are spread across many ridges upstream, while mangroves are found near the Salween Estuary.

The Salween River area has been known to support at least 140 species of fish, 7,000 species of plants, and up to 80 different types of endangered species in all. Minnows are by far the most prevalent fish found here. Some endemic (found nowhere else) fish species are *Hampala salweenensis* and *Hypsibarbus salweenensis*; up to 50 endemic fish species inhabit the biome.

Larger fauna in the Salween River biome include Siamese crocodiles, Asian small-clawed otters, and the fishing cat. The wild ox, wild donkey, and golden-eye monkey are ensconced along its banks. Turtles are an important component of the ecosystem, and a food source for many. Among the dozen different turtle species found in the Salween River are the Asian leaf turtle *(Cyclemys dentata)*, giant Asian pond terrapin (*Heosemys grandis*), and big-headed turtle (*Platysternon megacephalum*).

Threats

Mining, logging, and other disruptive human activities can undermine the high-relief landforms along the Salween River, adding to erosion and exacerbating the damage done by frequent landslides during the rainy season. Flooding, increased siltation and turbidity in the river, and fragmentation of habitats are the typical outcomes. Climate change impacts along the Salween River range from altered glacial melt regimes at the source, to heavier monsoon precipitation and wind across the main segments, to sea-level rise at the mouth. Additionally, some research suggests that dam construction is a net additive to global warming—apart from the destruction of forested areas it necessitates—by creating stagnant reservoirs conducive to algal blooms and rotting plants that release quantities of greenhouse gases.

William Forbes
Judy Flemons

Further Reading

Hedley, P., Michael I. Bird, and Ruth A. J. Robinson. "Evolution of the Irrawaddy Delta Region Since 1850." *Geographical Journal* 176, no. 2. (2010).

Twa, Saw Sein. "The Salween—My River, My Natural Belonging." *Watershed* 4, no. 2 (November 1998).

WWF Dams Initiative. *Rivers at Risk—Dams and the Future of Freshwater Ecosystems.* Washington, DC: World Resources Institute and World Wildlife Fund, 2004.

San Andres Archipelago Coral Reefs

Category: Marine and Oceanic Biomes.
Geographic Location: Caribbean Sea.
Summary: This group of coralline formations supports coral reef, mangrove, seagrass bed, algae, and forest habitats; it is protected in some measure by a system of conservation areas.

The San Andres Archipelago contains one of the most representative coral reefs in the Caribbean Sea, based on its development and complexity, which derive primarily from the strong marine current present in the archipelago and its distance from the mainland. While the total above-water surface area of the islands and immediate offshore waters of the archipelago is less than 45 square miles (116 square kilometers), it is the fundamental component of the vast Seaflower Biosphere Reserve of approximately 135,136 square miles (350,000 square kilometers). Under the dominion of Colombia, the full official name of the archipelago notes the three main islands: San Andres, Providencia, and Santa Catalina. The archipelago lies in the western Caribbean Sea, within 150 miles (240 kilometers) of Nicaragua and 480 miles (770 kilometers) from the Colombian mainland.

The average water surface temperature in the archipelago is 82 degrees F (28 degrees C), with a tropical dry climate. The mean air temperature is 81 degrees F (27 degrees C), moderated by the tradewinds. Besides the main islands, the archipelago also includes the atolls of Albuquerque, Courtown, Roncador, Quitasueño, and Serrana; and the reef banks of Serranilla, Bajo Alicia, and Bajo Nuevo. All of these areas are coralline structures created over old volcanoes in the early Cenozoic period. The same geologic, biologic, oceanographic, and meteorological processes have modeled the structures so the geomorphological and ecological features are common to all.

The archipelago has well-developed peripheral or barrier coral reefs in the marine current, generally on the side facing the current, and, offshore, wide pre-coral reef terraces. On the protected side, a more extensive barrier coral reef is present next to a lagoon terrace and a sedimentary shallow plane. In the lagoon basin, the depth can reach 66 feet (20 meters); nevertheless, columnar coral species can reach the surface. In protected sites, segments of barrier coral reef can be found where the lagoon basin is open or semi-open.

All of these coral reefs were modeled by sea-level oscillations during the cycles of ice age and thaw during the Pleistocene; by erosion and accretion processes in the modern era, or Holocene; and by long-term climate and current meteorological perturbations such as hurricanes.

Biodiversity

The coral species composition and morphology in the archipelago are generally determined by how open or closed a particular section of the lagoon basin is. The bottom can be dominated by seagrasses, algae, or coralline structures, depending on these factors, which in turn are shaped by three main elements: the marine current energy, the depth, and the presence of islands. The coralline formations in the archipelago are affected by marine current intensity that varies on a scale of 1–5; the average depth range is 33–39 feet (10–12 meters); and only two formations have islands that can influence the coral reefs with runoff, sediments, and nutrients that go to the lagoons. These factors have led these formations to hold from 6 percent to 28.5 percent coverage of coral species.

Six hermatypic, or stony, species associations are present in the archipelago, listed from more tolerant to marine current to less tolerant: red fouling algae (*Porolithon pachydermum*); fire coral (*Millepora complanata*) and colonial zoanthids (*Palythoa* spp.); elkhorn coral (*Millepora palmata*) and brain coral (*Diplora sfrigosa*); staghorn coral (*Acropora cervicornis*); finger coral (*Porites porites*); and boulder star corals (*Montastraea* spp.). A total of 57 hermatypic coral species are present here, together with incrusting calcareous algae. Several coral species present in the archipelago, such as column and cathedral corals, are absent in the continental reefs.

The islands and atolls provide habitat for 460 acres (186 hectares) of mangrove ecosystem, and for 18 resident and 76 migratory bird species, respectively, two of them endemic (found only here) and endangered, as well as many endemic subspecies. Four species of marine tortoises (*Caretta caretta*, *Chelonia mydas*, *Dermochelys coriacea*, and *Eretmochelys imbricata*) are present around the coral reefs, using the ecosystems for feeding and nesting. The relict forests of the islands are also important hot spots of biodiversity, with more than 360 species (roughly 75 percent native and 25 percent introduced); they also provide habitat for some endemic reptiles.

Besides the coral reef ecosystem, 4,942 acres (2,000 hectares) of seagrass beds occur within the archipelago, predominantly around the main islands. These, together with the coral formations, represent habitat for at least 270 species of fish, two of them endemic. In the coral reefs, the families *Pomacentridae, Labridae,* and *Scaridae* are dominant. The fish community tends to vary more based on reef types than on the structure of coral formations.

Environmental Threats

The main pressures on these ecosystems are represented by legal and illegal industrial fishing, including fishing of queen conch (*Strombus gigas*), Caribbean spiny lobster (*Panulirus argus*), spotted spiny lobster (*Panulirus guttatus*), and purple land crab (*Gecarcinus ruricola*). Artisanal fishing is relatively low-impact. Climate change, too, presents stark challenges to the coral reefs here, with heavier damage expected from storms, and especially from the rise in average seawater temperature, to which corals are particularly sensitive. Their *zooxanthellae* symbionts tend to disperse when seawater temperatures reach certain limits, exposing the coral polyps to predators, such as the crown of thorns starfish, and to disease; this is typically accompanied by the phenomenon of coral bleaching.

In part to protect the coral reefs from human damage, the Old Providence Natural National Park and the San Andres Archipelago Marine Protected Area were established. Along with the Seaflower Biosphere Reserve—which encompasses one-tenth of the Caribbean Sea—these reserves are policed to prevent overfishing and destructive recreational and tourist activities, as well as to monitor the health and balance of the marine and terrestrial species here.

José F. González-Maya
Amancay A. Cepeda
Diego Zárrate-Charry
Jan Schipper

Further Reading

Díaz, J. M. "Marine Biodiversity in Colombia: Achievements, Status of Knowledge, and Challenges." *Gayana* 67, no. 2 (2003).

Mejía, Luz Stella and Jaime Garzón-Ferreira. "Estructura de Comunidades de Peces Arrecifales en Cuatro Atolones del Archipiélago de San Andrés y Providencia (Caribe sur Occidental)." *Revista de Biologia Tropical* 48, no. 4 (2000).

Natural National Parks of Colombia. "Nature and Science Old Providence." http://www.parquesnacionales.gov.co/PNN/portel/libreria/php/decide.php?patron=02.02021503&f_patron=02.020215.

San Francisco Bay

Category: Marine and Oceanic Biomes.
Geographic Location: California.
Summary: Among the largest coastal estuaries in North America, San Francisco Bay is the site of landmark efforts to restore and preserve a complex and biologically diverse bioregion.

San Francisco Bay is both a bay and part of the largest coastal estuary in the western United States. Its definition as a bay derives from its position as an inland body of water directly channeled to the Pacific Ocean. This Pacific entrance, dubbed the Golden Gate by early European explorers, is bound by the Marin Headlands to the north and the Presidio of San Francisco on the south; it opens into the deep waters of the central bay. The salty waters of the bay mix with abundant freshwater from the deltas of the Sacramento River to the north and the San Joaquin River to the south.

Dams along these two major rivers have substantially altered important freshwater inputs into the estuary system. It is estimated that 40 percent of California's freshwater enters the San Francisco Bay before discharging into the Pacific Ocean. The combination of freshwater and saltwater in the estuary creates a relatively low salinity environment that is sensitive to both riverine and marine influences. It is a very nutrient-rich habitat today.

Covering a total area of approximately 1,600 square miles (4,149 square kilometers), the greater bay area features such contiguous components as San Pablo Bay, South San Francisco Bay, Suison Bay, Carquinez Strait, and the extensive delta areas of its two main tributary rivers. Via the rivers, the bay is connected to the ecosystems of California's great Central Valley, as well as the near-extinct Tulare Lake basin. A vital part of the bay area is its abundant wetlands, which support mollusks, crustaceans, amphibians, reptiles, and vast migratory bird populations. The marshes of the South Bay are high enough in salinity to support the commercial development of salt flats.

The maritime climate of the bay area is controlled by the southwestern flow of polar Pacific air, an effect that keeps water and land mass temperatures along the northern California coast cool and steady throughout the year. As a result of summer atmospheric inversions, the bay near the Golden Gate is characteristically shrouded in stratus cloud fog cover, particularly during the early part of the day during the summer months. These generally mild climate conditions and the voluminous interchange of marine and estuarine influences are the foundation of microclimates so extreme and diverse that unique classification systems have been developed to adequately describe some of the species that flourish within them.

Biota

The valleys and woodlands surrounding the bay receive an average rainfall of 20 inches (508 millimeters) per year, supporting a rolling landscape of shrubs, herbs, conifers, and broadleaf mixed forests of redwoods, oaks, maples, and alder. There is also extensive growth of eucalyptus, a nonnative species.

Phytoplankton are the primary producers of the food web that sustains the biodiversity of the bay. Elevated nitrogen levels and the turbidity of the bay's waters have had negative effects on phytoplankton production; the introduction of the Amur River clam further significantly reduced the biomass of phytoplankton, a condition that has continuing deleterious effects on the pelagic fish that depend on the biomass for survival.

Keystone species here include the primary consumers of phytoplankton—copepods, rotifers, ciliates, and flagellates—as well as the macro-invertebrates (mysids, shrimp, and amphipods), and planktivorous fishes such as sturgeon, chinook salmon, and American shad. Larger fish common to the bay include various saltwater, freshwater, and some anadromous species of shark, ray, sunfish, tuna, surfperch, halibut, sole, flounder, smelt, pipefish, lamprey, carp, bass, and catfish. The steelhead trout populations extending along the California coast are classified as endangered under the U.S. Federal Endangered Species Act. Frogs, toads, and salamanders are also common here, as are the leatherback sea turtle, the Pacific pond turtle, and a variety of lizards and snakes.

The diversity of habitat in the area sustains a thriving insect community including nearly 100 species of ants and bees; wasps, mud daubers, and yellowjackets; beetles, grasshoppers, and crickets; and dragonflies, moths and butterflies—all of which provide food for the millions of waterfowl that migrate to the bay via the Pacific Flyway of

A bird hatching from an egg in a nest by a salt marsh in San Francisco Bay. The South Bay Salt Pond Restoration Project plan aims to restore 50,000 acres of tidal wetlands in the area. (U.S. Geological Survey)

North America. Dabbling and diving ducks are common, as are the Canada goose, swans, terns, rails, egrets, the great blue heron, clappers, pelicans, red-tailed hawk, American kestrel, quail, dove, hummingbirds, kingfishers, ring-billed gulls, sparrows, goldfinches, and blackbirds.

Many of the aquatic mammals native to the region are at last growing in population, after staggering decimations during the California Fur Rush of the nineteenth century. Once plentiful communities of beaver, marten, mink, fox, weasel, and river and sea otters were driven to near extinction by global fur traders. Recovery is finally taking hold after more than 150 years. For example, the California golden beaver is recolonizing in the North Bay area; the North American river otter has similarly been noted. Other marine mammals that appear both offshore and in the bay proper include a range of whales, dolphins, and seals.

Common land mammals can be found both in protected park areas and in urban environments here. These include opossums, shrews, jackrabbits, coyotes, a variety of squirrels, and mice. In the higher-elevation, more heavily-wooded areas are found badgers, elk, deer, bobcats, mountain lions, and wild pigs. Bat communities are also common.

The San Francisco Bay biome is the site of one of the largest bayland restoration programs in the nation. The tidal flats and wetlands here are populated with sedges, rushes, cord grasses, and shrubs unique to salt marshes found along the banks of this tranquil estuary. These baylands provide essential nutrients that sustain shellfish and invertebrates, which in turn feed the fish and waterfowl that breed and spawn in its waters.

Human Impact and Conservation

Taken as a whole, the San Francisco Bay biome is one of the largest, richest—and heavily degraded—ecosystems in the world. It is also the site of landmark efforts to restore and preserve a complex and biologically diverse bioregion.

The urban, agricultural, and industrial geography of the San Francisco Bay biome was initiated in the early 19th century as fur traders recognized the bay and estuary for its ready access to contiguous rivers, and direct routes to the fecund valleys and mountains inland—ideal for commerce both before and after steamships and railroads were built. The bay was very quickly enlaced by early port settlements radiating from the San Francisco metropolitan core, all engaged in the trans-shipment of goods through the Port of San Francisco to markets in Asia, South America, and Alaska. The Oakland Pier in East Bay was built in 1863, accelerating patterns of settlement and trade. Today, the San Francisco Bay Area includes nine counties—Alameda, Contra Costa, Marin, Napa, San Francisco, San Mateo, Santa Clara, Solano, and Sonoma—and supports a 2010 census population of 7.15 million people.

The bay area is somewhat unique in its underlying intimate relationship between urban, industrial, and agricultural lands that surround it. Of a total surface area of 4.5 million acres (1.8 million hectares), nearly 2.5 million acres (1 million hectares) are open space, with some 1 million acres (405,000 hectares) held in public trust. This is the birthplace of early conservation and environmental movements that include the Sierra Club, Marin Conservation League, Save San Francisco Bay Association, San Francisco Estuary Project, State of California Coastal Conservancy, San Francisco Bay Area Conservancy, San Francisco Bay Conservation and Development Commission, the Audubon Society, the Bay Institute of San Francisco, and the Nature Conservancy. Other important science-based projects include the National Oceanic and Atmospheric Administration (NOAA) San Francisco Bay Watershed Database and Mapping Project and the Bay Area Ecosystems Climate Change Consortium.

Since the Gold Rush era of the mid-19th century, the San Francisco Bay biome has sustained unprecedented physical alteration and pollution, the sum of which resulted in the destruction of perhaps 90 percent of the system's natural baylands. At the same time, dams along the two major rivers substantially altered important freshwater inputs into the estuary system. The National Estuary Program was established by the U.S. Congress in 1987 as an amendment to the Clean Water Act; this program identified the San Francisco Bay as an estuary of national significance, and a Comprehensive Conservation and Management Plan was established in 1993, enumerating 145 actions to improve and restore the bay's estuarine system.

In 1999, the Baylands Ecosystems Habitat Goals was published, with the recommendation that the bay's tidal marshes should be increased by 100,000 acres. Referring to the 1993 Wetlands Conservation Policy, a benchmark was established for defining desirable tidal marsh and tidal flat sizes. These were defined as the percent similarity between their historical and present-day distribution (plus-or-minus 25 percent, according to the range of sizes in each category). To date, the marshes have reached approximately 50 percent of this goal, while the flatlands are close to realizing their 30,000-acre benchmark. The South Bay Salt Pond Restoration Project is an ongoing effort to restore 50,000 acres of tidal wetlands. Climate change and anticipated changes in sea level are ecological factors that will continue to challenge the health and diversity of the San Francisco Bay biome in future decades.

Victoria M. Breting-Garcia

Further Reading

Beidleman, Linda H. and Eugene N. Kozloff. *Plants of the San Francisco Bay Region: Mendocino to Monterey*. Berkeley and Los Angeles: University of California Press, 2003.

Hart, John and David Sanger. *San Francisco Bay: Portrait of an Estuary*. Berkeley: University of California Press, 2003.

Schoenherr, Allan A. *A Natural History of California*. Berkeley: University of California Press, 1992.

Vance, James E. Jr. *Geography and Urban Evolution in the San Francisco Bay Area*. Berkeley: Institute of Governmental Studies, University of California, 1964.

San Lucan Xeric Scrub

Category: Grassland, Tundra, and Human Biomes.
Geographic Location: Mexico.

Summary: This ecoregion retains much of its original flora and fauna because of geologic isolation. Notably, the area has been established as an Endemic Bird Area.

The San Lucan Xeric Scrub biome, sometimes referred to as the Cape Region, is located at the southern tip of the Baja California Peninsula of Mexico. It covers an area of approximately 1,500 square miles (3,900 square kilometers). This diverse landscape of granitic mountains, valleys, arroyos, and plateaus is covered with a variety of xeric (or low-moisture) vegetation, and is part of the greater ecoregion complex of the Sonoran-Baja Deserts. Plants and animals of this region evolved isolated from the greater landmass before joining the Baja Peninsula. In this sense, many species found here have evolved independently from others found in adjacent areas. Hence, the biome is sometimes referred to as an island of vegetation.

The average annual high temperature is 85 degrees F (30 degrees C); the average annual low is 64 degrees F (18 degrees C). Annual precipitation is about 16 inches (40 centimeters). This warm, arid climate supports a variety of wildlife, about 10 percent of which are endemic, or found nowhere else on Earth. The region supports a number of endemic birds, for example, and is listed as an Endemic Bird Area by BirdLife International.

The Cape Region contains the highest proportion of intact xeric scrub—315 endemic species—of any Mexican state, due primarily to its difficult topography. This ecoregion includes the plateaus between the coast and the lower limits of the dry forests that begin at around 800 feet (250 meters). There are two national parks here dedicated to protecting the xeric scrub and other flora and fauna: Bahía de Loret, with 798 square miles (2,070 square kilometers), established in 1996, and the much smaller Cabo Pulmo, with 27 square miles (70 square kilometers), established in 1995.

Flora and Fauna

The xeric plant communities that dominate the San Lucan ecoregion include: the cactus *cholla*, elephant tree (*Bursera microphylla*), *Lysiloma divaricata*, organ pipe cactus (*Stenocereus thuberii*), mala mujer (*Cnidoscolus angustidens*), *Yucca* spp., and barrel cactus (*Ferocactus* spp.). Other plants found here include the boojum tree (*Fouquieria columnaris*), date palm, giant coreopsis, ponderosa pine, the grass hairy grama (*Bouteloua hirsuta*), and spiderwort (*Commelina coelestis*).

An amazing variety of animal wildlife inhabits this hot, dry ecoregion. In total, 238 species have been identified here. Mammals include: Mexican long-nosed bat (*Leptonycteris yerbabuenae*), ring-tailed cat (Bassariscus astutus), black jackrabbit (*Lepus insularis*), desert kangaroo rat (*Dipodomys deserti*), mule deer (*Odocoileus hemionus*), coyote (*Canis latrans*), bobcat (*Lynx rufus*), cougar (*Puma concolor*), American badger (*Taxidea taxus*), gray fox (*Urocyon cinereoargenteus*), and kit fox (*Vulpes macrotis*).

Baja California has great biodiversity along its coasts, including populations of California sea lions and gray whales. The region has been called the World's Aquarium by conservationists, as the Gulf of California, Sea of Cortez, and Baja California's Pacific Ocean shores are home for up to one-third of Earth's marine mammal species.

There are 199 species of birds identified in the San Lucan Xeric Scrub biome, with two of these endemic. Some avifauna found in this ecoregion include: Cooper's hawk, sharp-shinned hawk, golden eagle, great blue heron, short-eared owl, burrowing owl, ring-necked duck, cedar waxwing, great horned owl, red-tailed hawk, California quail, Costa's hummingbird, Xantus's hummingbird, and Belding's yellowthroat, a highly threatened species. There are 56 reptile and amphibian species in the San Lucan, including the rosy boa and San Lucan rock lizard, and a remarkable variety of spiders.

Human Impact

It is believed that people first settled on the Baja Peninsula over 11,000 years ago, and that the region was dominated by only a few Native American groups. Europeans did not reach the area until 1539. Population density is low; there are fewer than 700,000 inhabitants in the entire California Baja Sur, but impact to the land is still significant. Due to its unique nature, climate, and surrounding coastal areas, the San Lucan Xeric Scrub region has seen rapid growth in tourism. Adventure-tourism has affected desert and coastal sandridge ecosystems; off-road driving damages soils and vegetation that are not easily regenerated in this environment.

More immediate and broadly distributed threats include cattle ranching, fuelwood extraction, and exploitation of wildlife and plants by locals. The poverty and the dependency on natural resources for survival of local inhabitants poses another problem. Perhaps the largest concern for the region are the effects of climate change. A 2010 study—sponsored by the Autonomous University of Baja California Sur (UABCS) in conjunction with Mexico's Northwest Center for Biological Research, the National Polytechnic Institute, and the Center for Scientific Research and Higher Education of Ensenada—lays out some of the possible consequences of global warming in the ecologically fragile Baja California Peninsula. These include more frequent and stronger hurricanes, loss of vegetation and soils, accelerated desertification, and negative impacts on fisheries and terrestrial biodiversity.

Medani P. Bhandari

Further Reading

BirdLife International. "Endemic Bird Areas." http://www.birdlife.org/action/science/endemic_bird_areas/index.html.

Murphy, Robert. "The Reptiles: Origins and Evolution." In *Island Biogeography in the Sea of Cortez*, edited by Ted J. Case and Martin L. Cody. Berkeley: University of California Press, 1983.

Riemann, Hugo and Ezcurra, Exequiel. "Plant Endemism and Natural Protected Areas in the Peninsula of Baja California, Mexico." *Biological Conservation* 122 (2005).

San Mateo (California) Creek

Category: Inland Aquatic Biomes.
Geographic Location: North America.

Summary: A relatively small but prolific wilderness area adjacent to urban areas on the Pacific coast.

San Mateo Creek is the most pristine intact coastal stream in southern California. (It is not to be confused with the San Mateo Creek in the San Francisco Bay area.) This creek is approximately 22 miles (35 kilometers) long, and its watershed covers about 139 square miles (360 square kilometers) of relatively undeveloped terrain. In part, it serves as the boundary between Orange and San Diego counties. Its watershed also includes Riverside County, where its headwaters arise, draining from the Santa Ana Mountains of the Cleveland National Forest. From this roadless wilderness area, runoff from rain flows westward and travels down steep canyons, through an alluvial plain, and into the Pacific Ocean.

The San Mateo Creek ecoregion encompasses areas of both mediterranean climate, to the north, and semiarid climate to the south and east. As a result, it is often described as an arid mediterranean, as well as semiarid steppe, environment. Average annual rainfall is between 15 and 20 inches (38 and 50 centimeters).

The watershed contains two distinct topographical regions: the upper and the coastal. The upper region lies mostly within San Mateo Canyon Wilderness in the Cleveland National Forest, with elevations up to about 3,500 feet (1,070 meters); its lower reaches run through Camp Pendleton Marine Corps Base and San Onofre State Beach (where the popular surfing beach, Trestles Natural Wetlands Preserve, is located).

Following a decreasing elevation gradient, the ecosystem starts at the ridgeline with mixed pine-oak forests descending to chaparral, coastal sage scrub, and finally to riparian habitats. It is home to many endemic (not found elsewhere) species, some of which are federally protected under the United States Endangered Species Act. This biome, though small, is extremely valuable.

Flora and Fauna

The vegetative community of the San Mateo Creek watershed is spectacular. Rainwater is sufficient to maintain a highly diverse floral ecosystem; spring rains bring an abundance of wildflowers blooming throughout the sage and chaparral. Five riparian plant communities have been identified in the Trestles Natural Wetlands Preserve alone. These include: coastal sage scrub or coastal bush scrub (scrub wetland); willow woodland (forested wetland); sycamore-cottonwood (forested wetland); freshwater marsh (emergent wetland); and jaumea meadow (emergent wetland).

Some of the individual plants found here are: larger trees and shrubs including coastal live oak (*Quercus agrifolia*), coastal scrub oak (*Quercus dumosa*), lemonade berry (*Rhus integrifolia*), Christmasberry (*Heteromeles arbutifolia*), California sage (*Artemisia californica*), ceanothus (*Ceanothus* spp.), poison oak (*Toxicodendron diversilobum*), and greasewood or chamise (*Sarcobates* spp.).

Fauna found in the region include 139 bird species, 37 mammal species, 46 reptile and amphibian species, and seven species of fish. The range features black bear, mule deer, bobcat, coyote, gray fox, American badger, ring-tailed cat, long-tailed weasel, bats, spotted owl, golden eagle, western pond turtle, toads, frogs, lizards,

A Bell's vireo. The subspecies "least" Bell's vireo, which has less yellow or green in its feathers compared to other Bell's vireos, is listed as a species of concern in the San Mateo Creek area. (U.S. Fish and Wildlife Service)

rattlesnakes, coyotes, skunks, and mice; the elusive mountain lion is here but rarely seen.

The San Mateo Creek ecoregion is unique in that much of its wildlife is on either the federal or State of California Threatened and Endangered Species lists. Some of these listed species have become the source of widespread media attention—for example, the California gnatcatcher (*Polioptila californica*), an 0.35-ounce (10-gram) bird that was not even a recognized taxon until 1993, when it was listed as threatened on the United States Threatened and Endangered Species list. The bird was featured on the cover of the January/February 1995 issue of *Audubon* magazine, with the headline, "Can this Bird Save California?" Because of legal battles between developers and conservationists, critical habitat for this bird still has not been designated.

Also of note is the proposed modification of State Route 241, a planned extension of an existing toll road that would terminate at San Onofre State Beach; road opponents fear the loss of important and unique habitat. Another species of concern is the southwestern willow flycatcher (*Empidonax trailli extimus*), whose breeding habitat is located in moist riparian vegetation, often containing willows. Another federally listed species is the least Bell's vireo (*Vireo bellii pusilus*), a subspecies impacted by habitat loss. Additionally, a small population of southern steelhead trout (*Oncorhyncus mykiss*) was found in the watershed; it survives because of the flow of freshwater, and is federally listed as endangered.

Human Impact

The early Native American inhabitants of the coast and the Santa Ana Mountains included the Kumeyaay, Luisaños, Cahuella, and Capeño. These groups fished the streams and found an ample food supply among the abundant plant life. The explorers Vizcaíno and Cabrillo reported that the native population did considerable burning of the brushlands, but the overall impact was probably not very great.

European inhabitants did not form a permanent presence until the mid-to-late 1700s, with the construction of the mission network and an extensive agricultural and irrigation system. The Spanish missions initiated changes to the rivers with the introduction of irrigation; they prospered until the separation of Mexico from Spain in 1821. By the mid-1800s, large portions of land had been opened to settlement by private ranchers. The land within the watershed was parceled out in the form of large Mexican land grants. Ranching became the predominant activity in the watershed.

Large numbers of cattle and sheep were brought in, and they grazed the grasslands of the lower San Mateo Creek drainage basin, drastically altering the native landscape. Widespread overgrazing throughout the area destroyed native vegetation; rancheros cut brush and trees and cleared underbrush with fires. The introduction of nonnative invasive plants displaced native grasslands, which probably was the single most destructive assault on the landscape.

In the late 1860s, gold miners from northern California flowed into the area, bringing about ecological destruction in the form of logging (for mine timbers) and mining for gold, lead, zinc, and silver. Subsequently, uncontrolled fire events threatened the water supplies of the surrounding region. In response, the California Forestry Commission voiced the necessity for special protection of the watershed to prevent fires and subsequent erosion. The Forest Reserve Act was signed by President Benjamin Harrison in 1891 to curb illegal timber cutting and mining. This Act established the boundaries of the Cleveland National Forest, which included a majority of the San Mateo upper watershed.

Today, the San Mateo Creek ecoregion supports more than 3 million nature visitors per year, mostly day picnickers and tent campers. The chief environmental concerns presently being studied and addressed for the ecosystem are: urban encroachment, groundwater depletion, erosion, military activities, the coming decommissioning of the San Onofre nuclear power plant, and the proposed toll road. Climate change also presents challenges, such as sea-level rise, higher temperatures, and ocean acidification. Impacts of these and related effects could include damage to trout and salmon population, streambank erosion and saltwater intrusion, increased wildfire frequency, and more invasive pests and disease.

Bob Whitmore

Further Reading

Lang, John S., Bruce F. Oppenheim, and Robert N. Knight. *Southern Steelhead* Oncorhynchus Mykiss *Habitat Suitability Survey of the Santa Margarita River, San Mateo and San Onofre Creeks on Marine Corps Base, Camp Pendleton.* Arcata: Coastal California Fish and Wildlife Office, 1998.

Palmer, Tim. *California Wild: Preserving the Spirit and Beauty of Our Land.* Minneapolis, MN: Voyageur Press, 2004.

San Mateo Creek Conservancy. "Trestles Natural Wetlands Preserve." http://trestleswetlands.org/Default.aspx.

Santa Marta Montane Forests

Category: Forest Biomes.
Geographic Location: South America.
Summary: This isolated mountain range, situated beside the Caribbean Sea in Colombia, supports a wide variety of flora and fauna with a high rate of endemism.

The Santa Marta Montane Forests biome spreads across a mountain range located in northern Colombia, but isolated from the Andes. The highest coastal mountain range in the world, the Santa Marta's tallest peak is at approximately 18,700 feet (5,700 meters) above sea level. The ecoregion has a triangular shape, with its northern edge running along the Caribbean Sea; it covers an area of approximately 4,600 square miles (12,000 square kilometers). Because of its unique location and altitude variation, this mountain range has all the possible tropical elevation steps, and therefore ecosystems, known for tropical America—from snow-capped peaks to lush, lowland rainforest.

Santa Marta montane forests begin above 1,650–2,620 feet (500–800 meters). The many isolated valleys and ridges along this gradient have led to high levels of endemism (species found nowhere else) and diversity, and it is one of the few areas in the Colombian region with a relatively low level of transformation of the natural coverage.

The average annual rainfall here is 80 inches (200 centimeters), although it can reach as high as 120 inches (300 centimeters) on the northern slopes. Typically, rainfall decreases as altitude increases. Rainfall peaks between September and December, and again between May and July. Temperatures range from averages of 81 degrees F (27 degrees C) at sea level to 43 degrees F (6 degrees C) or less in the highest areas. In general, the highest temperatures occur in the months of April, May, and June.

Flora and Fauna

The Santa Marta Montane Forests biome is recognized globally for its biological and cultural diversity, high levels of endemism, species richness, and the isolation of this massif from the Andean region. The ecoregion is home to at least 1,046 species of vertebrates and more than 3,000 species of vascular plants, which present high values of endemism at altitudes above 5,280 feet (1,610 meters).

The greatest diversity of plants can be found in the zone between 3,820 and 8,200 feet (1,000 and 2,500 meters). Some of the families of vascular plants of particular interest are the *Melastomataceae, Bromeliaceae, Asteraceae,* and *Lamiaceae.* Angiosperm families with the most genera include: *Asteraceae, Orchidaceae, Leguminosae,* and *Rubiaceae.* Among both of these listings are a large number of endemics.

The number and variation of fauna species in the Santa Marta forests is remarkable; over 600 animal species have been identified here. Mammals include the red howler monkey, lemurine night monkey, Hoffmann's two-toed sloth, ocelot, white-fronted capuchin, bicolor-spined porcupine, nine-banded armadillo, long-tailed weasel, neotropical otter, giant anteater, South American tapir, white-lipped peccary, gray fox, many small mammals, and a very high number of bat species.

More than 200 species of birds have been identified here, such as the endangered Santa Marta sabrewing, Santa Marta bush tyrant, and Santa Marta parakeet; the critical-listed blue-knobbed curassow; and blue-headed and red-billed parrots, white-tipped quetzal, golden-green woodpecker, rufous-tailed jacamar, laughing falcon, tropical screech-owl, black-crowned night-heron, as well as various hummingbirds. At least 20 amphibian species and nearly 40 reptile species have been identified.

Human Impact

This montane ecoregion is important globally based not only on the high biodiversity of ecosystems within its boundaries, but also on the cultural diversity that can be found here. Four indigenous peoples inhabit this mountain area and base their knowledge and traditions on their environment; these are the Kogui, Ika, Wiwa, and Kankuamo. In addition to these four tribes, the River Rancheria basin on the eastern slope of the mountain range is inhabited by the Wayuu indigenous group, which is originally from the Guajira peninsula. These peoples are few in number.

The mountains of this sierra ecoregion previously held a settlement of pre-Columbian aboriginal communities that left signs of their presence in various areas. Groups of Europeans have made settlements here since the late 19th century. The area covered by montane forests has declined drastically during the last 50 years. In some cases, it is estimated that the reduction in the original forest is between 70 and 80 percent.

The Santa Marta montane forests have also been affected over the last five decades by many political, economic, and social conflicts that have led to chaotic and uncontrolled use of the land and natural resources. Over time, this has transformed the natural covers and modified the original ecosystems, threatening all the components of its biodiversity. Also, the sociopolitical situation has limited the development of research and effective protection of the natural areas.

Additionally, climate change is forcing many species here to adapt, move, or perish. Habitats of some birds of the Santa Marta montane forests, for example, are projected to shrink in size by as much as one-third or more under future global warming scenarios. Many species will be under duress as temperatures rise and precipitation becomes less predictable.

Based on the importance of the region, different categories of protection have been established, and there are several regional processes of conservation that seek to reduce the increasing pressures caused by the expanding agricultural frontier. Because of the importance in terms of biodiversity, the area has received national and international attention. Two national parks have been established here, Sierra Nevada de Santa Marta National Park and Tayrona National Park; the ecoregion was designated as a Biosphere Reserve by the United Nations Educational, Scientific, and Cultural Organization (UNESCO), and is currently being considered for World Heritage Site status.

*One of the over 200 species of birds found in the Santa Marta Montane Forests region, this snowy egret (*Egretta thula*) was seen in Tayrona National Park in 2011. (Wikimedia/Biusch)*

Diego Zárrate-Charry
José F. González-Maya
Jan Schipper

Further Reading

Carriker, Melbourne R. *Vista Nieve: The Remarkable True Adventures of an Early Twentieth Century Naturalist and His Family in Colombia, South America.* Rio Hondo, TX: Blue Mantle Pubs., 2001.

Strattersfield, A. J., M. J. Crosby, A. J. Long, and D. C. Wege. *Endemic Bird Areas of the World: Priorities for Biodiversity Conservation.* Cambridge, UK: BirdLife International, 1998.

Velasquez-Tibata, Jorge, Paul Salaman, and Catherine H. Graham. "Effects of Climate Change on Species Distribution, Community Structure, and Conservation of Birds in Protected Areas in Colombia." *Regional Environmental Change* (July 2012).

São Francisco River

Category: Inland Aquatic Biomes.
Geographic Location: South America.
Summary: The longest river contained entirely within Brazil, the São Francisco is considered the river of national unification; it serves as a lifeline for agriculture and economic development in an area often beset by drought.

The São Francisco is the longest river contained entirely within Brazil, and the fourth-longest river in South America. It originates in the Serra de Canastra mountain range in the states of Minas Gerais and Bahia at about 2,400 feet (730 meters) in elevation, and winds north and east through the countryside before emptying into the South Atlantic Ocean. The total length of the São Francisco is nearly 1,900 miles (3,000 kilometers), and its watershed occupies more than 240,000 square miles (630,000 square kilometers).

The states of Alagoas, Pernambuco, and Sergipe also provide drainage into the São Francisco basin, bringing water from 168 rivers and streams, of which 90 are on the right bank and 78 on the left bank. Principal tributaries of the river include the Abaeté, Carinhanha, Das Velhas, Jequitaí, Paracatu, Corrente, Salitre, and Verde Grande Rivers.

Much of the São Francisco river basin is arid or semiarid; with its average water flow of 100,000 cubic feet (2,830 cubic meters) per second, the river provides a lifeline for the region known as Brazil's "drought polygon." This dry zone receives about 15–30 inches (37–75 centimeters) of precipitation a year, whereas much of eastern Brazil receives 40–70 inches (100–180 centimeters) annually. The Sao Francisco's flow is perennial, although its water level can change dramatically over the course of the year. The upper river basin—the Bahia interior forests—have average annual temperatures of 64–72 degrees F (18–22 degrees C). The lower river basin, primarily a semidesert ecoregion, has an average maximum temperature of 92 degrees F (33 degrees C) and average minimum temperature of 66 degrees F (19 degrees C).

Flora and Fauna

The upper river basin of the São Francisco is primarily savanna and forest; hardwoods grown in this area include the vinhatico, jacaranda, and the Brazilian cedar. The upper-middle basin is an agricultural region in which cotton, beans, rice, and corn (maize) are grown. The region also produces pineapples, potatoes, maté tea, melons, coffee, castor and cottonseed oils, and sugarcane (grown here mainly for rum). Other plant types growing in this region include the aloe plant, the cochineal cactus, and the vanilla plant.

Also in the middle river basin, stunted, thorny forest is typical; dominant trees in this area include the *barriguda*, the *catingueiras*, and the *juremas* (a palmlike tree). Many types of cacti and bromeliads also grow in this zone, along with rubber and cashew trees. Along the riverbanks in the lowlands here, cassava (manioc), corn, beans, and melons are cultivated in shallow waterbeds. On the coastal lowlands, rice and sugarcane are grown. The dry lowlands of the basin (or the *sertão*) are used largely for livestock grazing, with herds mainly featuring cattle, goats, sheep, and donkeys.

Non-domesticated mammal species that may be found in the upper, forested river basin include Coimbra's titi monkey, blond titi monkey (critically endangered), southern masked titi monkey, maned three-toed sloth (endangered), maned wolf, prehensile-tailed porcupine, ocelot, oncilla, lion tamarin (an endangered monkey species), ring-tailed coati, Brazilian gracile opossum, South American tapir, white-lipped peccary, and a variety of small mammals such as rodents and many bat species. There are several hundred bird species identified here, as well as many frogs (especially treefrogs), turtles, lizards, and snakes.

The lower, semiarid portion of the watershed supports smaller faunal communities. Mammals include several marmoset species, hairy-rumped agouti, seven-banded armadillo, several opossum species, tapir, peccary, mice and other rodents, and bats. There are also many birds, amphibians, and reptiles here. The river itself sustains more than 140 identified fish species, including anchovy, knifefish, catfish, tetra, and piranha. Although fish are plentiful, and there is an extensive fish-farming economy, damming of the river and use of the water for irrigation and hydroelectricity has greatly diminished fish population numbers.

Human Impact

The São Francisco is navigable for about 1,100 miles (1,800 kilometers) of its total length; a railroad line has been constructed to carry goods and passengers around the Paulo Alfonso Falls; other breaks include the rapids of Pirapóra in Mina Gerais, and on the border of the state of Bahia. In the 19th century, the São Francisco served as an important means of migration from the southeast to the northeast in Brazil; it was at this time the river achieved its reputation as the "river of national unification."

Because many tributaries of the São Francisco flow intermittently, based on the annual cycle of rain and drought, water is stored in reservoirs to assure a more consistent supply; however, this causes accelerated loss of water due to evaporation. The largest of these reservoirs is the Sobradinho Reservoir, located in the state of Bahia; it has a storage capacity of over 8 cubic miles (34 cubic kilometers) and a surface capacity of more than 1,500 square miles (4,000 square kilometers), making it the twelfth-largest artificial lake in the world.

The primary modern uses for the waters of the São Francisco river are generation of hydroelectricity and irrigation for agriculture. Demand for both is expected to increase in the future. The local development authorities project an increase of at least 2,000 square miles (500,000 hectares) in irrigated agricultural areas within the river basin. Somewhat counterintuitively, in relatively dry years, demand is expected to be greatest during the wet (rather than dry) season because more crops are planted during the wet season, and because of greater variability of rainfall during the wet season.

The São Francisco also provides most of the electricity used in northeastern Brazil; as of 2006, hydroelectric plants in the São Francisco River and its tributaries had an installed capacity of 7,800 megawatts, with plans to expand hydroelectric capacity to over 26,000 megawatts. The Paulo Alfonso Dam, built in 1955, now provides electric power for the whole of northeastern Brazil. Another dam with a large hydroelectric plant serving the region is the Tres Marias in Minas Gerais, built in 1961.

Development and economic activities in the river basin have caused environmental deterioration of the river. In Bahia, the chief threats are deforestation, discharge of raw sewage, and construction of dams. In Pernambuco, desertification is well advanced; and in Alagoas, deforestation has almost completely cleared the land. (As of 2006, an estimated 95 percent of the Atlantic rainforest of this region was already destroyed.)

The largest government-funded project on the river is currently underway—the diversion of 1.4 percent of the São Francisco's water to the some 12 million people who live in the dry *sertão* lowlands. The project includes 435 miles (700 kilometers) of canals, tunnels, and several dams. Construction began after protest in 2007; it is expected to displace almost 1 million people and to take 20 years to complete. Impacts from climate change will have to be taken into consideration, as evaporation rates will likely increase in the arid regions, even as water usage demand also increases.

Sarah Boslaugh

Further Reading

Eastham, J., M. Kirby, and M. Mainuddin. "Water-Use Accounts in Challenge Program on Water and Food (CPWF) Basins: Simple Water-Use Accounting of the São Francisco Basin." *CPWF Working Papers* 10 (2010).

Maneta, M. P., M. Torres, W. W. Wallender, S. Vosti, M. Kirby, L. H. Bassoi, and L. N. Rodrigues. "Water Demands and Flows in the São Francisco River Basin (Brazil) With Increased Irrigation." *Agricultural Water Management* 96, no. 8 (2009).

Osava, Mario. "Environment: River of National Integration is Dividing Brazil." Inter Press Service News Agency. http://www.ipsnews.net/2005/01/environment-river-of-national-integration-is-dividing-brazil.

Tortajada, Cecilia. "Sao Francisco Water Transfer." United Nations Human Development Office. http://hdr.undp.org/en/reports/global/hdr2006/papers/tortajada%20cecilia.pdf.

Sargasso Sea

Category: Marine and Oceanic Biomes.
Geographic Location: Atlantic Ocean.
Summary: An open ocean area surrounded by strong currents, this biome is the site of important marine research including biogeochemical and prokaryotic meta-genome studies.

The Sargasso Sea is a part of North Atlantic Ocean that overlaps the northern part of the Bermuda Triangle. The name is from the abundant *Sargassum* seaweeds (dominated by *S. natans* and *S. fluitans*) floating on the surface water in the region. They were reported by Christopher Columbus and his crew during their expedition to the New World. The Sargasso Sea is a slow-moving gyre area separated from the rest of the Atlantic Ocean by several surrounding strong currents: the Gulf Stream, North Atlantic, Canary, and the North Atlantic Equatorial. Found here is a large expanse covered with yellow *Sargassum*, plastic wastes, and occasional shipwreck derelicts (the source of many legends).

The Sargasso Sea is one of the few areas of the world lacking coastlines but designated as a sea; it is approximately 700 miles (1,100 kilometers) wide and 2,000 miles (3,200 kilometers) long, covering an area of at least 1.35 million square miles (3.5 million square kilometers) of water.

The water in the Sargasso Sea is warm, maintaining a high salinity concentration of around 36 percent; the temperature in the euphotic (sunlight-penetrated) zone averages up to 72 degrees F (22 degrees C). For these reasons, even though there are several species of plankton and massive amounts of seaweed floating on the water surface, the Sargasso Sea is still not nutritious enough to attract large communities of fish. The factors of low wind, low nutrients, and high salinity are chief reasons why the Sargasso Sea is sometimes considered a "desert among oceans."

Flora and Fauna

The *Sargassum* weed is known to harbor epiphytes (micro and macro), diverse fungi, over 100 invertebrate species, dozens of fish species, and about four seaturtle species—one of which is the threatened green seaturtle (*Chelonia mydas*). The abundant floating vegetation of *Sargassum* provides a habitat for several endemic (found nowhere else) species of marine animals, including Sargassum pipefish (*Syngnathus pelagicus*), Sargassum snail (*Litiopa melanostoma*), slender Sargassum shrimp (*Latreutes fucorum*), and the Sargassum crab (*Planes minutes*). In addition, the deeper reaches that feature drift algae provide critical spawning sites for major eel species including American eel (*Anguilla rostrata*) and European eel (*A. Anguilla*).

This generally nutrient-poor, or oligotrophic, sea has been one of the best-characterized oceans in its physical and biogeochemical properties, especially at the Bermuda Atlantic Time-series Study (BATS) site. Since the late 1950s, the Sargasso Sea has been important research location for marine biogeochemistry. More recently, the first major meta-genome research was carried out in the Sargasso Sea by J. Craig Venter and his colleagues. They have found many new lineages for the clade (or family) of SAR11 bacteria, which later were shown to constitute a major proportion of all

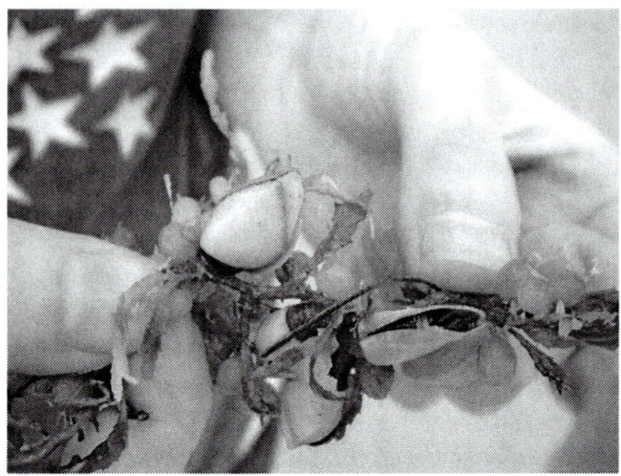

Sargassum provides habitat for more than 100 invertebrate species. These gooseneck barnacles were found living on the seaweed, which gave them a base from which to filter water for food. (U.S. Fish and Wildlife Service/Becky Skiba)

marine prokaryotes (both bacteria and *Archaea*) from the major oceans. This bacterium can surprisingly replicate efficiently in a low-nutrient environment, and is one of the smallest self-replicating cells found.

Other microbes sustained here include cyanobacteria, luminous bacteria, *Berkholderia,* and *Shewanella.* One of the cyanobacteria, *Prochlorococcus marinus,* though among the smallest photosynthetic organisms known, is a major phototroph in the ocean and greatly impacts the carbon cycle. *Berkholderia* was originally thought to be only a terrestrial bacteria. When this new, similar species was first discovered in the ocean, it was found to contain similar DNA and genes, suggesting genomic transfer. Terrestrial *Berkholderia* bacteria are known to be able to biodegrade a very toxic class of industrial product, the polychlorinated biphenyls (PCBs). Because of this ability, the bacteria have an important ecological and commercial potential for bioremediation, and sequencing this genome may play a major role in environmental protection.

Human Interaction

Beginning with the establishment of Hydrostation S in 1957, the time-series biogeochemical sampling between November 1957 to April 1960 by D. M. Menzel and J. H. Ryther effectively set the foundation for the modern science of marine biogeochemistry. They defined the seasonal cycle of primary productivity, and proposed many profound topics related to global biogeochemistry and global climate change. More recently, genomics research has allowed scientists to find evidence of genes that form the basis of the key chemical processes of ocean-dwelling microbes.

This genetic research has shed light on many new, and as yet unknown, micro-organisms. The Sargasso Sea, once the source for many legends and science fiction, has been a self-supporting environment for unique marine biota including some endemic fish and shells. It has also now been identified as a global hot spot of microbial diversity.

The projected effects of global warming on the Sargasso Sea biome are difficult to predict. It is known that sea surface temperature here has been more or less constantly increasing over the last 300 years; researchers tend to see this as a return to the historic norm, however. Global sea-level rise will affect this oceanic area, but consequences to its biota are unclear. More crucial will be any changes that precipitate from alterations in the complex suite of temperature, density, salinity, and velocity variables of the surrounding currents that form the gyre that sustains the Sargasso Sea.

Sanghoon Kang

Further Reading

Giovannoni, Stephen J., et al. "Proteorhodopsin in the Ubiquitous Marine Bacterium SAR11." *Nature* 438 (2005).

Lipschultz, Fredric, et al. "New Production in the Sargasso Sea: History and Current Status." *Global Biogeochemical Cycles* 16, no. 1001 (2002).

McKenna, Sheila and Arlo Hemphil. "The Sargasso Sea." Global Ocean Biodiversity Initiative. http://www.gobi.org/Our%20Work/rare-2.

Morris, Robert M., et al. "SAR11 Clade Dominates Ocean Surface Bacterioplankton Communities." *Nature* 420 (December 2002).

Venter, J. Craig, et al. "Environmental Genome Shotgun Sequencing of the Sargasso Sea." *Science* 304 (2004).

Saskatchewan River

Category: Inland Aquatic Biomes.
Geographic Location: North America.
Summary: This major freshwater body spans the Canadian prairies; the biome incorporates a broad scope of habitats and species, many now threatened by human activity.

The Saskatchewan River is a major freshwater biome in Canada, sustaining a variety of ecosystems and providing habitat for a rich diversity of plants and animals. Originating in the Columbia ice fields of British Columbia, the glacial water combines with snowfall meltwater and drains in two main branches that flow east across Alberta and Saskatchewan, then empty into Lake Winnipeg, Manitoba.

Geography and Climate

The North Saskatchewan River is 800 miles (1,300 kilometers) long, and the South Saskatchewan River is 865 miles (1,400 kilometers) long. The two branches merge, forming the Saskatchewan River, about 25 miles (40 kilometers) east of Prince Albert, Saskatchewan. The confluence is known as the Saskatchewan River Forks. The Plains Cree (*nēhiyawēwin*) refer to the river as *kisiskaciwani-sipiy*, meaning *swift flowing river*, describing the speed and power of the stream. The volume of this river averages about 22,400 cubic feet (634 cubic meters) per second.

Natural features of the river include shifting sandbars (both exposed and submerged), gravel, large stones, oxbows, lakes, rapids, islands, tree and shrub roots along shorelines, and fallen logs, all providing habitat for invertebrates, fish, birds, small and large mammals, and a variety of plant life. While a river brings to mind a body of water flowing along a channel, it is much more than that. The Saskatchewan River system is part of a watershed covering approximately 130,000 square miles (336,000 square kilometers) and filled with tributaries, forests, agricultural lands, delta areas, and marshes.

Most of the watershed is within a temperate, continental climate zone. The mean annual temperature ranges from approximately 36 to 43 degrees F (2 to 6 degrees C). Total annual precipitation ranges from 11 to over 32 inches (28 to over 80 centimeters) as one moves westward into the mountains.

Biodiversity

Invertebrates are important parts of the Saskatchewan River biome because they are food sources for fish and birds and are a good measure of the health of an ecosystem. The diversity and numbers of invertebrates is usually a better indicator of water quality than random water sampling. Invertebrates found in the Saskatchewan River system include water boatmen, roundworms, caddisfly larvae, stonefly nymphs, mayfly nymphs, bristle worms, midge larvae, and damselfly nymphs, as well as various clams, leeches, snails, beetles, crayfish, dragonflies, black flies, water fleas, fairy shrimp, and numerous other small organisms. Mosquitoes lay their eggs in standing water and are abundant in this ecoregion as an important food source for many birds and bats.

Minnows occupy lakes and rivers across Canada. Some common types found in the Saskatchewan River system are fathead minnows and brassy minnows. Larger fish, such as sauger, rainbow trout, and northern pike, feed on minnows. Like invertebrates, minnows are sensitive to water quality and changes in their environment. Increased pollution from prairie agriculture, wastewater effluent from human habitation, and other chemical releases are known to adversely affect minnow reproductive systems.

The Saskatchewan River biome is home to brook trout, burbot, cisco, goldeye, lake sturgeon, lake whitefish, longnose sucker, mooneye, northern pike, rainbow trout, sauger, shorthead redhorse, walleye, white sucker, yellow perch, and Atlantic salmon. The fish vary in number and habitat preference along the river, with some preferring shallower waters, some occupying deeper waters, and still others occupying pools of less turbulent water. The shorthead redhorse, for example, is sensitive to water quality and favors clear shallow water with a good sand or gravel bottom, whereas suckers are able to thrive in poor-quality water. Suckers are among the most prolific species in the ecoregion.

The river valleys of the Saskatchewan River system provide rich habitat for a great variety of birds. Some of the larger species include the peregrine falcon, Swainson's hawk, bald eagle, ruffed grouse, great horned owl, great blue heron, osprey, American white pelican, common merganser, Canada goose, Franklin's gull, and trumpeter swan. Smaller birds include the pileated woodpecker, blue-headed vireo, blue jay, magpie, crow, raven, swallow, black-capped chickadee, white-breasted nuthatch, house wren, American robin, Swainson's thrush, gray catbird, European starling, and cedar waxwing.

Some birds that typically migrate to warmer wintering grounds may stay in areas where the river does not freeze, mainly as a result of human influences. The Canada goose, for example, can sometimes be found year round in urban areas. In the Saskatchewan River delta, an estimated half-million breeding waterfowl make their way from the Arctic and northern boreal areas to renew the life cycle of their species. Other birds migrate from parts of North, Central, and South America each year to their particular territory along the river.

Large mammals that live in the watershed include the moose, white-tailed deer, mule deer, elk, black bear, and wolf. The animals are not easily seen during warm seasons, preferring to keep away from human habitations. Fruit and berry supplies growing along the river provide a source of food for bears in late summer, requiring extra caution for humans hoping to collect the same harvest. Smaller mammals—such as beavers, otters, fishers, minks, and coyotes—occupy small territories in and around the river.

The ecology of the Saskatchewan River system changes as it crosses through various zones. The mid-boreal lowland ecoregion occupies the northern section of the Manitoba Plain, from the eastern shore of Lake Winnipeg to the Cumberland Lowlands in Saskatchewan. The boreal transition ecoregion extends from southern Manitoba to central Alberta. The aspen parkland ecoregion extends in a broad arc from southwestern Manitoba, and northwestward through Saskatchewan to its northern apex in central Alberta. Moist mixed grassland is an ecoregion comprising the northern extension of open grasslands in the interior plains of Canada.

Each of these ecoregions has variations in climate, precipitation, vegetation, soils, and wildlife. The boreal transition ecoregion, for example, encompasses the area where the North Saskatchewan and South Saskatchewan rivers meet. Here is found a mix of deciduous boreal forest area and agricultural lands. Tall trembling aspen and balsam poplar provide thick canopy, with smaller shrubs filling in the understory. White spruce and balsam fir are the climax species, but have been reduced because of fires.

Human Impact

Traditionally, many indigenous peoples relied on native flora and fauna for survival. Many native plant and animal species of the biome are no longer found, however, having been replaced by imported, invasive, or stocked species. The Saskatchewan River is an important source of freshwater that humans and all other living things rely on for life.

The river is also an important source of agricultural irrigation, power generation, and recreation for people in Canada. Two hydroelectric power plants in Saskatchewan and one in Manitoba provide electricity. Hydroelectric dams have changed some areas and upset some habitats through the creation of reservoirs that flood large tracts of land. Silt and sediments removed at the dam sites reduce deposits on the downstream side, creating deeper channels and faster water flow. Chemical pollutants and wastes make consumption of resident fish risky. Intensive agriculture over many generations has depleted soil nutrients; farmers now use nontraditional methods of farming with greater amounts of fertilizer that eventually wash into the river, upsetting balanced ecosystems.

Changing climate conditions will bring changes to the Saskatchewan River system, also. Less annual snowfall in mountain valleys and less precipitation across the prairies will mean the bodies of water forming the Saskatchewan River will be drawn down more to support crops, urban landscaping, golf courses, industrial development, and human populations. This will challenge habitats deprived of their life-currents. Simultaneously, more precipitation in some areas of the watershed and earlier spring snowmelt will bring flooding and temporarily higher river levels that will also challenge watershed ecosystems with greater erosion, stream turbidity, and habitat disruption.

Yvonne N. Vizina

Further Reading

Casey, Alan. "Climate Prosperity." *Canadian Geographic*, October 2010. http://www.canadiangeographic.ca/magazine/oct10/south_saskatchewan_river.asp.

Hallstrom, Tim and Chris Jordison. *Fish Species of Saskatchewan.* Regina, Canada: Saskatchewan Watershed Authority, Fish and Wildlife Development Fund, 2007.

University of Saskatchewan. "Ecoregions of Saskatchewan." http://www.usask.ca/biology/rareplants_sk/root/htm/en/researcher/4_ecoreg.php.

Saya de Malha Bank

Category: Marine and Oceanic Biomes.
Geographic Location: Indian Ocean.
Summary: The largest submerged ocean banks in the world provide resources for fish and fishers.

Saya de Malha Bank is the largest submerged ocean bank in the world. Located in the western Indian Ocean, it covers an area of more than 15,600 square miles (40,500 square kilometers), representing one of the largest shallow tropical marine ecosystems on Earth. The bank is part of the underwater Mascarene Plateau, an underwater ridge east of Madagascar connecting the Seychelles and Mauritius, and is composed of the North Bank (or Ritchie Bank) and the much larger South Bank.

Because of its remoteness, the Saya de Malha Bank hosts some of the least explored tropical marine ecosystems. This highly productive oasis in the Indian Ocean sustains and represents an important food source for reef and pelagic fish. Diverse ecosystems such as coral reefs, seagrass beds, and shallow banks provide stepping-stone connectivity for species across the entire Indian Ocean. The great majority of the bank areas are located in international waters, and there is general concern regarding the sustainability of fish stocks targeted both by regional and international fisheries.

The banks were baptized around 500 years ago by Portuguese sailors who were exploring a new trade route to India. The origin of the name *Saya de malha*, ancient Portuguese words for military garb, apparently derives from the association of its impenetrable properties and the hazard that such mesh of shallow banks poses to navigation.

This shoal complex is situated in the great Mascarene Plateau, formed 20–40 million years ago by the volcanic hot spot of Réunion, which is still active today. These banks likely were islands that sank below the ocean surface as recently as 18,000 to 6,000 years ago. These days, the banks consist of a series of narrow underwater shoals with sandy bottoms, with depths varying from about 165 to 200 feet (50 to 60 meters), covered in major part by seagrass, surrounded by a shallower coralline rim that slopes to around 500 feet (150 meters) in depth. Completely detached from land boundaries, Saya de Malha Bank is surrounded by fairly sharp drop-offs up to 6,500 feet (2,000 meters) deep.

Biodiversity

The component banks of this biome are the tops of large seamounts, providing an extensive diversity of ecosystems throughout the depths and sustaining a great diversity of marine life. The plateau acts as a barrier to latitudinal water flows in the western Indian Ocean. The result is an upwelling system bringing deep, cold, and nutrient-rich waters to the surface, fueling primary productivity. In the deeper areas, impinging currents sustain food chains of small invertebrates and long-lived fish and coral species.

The banks represent one of the largest shallow tropical marine ecosystems, hosting the most extensive oceanic seagrass area in the world. Seagrass beds are among the most productive aquatic ecosystems; here, they cover 80 to 90 percent of the bottom—15,400 square miles (40,000 square kilometers). They provide a unique feeding area to one of the most numerous populations of the threatened green seaturtle (*Chelonia mydas*) in the western Indian Ocean. The remaining seafloor is covered by a diverse range of coral species and sandy areas.

The Saya de Malha Bank attracts significant numbers of tuna and billfish, and provides habitat to the world's largest fish, the whale shark (*Rhincodon typus*). The area is also one of the rare breeding grounds for the blue whale (*Balaenopterus musculus*), the largest animal on Earth.

A limited number of oceanographic surveys have been conducted in the area, and many of the species and habitats of this remote region remain poorly documented. Great expeditions in the area range from the British Royal Navy surveys in 1838 to the Russian (1960s to 1980s) and Japanese fisheries expeditions, and the recent Agulhas and Somali Current Large Marine Ecosystem (ASCLME) surveys.

*Visitors to an aquarium in Atlanta, Georgia, are dwarfed by a whale shark (*Rhincodon typus*), which is considered the largest fish in the world. The Saya de Malha bank's 15,600 square miles (40,500 square kilometers), which represent one of the largest shallow tropical marine ecosystems on Earth, offer habitat for whale sharks and blue whales. (Wikimedia/Zac Wolf)*

Human Impact

Most of the Saya de Malha Bank is in international waters, with Mauritius holding special fishing rights. Fishing is mostly by hand line, from dories deployed by mother ships based in Mauritius. International fleets operate deep-water fisheries, with purse-seiners and drifting long-liners targeting tuna species.

Saya de Malha provides the largest area for bank fishing on the Mauritius-Seychelles ridge. Most of the hand-line catch consists of *Lethrinus mahsena,* a reef fish locally known as dame berri. This fishery started in the 1950s, with a current catch of about 5,500 tons (5,000 metric tons) per year, and the Mauritius government is attempting to reduce the total allowable catches to sustainable levels.

Tuna fisheries have operated on an industrial scale since the 1950s in the southwestern Indian Ocean. Japanese, Taiwanese, and Korean fleets were later joined by fleets from Mauritius, France, and China. Estimates of Mauritius landings exceed 22,000 tons (20,000 metric tons) per year. The tuna fishery is an important industry worldwide, and the lack of information about it constitutes serious risk to the sustainability of open-water ecosystems. The Indian Ocean Tuna Commission is endeavoring to implement fishery observer programs.

In 2002, the region was proposed as an International Biosphere Reserve by the Lighthouse Foundation. In 2010, it was recommended as a High Seas Marine Protected Area by Greenpeace. The remoteness of the bank has protected the ecoregion from some threats (just as it has prevented much research and data collection). So far, prospecting for metals, oil, and gas beneath the seabed has returned weak results.

Climate change is also a major threat for the carbon-dominated food webs of the shallow banks. Accelerated absorption into sea water of carbon dioxide from the atmosphere can have deleterious effects, such as on the reproduction or shell formation in mollusks and crustaceans.

José Nuno Gomes-Pereira
Gisela Dionísio

Further Reading

Ardron, Jeff, et al. "Defining Ecologically or Biologically Significant Areas in the Open Oceans and Deep Seas: Analysis, Tools, Resources and Illustrations." International Union for Conservation of Nature. http://www.protectplanetocean.org/docs/GOBI_Report_2009.pdf.

Payet, R. "Research, Assessment and Management on the Mascarene Plateau: A Large Marine Ecosystem Perspective." *Philosophical Transactions of the Royal Society* 363 (2005).

United Nations Environmental, Scientific and Cultural Organization (UNESCO): World Heritage Convention. "Saya de Malha Bank, Mascarene Plateau." http://www.vliz.be/projects/marineworldheritage/sites/2_Masc%20Plateau_S%20Malha.php?item=The%20Indian%20Ocean.

Scandinavian Coastal Conifer Forests

Category: Forest Biomes.
Geographic Location: Europe.
Summary: These forests are a unique habitat for Norway spruce and Scots pine, stands of which provide valuable habitat for populations of seabirds and endangered species of lichens.

This temperate, palearctic region extending along the western Norwegian coastline of Scandinavia, between Lindesnes and Senja, supports an extended coastal conifer forest. Norway is located within a much larger green belt referred to as the boreal forest, or taiga, which is one of Earth's largest biomes; located just below the Arctic Circle, it encircles the northern parts of North America, Europe, Russia, and Asia.

Geography and Climate

The coastal topography is deeply etched as a result of successive glacial formations in the Pleistocene period. Jostedalsbreen National Park is situated at the Norwegian boundary of the Scandinavian Ice Sheet, the last formation of a Pleistocene glacial mass that covered Great Britain, Germany, Poland, and Russia for over 2.5 million years, ending only 11,700 years ago.

Fluted moraines, drumlins, bedrock troughs and deep saltwater inlets, bound by steep cliffs and verdant parallel-sided valleys, splay in ragged, fern-like patterns all along the fringes of the peninsula. Called fjords, these spectacular formations of sea and cliff lands provide a unique habitat for boreal conifers and various species of wildlife. Hundreds of rocky islands provide a variety of stippled landscapes and barren landmarks that include some of Europe's most important rookeries for colonies of seals and seabirds.

Flora and Fauna

This subarctic region is populated by fairly homogenous forests of spruce, fir, and birch. Intermediate post-fire successional stands of various deciduous trees are also common, usually in close proximity to inland bodies of water; important species include oaks, ashes, limes, and elms. Scandinavia is noted for its forests of Norway spruce (*Picea abies*) and Scots pine (*Pinus sylvestris*). Spruce forests grow in the central region of Norway; this is the only region in western Europe where these trees can be found in such dominance.

The Norway spruce is one of the world's hardiest, fast-growing evergreens. Standing full-grown at a height over 100 feet (30 meters), its luxuriant, deep green branches hang low with cones 4–8 inches (10–20 centimeters) in length. It is a time-honored favorite choice for Christmas trees. Norway spruces in old-growth forests are hundreds of years old. Scots pines, too, can be dated with ages in excess of 800 years. Older trees with maximum heights survive fires well; over time, boreal forests with high fire frequency develop multiple clustered diameter distributions, in contrast with old spruce forests where highly skewed diameters are noted.

Scandinavian forest grounds are scantily covered with ferns and mosses, with only a minor representation of shrubs and herbs where soils are moist. These forests are unique for the diversity of epiphytic species (those that grow on other plants) such as mosses, liverworts and lichen species that thrive on the detritus of old-wood forest stands. Lichens are living fungi whose structures are conformed to take advantage of life forms that produce food by photosynthesis. Common partners are algae and cyanobacteria, sometimes known as blue-green algae. These organisms create unique symbiotic growth forms determined by environment and climate. They are important soil stabilizers, protecting against erosion, and they are fundamental sources of fixed nitrogen.

Lichens are fragile and they are particularly sensitive to air-borne pollutants. Their presence or absence is a valuable indicator of the relative health of a particular ecoregion. Endangered species of coniferous lichens include the *Lobarion pulmonariae* community—a species that only grows on Norway spruce. *Pseudocyphellaria crocata*, *Uslea longissima*, and *Letharia vulpina* are other endangered lichens endemic to specific boreal forest habitats.

Conifers are an important food source for insects, small mammals, migratory birds, and grazing livestock. Large mammals that may be found here include moose, elk, reindeer, red deer, wolverine, weasel, arctic fox, and red fox. Small mammals include rodents such as mouse, shrew, and various bat species. Approximately 200 bird species have been identified here, including gulls, shrike, loons, merganser, owls, rough-legged hawk, peregrine falcon, white-tailed eagle, and various woodpeckers.

Human Impact

Ancient longhouses, farmsteads, churches, and early villages constructed from wood, reed, brick, and stone—and brightly decorated with delicate woodwork relief and inlay—continue to provide fresh inspiration for contemporary architects and designers. Early northern European homemakers here uprooted peats and sods for fuel, for roof cover, and for home insulation. Some natural homesteads provided unique ecological habitats for bugs and birds; they were often reconstructed and maintained for many generations.

Today's green roof technologies and structures here often incorporate centuries-old practices, and are an important feature of home and commercial construction in many countries world-wide. Iron ore was another important early natural resource; its mining and production required plentiful fuels for construction and smithing, which the forests amply supplied.

Over the centuries, there was increased demand for trade in particular commodities, of which sawn timber was a priority. Norwegian forests have been exploited intensively for export of roundwood, sawn timber, wood pulp, and wood tar for hundreds of years. All forest areas suffer from clear-cutting, plantation forestry, and habitat fragmentation as a consequence of urbanization, grazing, agriculture, and fire; there has been widespread destruction of important primeval habitats and ecosystems.

Norway's productive forests are limited by geography and climate; they represent only a quarter of the country's total land area. Over half of the land is dominated by valuable marshland, mountains, and nonproductive forest. In recent years, the planting of forest on areas that previously were used for farmland, and more sophisticated management of forest production, along with extensive planting of trees, has helped to increase the biomass quantities of Norway's forests.

In keeping with the United Nations Year of the Forest 2011 initiative, Norway hosted the Forest Europe Ministerial Conference on the Protection of Forests in Europe. Ministers and representatives of 42 European countries, six outside countries, and 29 international organizations convened to approve negotiations for legally binding protection measures to safeguard the biodiversity of Europe's expansive forest lands, recognizing that forests can have significant impacts on future climate trends, both positive and negative.

Victoria M. Breting-García

Further Reading

Esseen, Per-Anders, Bengt Ehnström, Lars Ericson, and Kjell Sjöberg. "Boreal Forests." *Ecological Bulletins* 46 (1997).

Helle, Knut, ed. *The Cambridge History of Scandinavia*. Cambridge, UK: Cambridge University Press, 2003.

Korsmo, Harald. "Conserving Coniferous Forest in Norway: A Critical Time for International Environmental Obligations." *Ambio* 20, no. 6 (1991).

Speer, Brian and Ben Waggoner. "Introduction to Lichens: An Alliance Between Kingdoms." University of California Museum of Paleontology. http://www.ucmp.berkeley.edu/fungi/lichens/lichens.html.

Scheldt River

Category: Inland Aquatic Biomes.
Geographic Location: Europe.
Summary: A heavily urbanized river biome and one of the last river-tide systems in Europe, the Scheldt encompasses valuable estuary wetlands but is threatened by widespread pollution.

The Scheldt River flows from northeastern France, cuts across northwestern Belgium, and after branching out as a great estuary across the border in southwestern Netherlands, flows into the North Sea near Vlissingen. The river springs in Gouy, France, at an altitude of about 300 feet (90 meters), and flows some 220 miles (354 kilometers) to its mouth, with the entire basin covering about 8,500 square miles (22,000 square kilometers). Several tributaries empty into the Scheldt River, including the Leie, Haine, Dender, and Rupel.

The river's name finds its origin in the old English word *sceald,* meaning shallow. The river is referred to as the *Schelde* in Dutch and *Escaut* in French. Different sections of the river have different localized names. From the source to the city of Ghent, Belgium, it is referred the Bovenschelde; between Ghent and the Belgian-Dutch border, it is called the Zeeschelde, reflecting its broad spread as it approaches the sea; and further downstream, it is dubbed the Westerschelde.

The Scheldt River is considered to be both a rainwater-fed river and a tidal river, with precipitation controlling water levels downstream to Ghent, while tides control the water levels from Ghent to the North Sea. Even though the Scheldt

River is one of the smallest rivers in Europe, its economic production places it in the top 10 of the world, with a value is similar to that of the Yangtse, Amazon, and Nile.

The climate here is temperate maritime, characterized by relatively warm summers and mild winters. January is the coldest month at about 37 degrees F (2.5 degrees C) on average, and July the warmest with an average of 63 degrees F (17 degrees C). In coastal areas, the proximity of the sea tends to provide lower temperatures in the summer and somewhat warmer temperatures in the winter. Because of its limited territory, the climate within the basin varies very little.

Biodiversity

The Scheldt is one of the few rivers left in Europe with an intact tide-river system comprising major saltwater and freshwater segments. The western Scheldt estuary system forms a unique hot spot for nature, with a high diversity of bird species migrating through each year. The estuary is a multi-channel system with several tidal flats both across the middle streams and along the edges.

These tidal flats are inhabited by a rich diversity of bottom-dwelling organisms, including ranges of mollusks and crustaceans, on which migratory birds feed. Mussels, snails, and oysters are also commercially farmed here. Freshwater and saltwater species can coexist as neighbors in this ecosystem; the high productivity of this estuary creates a the foundation of a complex food web. Closest to the sea, seals and porpoises often prey.

Human Impact

This river has a long and turbulent history with much human intervention to facilitate shipping and habitable land creation; it is truly an urban river. The first record of the river was from the Romans, who used it for shipping. At that time, the southwestern province of Zeeland (now The Netherlands) was almost completely underwater, which resulted in thick peatlands and clay deposits. The first modest diking of the river started in the 10th century, making parts of Zeeland habitable.

Throughout the 12th century, the Scheldt River was a smallish river with minor branches. Major flood events have altered its courses many times over the centuries, connecting and disconnecting various channels, breaching dunes, overwhelming marshes, and silting up channels—resulting in great habitat destruction at times.

Currently, the Scheldt River is left largely untouched for the first 10 miles (16 kilometers) from its source. The following 85 miles (137 kilometers) are partly canalized. The remaining length of the river mainly functions as an estuary, with the influence of the North Sea noticeable as far inland as the city of Ghent.

The river basin is inhabited by 10 million people. Intense use of the Scheldt estuary has resulted in severe pollution, eutrophication, and morphological changes threatening the natural refuges in this ecosystem. The Scheldt estuary is a uniquely polluted ecosystem due to erosion, oxygen depletion, and toxic heavy-metals inputs.

Erosion fills the river with sediment, creating a turbid environment. In addition, erosion contributes a substantial amount of nutrient loading (from adjacent agricultural areas) to the river. High nutrient concentrations in turn result in algal blooms, depriving the water of oxygen and threatening higher life forms, such as fish.

Since the late 1970s, awareness about the interaction between ecology and long-term economical health has become prevalent, and a push is being made to find a balance here. The future of the Scheldt River is unclear. Recovery and conservation, in terms of river health, would require a focus on three main points: returning oxygen to the river, removing toxic elements, and increasing the floodplain area to prevent the river from filling with sludge.

Climate change may impact the river by means of exacerbating hydrological extremes of low-flow and flooding along catchments in the basin. The lower Scheldt River in Belgium is particularly prone to flood risk, as coastal level changes include both sea-level rise and storm surge changes. Saltwater intrusion is also a mounting challenge as sea levels rise. Until now, the trend has been wetter winters and drier summers.

Peter Baas

Further Reading

Baeyens, W., B. van Eck, C. Lambert, R. Wollast, and L. Goeyens. "General Description of the Scheldt Estuary." *Hydrobiologia* 366 (1997).

Gilbert, A., M. Schaafsma, L. De Nocker, I. Liekens, and S. Broekx. *Case Study Report—Scheldt.* Brussels, Belgium: AquaMoney, 2007.

Lefebure, Arnould. *International Scheldt River Basin District: Management Plan Roof Report.* Antwerp, Belgium: International Scheldt Commission, 2009.

Scotia Sea Islands Tundra

Category: Grassland, Tundra, and Human Biomes.
Geographic Location: South Atlantic and Antarctic Oceans.
Summary: This harsh environment supports significant seabird, seal, and penguin populations that have slowly recovered from losses due to the commercial sealing and whaling industries.

The Scotia Sea Islands are comprised of the South Georgia, South Sandwich, South Orkney, South Shetland, and Bouvet Islands. All are territories of the United Kingdom except for Bouvet Island, which is a Norwegian possession. The islands' terrain ranges from cold, inhospitable rocky cliffs and ice sheets to lower, warmer areas heated by volcanic activity. These islands share environmental and biological characteristics, but also present their own unique habitats.

South Georgia Island is one of the largest and most biodiverse of all the islands and has no permanent human settlements, but the area does have several scientific observatories. The most significant disruptions to the ecosystem have come from the commercial sealing, whaling, and fishing industries. These island groups are dispersed over a wide area, but in general are some 1,000 miles (1,600 kilometers) east of the southern tip of South America, and 600–800 miles (965–1,300 kilometers) northeast of the Antarctic Peninsula.

The Scotia Sea Islands lie in the South Atlantic and Southern (Antarctic) Oceans between South America and Antarctica. The islands are located south of the Antarctic convergence, a biological marker cordoning off the cold Antarctic waters and their harsh environments. The Scotia Sea is approximately 350,000 square miles (906,496 square kilometers) in area; it was named for a ship of the 1903 Scottish National Antarctic Expedition that visited the region.

The Scotia Sea Islands tundra is linked with the Antarctic and sub-Antarctic climate zones. Many of the islands are partially or fully covered in ice fields, snow, or glaciers either permanently or for much of the year. A cold, stormy, windy climate and widespread cloud cover dominate, with little seasonal variation.

The Scotia Sea Islands sit above the Scotia Ridge and are of mixed continental and volcanic origins; the volcanic islands here represent the Antarctic region's only undersea volcanic arc. The South Sandwich Islands consist of 12 main islands and

These whaling boats could still be seen lying abandoned on the shore of South Georgia Island in 1994. In the 20th century, the whaling industry decimated cetacean populations in the area. (NOAA/Lieutenant Philip Hall)

several smaller islets; they have experienced relatively recent volcanic activity. Bouvet Island, small and isolated, contains an inactive volcano. The South Orkney Islands consist of four main islands, several smaller islets and offshore rocks, and a few coastal freshwater lakes. The South Shetland Islands feature 11 main islands. The larger, mountainous South Georgia Island has an area of 1,450 square miles (3,755 square kilometers), as well as several offshore rocks.

Biodiversity

The harsh environment of the Scotia Sea Islands ecosystem supports only primitive flora and a few native species of land birds. South Georgia Island has one of the region's most diverse habitats; the main flora of this tundra ecosystem consists of various mosses, lichens, algae, and liverworts. Other vegetation types include Antarctic hairgrass, fescue grass, coastal tussock grasslands, peat bogs, and feldmark.

There is an absence of native land mammals, reptiles, and amphibians. Invertebrate populations include arthropods, earthworms, mollusks, spiders, beetles, flies, and nematodes. Nonnative species successfully introduced into the ecosystem include only rats and reindeer, the latter growing to a population of several thousand since their introduction in the early 20th century. These and other, less successful, exotic species continue to alter the native habitats.

The deep ocean waters surrounding the islands support seabird, penguin, seal, and whale populations; they also harbor productive kelp beds, fish, and squid. Several species of seabirds, penguins, and seals visit the islands regularly, with some species establishing onshore breeding grounds and rookeries. Seabirds include several species of albatross, pigeon, petrel, prion, shag, and tern. Penguin species include the king, chinstrap, macaroni, gentoo, Adelie, and rockhopper penguins. Seal species include the Antarctic fur, sub-Antarctic fur, leopard, Weddell, southern elephant, and crabeater seals. South Georgia Island houses two endemic (found nowhere else) bird species: the South Georgia pintail duck and the South Georgia pipit.

Effects of Human Activity

Although the harsh environment of the Scotia Sea Islands has discouraged permanent human habitation, humans have significantly affected the ecosystem. The British and United States sealing industries flourished in the region in the 18th century, hunting the Antarctic fur seal for its hide and the elephant seal for its oil-rich blubber. The populations of both types of seals were decimated.

The whaling industry led to declining cetacean populations in the 20th century. Seal and whale populations slowly recovered, but faced new challenges and population declines in the late 20th and early 21st centuries. The commercial fishing industry threatens seals, as well as seabirds such as the albatross, through food competition, entanglement in fishing nets and lines, ocean debris and pollution, and overfishing.

Scientific interest in the region, most notably South Georgia Island, has resulted in the establishment of a British Antarctic Survey base, fisheries laboratory, and weather station on tiny Bird Island, off the tip of South Georgia. Disputes between Argentina and Great Britain over regional possessions such as the Falkland Islands led to the establishment of a British military station on South Georgia Island.

National and international legislation has protected the ecosystem's seal populations since the mid-20th century. Examples include the Convention for the Conservation of Antarctic Seals and the Antarctic Treaty. The environmental movement of the late 20th century furthered efforts to protect the ecosystem, including the preservation of South Georgia, South Sandwich, and Bouvet Islands as nature reserves. There is limited tourism to South Georgia Island, most of which is restricted by permit.

While the climate of the region is clearly cold and inhospitable to a wide range of species—plant, fish, and mammal—warming air and sea temperatures resulting from global climate change have been recorded in this remote corner of the world. The amount and thickness of ice in the frozen waters has declined. The long-term impact as far as precipitation and temperature is unclear, but sea-level rise certainly poses challenges to the littoral, intertidal, and coastal habitats around the Scotia Sea Islands.

Marcella Bush Trevino

Further Reading

Coad, Lauren, Neil D. Burgess et al. "Progress Towards the Convention on Biological Diversity Terrestrial 2010 and Marine 2012 Targets for Protected Area Coverage." United Nations Environment Programme World Conservation Monitoring Centre. http://www.unep-wcmc.org/progress-towards-the-cbd-terrestrial-2010-and-marine-2012-targets-for-protected-area-coverage_272.html.

Procter, D. and L. V. Fleming, eds. *Biodiversity: The UK Overseas Territories.* Peterborough, United Kingdom: Joint Nature Conservation Committee, 1999.

Watson, G. E., J. P. Angle, and P. C. Harper. *Birds of the Antarctic and Sub-Antarctic.* Washington, DC: American Geophysical Union, 1975.

Sechura Desert

Category: Desert Biomes.
Geographic Location: South America.
Summary: This coastal desert along the Pacific coast of Peru and Chile is prone to flooding from river surges and Pacific storms; the biota also depends on moisture delivered by coastal fogs.

The Sechura Desert is located north of the Atacama Desert in a narrow, longitudinal strip along the Pacific coast of South America, and extending inland to the foothills of the Andes Mountains. The entire stretch of the arid coastal desert from northwest Peru to northern Chile is commonly referred to as the Sechura Desert. Within Peru, the term is limited to the country's northern coastal desert, ending with the city of Piura and its surrounding tropical-dry forest ecoregion. The climate of this desert biome varies little because of the moderating influence of the Pacific Ocean. The summer average range is 77–100 degrees F (25–38 degrees C); the range in winter is 61–75 degrees F (16–24 degrees C). Summer lows are comparable to winter highs, around 75 degrees F (24 degrees C). While arid by most standards—annual rainfall averages about 3 inches (8 centimeters), the Sechura is nearly two orders of magnitude (100 times) wetter than the neighboring Atacama, generally accepted as one of the driest places on Earth. The Sechura also derives groundwater recharge from the many rivers along its edges and those that cut across its main body.

The Sechura Desert is filled with rocks degraded by weathering, erosion, and the formation of salt crystals that remain even after it rains. The soils are dry, low in organic matter, and more saline than the average soil of less arid regions. The desert district receives moisture via fog that forms over the nearby ocean. As in other coastal deserts, the biota has adapted to take advantage of these fog zones.

Fossils of mollusks and marine mammals from the Miocene are frequently found in the sedimentary deposits of sandstone in the center of the desert. Some rivers cross the desert, and human settlements have developed along them for thousands of years. The town of Sechura was relocated to its current location when a tsunami wiped out the original town in 1728. In 1998, the El Niño rain-fed floods temporarily turned part of the Sechura into Peru's second-largest lake, more than 10 feet (3 meters) deep and 90 miles (145 kilometers) long, punctuated by small islands of sand and clay.

Vegetation

Before the river valleys of the Sechura were converted to agricultural land, they supported large plant populations of *Acacia macracantha, Salix humboldtiana, Schinus molle, Sapindus saponaria, Muntingia calabura,* and semiwoody shrubs. Today, these rivers are diverted for intensive irrigated agriculture, and the city of Piura—one of the five largest in Peru—depends on the Sechura agricultural region, which is prone to flooding due to the proximity to so many bodies of water. In addition to floods from river surges, especially after wet springs or when snow melts rapidly in the Andes, the Sechura regularly experiences storms from the Pacific Ocean and floods at least briefly in most El Niño years.

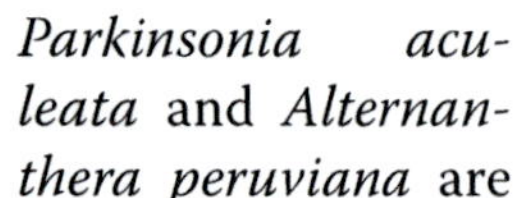

Chilean flamingos (Phoenicopterus chilensis) in flight in the Atacama Desert region just south of the Sechura Desert. Both deserts provide valuable habitat for the endangered birds. (Thinkstock)

Most of the vegetation in the Sechura region grows near the rivers. Squash and peanuts are indigenous to the bottomlands and have been cultivated for thousands of years. The carob tree, palo verde, salta grama grass, mangrove tree, giant reed, ditch reed, wild tomato, Peruvian daffodil, Peruvian papaya, and divi-divi are all found in the area. In the Piura vicinity, a variety of the mesquite tree called the algarrobo is endemic, that is, a species found nowhere else.

There are many species of pea-family algarrobo or bayahonda trees and shrubs, including *Prosopis abbreviata, P. affinis, P. alba, P. chilensis, P. fiebrigii, P. flexuosa, P. juliflora, P. kuntzei, P. nigra, P. pallida, P. rojasiana, P. rusciflolia, P. strombulifera,* and P. *tamarugo.* The Piura region also is home to the oldest limes in South America; a variety of endemic species of orchids; bananas; coconuts; rice; and mangos, both wild and commercially cultivated.

Advective fog (called *garua* in Peru and *camanchaca* in Chile) that forms over the Pacific, with its cold seawater upwelling zones along the coast, carries moisture to the southwestern hills and ridges of the Sechura, leading to the development of fog-zone plant communities called *lomas,* for *small hills.* Older botanical literature sometimes refers to the lomas as the fog belt, fertile belt, desert meadows, or fog oases, but in contemporary literature, the Spanish-language term is preferred. There are 40 discrete lomas in the Peruvian portion of the Sechura, and even more in the Chilean stretch, all of them fairly small. Lomas are extremely localized discrete communities of plant life, like islands surrounded by otherwise-arid desert.

Altitude affects the variety of plant life to some degree. Low shrubs are common below 328 feet (100 meters); cactuses thrive at higher elevations. *Parkinsonia aculeata* and *Alternanthera peruviana* are common, as are grasses that serve to stabilize the dunes: *Distichlis spicata* and *Cryptocarpus pyriformis* in the west, nearest the coast; and *Capparis scabrida* and *C. avicenniifolia* farther inland. Nearest to the Andes foothills, terrestrial bromeliads and columnar cactuses are common, as are dwarf trees such as *Eriotheca discolor, Bursera graveolens,* and *Acacia huarango.*

Fauna

Animal biodiversity is generally low in the Sechura, but it is an important corridor for migratory birds such as Baird's sandpiper (*Calidris bairdii*) and the Arctic sanderling (*Crocethia alba*). Distinctive short-range species include the pizarrita slender-billed finch (*Xenospingus concolor*) and Raimond's yellow-finch (*Sicalis raimondii*). Endangered birds here include the Chilean flamingo (*Phoenicopterus chilensis*), red-fronted coot (*Fulica rufifrons*), and white-winged guan (*Penelope albipennis*).

Among characteristic mammals here, the Sechuran fox (*Lycalopex sechurae*) is thought to be the smallest of the false fox canids, also called zorro. About the size of a cat, the Sechura fox is 20–30 inches (508–762 millimeters) long, with a 12-inch (305-millimeter) tail, and weighs 6–10 pounds (3–5 kilograms). Adapted for feeding on desert seed pods, capers, fruit, insects, and the occasional bird egg or small rodent, this nocturnal scavenger stays underground by day. Considered an at-risk species, the Sechuran fox has some protection in Ecuador and Peru.

The wild guinea pig known in Peru as the cuy has been part of the ecosystem since at least the time of the Moche civilization 2,000 years ago, which used it as a food source and may have introduced it from another region. Reptiles typical of the biome include Roedinger's lance head snake (*Bothrops roedingeri*) and Sechura Desert coral snake (*Micrurus ischudii*).

Environmental Threats

Major threats to the ecosystem include urban expansion and more damaging forms of commercial agriculture. As city areas swell, the remaining agricultural lands on which they depend are forced to use increasingly intensive methods, which can sometimes disrupt or destroy existing habitat in order to plant and irrigate crops. Agriculture also can attract new pests, while pesticides and fertilizers can harm microflora and microfauna, as well as riverine life. Opening carob groves to livestock can lead to overgrazing and invite new nonnative plant species. Hunting of some animals, principally ducks, periodically puts their populations at risk.

Rising temperatures, increased rainfall, and the potential for more flooding in the Sechura Desert is projected to interrupt plant growth cycles for some species, and to limit normal activity for some fauna. These effects of climate change may also encourage more invasive species as habitat ranges shift.

Bill Kte'pi

Further Reading

Cooke, Ronald, Andrew Warren, and Andrew Goudie. *Desert Geomorphology.* New York: Taylor and Francis, 1992.

Huey, Raymond B. "Parapatry and Niche Complementarity of Peruvian Desert Geckos." *Oecologia* 38, no. 3 (1979).

Huey, Raymond B. "Winter Diet of the Peruvian Desert Fox." *Ecology* 50, no. 6 (1969).

Laity, Julie J. *Deserts and Desert Environments.* Hoboken, NJ: Wiley-Blackwell, 2008.

Seine River

Category: Inland Aquatic Biomes.
Geographic Location: Europe.
Summary: One of France's major rivers, the heavily polluted Seine is at last seeing concentrated efforts to return its ecological health to previous habitat-supporting levels.

One of the five major rivers of France, the Seine flows through the capital, Paris, winds across northern plains, and joins ultimately with the waters of the North Atlantic Ocean. The Seine runs generally east-west, beginning as a stream on the Langres Plateau, passing through the Burgundy and Champagne regions, then Paris and the estuary segment, and empties into Baie de Seine, an extension of the English Channel. While the river is 482 miles (776 kilometers) long, it repeatedly loops back on itself, traveling only 250 miles (402 kilometers) by straight measure.

The Seine has a very gentle current. The river begins 1,545 feet (471 meters) above sea level and drops to 800 feet (244 meters) approximately 30 miles (48 kilometers) from its source. By the time it reaches Paris, the Seine is just 80 feet (24 meters) above sea level, and still more than 125 miles (200 kilometers) from the Channel.

The Seine River biome includes such main tributaries as the Aube, Yonne, Oise, and Marne Rivers. The Seine flows through a large farming region referred to as the Île-de-France, which has extremely fertile soil, intensively used for crops such as barley, wheat, corn, and many produce types, as well as vineyards. The river also provides the city of Paris with half of its drinking water.

Biodiversity

Efforts to clean the Seine have resulted in a return of salmon to the river after decades of absence resulting from river pollution that began in the late 1800s and killed off the once-large population. Today, Atlantic salmon are on the European Union's endangered species list. The cleaner water has resulted in a rebirth of lively fish habitat in the river, which now includes some 32 species. However, because the river remains polluted, only hardy fish species are here in abundance; these include roach, carp, chub, and bream. Trout and ombre also are native to the Seine, but can live only in the cleaner waters of the upper river.

Terrestrial fauna of the Seine River area include beavers, gray squirrel (*Sciurus carolinensis*), red squirrel (*Sciurus vulgaris*), least weasel, long-tailed weasel (*Mustela nivalis*), mink, skunk, raccoon, and muskrat. Avian species are represented here by duck, heron, cormorant, Merlin (*Falco columbarius*), killdeer (*Charadrius vociferus*), spotted sandpiper (*Actitis macularius* and *A. macularia*), Franklin's gull (*Leucophaeus pipixcan*), herring gull (*Larus argentatus*), and mourning dove (*Zenaida macroura*).

The characteristic flora of the Seine, rendered artistically in the 19th century by renowned Impressionist painters such as Claude Monet, Edouard Manet, Auguste Renoir, and Vincent van Gogh, include chestnut trees (*Castanea* spp.), linden (*Tilia* spp.), and balsam poplar (*Populus balsamifera*), mixed with dense clusters of European mistletoe (*Viscum album*).

Impact of River Modifications

While there have been several positive effects of human involvement, especially through recent restoration efforts, the negative impact includes high pollution levels from industrial discharge, agricultural runoff, and sewage dumping. Mass fish die-offs have sometimes resulted. Parisian officials and other governmental agencies are working together to lower the pollution levels and restore the Seine, making the river safer for fish and people.

Paris is near the point of convergence of several of the Seine's tributaries, and a system of canals, locks, and dams were built to control water flow and improve navigation. Reservoirs were added for storage of drinking water, as well as flood protection. In the past, floods caused the river to overflow from heavy rainfall; much of this seasonal flooding has been lost, with concomitant effects on floodplain vegetation and fauna. Deforestation and marshland drainage to maximize space for farming also have adversely affected the wetlands. The redirecting of the natural flood zone, in conjunction with the canal-and-dam system, is seen as a culprit in decreased diversity of fish species, harshly affecting populations of eel, sturgeon, and salmon here.

More than 50 types of pollution are monitored in the Seine; some areas have recorded over 100 times the European Union safe-swimming limit for certain strains of bacteria. Toxic algae blooms are a periodic concern, as is sewage overflow during heavy rains. From time to time, government agencies have resorted to pumping bubbles directly into the Seine to counteract low-oxygen zones and attempt to save fish. Some efforts are questionable: About 20 species of nonnative fish were introduced into the Seine, in a gesture to support species diversity, but the long-term effects on habitat and native species is unknown.

Climate change could complicate the river clean-up programs. Warmer temperatures foster algal spread and tend to augment riverine and groundwater nitrate concentrations. Warming also alters the suitability of various river habitats for certain species. Gudgeon, minnow, and stone loach have prospered in recent years, while grayling, brown trout, nase, and dace populations have been undercut; some of this change is allayed to damming, however, and it is not clear how much is due directly to temperature, precipitation, and runoff changes. Also, many individuals of some fish species have suffered from average size reduction, as harsh conditions multiply.

The Syndicat Interdepartemental pour l'Assainissement de l'Agglomeration Parisienne (SIAAP) has been working since the 1970s to clean up the Seine by creating ways to treat wastewater and permanently raise the oxygen levels to prevent further mass fish die-offs. SIAAP recently installed a massive station to treat wastewater by removing excess ammoniacal nitrogen. The facility, on the riverbank at Acheres, is one of the largest wastewater treatment sites in the world, capable of processing 2.2 million cubic yards (1.7 million cubic meters) of wastewater per day from the metropolitan Paris area.

Elizabeth Stellrecht

Further Reading

Bendjoudi, H., et al. "Riparian Wetlands of the Middle Reach of the Seine River (France): Historical Development, Investigation and Present Hydrologic

Functioning. A Case Study." *Journal of Hydrology* 263 (2002).

Billen, Giles, et al. "The Seine System: Introduction to a Multidisciplinary Approach of the Functioning of a Regional River System." *Science of the Total Environment* 375 (2007).

Boët, Philippe, et al. "Multiple Human Impacts by the City of Paris on Fish Communities in the Seine River Basin, France." *Hydrobiologia* 410 (1999).

Lafite, Robert and Louis-Alexandre Romana. "A Man-Altered Macrotidal Estuary: The Seine Estuary (France): Introduction to the Special Issue." *Estuaries and Coasts* 24, no. 6 (2001).

Mauch, Christof and Thomas Zeller. *Rivers in History: Perspectives on Waterways in Europe and North America.* Pittsburgh, PA: University of Pittsburgh Press, 2008.

Rosenblum, Mort. *The Secret Life of the Seine.* Reading, MA: Addison-Wesley, 1994.

Senegal River

Category: Inland Aquatic Biomes.
Geographic Location: Africa.
Summary: Traversing four countries, this river and its wetlands support a wide array of animal and plant life; habitats are threatened by poor resource management and widespread poverty.

Located in West Africa, and flowing through four countries—Guinea, Mali, Mauritania, and Senegal—the Senegal River is more than 1,000 miles (1,609 kilometers) long. It is formed in the southwestern section of Mali by the confluence of the Bafing and Bakoy rivers, which flow from Guinea's Fouta Djallon highlands. Like other West African rivers such as the Niger, Volta, and Gambia, the Senegal takes a meandering course that traverses a variety of climates with a contrast of dry and rainy seasons. The Senegal River runs generally northwest in its upper reaches; it forms the border of Senegal and Mauritania, crosses the Talari Gorges, the Gouina Falls, and then tidal marshes as it nears the coastal city of Saint-Louis in northwest Senegal, where it pushes around a large, built-up sand spit, Langue de Barbarie, and flows into the North Atlantic Ocean.

The Senegal's chief tributary is the Falémé River, which flows along Senegal's border with Mali. The Karakoro and Gorgol rivers also are major tributaries. Geological changes over time have created a vast patchwork of channels, floodplains, and islands that dot the delta of the Senegal River. Some of the islands of the delta are topped by inland dunes, the largest of which are the Zairé and the Birette.

Upstream, the Senegal is tidal for about 300 miles (483 kilometers). During the rainy season, when the waters are swollen, travel to Mali by water is possible. On both the Senegalese and Mauritanian sides of the river, much of the population makes its living from subsistence farming and fishing; the floodplain is also an ideal place for growing rice.

Biodiversity

The Senegal River region is home to numerous seabirds and waterbirds; Delta de Saloum National Park and Iles de Madeleine National Park are two important bird sanctuaries. Some of the birds found in these preserves are the greater and lesser flamingo (*Phoenicopterus ruber* and *P. minor*), pink-backed pelican (*Pelecanus rufescens*), cormorant (*Phalacrocoracidae*), reed cormorant (*Microcarbo africanus*), red-billed tropicbird or boatswain bird (*Phaethon aethereus*), slender-billed tern, common tern (*Numenius tenuirostris*), little tern (*Sternula albifrons*), and a variety of parrots. The slender-billed tern and royal tern are considered to be of international significance. Many of the parrots, and some other species here, are part of one the largest pet-bird export industries in the world.

Senegal's Djoudj National Park, a United Nations Educational, Scientific, and Cultural Organization (UNESCO) site, hosts 366 bird species, including garganey (*Anas querquedula*), shoveler (*A. clypeata*), pintail (*A. acuta*), black-tailed godwit (*Limosa limosa*), great white pelican (*Pelecanus onocrotalus*), and avocet (*Recurvirostra avosetta*). Approximately 3 million migratory birds fly through the protected areas in the Senegal River each year.

Mangroves provide important habitat for some bird species; the ecoregion contains some of the most northern mangroves on the African continent. This habitat area forms the northern limit for the west African dwarf crocodile (*Osteaoaemus tetraspis*) and west African manatee (*Trichechus senegalensis*). Mammals in the river basin include the warthog (*Phacochoerus africanus*). Several species of crocodile and gazelle have been successfully reintroduced into the area.

More than 800 plant species have been identified within the Senegal River basin in the area formed by Lac de Guiers and the lower Senegal River. It is a biodiverse wetland, and 33 of the species are endemic, that is, found nowhere else. Vegetation along the river's banks ranges from semidesert Sahelian grassland and shrubland in the north, to progressively moister Guinea savanna in the south. In the wetter parts, seasonally inundated swamp forests line the rivers. Floodplain vegetation includes perennial grasses and sedges, and reed mace (*Typha domingensis*) in the main channels.

One of the effects of the construction of the Maka-Diama Dam has been an increase in malaria in the Senegal River area. This man was carrying home mosquito nets for his family in Senegal in March 2010. (USAID/Debbie Gueye)

Due to damming of the river, aquatic plants such as *Typha australis, Pistia startioles,* and *Salvinia molesta* have proliferated in the river's distributaries and in the irrigation canals. Their growth reduces flow velocities and encourages insects and disease that displace other species, reduce fish production, and impede fishing as a commercial enterprise.

Environmental Threats

Various habitats in the Senegal River biome have been threatened by dam construction, depletion of native biota due to irrigation, overgrazing, and some areas of desertification. The Senegal River lies within the Sahelian zone of Africa, a wide band where the effects of climate change are apparent in drought-impacted areas, extensive of deforestation and desertification. The entire region has been beset by high population growth, rapid and uncontrolled urbanization, pervasive and widespread poverty, lack of sustainable development, and ethnic tensions. All these problems make it extremely important to engage in responsible management of the river's resources.

The country of Senegal has traditionally joined with Guinea, Mali, and Mauritania to manage the Senegal River watershed, but not without difficulty. In 1972, the Organization to Enhance the Senegal River (*Organisation pour la Mise en Valeur du Flueve Sénégal* or OMVS) replaced the Organization of Senegal River States, but Guinea refused to accept the agreement. Guinean officials finally came on board in 2005.

The Manantali Dam, which serves as a reservoir, is located in Mali. A second dam, Maka-Diama, is situated along Senegal's border with Mauritania; its purpose is to prevent saltwater from flowing into upstream waters. While the benefits from the dams have been extensive to the human population, changing the natural flow into other waterways has created problems for various ecosystems.

Despite the need to cooperate over shared resources, tension has been particularly high in areas that were affected by the building of dams. A glaring example is Lac de Guiers, a shallow, brackish body of water that is one of the most important lakes in Senegal. Before the Diama Dam was built, low-saline water flowed into the lake, but that flow has been stabilized. Because the lake is used for fishing, supplying drinking water for humans and animals, and for crop irrigation, that shift had a serious effect on the local population as well as on the lake's ecosystems. As saltwater was prevented from following its normal course, parasitic growth became a major health problem, resulting in the spread of several bacterial diseases. Also since the dam was built, malaria has become a year-round problem in the Senegal River area. On the other hand, incidences of waterborne diseases that contribute to malnutrition have decreased because of the more controllable hydraulic conditions in the upper basin.

Although the use of chemical fertilizers was minimal in the past, the Senegal River is now being contaminated by heavy metals and pesticides, including discharge from sugar plantations and dissolved solids that contaminate the waters during the rainy season, as well as from runoff from hydropower plants. All remain ongoing threats that are being addressed throughout the river basin community. Working with funding from international organizations, OMVS has created an environmental impact program to study and implement ways to mitigate negative effects on this environment.

ELIZABETH RHOLETTER PURDY

Further Reading

Godana, Bonaya Adhi. *Africa's Shared Water Resources: Legal and Institutional Aspects of the Nile, Niger, and Senegal River Systems.* London: F. Pinter, 1985.

N'Diaye, El Hadji Malick et al. "Dam Construction in the Senegal River Valley and the Long-Term Socioeconomic Effects." *Knowledge, Technology, and Policy* 19, no. 4 (2007).

Park, Thomas K., ed. *Risk and Tenure in Arid Lands: The Political Ecology of Development in the Senegal River Basin.* Tucson: University of Arizona Press, 1993.

Stuart, Simon N., et al. *Biodiversity in Sub-Saharan Africa and Its Islands: Conservation, Management, and Sustainable Use.* Gland, Switzerland: International Union for Conservation of Nature (IUCN), 1990.

Venema, Henry David, et al. "Evidence of Climate Change in the Senegal River Basin." *International Journal of Water Resources Development* 12, no. 4 (1996).

Serengeti Volcanic Grasslands

Category: Grassland, Tundra, and Human Biomes.

Geographic Location: Africa.

Summary: The Serengeti Volcanic Grasslands is an exceptional ecosystem comprised of vast plains of short grasses that are home to migratory herds of ungulates and many bird species; it is also a magnet for ecotourists.

The Serengeti volcanic grasslands is a 9,650-square-mile (25,000-square-kilometer) area straddling the borders of Tanzania and Kenya that gets its name from the Maasai word *Siringit,* meaning *endless plains.* The ecosystem is defined by the movement patterns of the migratory ungulates, which roam the plains in search of food and water, stimulated by local rainfall. The grasslands of the Serengeti Plains are bordered in the southeast by the Ngorongoro Mountains and the active Oldonyo le-ngai volcanic mountain, as well as the Great Rift Valley. To the south are the Sukuma plains. To the west is Lake Victoria, Africa's largest inland water body; in the north are the Mau highlands, Siria escarpment, and Mara River in Kenya and Tanzania.

The Serengeti volcanic plains are mostly contained within protected areas. The Tanzania portion of the region is located in the Serengeti National Park and Ngorongoro Conservation Area, where the pastoral human community resides and mixes with friendly wildlife.

The Ikorongo, Grumeti, Maswa, and Kijereshi game reserves are used for sustainable tourist hunting to support ecosystem conservation and the subsistence communities. The Loliondo game controlled area and the Ikona and Makao wildlife management districts are owned by the communities for wildlife conservation and socioeconomic development. In Kenya, the Serengeti volcanic plains fall within the Maasai Mara National Reserve and surrounding group ranches.

The Serengeti volcanic grasslands are a result of Pliocene- and Pleistocene-age volcanic eruption of the Ngorongoro Mountains. Aerial ash and debris from these volcanic highlands was blown westward to form the Serengeti Plains, producing a core area of basic, mineral-rich deposits supporting this very fertile ecosystem. In other areas, the soils are dominated by black clay soils, or by highly saline, alkaline, and shallow sandy loam soils. These latter tend to be less fertile.

Biodiversity

Supported by the rich volcanic soils, the Serengeti grasslands proliferated and expanded, with vast spreads of short grasses; oat grass (*Thermeda triandra*), *Cynodon* spp., and *Chloris* dominate many areas. There also are *Sporobolus* and sedges such as *Kyllinga*. In places, the plains are interrupted by open woodlands, with *Acacia* and *Combretum* quite typical.

These grass-based lands attract millions of animals; among the leading fauna here are more than 1.5 million migratory wildebeest, 600,000 plain zebras, and in excess of 300,000 Thompson and Grant's gazelles. These huge herds of ungulates, or hooved animals, support diverse carnivore populations of lion, hyena, cheetah, wild dog, and jackal. The population of these ungulates is so large that by the grazing, breeding, calving, and migration patterns, it in great measure determines the structure and function of the environments around it.

The plains are home to such endangered and threatened species as eastern black rhino (*Diceros bicornis michaeli*), pangolin or scaly anteater (*Manis* spp.), African elephant (*Loxodonta africana*), African wild dog, and cheetah (*Acinonyx jubatus*).

The Serengeti Volcanic Grasslands biome is also home to diverse invertebrate communities, among the most notable of which is the harvester termite, which plays a vital role in nutrient cycling across the plains. Others include the dung beetles (*Scarabidae* spp.), well known for cleaning the mess, that is, breaking down carrion and waste, left by other animals—and thereby dispersing and recycling nutrients throughout the biota here.

Serengeti is home to a diverse community of raptors, notably vultures, kites, and eagles; each these groups is sustained by the presence of enough prey species. Other iconic fauna here include crocodile, giraffe, topi, hartebeest, eland, and various snakes.

Birds of the Serengeti include many vibrantly-colored species, such as black-headed heron (*Ardea melanocephala*), purple-grenadier (*Uraeginthus ianthinogaster*), rufous-railed scrub-robin (*Cercotrichas galactotes*), yellow-throated sandgrouse (*Pterocles gutturalis*), Fisher's lovebird (*Agapornis fischeri*), yellow-collared lovebird (*Agapornis personatus*), Kori bustard (*Ardeotis kor*), Madagascar bee-eater (*Merops superciliosus*), African hawk-eagle (*Hieraaetus spilogaster*), violet-backed starling (*Cinnyricinclus leucogaster*), ostrich (*Struthio camelus*), and flamingo.

Human Activity and Threats

The grasslands are vital to the economies in East Africa, as for instance, more than half of all tourists visiting Tanzania are drawn to the Serengeti. The plains are inhabited by diverse ethnic groups, most notable of whom are the Maasai, who inhabit the eastern swaths of the region in both Tanzania and Kenya. The Maasai are culturally nomadic pastoralists whose lives depend on moving their livestock—mostly cattle, goats, and sheep—following available pasture.

Other ethnic groups include the Bushmen group the Hadzabe in the southeastern portion of the region, who are mainly hunters and gatherers. Other peoples include the Kurya, Ikoma, Ikizu,

Isenye, Nata, and Sukuma, who mostly inhabit the northern, western, and southwestern reaches of the ecosystem. They are mainly agro-pastoralists, practicing both livestock herding and small-scale crop farming.

Despite their apparently endless bounty, the Serengeti volcanic grasslands are an ecological island in a rising sea of humanity. Located within Serengeti National Park in Tanzania and Maasai Mara National Reserve in Kenya, parts of the northern reaches have been converted to mechanized agriculture, while a great deal of the southwestern portion around the Maswa Game Reserve in Tanzania has been turned into cotton fields. Rising threats and damage are coming from increased poaching, human-livestock disease transmission, heavier road traffic, and political interference in protected areas. Conflicts also have escalated over rights to and control over the rising population of elephants that move in and out of the reserves, causing havoc in some settlement areas.

Climate change has made rainfall unpredictable, impacting the migrating patterns of local wildlife. As a result, drought periods have increased and many animals have died, especially among the grazers. Altering migration pattern and drier conditions have triggered disease outbreaks, and the increase of pests such as ticks are leading to an imbalance across habitats here. If the changes intensify as expected, there is expanding potential for species extinctions.

Alex W. Kisingo

Further Reading

Gottschalk, Thomas. "Birds of a Grumeti River Forest in Serengeti National Park, Tanzania." African Bird Club. http://www.africanbirdclub.org/feature/grumeti.html.

Mburia, Robert. "Climate Change and Species Loss." Climate Emergency Institute. http://www.climate-emergency-institute.org/species_loss_robert_m.html.

Sinclair, A. R. E., Craig Packer, Simon A. R. Mduma, and John M. Fryxell, eds. *Serengeti III: Human Impacts on Ecosystem Dynamics.* Chicago, IL: University of Chicago Press, 2008.

Serra do Mar Coastal Forests

Category: Forest Biomes.
Geographic Location: South America.
Summary: The Serra do Mar coastal forests form a key part of Brazil's critically threatened Atlantic Forest ecoregion. Very high species diversity and high rates of endemism make their conservation of paramount importance.

The Serra do Mar Coastal Forests biome stretches 1,500 miles (4,000 kilometers) along the coast of Brazil and also includes parts of Paraguay and Argentina. Moisture-laden clouds from the Atlantic Ocean keep the forest wet year round, and most tree species are evergreen. Largely a lush, tropical rainforest, Serra do Mar is referred to as part of the Atlantic Forest or *Mata Atlântica.* The climate is subtropical with a mean annual rainfall range of 55–157 inches (1,400–4,000 millimeters), and no dry season.

When Brazil was colonized in 1500, the Atlantic Forest was approximately 24.7 million acres (10 million hectares). At present, roughly just 5 percent of the original forest remains, all in isolated fragments and mostly in the Serra do Mar Coastal Forest itself. Although the majority of Brazilian cities, metropolitan areas, and about half the national population resides on the Atlantic coast, the forest has survived because the land, marked by steep terrain, is unfavorable for agricultural development. Currently only about 37 percent of the remaining forest is protected.

Despite its greatly diminished size, the Serra do Mar remains an eco-rich region containing numerous endemic (found nowhere else) species, many of which are threatened with extinction. In excess of half the tree species, and 92 percent of the amphibians, are endemic to this forest. Approximately 40 percent of Serra do Mar's vascular plants, and as many as 60 percent of its vertebrates are endemic to the forest. More than 140 threatened terrestrial species common to Brazil are found in Atlantic Forest. Thirty-five of Paraguay's threatened species take refuge here, and

22 of Argentina's threatened species dwell in that country's interior forest portion.

Biodiversity

New species are continually found here; in fact, more than 1,000 new flowering plants were discovered in the Serra do Mar between 1990 and 2006. Among the local plants are flowering Cassia (*Cassia fistula*); *Tibouchina*—a genus of about 350 species of neotropical plants in the *Melastomataceae* family—orchids (*Orchidaceae*); a family of monocot flowering plants called *Bromeliaceae*; many local varieties of the myrtle (*Myrtaceae*) family; and true laurel (*Lauraceae*).

The forest is divided into habitat types: the lowlands or coastal Atlantic forest, which comprises mainly tropical moist broadleaf forest; deciduous and semi-deciduous forest that extends across mountain foothills and slopes; and the forests in the cooler south, which are dominated by Parana pine (*Araucaria angustifolia*) or laurel. The Serra do Mar also encompasses mangrove forests, high-altitude grasslands or *campo rupestre*, coastal forests, and scrub on sandy soils, called *restinga*.

The Serra do Mar Forest biome is home to approximately 350 recorded bird species, and it is an important breeding ground for harpy eagle (*Harpia harpyja*), red-tailed Amazon (*Amazona brasiliensis*) and black-fronted piping guan (*Pipile jacutinga*), among many others.

Among the endemic mammal species are the endangered woolly monkey (*Brachyteles arachnoides*) and the critically endangered black-faced lion tamarin (*Leontopithecus caissara*). Many endemic species are so rare that their conservation status remains unknown, such as the officially data-deficient Ihering's three-striped opossum (*Monodelphis iheringi*).

The Serra do Mar Coastal Forest biome also hosts populations of the endemic and critically endangered golden frog (*Brachycephalus pernix*), and a newly identified species of blonde capuchin (*Cebus queirozi*).

Other noteworthy species of the ecosystem include jaguar, ocelot, bush dog, La Plata otter, 20 bat species, and a number of endangered primates, notably muriqui and brown howler monkey.

Tropical vegetation in the Serra do Mar Forest along the coast of Parana State, Brazil. Over 50 percent of the tree species and 92 percent of the amphibians in the forest are found nowhere else. (Wikimedia/Ângelo Antônio Leithold)

Environmental Threats

The Serra do Mar Forest biome faces various environmental and ecological threats, among them deforestation, diseases that can impact wildlife populations, and climate change. Global warming effects in this part of South America are anticipated to include higher average rates of precipitation. The impacts are expected to result in heavier storms that will generate more frequent and severe floods and landslides, each of which can lead to fragmented habitat.

Deforestation in riparian zones has been shown to decrease the amphibian population in the Brazilian Atlantic Forest segment, and undoubtably affects many other areas and taxa as well. If new, more lax forestry regulations are implemented, restoring deforested areas will no longer be required. However, there are private landowners restoring their own forests. Also impacting the forest are urban sprawl and its associated pollution and acid rain, as well as logging and poaching.

Renata Leite Pitman

Further Reading

Becker, C. G. and K. R. Zamudio. "Tropical Amphibian Populations Experience Higher Disease Risk in Natural Habitats." *Proceedings of the National Academy of Sciences of the United States of America* 108, no. 24 (2011).

Loyola, R. D., U. Kubota, G. A. B. da Fonseca, and T. M. Lewinsohn. "Key Neotropical Ecoregions for Conservation of Terrestrial Vertebrates." *Biodiversity and Conservation* 18 (2009).

Morellato, L. P. C. and C. F. B. Haddad. "The Brazilian Atlantic Forest." *Biotropica* 32, no. 4 (2000).

Olson, D. M. and E. Dinerstein. "The Global 200: Priority Ecoregions for Global Conservation." *Annals of the Missouri Botanical Garden* 89, no. 2 (2002).

Webb T. J., F. I. Woodward, L. Hannah, and K. J. Gaston. "Forest Cover–Rainfall Relationships in a Biodiversity Hotspot: The Atlantic Forest of Brazil." *Ecological Applications* 15, no. 6 (2005).

Severn Estuary

Category: Marine and Oceanic Biomes.
Geographic Location: Europe.
Summary: This robust environment for migrating fish and birds holds huge potential as a source for wind- and tide-driven electrical power generation that could supply Britain with up to 5 percent of its energy needs.

The Severn Estuary, stretching from the mouth of the River Severn to the Bristol Channel on the Irish Sea, forms the marine boundary between Wales and southwestern England. The Severn is the longest river in Great Britain; the estuary is actually the mouth of three other rivers as well: the Wye, Usk, and Avon. Various other smaller rivers also are tributaries to the estuary.

The estuary is funnel shaped; it has the second-highest tidal ranges in the world, generally reaching 49 feet (15 meters). This has helped the biome become a focus for tidal energy ideas. The underlying geology of the estuary basin consists of rock, gravel and sand, which help produce strong tidal streams as well as high turbidity that makes the water brown.

The climate of Severn Estuary is oceanic, typically cool in winter with warmer summers and rainy year round, especially in winter. Temperatures rise from an average cool range of 34–39 degrees F (1–4 degrees C) in winter to average highs of 64–72 degrees F (18–22 degrees C) in summer. Annual rainfall averages 39 inches (1,000 millimeters), with as much as 79 inches (2,000 millimeters) in higher elevations. The Estuary is located in the second windiest region of the United Kingdom, with the winds coming mostly from the southwest and northeast. Due to global-warming-driven changes in climate, governmental organizations are predicting the area will become the hottest region in the United Kingdom (UK).

Biodiversity

The Severn Estuary biome is an important site for wintering and wading birds that migrate through the area, supporting more than 10 percent of Britain's wintering population of dunlin, as well as significant numbers of Bewick's swans, European white-fronted geese, and wigeon. The region also supports wintering populations of gadwall, shoveler, and pochard. Over winter, the area regularly supports about 85,000 waterfowl overall, including shelduck, teal, grey plover, lapwing, redshank, and curlew. Whimbrel and ringed plover pass through in large numbers on migration, the former particularly in spring.

More than 110 species of fish, including seven different species of migratory fish such as Atlantic salmon, common eel, Allis shad, and sea trout pass through the estuary—more than any other British estuary. Several rare species of river lamprey, sea lamprey, and twaite shad use the estuary, which also serves as a nursery for juvenile fish that feed heavily on diatoms and plankton.

Saltmarsh, located in the upper parts of intertidal mudflats, is found all along the estuary's fringes in the Avon area. Only a limited number of salt-tolerant plants can grow in these conditions, including nationally scarce species of slender hare's-ear, sea clover, and bulbous foxtail. Saltmarsh, a haven to gastropods, mollusks, and insects, provides important feeding and roosting areas for waterfowl and waders. The harsh, wind-whipped conditions of the estuary are well-tolerated by alder and oak, the most dominant tree species here.

Power Generation

The Severn Estuary provides a valuable power-generation opportunity that has been contemplated for decades; the tidal power potential here has been reliably quantified at 8,640 megawatts when the tide flows, a viable opportunity to generate as much as 5 percent of the UK's power needs. Various environmental issues, however, and underlying site preparation, construction, and operating costs have kept progress sidelined for years. The issue came to the fore in January 2008 when the Severn Tidal Power Feasibility Study began; its mandate: to assess all tidal range technologies, including barrages, lagoons, and others. Some of the fresh urgency stems from the UK's need to address its carbon footprint goals under its responsibility to European Union (EU) renewable energy targets; its substantial and growing problems with its aging nuclear power plant infrastructure, along with that industry's waste storage challenges; and the competitive gains the country could make regarding this emerging global technology.

The UK government is committed to generating 20 percent of its energy needs from renewable sources in the near future; tidal generation in the Severn is more likely to be part of the mid- to long-term electricity solution. Wind power installations here, meanwhile, could form a portion of the short-term answer.

Environmental Issues

The wetlands of the estuary are a Ramsar preservation site; the biome is also recognized as a Special Protection Area (SPA) under the EU's directive on the conservation of wild birds. The estuary has further been deemed a Special Area of Conservation (SAC) under the EU Habitat Directive, and part of the estuary has been designated a Site of Special Scientific Interest (SSI).

The estuary and its coastal area supports major Atlantic Ocean-connected ports and key cities including Cardiff, Bristol, and Gloucester. The beaches and undeveloped coastline, including low-lying wetlands and cliff scenery, provide recreational opportunities that must be balanced against the need to preserve and protect wildlife habitats. The ecoregion is managed through the Severn Estuary Partnership program, established in 1995 as a linkage of organizations and individuals committed to caring for the estuary.

The estuary is chronically threatened by natural processes such as coastal erosion from the high tidal range; strong currents and storm surges; and development pressure along the shoreline; as well as environmental threats from industrial, chemical, and domestic sewage discharge. Policies, programs, and projects to minimize the pollution in this ecoregion are ongoing.

Rising temperatures also are a concern; their effects on habitat are being monitored. Severe storms have already caused substantial coastal erosion and recurring floods. One of the biggest threats to the estuary comes from rising sea levels. More than three-quarters of the Severn Estuary mudflats, it is estimated, could vanish during this century as sea levels rise due to global warming, which would have a huge impact on biodiversity here. This impact would compel changes in the distribution of native species, and would enable some nonnative species to become more common. Scientists have been collecting evidence over the past 30 years of animals occurring outside their usual or expected ranges here, including the northern expansion of marine mollusks, plants, migratory birds, and fish.

Medani P. Bhandari

Further Reading

Edwards, S. D., P. J. S. Jones, and D. E. Nowell. "Participation in Coastal Zone Management Initiatives: A Review and Analysis of Examples from the UK." *Ocean & Coastal Management* 36 (1997).

Holgate-Pollard, D. "Coastal Management: The Policy Context." *Marine Environmental Management Review of 1995 and Future Trends* 3, no. 7 (1995).

Knowles, Steve and Louise Myatt-Bell. "The Severn Estuary Strategy: A Consensus Approach to

Estuary Management." *Ocean & Coastal Management* 44 (2001).
Sorensen, J. "National and International Efforts at Integrated Coastal Management: Definitions, Achievements and Lessons." *Coastal Management* 25 (1997).

Shark Bay

Category: Marine and Oceanic Biomes.
Geographic Location: Australia.
Summary: This World Heritage Area is known for its rare stromatolite community, extensive seagrass beds, and diverse shark population.

Shark Bay lies about 500 miles (805 kilometers) north of Perth, in western Australia, and is renowned for its wide variety of sharks and for its environmental significance. In 1991, Shark Bay became a World Heritage Area due to its natural heritage values, which include the Bay's stromatolite (single-cell cyanobacteria that date back to the Earth's creation) community, and otherwise diverse animal life. The biome also is widely known for seagrass beds that are among the largest in the world; the seagrass habitat provides food and shelter to numerous marine life species.

Shark Bay is located near the northern limit of the transition zone between temperate and tropical environments; the surface temperature of the water averages 70–79 degrees F (21–26 degrees C). The area has a semiarid-to-arid climate, with hot, dry summers and mild winters. The rainy season occurs during winter, with annual rainfall averaging 8–16 inches (203–406 millimeters). The western portion of Shark Bay receives more rain than the eastern side, which has semidesert conditions, while the western region is dry mediterranean. The average depth of the water in the biome is 30 feet (9 meters); the deepest sections are approximately 95 feet (29 meters).

The Bay also holds historical significance, as it is the earliest recorded site of European exploration on the western Australian coast. Many of the islands here are named after these explorers, such as Dirk Hartog Island (for a Dutch seaman of 1616). William Dampier gave Shark Bay its name in 1699, when he and his crew were mapping the area.

Biodiversity

So named for some of its more prevalent inhabitants, Shark Bay is home to 28 species of shark and six types of ray. The bay has one of the largest populations of tiger sharks in the world. Tiger sharks inhabit inshore waters, deeper offshore areas and reefs, and are the top predator in this marine ecosystem. Their main prey are sea snakes and dugongs, or sea cows, and they also feed on green sea turtles, and various fish, as well as smaller sharks and rays.

Shark Bay is one of the few locations in the world where living stromatolites are found. Since stromatolites have such a long biological history, they serve as an example of Earth's original life-evolutionary progress. In fact, they are theorized to date back to the Snowball Earth period of the Ediacaran era, more than 650 million years ago. Extremely fragile, and growing only a few tenths of an inch (millimeters) per year, these blue-green algae nevertheless build up rocklike structures as they capture and bind together tiny grains of silt and sediment.

The stromatolites of Shark Bay are found in Hamelin Pool, a semi-enclosed, hypersaline embayment that has twice the salt concentration in its water as the marine norm nearby. The elevated salinity level is a result of limited tidal exchange from the sea, along with high evaporation. Due to their delicate nature, stromatolites are easily destroyed by other creatures that feed on them or simply crush them underfoot. The salinity levels and closed nature of Hamelin Pool provide an excellent environment for the stromatolites to thrive without disturbance, as few other creatures can survive for long in the pool.

Shark Bay also is well known for its wildlife diversity. Shark Bay Marine Park—established in 1990 under the Conservation and Land Management Act within the World Heritage site—is a protected and preserved area that enables animal life to flourish without much human interference. It is home to thousands of sharks and to numerous other aquatic and terrestrial animals. Marine animals thrive in its vast seagrass beds, which span 2,500 square miles (6,475 square kilometers) of the ocean floor. The seagrass beds provide food, shelter, and protection to 60 species of fish and small gastropods, and acts as a food source for the population of the threatened dugong.

While the dugong is somewhat rare in most parts of the world, Shark Bay is home to 10,000 individuals, which is about one-eighth of the world's dugong population. Even though the dugongs are a food source for the area's sharks, they manage to hide and thrive in the dense thickets of the seagrass beds. While seagrass is the dominant marine habitat, plenty of creatures live in the sand and open water as well, such as dolphins, whale sharks, sea turtles, migrating humpback whales, lobsters, shrimp, coral, sea snakes, and more than 300 species of fish.

Rocklike clusters of stromatolites growing in Shark Bay, one of the few places where these organisms, which may date back 650 million years, still thrive. (Thinkstock)

Several members of Shark Bay's animal population live on land; these include wallabies, bandicoots, kangaroos, emus, and mice. More than 100 species of amphibians and reptiles live on the islands, including frogs, lizards, snakes, skinks, and geckos. Shark Bay provides a safe environment for such threatened species as the Shark Bay mouse, banded hare-wallaby, rufous hare-wallaby, western barred bandicoot, and burrowing bettong.

Conservation Efforts

Because of Shark Bay's population of endangered and native animals, Western Australia's Department of Environment and Conservation is striving to conserve and protect the Bay for future generations. The Shark Bay World Heritage Area covers approximately 13,670 square miles (35,405 square kilometers), with two-thirds of it a marine habitat. The Australian government is working to remove nonnative animals such as goats, foxes, cats, and rabbits, which threaten both the endangered species and many other native ones.

The government is managing human use of Shark Bay, with strict laws concerning fishing, boating, and camping to further preserve the environment of the animals living in the area. While tourism is encouraged, to bring in funds for the preservation of Shark Bay, the Australian government is strategically developing specific sites and areas that will have the least effect on wildlife.

Climate change is impacting Shark Bay. Warming temperatures and increased river flooding from heavier rain events is altering the composition of microbial species in the Bay, and upsetting the ecosystem of the stromatolites. In addition, the increasing amounts of sediment washing into the Bay inflicts damage on parts of the seagrass meadows. These habitats are complex; the effects will have to be monitored closely to gain information on how to help sustain them.

Elizabeth Stellrecht

Further Reading

Berry, P. F., et al., eds. *Research in Shark Bay: Report of the France-Australe Bicentenary Expedition Committee.* Perth: Western Australian Museum, 1990.

McCluskey, Paul. *Shark Bay World Heritage Property Strategic Plan 2008–2020.* Perth: Australia Department of Environment and Conservation, 2008.

Richards, Jacqueline D. and Barry Wilson, eds. *A Biological Survey of Faure Island, Shark Bay World Heritage Property, Western Australia.* Perth: Western Australian Museum, 2008.

Walker, Gabrielle. *Snowball Earth: The Story of the Great Global Catastrophe That Spawned Life as We Know It.* New York: Crown Publishers, 2003.

Shetland Islands Intertidal Zone

Category: Marine and Oceanic Biomes.
Geographic Location: Europe.
Summary: The diverse and rugged sub-Arctic coastline of the Shetland Islands is exposed to extreme wind and waves, and is home to populations of hardy plants, birds, and marine life.

The Shetland Islands is an archipelago of about 100 islands around 106 miles (170 kilometers) north of mainland Scotland. Shetland covers an area of 567 square miles (1,468 square kilometers), with a coastline of 1,679 miles (2,702 kilometers). The largest island is known as Mainland, and is one of the 16 inhabited islands, with a total population of about 22,000 people. Shetland was formerly called *Hjaltland,* possibly derived from the Old Norse word *hjölt,* meaning "the handgrip of a sword," reflecting the island chain's resemblance to a sword.

Shetland is sub-Arctic, but the strong influence of the North Atlantic Drift provides an oceanic climate, with long, mild winters and short, cool summers. The mean maximal temperature range is 41–57 degrees F (5–14 degrees C), and the sea temperatures vary from 45–55 degrees F (7 to 13 degrees C). In general, the climate is windy and cloudy. Fog is common during summer. The prevailing wind is southwesterly; storms in Shetland can be exceptional. Gusts of more than 124 miles per hour (200 kilometers per hour) have been recorded, and the islands have five times the number of gale days as mainland Scotland. Huge seas batter the west coast, creating storm beaches up to 59 feet (18 meters) above sea level.

The archipelago has a complex geology, with many faults and fold axes. During the ice ages, glaciers entirely covered the islands, as shown by the numerous voes (glaciated valleys drowned by rising sea levels). One intertidal geological feature of interest is the largest active tombolo (mound or sandbar) in the United Kingdom, which connects St. Ninian's Isle to Mainland.

Intertidal Habitats and Species

Shetland's coastline experiences a broad range of environmental conditions. The west coast is extremely exposed to waves, with warmer water from the North Atlantic Drift, whereas the upper reaches of the voes (small bays or narrow creeks) are almost completely sheltered, and the east coast is washed in cool water from the North Sea. This variety of conditions, combined with the long and complex coastline, creates a broad range of intertidal habitats.

Much of the coastline is rugged and rocky, with many high cliffs and a heterogeneous range of features, such as boulders, ledges, crevices, gullies, and rock pools. These features support many types of seaweed and diverse fauna, including a great variety of limpet (order *Patellogastropoda*), which are aquatic snails with conical shells; winkle (*Littorina littorea*), a small edible sea snail species; and dogwhelk (*Nucella lapillus*), a predatory sea snail also called Atlantic dogwinkle; as well as widely varied species of mussel, barnacle, anemone, and crustaceans such as crabs.

Well above the tidal zone, and out of reach of grazing sheep and human interference, coastal cliffs here support sizable breeding populations of seabirds including fulmar, gannet, storm petrel, shag, kittiwake, guillemot, razorbill, and puffin. Some of these birds interact heavily with the intertidal zone, preying upon its fauna.

At the back of numerous bays in the Shetland are beaches of golden sand. Salt marshes dominated by fine grasses are found at the head of some voes, and there are several expanses of intertidal mud. These areas can be rich in marine invertebrates, providing important foraging areas for waders and seabirds.

A total 274 species of benthic algae are recorded in Shetland. One is a rare, free-living form of knotted wrack (*Ascophyllum nodosum mackii*), which occurs in sheltered water and grows in spherical masses up to 24 inches (600 millimeters) in diameter, giving it the name crofter's wig (a croft being a small farm). The complex structure of the seaweed provides habitat for many small invertebrates.

Larger marine life here includes various trout varieties, mackerel, ling, haddock, halibut, porbeagle shark, scorpionfish, butterfish, assorted crabs, and urchin. All the land animals in the Shetland Islands—including otters, mice, and rats among those that frequent the tidal zones—were introduced by humans.

Shetland is largely treeless, but since the ground has an abundance of peat deposits, the islands likely did have a warmer climate at some point in its history. Local vegetation and habitat that is often washed in spray from the high-tide intertidal wave action includes blanket bog dominated by heather, cotton grass, dwarf shrubs such as crowberry and bilberry, and sphagnum moss. There are various greens such as heather, grasses and sedges, mosses, liverworts, ferns, and lichens

Human Settlement and Activity

People have lived in Shetland for more than 6,000 years. Stone Age hunter-gatherers harvested species that they found locally, leaving the remains in archaeological middens (rubbish dumps). These middens reveal a range of intertidal species, including oysters, mussels, limpets, crabs, and cockles, as well as fish and seabird remains. Limpets are abundant on the rocky shores and are particularly common in middens. Limpets also are important to aquaculture, primarily for salmon and mussels to feed upon, while Pacific and European oysters are often grown intertidally on racks.

In the U.K., the intertidal zone, known generally here as *foreshore*, is deemed to be owned by the Crown, although shores in Shetland are notable exceptions, where Udal Law generally holds. Udal Law is a near-defunct Norse system in which individuals have rights of ownership of the foreshore. This law has been upheld intermittently by Scottish courts, and has important implications for construction of pipelines, cables, or coastal defenses—as well as conservation efforts.

Discovery of North Sea oil provided a major economic boost—and the dangers that come with it. A large terminal at Sullom Voe processes and stores oil from the North Sea and Russia. The oil tanker *Braer* ran aground on the south coast of Mainland in January 1993, releasing 93,696 tons (85,000 metric tons) of crude oil. The spill had short-term effects on fish, shellfish, marine mammals, and seabirds, but fortunately long-term effects on intertidal species were less than anticipated.

The impact of climate change seems to pose some positives for the Shetland Islands. As the sea waters warm, fewer storms could hit the islands, moving at least some of the harshest weather further south to southern England and France. However, warming temperatures could spell more coastal flooding for the archipelago. The increase in sediment flushed into the tidal areas, as well as erosion of some of the more sensitive zones, would create difficulties for many of the intertidal faunal communities here.

Angus C. Jackson

Further Reading

Baxter, J. and S. Mathieson. *Broad Scale Habitat Mapping of Intertidal and Subtidal Coastal Areas: Busta Voe and Olna Firth, Shetland.* Wallsend, Scotland: Entec, 1996.

Johnston, Laughton. *A Naturalist's Shetland.* London: A & C Black Publishers Ltd., 2002.

Scottish Natural Heritage. "Natural Heritage Futures: Shetland." http://www.snh.gov.uk/docs/A306317.pdf.

Shu Swamp

Category: Inland Aquatic Biomes.
Geographic Location: North America.
Summary: Shu Swamp is one of the few extant and intact red maple-blackgum swamps on Long Island, New York, supporting native brook trout and old-growth stands of tulip tree and American beech.

Situated a mere 12 miles (20 kilometers) from New York City, Shu Swamp is one of the few remaining intact red maple-blackgum swamps on Long Island. The name *Shu* is a corruption of the Dutch word *sheogh,* which means "cascading waters." This type of wetland can be found along the North Atlantic coast of the United States from New Jersey to New Hampshire. In New York State, red maple-blackgum swamps are restricted to 20 or 30 sites, primarily on Long Island.

The landscape surrounding Shu Swamp consists of low-lying hills along the intricate coastline of Long Island Sound, remnants of a glacial moraine that formed the island during the last ice age. The climate today is temperate, with warm, humid summers and cool winters. Shu Swamp is fed by Beaver Brook, which originates from seeps and springs in the low hills; flows through a series of small ponds; and eventually reaches the heavy clay soils of Mill Neck Valley, where a high water table has resulted in the formation of the wetlands. The swamp marks the transition between upland oak-tulip tree forests and the tidal marshes of Mill Neck Creek, which drain into Cold Spring Harbor and Oyster Bay, one of several embayments on Long Island Sound.

There are many ecological links among these neighboring habitats, but Shu Swamp's unique geology and plant community clearly distinguish the wetland as a discrete ecosystem. Although it has been highly modified by suburban development, the 59 acres (24 hectares) of land that comprise Shu Swamp are protected as the Charles T. Church Nature Preserve.

Vegetation

In addition to hosting red maple (*Acer rubrum*) and blackgum (*Nyssa sylvatica*)—also known as tupelo—as its constituent tree species, Shu Swamp is renowned for containing approximately 30 acres (12 hectares) of old-growth tulip tree (*Liriodendron tulipifera*) and American beech (*Fagus grandifolia*) forest. In particular, the tulip trees are thought to be candidates for the oldest and largest trees on Long Island, and they are certainly among the tallest trees in the eastern United States, reaching more than 148 feet (45 meters). The Shu's tulip trees are generally 150–200 years old, while the largest specimens, reaching nearly 5 feet (1.4 meters) in diameter, are estimated to be 350–600 years old.

More than 300 vascular plant species have been identified in Shu Swamp, including two state-threatened species, American strawberry bush (*Euonymus americanus*) and sweetbay magnolia (*Magnolia virginiana*). The midstory is dominated by sweet pepperbush (*Clethra alnifolia*), while the inundated areas are dominated by skunk cabbage (*Symplocarups foetidus*), a low-growing plant. Skunk cabbage emits a foul odor to attract pollinators such as flies; it is notable for its ability to generate heat and thereby melt its way through frozen ground and flower during the winter.

Fauna

Shu Swamp provides important habitat and breeding ground for many bird species, including red-tailed hawks *(Buteo jamaicensis),* wood ducks (*Aix sponsa*), Virginia rails (*Rallus limicola*), ovenbirds (*Seiurus aurocapilus*), and swamp sparrows (*Melospiza georgiana*). In addition, the adjacent tidal wetlands and Beaver Lake attract ospreys (*Pandion haliaetus*), herons, and high numbers of waterfowl.

The swamp is notable for supporting one of the few spawning populations of native brook trout (*Salvelinus fontinalis*) on Long Island. Although not endangered, many populations of brook trout—New York's official state fish and Long Island's only native salmonid—have been reduced and/or lost due to habitat destruction and the introduction of competitively superior fish species. There also are also reports of the rare American brook lamprey *(Lampetra appendix)* spotted in the clear-water streams of Shu Swamp. The waters downstream

*The invasive plant species garlic mustard (*Alaria petiolata*) has been found in the Shu Swamp among the over 300 vascular plant species and two state-threatened species. (U.S. Fish and Wildlife Service/Steve Hillebrand)*

where the swamp connects to Beaver Lake have been invaded by introduced Asian carp (*Cyprinus carpio*), which grow to nearly 3 feet (1 meter) long and are known to negatively affect native fish populations elsewhere.

Environmental Threats

The greatest threats to Shu Swamp come from the high degree of suburban development in the surrounding area. In general, wetlands are particularly sensitive to runoff from roads, households, and golf courses, and may be affected by offsite pollutants flowing downstream through the watershed. Located in one of the most densely populated regions of the United States, Shu Swamp is affected by human pollution. Fortunately, several small nature preserves upstream of the Shu, such as Coffin Woods and Upper Francis Pond Preserve, provide habitat corridors for some of the stream catchments running into Beaver Brook.

Other threats to the habitat include invasive plant species such as garlic mustard (*Alaria petiolata*) and English ivy (*Hedera helix*), which may outcompete native plant species. Another concern is that the marshland may be in danger of "drowning" in the face of sea-level rise driven by global warming. Saltwater intrusion, if permitted to reach substantial portions of Shu Swamp, would have devastating impacts on both native vegetation and animal life. Despite these threats, active management by the North Shore Wildlife Sanctuary and other volunteer groups continues to maintain Shu Swamp as one of the finest examples of red maple-blackgum swamps in the region.

CLAY TRAUERNICHT

Further Reading

Greller, Andrew M. "A Classification of Mature Forests on Long Island, New York." *Bulletin of the Torrey Botanical Club* 194, no. 4 (1977).

Johnston, Carol. "Spring Wildflowers at Shu Swamp Preserve." *Long Island Botanical Society Newsletter* 6, no. 3 (1996).

Karpen, Daniel. "Old Growth Forests on Long Island, New York." *Long Island Botanical Society Newsletter* 10, no. 3 (2000).

New York Natural Heritage Program. "Online Conservation Guide for Red Maple–Blackgum Swamp." http://www.acris.nynhp.org/guide.php?id=9900.

Siberian Broadleaf and Mixed Forests, West

Category: Forest Biomes.
Geographic Location: Europe.
Summary: The western Siberian broadleaf and mixed forests lie at the juncture of the tundra and

mixed forest regions north of Kazakhstan. Also known as hemiboreal forest, the biome is rich in animal and plant diversity.

The Siberian broadleaf and mixed forests define a slender ecoregion that lies at the southern edge of the broad Siberian taiga biome, just north of Kazakhstan. The Russian taiga is the largest forest in the world; forests cover 48 percent of the Siberian landscape and represent 20 percent of all of Earth's forested lands. The Siberian Broadleaf and Mixed Forests biome is also known as the western Siberian hemiboreal forest, because the area is located halfway between the temperate and boreal zones. This area encompasses the western edge of the North Asian hemiboreal region.

Hemiboreal forests are a relatively new ecological designation, as they refer to specific conditions and habitats common to boundary communities, and forests located at the juncture of boreal tundras and temperate-zone forests.

The Siberian hemiboreal forests lie at the southern edge of the Euro-Siberian biogeographic region that falls inside the Palearctic ecozone, a vast belt that runs across North America and Eurasia, including lands in Canada and Scandinavia. This expansive ecozone is characterized by temperate broadleaf and mixed forests, and by conifer forests. Because North America was once connected to Eurasia via the Bering Strait land bridge, these areas have similar plant and wildlife habitats. Hemiboreal forests are the remnants of broad tracts of forests that flourished long ago; many did not survive the last periods of ice-age glaciation. The current climate supports a rich and diverse ecology here, despite clear-cutting, frequent fires, and poaching.

Vegetation

The western Siberian landscape is varied, with broad overlaps of particular types of vegetation. From north to south, the landscape transitions through five bioclimatic zones: tundra, forest-tundra, taiga, forest-steppe, and steppe. The complexity of this distribution creates integrated biophysical relationships that are difficult to describe even with the use of sophisticated, high-resolution radiometric technologies.

Biophysical relationships are further complicated during wide-scale events such as fires. The quantification of the processes of secondary succession stages, and the levels of carbon dioxide and water vapor available for heterotrophic respiration, are vital indicators of net biome productivity. The structure, age, and composition of regenerating areas are important to identify during recovering succession periods.

Fragmentation of Siberian hemiboreal landscapes due to fire and other causes, with subsequent plant dispersal and the disconnection of native ecosystem habitats, has accelerated the decline of biodiversity across the ecozone. Native forests were originally abundant. Birches dominated northern reaches. In more temperate southern areas, Siberian firs and Siberian spruces shared the biome as dominant tree species, along with tilias—a genus with 30 different trees known by various names including lime, linden, and basswood. Frequent fires have altered this distribution of species, often replacing native canopies with secondary successions of birches and aspens.

The hemiboreal forests provide habitats for more than 400 other plant species, including shrubs and nemoral (wood or grove) grasses. Taiga and steppe species are commonly found along the western Siberia forest corridor.

Fauna

Bird types in these Siberian forests include white partridge, snowy owl, gulls, and loons. Geese, swans, and ducks migrate into the region—accompanied by mosquitoes, gnats, and other insects. Indigenous birds include the demoiselle crane, steppe eagle, great and little bustard, finches, kestrels, and other falcons.

The taiga forest zone is home to such large mammals as European elk, brown bear, reindeer, lynx, and sable. Forest birds here include owls and the nightingale. The broadleaf forests also contain wild boars, deer, wolves, foxes, minks, snakes, lizards, and tortoises. The forests of southeastern Far Eastern Russia are home to Siberian tigers (*Panthera tigris altaica*), also known as the Amur tiger—among the world's largest cats—as well as the Amur leopard, bears, musk, and various deer species. Rodents such as marmots; hamsters; and five species of suslik, a type of ground squirrel, inhabit the steppes.

Endangered animals include *Coregonus tugun*, a ray-finned fish of the Lena River basin; Siberian crane or Siberian white crane (*Grus leucogeranus*); tawny owl or brown owl (*Strix aluco*); and a subspecies of the European beaver, *Castor fiber pohlei*. The Siberian tiger is a critically endangered species impacted by habitat encroachment from logging and poaching.

Environmental Threats

Accurate topographical surveys of western Siberia land cover are essential to climate studies and efforts to simulate potential feedback relative to global warming. Land-atmosphere exchanges of energy, water, carbon dioxide, and other greenhouse gases are relative to the composition and structure of a given ecoregion. Current scenarios suggest that boreal species will migrate to the north as temperatures rise. Some data suggests that woody vascular species already are found in taiga areas.

Some scientists believe that if this expansion continues, the atmosphere will react with corresponding declines in albedo (visual surface brightness when viewed with reflected light) and increased absorption of solar energy, leading to additional heating of the atmosphere. These transitions are expected to have a profound affect on the Siberian landscape. It is predicted that by 2080, more than half of Siberia will be covered with forest-steppe and steppe ecosystems. New temperate broadleaf forests will contribute to fuel loads and add to subsequent fire hazards.

Wetlands are notoriously difficult to represent in geographic data profiles, particularly data sets derived from satellite imagery. It is estimated, however, that roughly 20 percent of the western Siberian forest lands are swamplands, including riparian stretches of the Irtysh River basin. Extensive lakes and wetlands in western Siberia contribute to the accumulation of organic peat carbon.

Ice-core records of atmospheric methane in the western Siberian lowlands suggest that accumulations have accelerated dramatically during the Holocene period, creating a vast, long-term carbon dioxide sink and global methane source. These reserves represent roughly 26 percent of all terrestrial carbon accumulated since the last glacial maximum. During periods of global warming, peats will decompose, releasing carbon dioxide into the atmosphere. However, these processes are difficult to assess quantitatively, given the paucity of information about the depth, age, and content of Siberian peat accumulations.

Conservation Efforts

Regrettably, the lands of the western Siberian hemiboreal forest are not included in the Russian system of protected lands known as *zapovedniks*. Closely related areas include the Orenburgsky Zapovednik at the far western Kazakh border, and the Olekminsky Zapovednik of central Siberia. The Pleistocene Park project, the brainchild of researcher Sergey A. Zimov, is of interest to climatologists worldwide. This project is the result of intensive research on a vast ecosystem that stretched across the northern latitudes of Canada, Europe, and China at the end of the Pleistocene era. One grassy corner of eastern Siberia survived the onslaughts of glacier advances and was the center of a highly diverse community of coexisting mammals, including mammoths, bison, horses, reindeer, musk oxen, elk, moose, saiga antelope, and yaks.

In 1980, Zimov founded the Northeast Science Station in Cherskii in the Republic of Sakha (Yakutia), and in 1989, he initiated the Pleistocene Park Project. On a 62-square-mile (160-square-kilometer) plot of grassland, project partners plan to reintroduce Pleistocene mammals to the Siberian plains. The effort will re-create the ecological conditions that existed just before the Holocene era.

It is the assumption of the project directors that restoring that ecology could effectively cool the atmosphere, mitigating the escalating feedbacks of global warming, including the migration of species northward. Genetic cloning techniques could even make it possible to reintroduce the great mammoths native to Eurasia 10,000 years ago. Project plans include a fivefold expansion of this grassland habitat.

Victoria M. Breting-Garcia

Further Reading

Frey, Karen E. and Laurence C. Smith. "How Well Do We Know Northern Land Cover? Comparison of Four Global Vegetation and Wetland Products With a New Ground-Truth Database for West Siberia." *Global Biochemical Cycles* 21 (2007).

Laletin, Andrei P., Dmitry V. Vladyshevskii, and Alexei D. Vladyshevskii. "Protected Areas of the Central Siberian Arctic: History, Status, and Prospects." USDA Forest Service Proceedings. http://www.fs.fed.us/rm/pubs/rmrs_p026/rmrs_p026_015_019.pdf.

Meroni, M., et al. "Carbon and Water Exchanges of Regenerating Forests in Central Siberia." *Forest Ecology and Management* 169 (2002).

Rodriguez, Antonio Rigueiro, et al. *Agroforestry in Europe: Current Status and Future Prospects.* Berlin: Springer Science & Business Media B.V., 2009.

Smith, L. C., G. M. MacDonald, A. A. Velichko, D. W. Beilman, O. K. Borisova, K. E. Frey, et al. "Siberian Peatlands: A Net Carbon Sink and Global Methane Source Since the Early Holocene." *Science* 303 (2004).

Tchebakova, N. M., E. Pafenova, and A. J. Soja. "The Effects of Climate, Permafrost and Fire on Vegetation Change in Siberia in a Changing Climate." *Environmental Research Letters* 4 (2009).

Zimov, Sergey A. "Pleistocene Park: Return of the Mammoth's Ecosystem." *Science* 308 (2005).

Siberian Coastal Tundra, Northeast

Category: Grassland, Tundra, and Human Biomes.
Geographic Location: Eurasia.
Summary: This land of marshes and permafrost hosts anadromous fish, migratory birds, and reindeer; its marshes and seabeds hold vast amounts of carbon dioxide and methane, both important catalysts of global climate change.

The northeastern Siberian coastal tundra stretches from the eastern Siberia and Laptev Seas along the far-northeastern maritime border of Eurasia just above the Arctic Circle, and extends across the Chukotka Peninsula that borders the Bering Strait just west of Alaska. It is an ecosystem within the larger, Arctic Tundra biome, which extends around the Arctic Circle and includes the far-northern lands of North America and Northern Europe. Similar landscapes can be found in the Antarctic.

Sometimes, the tundra is referred to as a borderland of the larger taiga biome. The Siberian taiga is a vast boreal forest ecosystem that extends 1.5 million square miles (3.9 million square kilometers) across eastern and central Russia. However, both have distinct biological features. The tundra covers nearly one-fifth of Earth's lands, and is noted for its frigid temperatures and scant precipitation. At its extremes, the landscape becomes a polar desert.

These harsh conditions progressively limit the biodiversity of species and subspecies that can survive exposure at northern climatic extremes. Mats and cushions of mosses, lichens, willows, sedges, grasses, and Arctic poppies populate in a permafrost landscape. In the far northeastern corner of the Siberian tundra, the lands of the Chukotka Peninsula are densely covered with tussock-dwarf shrubs and cotton sedge (*Eriophorum vaginatum*). Common plant species here also include cotton grass (*Eriophorum* spp.), sedges (*Carex* spp.), dryas (*Dryas punctata*), willows (*Salix* spp.), crowberry (*Empetrum* spp.), cranberry (*Vaccinium vitis-idea*), and mosses.

Biodiversity

Three natural zones—polar desert, tundra, and taiga, or forest tundra—stretch into a vast area of the Russian Arctic and sub-Arctic, encompassing an area of 3.9 million square miles (10 million square kilometers). Roughly 16 ethnic groups in a total human population of approximately 11 million are sparsely scattered in settlements across the region. Reindeer herding, hunting, and fishing are the staples of a subsistence culture thousands of years old. Despite the harshness of the environment, the coastal tundra provides dense habitats for anadromous (fish born in freshwater

Reindeer domesticated by the indigenous Nenet people on the Yamal Peninsula in Siberia. The nomadic Nenet have maintained a subsistence herding culture in the area for over 1,000 years, but are threatened by development of oil and gas resources and by climate change. (Thinkstock)

that live and grow in the seas) and freshwater fish, summertime migratory birds traveling along the Nearctic and Palearctic flyways, and herds of migrating wild reindeer (*Rangifer tarandus*). All of these provide important resources for indigenous subsistence cultures throughout Siberia—and prey for carnivores.

The polar bear is a key mammal whose survival is endangered by rapidly escalating climatic change that is adversely affecting its habitat. Other endangered species include the Arctic fox, caribou, rockhopper penguin, musk ox, and petrel. Among the fish in the region's rivers are perch, pikes, char, carp, and salmon.

The coastal tundra has an close relationship with the bodies of water that feed its treeless landscape; all of the central Asian rivers eventually flow northward out across the coast before flowing into the Arctic Ocean. Important rivers that migrate across the tundra include the Yana, Khroma, Indigirka, Alazeya, and Kolyma. This combined catchment area constitutes some of the most productive wetlands in Russia. Steller's and spectacled eiders (*Polysticta stelleri* and *Somateria fischeri*) are two of myriad waterfowl species that breed in this fertile marshland. Among the many others are red-throated loon (*Gavia stellata*), northern fulmar (*Fulmarus glacialis*), great white pelican (*Pelecanus onocrotalus*), northern gannet (*Morus bassanus*), and gray heron (*Ardea cinerea*).

Environmental Concerns

Siberia's rivers are critical pathways that carry toxic effluents out into the waters of the Arctic Circle. Researchers recognize the particular signature of each river on the Arctic basin, reflecting the chemical and biogeographical sources of the samples collected. Radionuclides (atoms with unstable nuclei) are of particular concern. They are residuals from the Khystym and Chernobyl accidents and are still monitored. Discharges of organic and inorganic carbons also are monitored for their effects on sensitive Arctic wetlands.

All of this is monitored via the cryosphere, Earth's ice cover—snow, ice, glaciers, sea ice, and permafrost. The geophysical dynamics of the Siberian Coastal Tundra biome are complex and sensitive to climatic and environmental variation, especially where permafrost soils serve as vast sinks holding carbon dioxide and methane. Sediment samples provide climatologists critical information about the effects of global warming on the tundra.

It is recognized that remobilization of even small amounts of the vast reserves of frozen methane can quickly accelerate a rise in temperature. Scientists estimate that more than 80 percent of the Eastern Siberian Arctic Shelf region is saturated with methane. Continuing fluxes in geothermal seawater heat act as a conductor force, bringing methane deposits up from Arctic permafrost-related seabeds and into the atmosphere. These effects are consistent with similar releases of methane and carbon dioxide in the world's oceans, and are important parameters for understanding the dynamics of future climate change in the Arctic.

In 1993, the Bolshoi Arktichesky (Great Arctic) Zapovednik was established, including 16,097 square miles (41,692 square kilometers) of the Taimyr Peninsula and the waters of the Karsk and Laptev Seas. These preserved lands provide valuable habitats for protected species, including the polar bear, reindeer, walrus, and beluga. Migratory birds are similarly protected.

VICTORIA M. BRETING-GARCIA

Further Reading

Anderson, Leif G., Kjell-Åke Carlsson, Per O. J. Hall, Elis Holm, Dan Josefsson, Kristina Olsson, et al. "The Effect of the Siberian Tundra on the Environment of the Shelf Seas and the Arctic Ocean." *Ambio* 28, no. 3 (1999).

Nuttall, Mark, and Terry V. Callaghan, eds. *The Arctic: Environment, People, Policy.* Boca Raton, FL: CRC Press, 2000.

Shakhova, Natalia, Igor Semiletov, Anatoly Salyuk, Vladimir Yusupov, Denis Kosmach, and Orjan Gustafsson. "Extensive Methane Venting to the Atmosphere from Sediments of the East Siberian Arctic Shelf." *Science* 327 (2010).

Sitch, Stephen, A. David McGuire, John Kimball, Nicola Gedney, John Gamon, Ryan Engstrom, et al. "Assessing the Carbon Balance of Circumpolar Arctic Tundra Using Remote Sensing and Process Modeling." *Ecological Application,* 17, no. 1 (2007).

Sierra Madre Pine-Oak Forests

Category: Forest Biomes.
Geographic Location: Mexico.
Summary: These forests represent a rich mixture of tropical and temperate species, some of them endemic.

The Sierra Madre Pine-Oak Forests biome is a montane ecosystem with high levels of biodiversity and endemism (species found nowhere else), and a convergence of species from North and Central America. It is possible to see jaguar stalking deer among lush oak forests, shrouded in orchids and other epiphytes (air plants). To the casual observer, these pine and oak forests of southern Mexico may not seem as lush as the tropical rainforests, and in plants that is certainly the case, but the Sierra Madres are a crossroad for many vertebrate fauna. Due to the strong elevational gradient, the region supports disproportionately high animal diversity.

This biome is made up of three geographically distinct ecoregions: Sierra Madre Occidental pine-oak forests, Sierra de la Laguna pine-oak forests, and Sierra Madre Oriental pine-oak forests. A total 27 species of conifer trees and 21 oak species have been recorded in these mountains, making them a center of diversity for these tree groups. However, oak and pine are highly desirable timber species, making this region one of the montane biomes most heavily-affected by human activity. Impacting the forest areas are logging, conversion to agriculture and pasture, and broadening human settlement.

The climate is temperate and humid; temperatures average 60–68 degrees F (16–20 degrees C), and annual mean precipitation is 27–157 inches (700–4,000 millimeters).

Vegetation

Oak forests gradually merge into pine forest moving upward along an elevational gradient. Oak forests dominate at 4,921–8,202 feet (1,500–2,500 meters), transitioning to cloud forests at 7,546 feet (2,300 meters), and pine-oak forests and some dispersed pine clusters live at still higher elevations. The most representative families are *Pinus, Quercus, Asteraceae, Fabaceae, Poaceae,* and *Euphorbiaceae.* In total, 12 main types of oak trees are found within these conifer forests.

The more humid forest areas contain cloud forests and numerous epiphytes, among them *Odontoglossum* spp., *Tillandsia prodigiosa,* and *Peperomia galioides*; scrub (*Eupatorium* spp. and *Ternstroemia pringlei*); and herbs (*Smilax moranensis, Spigelia longiflora* and *Salvia* spp.). These forests house a mix of vegetation with neotropical and boreal elements, including *Oreomunnea mexicana* and *Weinma*.

The pine-oak forests in the northern reaches of the biome contain many endangered species, including trees such as Hickels fir (*Abies hickelii*) and Mexican Cyprus (*Cupressus bethamii* var. *lindleyi*) as well as many species of endemic ferns and water lilies.

Fauna

Among the many natural phenomena in these forests is the migration of the monarch butterfly (*Danaus plexippus*). Monarchs migrate from Mexico to the United States and Canada, and over three generations they make the round-trip journey back to the Mexican highlands to breed en masse.

Birds are among the most diverse vertebrate taxa here, including more than 300 species from the north and south, including the Mexican jay (*Aphelocoma ultramarina*), Mexican chickadee (*Parus sclateri*), maroon-fronted parrot (*Rhynchopsitta terrisi*), thick-billed parrot (*R. pachyrhyncha*), Strickland's woodpecker (*Picoides stricklandi*), and Montezuma quail (*Cyrtonyx montezumae*). The imperial woodpecker (*Campephilus imperialis*), the world's largest woodpecker, was once widespread in this biome, but today is thought to be extinct due to loss of old-growth habitats.

Among the mammals found here are large predators such as the jaguar (*Panthera onca*) and gray wolf (*Canis lupis*)—this is perhaps the only part of the world where these predators overlap—as well as white-tailed deer (*Odocoileus virginianus*), Buller's chipmunk (*Tamias bulleri*), endemic Zacatecan deer mouse (*Peromyscus difficilis*), and rock squirrel (*Spernophilis variegatus*).

The number of reptile and amphibian species in this biome is quite high, including at least three species of rattlesnakes: rock rattlesnake (*Crotlaus lepidus*), twin-spotted rattlesnake (*C. pricei*), and ridge nose rattlesnake (*C. willardi*).

Environmental Threats

There are numerous threats to the native forest, including overgrazing, clearing of forests for timber and fuelwood, and human development. Because these forests are restricted to mountainous areas, there is growing concern that global warming will force the more temperature-sensitive species to move toward the summits until there is nowhere left to go. This is a special concern for the many species restricted to high-elevation areas, where slight changes in climate will have irreversible effects on restricted populations.

Jan Schipper
José-F. González-Maya

Further Reading

Barbour, M. G. and W. D. Billings, eds. *Forests of the Rocky Mountains. North American Terrestrial Vegetation.* Cambridge, UK: Cambridge University Press, 1988.

Brown, David E., ed. *Biotic Communities: Southwestern United States and Northwestern Mexico.* Salt Lake City: University of Utah Press, 1994.

Collar, N. J., L. P. Gonzaga, N. Krabbe, A. Madroño-Nieto, L. G. Naranjo, T. A. Parker et al. *Threatened Birds of the Americas. The ICBP/IUCN Red Data Book, Part 2, 3rd Ed.* Cambridge, UK: International Council for Bird Preservation, 1992.

Stattersfield, A .J., M. J. Crosby, A. J. Long, and D. C. Wege. *Endemic Bird Areas of the World, Priorities for Biodiversity Conservation.* Cambridge, UK: BirdLife International, 1998.

Sierra Nevada Forests

Category: Forest Biomes.
Geographic Location: North America.
Summary: The biologically diverse Sierra Nevada forests have been altered by intensive management practices that diminish forest health and resilience to wildfires.

The Sierra Nevada mountain range in the American West is renowned for forests with incredible biodiversity and spectacular scenery. Encompassed by peaks up to 14,000 feet (4,267 meters), river valleys, steep canyons, lakes, meadows, and rock outcroppings, the forests support remarkable plant and animal variety, and provide valuable natural resources and recreational opportunities for the region.

Logging, intensive forestry, wildfires, and urbanization threaten the Sierra Nevada forests, however, and also affect surrounding human communities. Though sizable forest segments are managed and protected by federal agencies, additional policy and community initiatives seek to improve forest health and maintain ecosystem services.

Vegetation

The Sierra Nevada forests are comprised of various forest types influenced by elevation, climate, and precipitation. Running northwest to southwest approximately 400 miles (644 kilometers) long by 50 miles (81 kilometers) wide, the eastern slope of the mountain range is much steeper and drier than the west. The towns and agricultural regions in the California Central Valley lead upwards into the chaparral and foothill woodlands on the west slope, followed at higher elevations by a mixed conifer forest dominated by ponderosa pine, sugar pine, Douglas fir, and white fir. Comprised of 27 species of conifers, the Sierra Nevada forest is considered one of the most diverse temperate coniferous forests in the world.

Above the mixed conifer forest, white fir, red fir and lodgepole pine species are prevalent. Just below the timber line, high-elevation pine species may be densely or sparsely present, including mountain hemlock, whitebark, foxtail, and limber pine. On the eastern slope, high-elevation areas of tundra and bare rock extend down to forests composed of lodgepole pine, red fir, and Jeffery pine. Below 6,000 feet (1,829 meters) on the eastern face, the open woodland of pinyon pine and western juniper descends into the desert scrub species of the Great Basin.

Biodiversity

These biologically diverse forests contain robust plant and animal assemblages, and perform vital ecological functions. Half of California's vascular plant species occur here, including 200 rare species and 400 that are endemic to the Sierra Nevada, that is, not found elsewhere. The forests support a

Giant Sequoia

As many as 75 groves of giant sequoia (*Sequoiadendron giganteum*) exist in the Sierra Nevada's mixed forest within a narrow band 5,000–8,000 feet (1,524–2,438 meters) in elevation. Giant sequoias are the largest trees on Earth, with some individuals recorded as more than 300 feet (91 meters) tall and 36 feet (11 meters) in diameter; some are as old as 3,200 years.

While one-third of the original giant sequoia groves have been harvested, some as recently as the 1980s, Giant Sequoia National Monument now preserves 33 of the remaining giant sequoia groves. Its wilderness areas are typically high-elevation areas that are protected from commercial logging, but still subject to intensive grazing.

Park visitors gaze up at giant sequoias in California. Giant Sequoia National Monument, established in 2000 by President Bill Clinton, protects 328,000 acres (1,330 square kilometers) of forest. (Thinkstock)

host of fauna as well, with 60 percent of California's vertebrate species found here, in addition to a diverse invertebrate population including many endemics and species with local distributions.

Among the animals found in this vast area, depending on elevation and climate, are black bear, mountain lion, mule and whitetail deer, gray fox, jackrabbit, California mouse, striped skunk, chickaree and chipmunk, and various varieties of squirrel.

Among the bird population are Bullock's oriole, screech owl, California and mountain quail, red-breasted sapsucker, wild turkey, turkey vulture, Audubon and Nashville warbler, white-headed woodpecker, and the rock and winter wren.

Amphibians and reptiles abound, including bullfrogs, California newt, Pacific treefrog, red-eared turtle, Gilbert's skink, and a range of snakes including coral-bellied ringneck, gopher, racer, sharp-tailed, Calley garter and western rattlesnake.

The Sierra Nevada forests are known for trout, including brown or German brown trout; golden trout, which is California's state fish; rainbow trout; and brook trout, which is often found in high-elevation alpine lakes and brooks.

Yosemite National Park is home to many creatures; wildlife biologists who monitor the park's fauna have identified threatened species as a way to ensure their long-term survival. Among those listed as under federal and/or state protection are: Sierra Nevada bighorn sheep, California wolverine, Pacific fisher, and Sierra Nevada red fox.

Environmental Threats

Research in the Sierra Nevada region reveals declining and at-risk plant and animal populations in these significant forest habitats, as well as scores of newly described plant species. The World Wildlife Fund has given this temperate coniferous forest a critical/endangered conservation status; the Sierra Nevada was included in the top 10 ecosystems needing protection from climate change in a 2010 report by the Endangered Species Coalition.

The ecosystem services of the Sierra Nevada forests are critical to the well-being of California residents. More than half of the water, timber, and energy from hydropower used in the state originates from this region, while its trees—particularly the old-growth forest stands—are able to capture and store carbon dioxide better than many other terrestrial ecosystem processes.

Logging, forestry practices, fire suppression, and urbanization are some of the ongoing threats to the biome. More than 60 percent of these mixed conifer forests have been altered. The move toward tree plantations and intensive forestry practices has simplified forest structure and composition, causing reduced genetic variability, impaired function, and diminished resilience to disturbance. The lack of genetic diversity from monoculture replanting after timber harvesting has made forests more susceptible to pests and disease such as bark beetles and fungal pathogens.

Fire suppression was implemented as human populations increased in the mountains and foothills, although fire was historically part of the natural cycle in the Sierra Nevada forests. Forest simplification and fire suppression together contribute to the greatly increased number and severity of catastrophic wildfires in the area. The decline in natural firebreaks such as old-growth patches or moist riparian areas, combined with increased fuels from the understory, lowers resilience to forest fire, and affects the health of forests and residents alike. Additionally, ozone and other chemicals from air pollution in coastal and valley cities have damaged many tree species and hinder their beneficial carbon sequestering function.

Conservation Efforts

An early advocate for Sierra Nevada preservation, environmentalist John Muir was instrumental in creating the celebrated Yosemite National Park in 1890, as well as the prominent conservation organization Sierra Club. Since then, the federal government has established nine national forests, four national parks, two national monuments, and 26 national wilderness areas that cover more than 15 million acres (6 million hectares), and provides varying degrees of protection for the Sierra's vast natural resources.

National forests account for the majority of the area, and the U.S. Forest Service administers multiple-use management of public lands that allows logging and grazing. The national parks share a high degree of protection for mid-elevation forests, and harbor the largest remaining blocks of relatively intact mixed conifer forests.

Sierra Nevada forest managers have faced the challenge of balancing multiple environmental perspectives and economic interests while addressing public safety. As a result, myriad partnerships concerning forest management and conservation have emerged. The Sierra Nevada Ecosystem Project, requested by Congress in 1992, was tasked with compiling an accurate, multidimensional ecosystem assessment to enable sustainable forest management and identify economic, ecological, and social linkages to inform future policy decisions. Following that initiative, the Sierra Nevada Forest Plan was created in 2001 and amended in 2004 and 2010, after significant public input, to improve protection of old forests, wildlife habitats, watersheds, and communities in the Sierra Nevada region.

In 2004, the U.S. Forest Service began the Forests with a Future campaign, aimed at protecting Sierra Nevada old-growth forests, wildlife, and communities from catastrophic wildfires by building public understanding and teaching tools to reduce wildfire damage. Numerous conservation coalitions also are actively engaged in contributing to forest management policies to ensure that biodiversity and ecosystem functions are preserved, especially as climate change affects forest and water resources, and vulnerable plants and animal populations. These collaborations and policy initiatives seek to maintain the comprehensive benefits derived from the Sierra Nevada forests now and in the future.

Nicole Menard

Further Reading

Murphy, Dennis D., Erica Fleishman, and Peter A. Stine. "Biodiversity in the Sierra Nevada." In *Proceedings of the Sierra Nevada Science Symposium*, edited by D. Murphy and P. Stine. Albany, CA: Pacific Southwest Research Station, Forest Service, U. S. Department of Agriculture, 2004.

Storer, Tracy I., Robert L. Usinger, and David Lukas. *Sierra Nevada Natural History.* Berkeley: University of California Press, 2004.

United States Forest Service. *Sierra Nevada Forest Plan Monitoring Accomplishment Report for 2011.* Washington, DC: Forest Service, 2012.

Simpson Desert

Category: Desert Biomes.
Geographic Location: Australia.
Summary: This desert has the world's largest parallel sand dune structure. Despite harsh conditions, it is home to diverse fauna and flora.

More than two-thirds of the Australian landmass is classified as arid or semiarid, and the Simpson Desert is one of the most arid parts of the continent. Areas of the Simpson have long-term annual average rainfall of 6 inches (140 millimeters), but the rainfall is far from consistent; it occurs in random bursts of thunderstorms every few years, with exceedingly dry periods in between. This limited rainfall and a temperature range of 32–122 degrees F (0–50 degrees C) mean that evaporation rates far exceed rainfall rates.

The Simpson covers 67,568 square miles (175,000 square kilometers) in the corners of the states of Queensland, South Australia, and the Northern Territory. It is bordered on the south by Warburton Creek and the Tarari Desert, on the west by the Finke and Todd rivers, and on the east by the Diamantina and Georgina rivers.

The Simpson Desert is characterized by vast dune fields. The red sand dunes run 99 miles (160 kilometers) in a north-south direction and can reach 131 feet (40 meters) in height. Linear sand dunes are created when bimodal winds (in this case, southeasterlies and southwesterlies) blow the sand particles. The sand dunes are of Pleistocene origin and have a thin, 7-foot (2-meter) layer of sand atop consolidated clays.

The watercourses of the region rarely flow, but when they do, the flows are slow-moving but large.

Although Warburton Creek has water in it every few years, Lake Eyre, into which the Warburton flows, has filled only three times in the 20th century. While arid above the ground, the Simpson Desert sits atop the Great Artesian Basin, a vast underground aquifer that receives water falling on the west of the Great Dividing Range thousands of miles (kilometers) away.

Flora

During wet years, the tops of the sand dunes are held together by vegetation, but more typically, huge sandstorms are whipped up by winds that are unchecked by topography. During decades of drought, the parallel red sand dunes are vegetated with isolated hummocks of sandhill canegrass and nitre bush. Both of these plants provide support and structure in the shifting sands of the dune tops, and within this structural support, small native mammals and reptiles make burrows for homes.

The swales are vegetated by acacia and hakea shrubs, with coolabah eucalypts lining the edges of the watercourses and lignum swamps filling the floodouts, areas flooded when the watercourses flow. Other hallmark plants of the biome include canegrass (*Zygochloa paradoxa* and *Triodia basedowii*), a hardy grass that grows on the tips of sand dunes; parrot bush (*Crotalaria cunninghamii*), with narrow, yellow-green flowers; and spinifex grass, which also grows on sand dunes.

Fauna

Despite being a highly inhospitable environment, the Simpson Desert originally supported diverse fauna, a recorded 34 mammal, 231 bird, and 125 reptile species. While arid Australia is an epicenter of mammalian extinctions worldwide, several unique species have persisted—possibly because introduced predators are less able to attain the densities needed to drive these small extant species extinct. Crest-tailed mulgara, for example, are carnivorous marsupial predators with a sandy-colored coat and a black crestlike tail; they inhabit burrows within the dune fields.

On the edge of the Simpson lives the kowari, another carnivorous marsupial, with a black, bushy tail. The kultarr is a smaller marsupial carnivore that stands upright like a tiny kangaroo and appears to hop when fleeing. The dusky and fawn-hopping mice are Australian native rodents that are endemic (found nowhere else) to this region; they hop along like tiny kangaroos as they search for seeds and insects.

Perhaps the most iconic species is the Eyrean grasswren, which is endemic to the Lake Eyre basin. It was recorded in the 19th century and then not observed again until 1976. The Eyrean grasswren feeds on seeds and small invertebrates that it collects while hopping around on the ground with its tail cocked.

Other birds of the dunes and swales are the whitefaces, zebra finches, emus, Australian bustards, and raptors such as whistling kites, brown falcons, and wedge-tailed eagles. When rainfall fills the watercourses and lakes of the region, birdlife erupts, and mass breeding events occur. Species of avocet, ibis, stilt, pelican, cormorant, and duck migrate in from thousands of miles (kilometers) away to use the short increase in resources for breeding.

Human Settlement

Aboriginal people have been living alongside water in and around the Simpson Desert for thousands of years. They were present when giant rhinoceros-size marsupials, Komodo-dragon-sized goanna, and 33-foot (10-meter) snakes and crocodiles roamed the region, tens of thousands of years ago. The tracks of *Euwinia grata,* a gigantic wombat, can be seen in the rocks along Warburton Creek.

Explorer Charles Sturt was the first European to see the Simpson, during 1844–46, but it was not until 1936 that Ted Colson became the first European to cross it entirely. Captain James Lewis's report of his expedition along Warburton Creek in 1874–75 described a large population of Aboriginal people sustaining themselves around isolated water holes on bountiful fish from the creek, supplemented with smaller marsupials like boodies and possums, and herbage like nardoo. These water holes persisted despite a lack of rain for several years previous.

Lewis reported on groups of Aboriginal people who had moved deep into the desert during rainy periods but could not return to the Warburton because the water sources had dried up in the intervening years. The desert was named after Alfred Allen Simpson, the president of the South Australian branch of the Royal Geographical Society of Australasia.

After the region was surveyed in the late 19th century, the Simpson Desert region was quickly opened to pastoralism. Vast cattle stations were created, but they could run stock only at low densities unless there was rain. Stations like Kalamurina and Macumba were attractive because they offered cattle routes with water sources from Alice Springs south to the Birdsville Track, and on to the markets in Adelaide and beyond. During years of floods or local rainfall, they also offered opportunities for agistment (care and feed for payment) and fattening before sales.

Conservation and Threats

The Simpson Desert has been reasonably well protected in the past by the Simpson Desert Conservation Park/Regional Reserve/National Park and Witjira National Park. In 2006, the Australian Wildlife Conservancy acquired the 2,550-square-mile (6,605-square-kilometer) Kalamurina Sanctuary, which links the conservation areas of the Simpson Desert with the Warburton Creek. This sanctuary allowed the fauna of the Simpson to access the life-giving water and vegetation of the creek's floodouts without competing with domestic livestock, which had been the case since Europeans began pastoralism in the region.

The parallel sand dunes of the Simpson Desert are the largest in the world, running 99 miles (160 kilometers) from north to south. They were shaped by the area's southeasterly and southwesterly winds. (Thinkstock)

Despite being covered by a conservation estate, the Simpson still faces severe threats, including exotic species. Early settlers to Australia introduced European rabbits, European red foxes, and feral cats, all of which have been linked to the decimation of Australian native mammals weighing 1.2 ounces (35 grams) to 12 pounds (5.5 kilograms). Noted conservationist Hedley Finlayson conducted fauna surveys of the region in the 1930s and again in the 1950s, and recorded the demise of many species of native mammals. He recorded mass declines of numbats, boodies, brushtail possums, lesser bilbies, and golden bandicoots in the region, and was the last person to record the existence of the desert rat kangaroo near Kalamurina Sanctuary.

Finlayson largely attributed the decline to the effects of introduced predators, in contrast with the activities and effects of such population-trimmers as the dingo, which is considered native, albeit only in the last few thousand years. Camels and rabbits, on the other hand, remove the precious vegetation, causing erosion and lowering the carrying capacity for native herbivores. There is a push within Australia to stop persecuting the dingo because of the protection it affords native wildlife through mesopredator suppression of foxes and cats.

Untapped bores tapping into the Great Artesian Basin also are a problem in the Simpson. Water from the White Bull Bore on Kalamurina Sanctuary comes out of the ground at 207 degrees F (97 degrees C), under enough pressure to drive it 34 miles (55 kilometers) upstream without any pumps. Untapped bores in the Simpson drain the Great Artesian Basin and create biospheres of vegetation damage around artificial water holes and unnatural wetland ecosystems, also providing permanent refuge for species such as the water rat that may not have survived otherwise in such dry places.

Climate change, too, poses threats to the natural balance of the Simpson Desert biome. Temperatures are projected to rise, and evaporation rates will follow suit, making water availability more difficult seasonally in some areas, and causing no-flow periods to be extended when drought conditions prevail year to year. Predicting changes in seasonal rainfall, flooding, aquifer recharge, and surface inundations, however, is far less practicable. What is clear is that plant and animal species will be pushed—depending on the severity of the warming and precipitation scenario—to adapt, migrate, or leave the most harshly arid areas altogether.

Matt W. Hayward

Further Reading

Burbidge, A. A. and N. L. McKenzie. "Patterns in the Modern Decline of Western Australia's Vertebrate Fauna: Causes and Conservation Implications." *Biological Conservation* 50 (1989).

Coenraads, R. R. and J. I. Koivula. *Geologica: Earth's Dynamic Forces.* Elanora Heights, Australia: Millennium House, 2007.

Dickilometersan, C. R., A. C. Greenville, B. Tamayo, and G. M. Wardle. "Spatial Dynamics of Small Mammals in Central Australian Desert Habitats: The Role of Drought Refugia." *Journal of Mammalogy* 92 (2011).

Finlayson, H. H. "On Central Australian Mammals. Part IV—The Distribution and Status of Central Australian Species." *Records of the South Australian Museum* 13 (1961).

Letnic, M. and F. Koch. "Are Dingoes a Trophic Regulator in Arid Australia? A Comparison of Mammal Communities on Either Side of the Dingo Fence." *Austral Ecology* 35 (2010).

Sitka Spruce Plantation Forest

Category: Forest Biomes.
Geographic Location: North America.

Summary: The first afforestation site in North America has tested the toughness and adaptability of the Sitka spruce species since the project began in 1805.

The original Sitka Spruce Plantation Forest biome is located on the Aleutian island of Amaknak (also called Umaknak), Alaska; it is the site of the oldest recorded afforestation project on the continent. The forest, known to researchers as The Forest or AHRS UNL-074, is a National Historic Landmark site and was naturally treeless until the early-19th-century experiment that planted the spruce here.

The Sitka spruce is a species named for the Alaskan town of Sitka—even though the tree is found all along the northwest coast of the United States. It has a straight conical shaped trunk and long downward flowing branches. The tree can grow up to 164 feet (50 meters) tall—or even to 300 feet (92 meters) in the most spectacular of the documented cases—with a trunk diameter of more than 6 feet (2 meters). The tree's bark is greyish brown with curved fissures and flaky plates as it grows, and the leaves are green flattened needles that are stiff, hard, very sharp, and grow individually. The spruce reaches maturity relatively quickly, and takes 40–60 years to reach its maximum timber potential, rather than, for example, 150 years for oak.

Amaknak is located 800 miles southwest of Anchorage, Alaska. The climate of the island, and all of adjoining Unalaska Bay, is cold maritime, with long periods of wind, drizzling rain, and fog. The mean annual temperature is about 40 degrees F (5 degrees C), and temperatures range from a cold of 32 degrees F (0 degrees C) in winter to a summer high of 52 degrees F (11 degrees C). The area has approximately 225 days of rain every year, making it one of the rainiest places in the United States.

Afforestation

The Sitka Spruce plantation forest here was started in 1805, when the Aleutian Islands were part of Russia. The Russian government supported a tree-planting program on Amaknak Island in order to establish a reliable timber industry and promote human settlement on the treeless island. The plantings consisted mainly of Sitka spruce (*Picea sitchensis*) that was transplanted from southeast Alaska's Kodiak Island to the Russian establishment of Unalaska here and on neighboring islands. Climatic and ecosystem conditions on the island proved generally too harsh for the forest to thrive.

After the area was purchased by the United States in 1867, the project attracted visiting botanists, who, throughout the 20th century, kept trying to develop forest on the island. Many waves of seedlings from the contiguous U.S. were transplanted to Unalaska through the mid-20th century, in the effort to combat erosion and to make the island more habitable for settlers.

The Sitka spruce was the most successful of the various planted species, and reforestation efforts tended to focus on the continuation of this particular conifer. It normally has a lifespan of 400–700 years. A handful of the original, 1805-planted trees have endured both the tough northern climate and soil, as well as contamination from human projects and pests such as bud moth (*Zeirapnera* spp.).

It is thought that, as some Sitka spruce on Amaknak and Expedition Islands have survived more than 200 years, and have spread some of their own seedlings, the species could become naturalized on those islands. Natural regeneration is indeed succeeding in some areas, with seed crops succeeding from plantings established in 1950–70. Successful spruce trees, however, need on the order of 670–700 growing degree days for pollination, fertilization, and embryo growth to occur properly, and an additional 625 growing degree days for germination after pollination.

Biodiversity

Sitka spruce can grow close together to make a very dense canopy that blocks out sun from penetrating to the woodland floor, making it difficult for other plants to grow beneath them. Sitka spruce forests at their maximum provide shelter from wind, rain, cold, and heat to biota below, so that such animals as deer and fox can find cover. Birds of prey, including goshawks and sparrowhawks, tend to find the spruce ideal for nesting and hunting. Birds such as the crossbill, tree creeper, coal tit, and siskin live and feed around Sitka spruce in many locations.

Prior to the establishment of the forests on Amaknak, native vegetation was primarily meadow with grasses, sedges, and herbs. The first plot of trees was planted among lyme grass (*Leymus* spp.). The volcanic soils, finely textured and rich in organic matter, were considered adequate to support rapid tree growth.

In other parts of the world, notably Iceland, arborists and other researchers have attempted to replicate the Sitka spruce afforestation program that was begun here on Amaknak. The scale of the life span of the species makes it difficult to assess the relative success of failure of such experiments in the near term.

Environment

Increased human population, exposure to industrialization, and construction on adjacent lands has altered the topography of the Sitka Spruce Plantation Forest biome here. Drainage systems on the site have not been updated, and with the frequent flooding, seepage from underground fuel tanks and a diesel fuel spill have contaminated water runoff and soils. These factors endanger the sites of the remaining original trees.

There are recent initiatives to arrange a pollution cleanup in and around the forests by both private land owners and public agencies. The U.S. Forest Service has recommended installation of a new drainage system and a public education program to inform the community of the ecological repercussions of construction and contaminants in the area.

Climate change poses other threats to the forest that could have both positive and negative effects. Milder temperatures could extend the growing season and curtail tree mortality. However, the warmer temperatures—and increased rainfall—could attract more pests and, with sea-level rise, worsen erosion problems.

Julie Kenkel

Further Reading

Alden, J. and D. Bruce. *Growth of Historical Sitka Spruce Plantations at Unalaska Bay, Alaska.* Portland, OR: U.S. Forest Service Pacific Northwest Research Station, 1989.

Dammert, Auri. "Dressing the Landscape: Afforestation Efforts on Iceland." *Unasylva* 52, no. 207 (2001).

Hermann, R. "North American Tree Species in Europe." *Journal of Forestry* 85 (1987).

Slums

Category: Grassland, Tundra, and Human Biomes.
Geographic Location: Global.
Summary: Slums are a challenge to cities around the world and a growing threat to a wide range of ecosystems.

It is estimated that one-third of the world's urban population lives in slums, and four out of 10 inhabitants of the developing world are classified as informal settlers. Globally, 1 billion people live in slums—and the figure will likely grow to 2 billion by 2030. Sub-Saharan Africa has the biggest share of slum dwellers, with more than 70 percent of its urban population residing in slums. The percentage of urban dwellers living in slums actually decreased from 47 percent to 37 percent in the developing world between 1990 and 2005. However, due to rising population, and the rise especially in urban populations, the number of slum dwellers is rising worldwide.

The word *slums* is used to describe a wide range of poor living conditions coupled with urban areas of low income or poverty. It is thought to have originated from the Irish phrase *'S lom é*, meaning a bleak or destitute place. Slums encompass deteriorated as well as informal settlements. The United Nations Human Settlements Programme defines a slum as a rundown area of a city characterized by substandard housing and squalor, and lacking security. In developed countries, *slums* could refer to housing areas that were once relatively affluent but deteriorated as the original dwellers moved on to newer and better parts of the city. In the developing countries, *slums* refers mainly to the sprawling informal settlements found in cities.

Depending on the cultural and language background, slums are also known as *bidonvilles, habitat précaire, shanty towns, favelas, skid row, barrio, ghetto, barraca, villa miseria, mabanda, mudun safi, chawls,* and so on. Normally, they are erected on land of high risk (such as steep slopes, floodplains, or garbage dumps) with unclear or informal property rights. Largely, the identification of an area as a slum is based solely on socioeconomic criteria and much less on racial, ethnic, or religious criteria—although in most developed countries of Europe and North America, they have become associated with foreigners lacking job skills or entitlements to live in those countries.

Most slums, fundamentally, are characterized by being inhabited by poor and socially underprivileged populations of the society. Buildings in slums may vary from simple shacks to permanent and well-maintained structures; it is not uncommon to find relatively well-built structures within slums. Those sound structures may be owned by successful slum-based business owners who have made their fortune from the plight of slum dwellers, through activities such as selling water. Slums tend to lack clean water, electricity, sanitation, and other basic services. They are areas of high rates of poverty, urban decay, poverty, illiteracy, and unemployment. Government authorities perceive them as breeding grounds for social problems such as alcoholism, crime, drug peddling and addiction, mental illnesses, and even suicide. In many developing countries, they are even seen as areas of high rates of disease due to unsanitary conditions, malnutrition, and lack of basic health care.

The low socioeconomic status of the residents is another common feature of slums. Many slum dwellers, therefore, employ themselves in the informal economy. In some cases, such as Kibera in Kenya or Mumbai in India, slums are places of thriving labor-intensive businesses such as garbage recycling, leather work, and various cottage industries. Common livelihood activities in the slums include domestic work, street vending, drug dealing, and even prostitution. In some slums, people recycle refuse of different kinds, from household garbage to electronics, for a living, selling either the odd usable goods or stripping broken goods for parts or raw materials.

Emergence

Most slums are formed as a result of rapid urban population increase due to several factors:

- *Rural-urban migration.* Many people move to urban areas in search of jobs and a better life. Lack of infrastructure in rural areas, and location of industries mostly within the urban centers in the developing countries, encourages those in the rural areas to move to urban centers. This in turn outpaces many urban facilities, especially housing.
- *Natural population growth.* Most slums occur in developing countries with high population growth rates. The rate of population growth outnumbers the growth of facilities in many cases, leaving the urban poor unable to cope and desolate.
- *Population displacement.* During conflicts or large-scale industrial development, people who have nowhere else to go settle informally within valleys, riverbanks, or hilly places usually least-preferred for urban development or settlement. In cases of conflicts, even people who once had good housing and sources of income may become desperate and lack the ability to afford decent housing.
- *Lack of income opportunities.* Large portions of the population are unable to meet housing and related costs, and are thus relegated to the periphery of towns or areas that are least suitable for settlement.

In some of the cities, having a high population combined with urban-specific transformation processes result in segregation implications, such as inner-city deterioration, gentrification, and counter-urbanization. Moreover, rapid urban

This shantytown occupied by Roma people in the city of Sophia, Bulgaria, illustrates two common problem of slums: flooding during rainstorms and narrow accessways that lead to blocked drainage and garbage buildup. Other problems related to poor infrastructure are lack of clean drinking water and high risk of fires. (World Bank/Scott Wallace)

population increases and the often-related spatial segregation of urban population segments on socioeconomic and ethnic grounds is fertile ground for the growth of slums in the developing world. Urban authorities also contribute to the emergence of slums through a general failure of housing and land markets to provide for the land and housing requirements of rapidly growing urban low-income populations in a timely fashion, and in sufficient numbers and locations.

In many cases, there exists apathy toward the slums, or even a tendency to assume their nonexistence. Indeed, lack of planning for long periods by the urban authorities also results in illegal occupation of urban lands and commensurate flouting of building regulations and/or of urban zoning prescriptions. An example of this wanton flouting would be a drainage line constructed across a river channel in the Mukuru slums in Kenya, hindering the flow of water and increasing the likelihood of flooding during heavy downpours. Considerable political and institutional inertia permits slums to expand to levels where their sheer magnitude overwhelms the capacity of existing institutional arrangements to effectively address their issues.

In many cases, lack of planning overshadows the political desirability of intervention. Further, attempts to formally address issues such as urban degeneration, explosive growth of informal housing, and illegal urban land occupancy are all too often ad hoc, marginal, and insignificant in relation to the scale and scope of the issues at hand. The nature of such interventions appears to indicate that the phenomenon of slums and the related problems are generally little understood, and that public interventions more often than not address symptoms rather than the underlying causes.

With the rise of the cities as the predominant and preferred residential locus of the majority of the world's population, the world is being faced with the reality that many large and medium-size cities are increasingly becoming areas of impoverished urban exclusion, surrounded by comparatively small pockets of urban wealth. This kind of inequality could lead to instability in many of the emerging cities.

Major Challenges

Slums by nature face myriad challenges. Some of these challenges are a result of the physical location of the slums. The growth or decline of slums may be closely linked to variations in the rural and urban economy and to related poverty levels. It is also a factor of demographics in terms of household formation rates, as well as the effectiveness of public interventions. Because slums are located in areas that are least habitable, and in many cases, residents pay no land rates to the government agencies, they remain a refuge for the poor or new immigrants still looking for sources of income. Slum residents tend to have low average incomes, high levels of unemployment, and relatively low levels of education. As a result, they are often stigmatized, leading to social discrimination.

There are exceptions such as Bangkok, Thailand, where only a minority of the slum dwellers are considered poor and stigmatization is less, and Havana, Cuba, where slum dwellers have secure tenure and access to the same social infrastructure as nonslum dwellers. The Havana experience is different due to the proactive attempt by the government to improve the general welfare of the slum dwellers.

The often-ignored isolation and victimization, difficult access to physical and social infrastructure, and generally higher incidence of violence and crime generate patterns of depressed urban areas where the inhabitants, despite their heterogeneity, find common interests on the basis of unsatisfied basic needs. In many developing countries, the proliferation of slums and their challenges are a result of these societal failures. In others, the problems may be a result of the mentality or the social orientation of the slum dwellers.

Underlying Problems

One of the leading causes of problems or challenges in slums is the lack of space. There are simply too many people within a small area of land. Slum residents are thus forced to use any small space available, regardless of its suitability in terms of exposure to floods, drainage channels, or need for open space for social amenities. As a result, housing structures may be constructed in inappropriate areas, such as blocking drainage channels, or in the path of floods.

Lack of space is also partly responsible for the small accessways common within the slums, which often double as walkways. Small accessways are in turn responsible for blockages, as their capacity to hold garbage is small.

The other major challenge within slums is lack of income-earning options. The majority of the slum dwellers are living below the poverty line and thus have limited possibilities to explore other alternatives. They are not able to put up more durable, flood-resistant buildings; neither are they able to acquire suitable land outside the danger of floods or overflowing drainages. They are also limited in terms of ability to coordinate responses to calamities such as floods or fire outbreaks because all their efforts are directed toward obtaining food for daily survival.

Slums, like other informal settlements frequented by the urban poor, receive little or no services from central or local governments. Due to lack of basic services in waste/garbage collection, there is often a huge accumulation of garbage, which flows into the drainage lines, leading to blockages. In cases where the garbage is emptied into a river channel, there are increased chances of blocked channels. Blocked river channels are known to cause back water flows, which can cause floods and even riverbank erosion. Slums also lack proper drainage facilities. The drainage channels are few and largely interspaced, which increases the possibility of blockage and stormwater damage during heavy rainfall.

Most slum residents are unaware of the possible solutions to some of the challenges facing them. In some cases, flood risks could have been avoided by simple solutions like clearing the blocked channels, building away from drainage lines, and avoiding construction of buildings close to riverbanks. This lack of awareness is also evident through some of the solutions adopted by the residents during floods, such as staying inside the flooded houses on top of beds, tables, or stools, oblivious of the imminent danger of increased volumes of water.

Poor maintenance of facilities within the slums is another cause for worry. There may be no schedule of maintenance of facilities like drainage systems. This lack of maintenance is also found in water pipes. It is not uncommon to see burst water pipes, making it easy for raw sewage to contaminate drinking water. This kind of contamination further reduces the availability of clean water within the slums. In many cases, the contamination of clean drinking water results in the spread of diseases such as cholera, dysentery, and typhoid. Lack of maintenance of the drainage way results in congestion or blockage of the drainage pipes or lines, creating fertile scenarios for flooding.

It should be noted that underlying problems cannot be solved internally within the slums and do require external intervention. The underlying problems are mostly the result of societal or economic failure and can best be solved through government policy intervention. The flood menace remains a big challenge to slums. In many cases, failures in drainage systems and blocked river channels increase the flood hazard and risk. Slum houses, due to the lack of space, are in many cases constructed close to the riverbanks, increasing the flood risk during heavy rains, and further disrupting riverine habitat.

In addition to flood risks, the threat of fire is also intensified in slum environments. Fires, indeed, are sometimes set by land owners or developers in order to clear areas for more profitable ventures. The environment in and around slums, therefore, is more likely to experience flooding and erosion, severe fire risk, and general degradation of the natural infrastructure.

Types

Generally, there are two main types of slums. On one hand are what could be considered slums proper, in many cases referring to inner-city residential areas that were laid out and built several decades ago in line with prevailing urban planning, zoning, and construction standards; over time, they have progressively become physically dilapidated and overcrowded to the point where they become the residential zone for the lowest income groups. These slums could retain some underlying order and even admired architecture.

On the other hand are shanties or spontaneous housing. In many cases, these informal settlements are the result of illegal or semilegal urbanization processes or unsanctioned subdivisions of land at the urban periphery. An example of this type of land invasion would be squatters who erect housing units without formal permission from the land owner using materials and building standards not in line with local building codes.

Global Action

The situation for slum dwellers varies widely. In democracies where slum dwellers are citizens with voting rights, politicians have an interest in improving living conditions in slums in an attempt to attract slum dwellers' votes and maintain political power. However, there have also been cases where governments have forced evictions and slum clearance programs because the slums were viewed as eyesores.

Wholesale urban renewal programs, slum regularization, upgrading, and community-based slum networking are increasingly being undertaken by city managers and governments. The United Nations Human Settlements Programme and similar global organizations have proposed administrative reforms for greater efficiency and reduction of corruption to permit the implementation of pro-poor social policies, with tangible successes in the areas of social housing, transportation, education, and public participation.

The Human Settlements Programme argues that it is possible to reduce the number of slums or improve living conditions for their residents by having citywide, rather than ad hoc, slum improvement, environmental improvement, land regularization, housing finance provision, and urban poverty reduction, as well as by forming partnerships with the private sector, nongovernmental organizations, and communities. There is also the need to combine these actions with true decentralization and empowerment of local governments. Further, the participatory involvement of both the slum residents and city managers in setting local priorities, participatory decision-making, and community-based involvement in implementation is being advocated.

Okeyo Benards

Further Reading

Davis, Mike. "Slum Ecology—Inequity Intensifies Earth's Natural Forces." *Orion Magazine*, March–April 2006. http://www.orionmagazine.org/index.php/articles/article/167/.

Floris, Fabrizio. *Puppets or People? A Sociological Analysis of Korogocho Slum.* Nairobi, Kenya: Pauline Publications, 2007.

United Nations Human Settlements Programme. "The Challenge of Slums: Global Report on Human Settlements 2003." UN-Habitat. http://www.unhabitat.org/downloads/docs/GRHS.2003.0.pdf.

Snake-Columbia Shrub Steppe

Category: Grassland, Tundra, and Human Biomes.
Geographic Location: North America.
Summary: This vast sagebrush steppe provides a northern transition from the cold Great Basin desert to the semiarid steppe of the northern intermountain west.

The Snake-Columbia Shrub Steppe biome comprises tablelands, intermountain basins, dissected lava plains, and scattered low mountains of the northern Great Basin. Extending from the Teton Range along the Idaho and Wyoming borderland west to the foothills of the Cascade Mountains, the area is a veritable sea of sagebrush steppe punctuated by concentrated areas of row-crop agriculture.

Wildlife here includes pronghorn antelope, greater sage-grouse, Brewer's sparrow, sage thrasher, and long-billed curlew. This classic western landscape has set the scene for plateau Native American cultures and European immigrants, and provided a ranching and farming heartland. The massive river systems here have been harnessed for irrigation and electricity, and now wind farms are becoming common sights.

Geography and Climate

The Snake-Columbia Shrub Steppe biome is located generally along the leeward side of the Cascade Mountains in Washington and Oregon, extending east across southern Idaho along the Snake River plain. Geologically, the Snake River traces the path of the North American Plate over the Yellowstone hot spot, located under Yellowstone National Park, a relatively thin portion of the Earth's crust known for volcanic activity. Most of the area is underlain by Miocene-age basalt flows covered with loess and volcanic ash—generating highly fertile soils. This is a region of smooth plains and deeply dissected terrain, with river-cut terraces along the Snake River.

Average annual temperatures vary from 40 to 55 degrees F (4 to 13 degrees C), with a growing season of approximately 160 days. Annual precipitation across the plateaus averages 6 to 20 inches (152 to 508 millimeters), but is much higher in isolated mountain areas. The least precipitation falls in the lowest valleys, and summers overall tend to be dry. However, surface water is abundant, largely from the Snake and Columbia River systems.

The Snake River running through Hell's Canyon, where its banks are lined with the grasses and low vegetation typical of the Snake-Columbia Shrub Steppe. (Thinkstock)

Crop irrigation is by far the largest use of water in the region. Groundwater is pumped from basalt aquifers and from deep soils along river valleys. The dominant soils in the region include the black, organic-rich soils called mollisols that are typical of Midwestern prairies. Aridisols, soils with relatively little organic material and characteristic of cool deserts, are also fairly common here.

Diverse Biota

Shrub steppe occurs in temperate latitudes, typically where semiarid conditions support a mix of grass cover and scattered shrubs. Most of the Columbia plateau is a steppe of sagebrush (*Artemesia* spp.), one of the most abundant and widespread types of vegetation in North America. Soils supporting sagebrush steppe in this region tend to be deep, often with a thin, fragile crust of algae, lichen, and moss. This shrub steppe is dominated by perennial grasses, including western wheatgrass (*Agropyron* spp.) and bluebunch wheatgrass, with several distinctive subspecies of big sagebrush and antelope brush. Areas with the deepest soils commonly support basin big sagebrush, but because these lands were among the only areas suitable for row-crop agriculture, most have been plowed.

Drier, windswept, or rockier sites tend to support Wyoming big sagebrush, often mixed with black sagebrush, low sagebrush, budsage, and spiny hopsage. More moist, or even poorly drained, sites might include silver sagebrush, bluegrass, muhly, wildrye, tufted hairgrass, and various sedges. Historically, the action of natural wildfire, as well as grazing by bison, probably maintained a patchy mosaic of open and closed shrubland among the grasslands. Shrubs may increase in density following heavy grazing and/or with suppression of wildfire. In many areas where surface disturbance has occurred, the invasive annual cheatgrass or other annual brome grasses are abundant.

On shallow, stony, or poorly drained clay soil, around valley margins, and along gentle slopes, dwarf-shrub steppe is commonly dominated by low sagebrush and close relatives such as early sagebrush and black sagebrush. On shallow basalt flows, scabland occurs in rocky patches, dominated by rigid sagebrush or buckwheat, or occasionally only by grasses and forbs. Because of poor drainage through basalt, these soils are often saturated from fall to spring by winter precipitation but typically dry out completely by midsummer. Total vegetation cover is typically low; annual plants may be seasonally abundant, but cover of moss and lichen is often high in undisturbed areas. On highly eroded volcanic ash and tuff, the harsh soil and high rate of erosion support sparse vegetation that often includes spiny hopsage, needlegrass, buckwheat, and bitterbrush.

At higher elevations, mountain big sagebrush steppe becomes dominant, extending up among woodlands and forests on ridgetops and mountain slopes. Other low sagebrushes, along with snowberries, juneberries, and currants, and mixed with a greater diversity of grass and forb plants are common in these areas.

Western juniper savannas pick up in abundance across a range of elevations on the western margins of the Columbia plateau, from southwestern Idaho, along the eastern foothills of the Cascades, and south into California. Where this vegetation grades into relatively moist forest or grassland habitats, these savannas become restricted to rock outcrops or escarpments with excessively drained soils. Throughout much of this range, fire exclusion and removal of fine fuels by grazing livestock have reduced fire frequency and allowed western juniper to expand into adjacent shrub steppe and grasslands.

The Columbia River system at one time sustained one of the largest salmon runs in the world. Today, the salmon have declined to less than a tenth of their former population, as a result of the effects of dams, diversions, overfishing, and degradation in surrounding uplands. More than 200 vulnerable plants and animals, including some 70 endemic (found nowhere else) plant species, are found here. The sagebrush steppe continues to support extensive herds of pronghorn antelope that still have seasonal migrations; some of the remaining core habitat for greater sage-grouse; and numerous birds of prey that nest here at higher densities than anywhere else on Earth.

Human Activity

The cultural history of the Snake River plain and Columbia plateau has been largely influenced by the main river systems. To the plateau Indian cultures such as the Nez Perce, salmon were central to life, although with the relatively recent introduction of horses by the Shoshone in recent centuries, hunting of bison became increasingly important. Throughout the early 1800s and before the development of railroads, explorers, trappers, and European-American settlers passed through this region along what became the Oregon Trail. Most of the area was homesteaded by the turn of the 20th century, at the point when dams began to be built throughout the region's river systems.

Today, this region is primarily a mixture of cropland and rangeland. Irrigated agriculture is most significant, with crops ranging from potatoes and peas to wheat and alfalfa. Grazing is the major land use in the drier parts of the region, with about one-third of the land being federally managed.

Environmental Concerns

Water erosion, wind erosion, surface compaction, and invasion of undesirable plant species are major resource-management concerns today. The recent growth in wind-energy development has increased the urgency for action to limit further fragmentation of wildlife habitat. With this in mind, the Western Governors Association and member states have begun to proactively provide wildlife habitat information to inform the planning decisions for renewable energy.

Climate change impacts may be felt in the arid region; nitrogen deposits, atmospheric carbon dioxide, and other changes will impact the grasslands. Climate change may impact the size of the sagebrush areas, for example, which in turn constricts grouse and other birds and mammals living in these habitats.

Patrick J. Comer

Further Reading

Cutright, Paul Russell. *Lewis and Clark: Pioneering Naturalists.* Lincoln: University of Nebraska Press, 2003.

Petersen, Keith C. *River of Life, Channel of Death: Fish and Dams on the Lower Snake.* Corvallis: Oregon State University Press, 2001.

Waring, Gwendolyn. L. *A Natural History of the Intermountain West: Its Ecological and Evolutionary Story.* Salt Lake City: University of Utah Press, 2011.

Snake River

Category: Inland Aquatic Biomes.
Geographic Location: North America.
Summary: The Snake River and its peoples are defined by fish, especially salmon, and the struggle continues to provide safe, clean habitat in the face of dams and climate change.

The mighty Columbia River in the Pacific Northwest region of the United States has more than 60 tributaries, and the Snake River is its largest and most important. The Snake River watershed of 173,984 square miles (280,000 square kilometers) is larger than the state of Idaho, with an average discharge of over 53,000 cubic feet (1,500 cubic meters) per second. The Snake River itself is approximately 1,078 miles (1,735 kilometers) long, and flows through forests, mountains and plains in Wyoming, central Idaho, southeast Washington, and northeast Oregon. Throughout its long history, volcanoes, flooding, and glaciers have shaped the river and its shores.

The Snake River plain was created by a volcanic hot spot beneath Yellowstone National Park, which holds the headwaters and origin of the river. Flooding as the glaciers retreated after the ice age created the current landscape, including eroded canyons and valleys. Mountains and plains are typical terrain along the river. The Snake River has more than 20 major tributaries, most of them in the mountains; Hells Canyon is the deepest river gorge in North America.

The Snake River is home to salmon and steelhead, which were central to the lives of the Nez Perce and Shoshone, the dominant tribal nations before the Europeans came. People have lived along the Snake River for over 15,000 years. The Snake River may have been given its name by the Shoshones, as a hand signal made by the Shoshones representing fish was misinterpreted by Europeans to represent a snake.

Flora and Fauna

The areas near the Snake River contain diverse flora, and the river basin itself was once home to broad shrub-steppe grassland. Riparian, wetland, and marsh habitats are found along the Snake River today, and thus a broad range of plants inhabit the area. At higher elevations, Ponderosa pine, Douglas fir, and other conifers are dominant. In drier areas, sagebrush and desert plants are the widely distributed dominant vegetation.

Salmon remain perhaps the most important animal species of the Snake River. Portions of the watershed have the largest, highest, and most intact salmon habitats in the continental United States. Anadromous steelhead, chinook salmon, and sockeye salmon are born in the Snake River watershed, swim downstream, grow and live in the Pacific Ocean, and return to the river to breed, spawn, and die. The river is also home to the bull trout, which overwinters there, and other year-round species of fish.

Through their life cycle and migrations, anadromous fish such as salmon and steelhead transfer nutrients from the oceans to inland rivers,

A researcher weighing a steelhead trout before tagging it at the Eagle Creek National Fish Hatchery in Oregon in 2010. The tracking tags help measure the success of restoration efforts. (U.S. Fish and Wildlife Service/Tess McBride)

tributaries, forests, and mountains. These fish also provide one of the primary foods of killer whales in coastal waters, and of the grizzly bear, gray wolf, wolverine, lynx, and other carnivores inland. More than 200 bird species also benefit from salmon, including birds of prey such eagles and falcons. Other notable fauna in the Snake River biome include several species of frogs, such as the inland tailed frog, Northern leopard frog, and Columbian spotted frog, as well as reptiles including the western toad.

Threats

Pressures began to mount upon salmon in the 19th century, with the establishment of commercial fisheries and canneries on the river. Up until the 1930s, native people continued to live on the river and fish year round, moving with the seasonal and migratory patterns of the fish.

The most significant impact occurred in the 20th century, with the widespread construction of hydroelectric dams along the Columbia River and all its tributaries, including four dams on the lower Snake River. The Lower Snake River Project resulted in four locks and dams—Ice Harbor Dam, Lower Monumental Dam, Little Goose Dam, and Lower Granite Dam—being constructed in Washington State. These dams became operational between 1961 and 1975. They are run-of-river facilities; this means the dams have limited storage capacity in their reservoirs, and water passes through the dam at about the same rate and volume as it enters the reservoir. The dams provide inexpensive hydropower, irrigation, navigation, and recreation to the region, but have had devastating impacts on fish and fish habitat.

Dams affect salmon and steelhead in many ways; they flood spawning areas, change historic river flow patterns, and raise water temperatures. Dams block the passage of fish between their spawning and rearing habitat, and the Pacific Ocean. If no artificial fish passage is provided, that blockage is permanent. Dams block the passage of both adult and juvenile fish. Adult fish passage has more recently been addressed through the creation of fish ladders.

The more significant problem is the need for juvenile fish to go through or around the dams. Juvenile fish that are drawn into the dam turbines can be killed or injured. The intense water pressure alone can kill the fish. If turbine passage is the only way past a dam, 10 to 15 percent of the fish that go through each dam's turbines will die. With that much mortality at each dam, fish that pass multiple dams—fish from the Snake River must pass through up to eight dams—have a statistically high probability of dying before they reach the ocean.

Environmental groups and scientists have requested the federal government to remove the four lower Snake River dams, arguing that mortality of juvenile fish would significantly decrease and that overall fish populations would recover after this removal. They also argue that these species are more likely to adapt to and survive climate change if these dams are removed. Salmon and steelhead require clear, cold water. The mountainous areas of the Snake River are likely to stay cool, and could be sanctuaries for salmon and steelhead populations throughout the region as global warming pushes up temperatures in the lower altitudes.

The dam removal could also be consistent with federal government obligations to native peoples in the Pacific Northwest. These groups signed treaties in the 19th century for rights to fish for salmon and steelhead in their usual and accustomed places. Government mainly failed in its obligation, as federal dams are the major cause in the decline of these fish. Therefore, the removal of these dams would be consistent with fulfilling federal treaty obligations.

Anticipated climate change impacts along the Snake River are likely be felt mostly in the form of reduced snowfall, which will decrease the amount of water runoff into the river system. This in turn may affect the availability of water, especially during the summer months, to keep habitats fully supportive of native plants and animals. Some research has found that recent sharp increases in the amount of zinc and other toxic metals in the upper Snake River streams—quite damaging to fish and other creatures—is apparently tied to warmer average air temperatures. Scientists are attempting to pinpoint the processes involved, but suspect lower levels of water to filter out metals is one aspect, as plants draw up more water when evaporation rates increase. In addition to lower filtration rates, more rock surface exposed to sunlight, air, and wind—because of lower stream levels—allows faster weathering and movement of metals into the water.

Magdalena Ariadne Kim Muir

Further Reading

Berwyn, Bob. "Global Warming May Be Upping Snake River Metals Pollution." *Summit County Voice,* September 12, 2012. http://summitcountyvoice.com/2012/09/12/global-warming-may-be-upping-snake-river-metals-pollution/.

Lovell, Mark D. and Gary S. Johnson. *Assessment of Needs and Approaches for Evaluating Ground Water and Surface Water Interactions for Hydrologic Units in the Snake River Basin.* Moscow: Idaho Water Resources Research Institute, 1999.

Slaughter, Richard A. and Don C. Reading. "Snake River Case Study: Institutions, Adaptation, the Prior Appropriation Doctrine, and the Development of Water Markets." Climate Impacts Group, University of Washington. http://cses.washington.edu/cig/res/sd/snakecasestudy.shtml.

Solomon Islands Rainforests

Category: Forest Biomes.
Geographic Location: Pacific Ocean.
Summary: The Solomon Islands rainforests, although home to a limited number of species of mammals, birds, and amphibians, have an unusually large proportion of endemic species.

The Solomon Islands is a sovereign state consisting of more than 1,000 islands located in the Pacific Ocean, east of Papua New Guinea. Most of the Solomon Islands are part of the Solomon Islands Rainforest biome. Because of the isolation of this island chain, and the rugged topography and the lushness of the habitats here, a full range of endemic (found nowhere else) species of flora and fauna live in these tropical rainforests.

The oceanic-equatorial climate here provides a mean temperature of 80 degrees F (27 degrees C) throughout the year, with few extremes of temperature, although the June-through-August period has slightly cooler weather than the rest of the year. While most of the Solomon Islands are within the Solomon Islands Rainforest biome, the Santa Cruz Islands group makes up part of the somewhat different Vanuatu Rainforests ecosystem.

Soil quality in the Solomon Islands rainforest varies greatly, from rich volcanic soil to infertile limestone. The climate here is considered tropical wet; the ecosystem is comprised chiefly of tropical lowland and montane forest habitats. Although most land on the islands lies below an elevation of 3,000 feet (914 meters), the predominant terrain type is forested hills.

Flora

Seven broad vegetation groups exist in the Solomon Islands, including coastal strand vegetation, freshwater swamp forests, mangrove forests, montane rainforest, two types of lowland rainforests, seasonally dry forest, and grasslands—this latter type found only on the large island of Guadalcanal. The lowland rainforest is the most common plant community.

The lowland rainforest canopy tends to be uneven, as frequent natural disturbances such as subsidence, fallen trees, and tropical storms disrupt it. Six diverse lowland rainforest types are spread about the Solomon Islands, resulting from a forest's location on the northern or western side of an island, its elevation, and the local level of disturbance.

A variety of common tree species exist here; among the most common are penaga (*Calophyllum kajewski*), kamani (*C. vitiense*), the fruit-bearing sumac (*Campnosperma brevipetiolata*), sea bean (*Maranthes corymbosa*), beabea (*Schizomeria serrata*), elephant apple (*Dillenia solomonensis*), the deciduous whitewood (*Endospermum medullosum*), and the eucalypt black wattle (*Gmelina mollucana*). Other important plants include the many endemic orchid and palm species.

Fauna

As true oceanic islands, the Solomon Islands are home to a high number of endemic vertebrates. The rainforests have fewer mammals than other nearby regions, such as New Guinea to the west, but there is a large gallery of bats here. The Solomon Islands rainforests are thought to contain fewer than 50 mammal species. At least half of these are endemic to the biome, including nine rodent species; 15 species of family *Pteropodidae,* or old-world fruit-eating bats; a free-tailed bat of genus *Molossus;* and the Solomons horseshoe bat (*Anthops ornatus*).

A similar pattern holds with avian species: relatively low diversity but high rate of endemism. The Solomon Islands rainforests are a haven to more than 40 families and subfamilies of birds, with perhaps 200 species. More than one-third of these are endemic to this ecosystem, a factor making the Solomon Islands rainforests a critical global area for bird conservation.

Among the endangered rainforest bird species found here are Woodford's rail (*Nesoclopeus woodfordi*), Makira moorhen (*Gallinula silvestris*), chestnut-bellied imperial pigeon (*Ducula rubricera*), white-eyed starling (*Aplonis brunneicapilla*), and imitator sparrowhawk (*Accipiter imitator*).

Environmental Threats

Direct human alteration of riverine and coastal areas, along with typically poor soils in some areas, have contributed to the depletion of lowland Solomon Island rainforests, and fragmentation of remnant coastal swamp vegetation and pandanus thickets. Some of the outlying coral atolls of the archipelago are in better native condition than the larger islands.

The rainforests are subject to tropical cyclones from November to April; these storms have proved to be sources of natural disturbance to flora and fauna, as have extreme droughts, which occur with some regularity, generally on a six- to 20-year cycle. However, climate change impacts upon this region, already prone to major storms and droughts, could push some habitats here beyond tipping points that have evolved over thousands of years.

Stephen T. Schroth
Jason A. Helfer

Further Reading

Gillespie, A. and W. C. G. Burns, eds. *Climate Change in the South Pacific: Impacts and Responses in Australia, New Zealand, and Small Island States, Vol. 2.* Dordrecht, Netherlands: Kluwer Academic Publishers, 2000.

McKibben, B. *Deep Economy: Economics as if the World Mattered.* London: Oneworld Publications, 2007.

Wolff, T. "The Fauna of Rennell and Bellona, Solomon Islands." *Philosophical Transactions of the Royal Society of London, Series B, Biological Sciences* 255 (1969).

Sonoran Desert

Category: Desert Biomes.
Geographic Location: North America.
Summary: Bracketed by temperate regions to the north and the tropics to the south, the Sonoran Desert is home to a biota with high levels of biodiversity from distinct backgrounds.

About 60 million years ago, what is now the southwestern United States and northwestern Mexico contained a relatively wet forest that was composed of both tropical and temperate species. Deserts were nowhere to be found. However, when conditions gradually shifted from wet to dry about 30 million years ago, some species adapted to these changing conditions toward what would later make up the Sonoran Desert biome of today.

Climate and Origins

The connection of desert species to the temperate and tropical forests from which they originally evolved is clearly visible in the flora of the Sonoran Desert. The five species of columnar cacti that impart much of the character of this region, for example, originated in tropical latitudes, in the remarkable Tehuacán valley of central Mexico. Additionally, more than half of the plants that currently exist in the desert are annuals. Annuals are plants that complete their life cycle in one year and take advantage of temporary resources, and here in the arid Sonoran Desert climate, that especially means rain. These species are about equally divided between those of temperate origin, which emerge in winter, and those of tropical origin, that emerge during summer. Yet, it took millions of years of evolution and shifting climatic conditions for the formation of the desert that we know today. In fact modern desert communities did not appear until just 8,000 years ago. The Sonoran Desert is a new thing in geologic terms.

The Sonoran Desert exists in a continuum of rainfall seasonality, with two peaks of rain: one in the summer and one in the winter. Winter precipitation comes to the region from the Pacific Ocean, which brings cool, soaking rains. The amount of winter rain tends to be modulated by the La Niña and El Niño oceanic-driven climatic cycles, with more winter rain during the warm El Niño phase than in the cool La Niña period.

In the summer, the heating of the land that occurred during the late spring draws in moisture from the Gulf of California, which creates huge, powerful thunderstorms with bursts of rain and lightning that can cool a hot summer day in minutes. This summer rain system is called the Mexican monsoon. Additional but less reliable summer rain comes in the form of tropical hurricanes in the eastern Pacific that form off the west coast of southern Mexico.

Biotic Response

Throughout the year, this seasonality is vividly apparent. Summer rains bring a flush of activity to species that originated in tropical latitudes. Cacti release their crowns of multicolored flowers, which are visited by pollinating bats. Elephant trees and various legume trees and shrubs set forth beautiful displays of foliage and flowers. Frogs dormant the rest of the year are awoken by ephemeral pools of water; they fill the desert night with a chorus of croaking.

But on a cool winter day after a gentle rain, none of what comes to life in the summer is visible. Now it is the species of temperate origins that capture the attention. Ferns that in the dry months are mysteriously absent emerge from rocky cracks and blanket previously barren slopes with a bright green.

Unexpectedly cold winter nights bring freezing temperatures that once every few decades are severe enough to kill back vast numbers of tropically derived species, keeping the balance between the temperate and tropical nature of the desert. Spring seasons following fortuitous winter rains erupt in fields of wildflowers that carpet the desert floors in yellow, blue, pink, and white. It is this climatic complexity that makes the Sonoran Desert a composition of such divergent life forms and leads to high levels of biodiversity.

Regional Habitats

The Sonoran Desert is not a uniform landscape; it is composed of several distinct regions. The saguaro-palo verde studded landscape defines the northern Sonoran Desert. The Lower Colorado Valley in the vicinity of the Colorado River and its delta is the most arid corner of North America. The diminished Colorado River Delta still serves as critical habitat for many migrating birds, although the human diversion and reduction of freshwater flow has severely degraded this once great ecosystem.

Just west of the delta, the Gran Desierto hosts the largest complex of sand dunes on the continent, remnants of the land forms that used to fill the Grand Canyon. Here too, the Pinacates resembles a lunar-like landscape of lava flows and massive craters formed by relatively recent localized volcanic activity, 1.7 million to 16,000 years ago.

The western coast of Sonora, Mexico, is heavily influenced by the buffering maritime influence of the Gulf of California and unpredictable, scant rainfall. A landscape of tall cacti and succulent trees with swollen trunks, and a low layer of shrubs interspersed at regular intervals, hugs the coast for miles and is the vegetation found on the individually unique and pristine islands of the Gulf of California.

Inland of this coastal fringe, the plants become denser and are characterized by a multitude of semi-deciduous tree species and various columnar cacti. Further east or south, the vegetation thickens progressively until the flat, arid valleys of bare ground, occasional shrubs, and scattered stands of trees fade away entirely. Instead, on the eastern side of the state of Sonora is the towering Sierra Madre Occidental and its deep *barrancas*, or canyons, filled with lush tropical vegetation, here at its northernmost extent in the New World.

A close-up of the hot pink blossoms of the beavertail cactus (Opuntia basilaris) *in the Sonoran Desert. Many types of cacti bloom in the desert from March to June. (Thinkstock)*

Flora and Fauna

Many plants thrive in the Sonoran Desert. The endemic (found nowhere else) saguaro cactus (*Carnegiea gigantea*) is perhaps the best-known plant found here, but others are vital to support local wildlife. Such plants include beavertail (*Opuntia basilaris*), cholla (*Cylindropuntia* spp.), prickly pear cacti (*Opuntia* spp.), legume trees such as iron wood (*Olneya* tesota) and palo verde (*Parkinsonia* spp), diverse grasses, and others. Cacti provide nourishment to many animals and keystone legume trees provide shade to animals and plants alike. From March to June, many cacti produce white, red, pink and yellow flowers, which dot the desert landscape with vivid color.

Other plants of interest include the only regional endemic palm here, the California fan palm, which is found in the Colorado Desert section of the Sonoran Desert and into Baja California. It is also found at several oases, including the one in Joshua Tree National Park.

The Sonoran desert hosts a great diversity of mammals, reptiles, and other animals. The black Mexican king snake, Arizona night lizard, desert iguana, and desert box turtle are among the most widespread reptiles. Coyotes and desert bighorn sheep rank among the hallmark mammal types. Many birds are endemic to the Sonoran Desert, such as the greater roadrunner, Gila woodpecker, and burrowing owl.

Threats

In the latter half of the 20th century, largely due to intensive irrigation and the widespread use of air-conditioning, the populations of towns in the Sonoran Desert multiplied from a few thousand individuals to form the three major population centers of the region: Phoenix, Arizona, with a population of 4 million; Tucson, Arizona, population 1 million; and Hermosillo, Sonora, population

800,000 (numbers are from 2010). Each of these urban centers continues to expand at nationally high levels, welcoming in new residents from distant regions. The pace of urban growth threatens to undermine the natural infrastructure of the Sonoran Desert biome.

As is expected in a desert where water is the principal and scarcest resource, dramatic hydrological projects of enormous scale have been erected. Vast canal systems channel water from the Colorado River in Arizona and the rivers of the Sierra Madre across the desert to the sprawling cities. Unregulated pumping of the ancient aquifers that lie below the sediment-filled valleys, dating from the prehistoric times when forests grew in the lowlands, fuel continued growth. With the expansion of desert cities comes the loss and fragmentation of habitat due to draw-down of water, construction, pollution, and other collateral effects of urban development.

A related threat to the biodiversity and economic functioning of the Sonoran Desert is an onslaught of exotic plant and animal species that were introduced in the last century, many of which have become invasive, aggressively spreading in recent decades. None seem more threatening than buffelgrass, which introduces a novel fire regime to the previously fire-exempt desert, with deadly consequences for Sonoran Desert plants that are not adapted to survive fire.

The Sonoran Desert relies upon its uniquely consistent precipitation cycle to support the distinct plant and animal biota of the region. Climate change is expected to cause increased warming in this area, with reduced annual precipitation and faster evaporation, but larger and more dramatic pulse type rain events. With increased warming some species, especially those already living at the highest elevations, may have nowhere else to go; if they cannot adapt, they may become extinct. Other species will likely have to adapt to take advantage of the pulses of resources to withstand the longer periods of drought.

Benjamin Theodore Wilder
Pedro P. Garcillán
Brigitte Marazzi

Further Reading

Felger, Richard S. and Mary Beck Moser. *People of the Desert and Sea: Ethnobotany of the Seri Indians.* Tucson: University of Arizona Press, 1985.

Shreve, Forrest and Ira L. Wiggins. *Vegetation and Flora of the Sonoran Desert.* Palo Alto, CA: Stanford University Press, 1964.

Turner, Raymond M., Robert H. Webb, Janice E. Bowers, and James R. Hastings. *The Changing Mile Revisited.* Tucson: University of Arizona Press, 2003.

South China Sea

Category: Marine and Oceanic Biomes.
Geographic Location: Asia.
Summary: Unparalleled biodiversity is the rule here, but there is a race on between fossil fuel extraction, international political tension, and those cooperating to restore, balance, and conserve the ecosystem.

Awash in the waters of the Indian and Pacific oceans, the South China Sea contains up to one-third of the world's marine biodiversity, which gives the region a substantial renewable resource base—but one that by some accounts is slowly crumbling. Sealed below its floor is an unusually large amount of crude oil and natural gas, which has sparked international conflict over extraction rights. Conservation and preservation of the sea's unique and diverse ecosystems may prove to be the key that opens the way for cooperation among the people of the region.

Geology and Origins

The South China Basin is the deepest part of the sea in this region. It is surrounded by three significant geomorphologic features: a segment of Eurasia's continental shelf in the west; the Reed Tablemount and Manila Trench to the east; and the Sunda Shelf to the south. Around 30 million years ago, a vast block of rock called the Reed Tablemount slowly separated from the Eurasian continent. In between, upwelling magma cooled and hardened to compose the basin floor, a process that ended 17 million years ago. The Spratly Islands sitting atop the Reed Tablemount are mostly uninhabited, but are the most politically contested archipelago in the world.

To the east, this submarine plateau drops into the Manila Trench, an abysmal habitat where marine life remains mostly a mystery. As the Eurasian tectonic plate slowly moves below the Philippine plate, the upper layer of the Earth's crust cracks and fissures, giving magma from below an opportunity to flow to the surface and create volcanic islands.

In the warm tropical and subtropical climates, corals colonize these mountains to form coral reefs. When the forces of erosion cause rock to disappear below the water's surface, the remaining living rings of coral create an atoll surrounding a lagoon. The coral reefs and atolls stretching from the Philippines island of Palawan down to northern Borneo form the base of southeast Asia's Coral Triangle, which consists of a network of protected areas that tapers toward Australia. Environments like these are mainsprings of marine biodiversity—and extremely rich hatcheries for both the Indian and Pacific Oceans.

Biodiversity

Much of the South China Sea's biodiversity springs from the continental and Sunda shelves here, which along with the Tablemount endow it with general shallowness. Along the coastal rim, terrestrial biomes such as tropical rainforests and savannas interact with inland aquatic biomes such as the Mekong, Red, and Pearl rivers, with marine estuaries creating life zones bristling with myriad species of plants and animals.

Among the most animated transition zones along the coast of the South China Sea are mangrove forests, in which gentle warm-water sea currents bring salt to the mouths of freshwater rivers and streams. Such areas reproduce exceptionally high levels of biomass that in various ways become dynamic natural resources. Mangroves protect coastlines from erosion and the added destructive effects of recurrent typhoons, which usually form between May and September. The region supports about 30 species of mangrove, and has extensive swaths of seagrass beds in the intertidal zones around its perimeter—yet another rich area providing shelter, breeding and feeding grounds for myriad mollusk, crustacean, fish, and marine mammal species. However, progressively more such habitat is being lost to industrial pollution, agricultural runoff, coastal development, aquaculture, and overfishing.

The Sunda Shelf and Coral Triangle have some of the world's most spectacular coral reefs, which teem with species of mollusks, crustaceans, sea snakes and turtles, but they, too, are succumbing to environmental threats. Reef raiding, blasting, and poisoning have depleted these creatures; subsistence and commercial fishers compete for depleted fish stocks, while demand for seafood is at an all-time high. The seagrass and soft bottom beds suffer similar damage, as trawling ships scrub the floor of all that can and cannot be eaten by human beings. International bodies have made some headway in outlawing such practices and introducing more sustainable fishing regimes, but the struggle is by no means won.

Human Factors

The territories of Brunei, Cambodia, China, Indonesia, Malaysia, Philippines, Thailand, and Vietnam extend into the South China Sea. Approximately 270 million people live in or near the coast of the South China Sea, a population expected to double by 2040. The economies of most of these countries depend heavily on international trade in manufactured and agricultural goods. Because shipping is the least expensive form of transport, the sea is central to intraregional trade.

Further, it is also one of the world's major global shipping lanes, because its waters funnel through the Straits of Malacca, which for hundreds of years has been a significant passageway for people and goods traveling east and west. With the opening of the Suez Canal in 1869, and the more recent rise of China and Japan as first-rate economies, the interaction among Europe, southwest Asia, and Africa with the countries of the South China Sea has never been greater.

Fossil Fuels

Civilization's dependence on petroleum puts at risk the localized and overall South China Sea biome—whether the threat comes from military conflict over access to these resources, production accidents such as pipeline blowouts and spills, contemporary global warming that will degrade long-established ecosystems as the sea-level rises and seawater acidity increases out of balance, or plastic containers and packaging that pour from coastlines to foul breeding areas of vulnerable marine fauna.

The continental shelves surrounding the basin hold about 7.7 billion barrels (1.2 cubic kilometers) of proven oil reserves, and a consensus estimate is that a total of 28 billion barrels (4.5 cubic kilometers) may be available. Natural gas reserves are estimated to total around 266 trillion cubic feet (7,500 cubic kilometers). The presence of fossil fuels and the richness of marine biodiversity are mainly the result of the sea's unique geomorphology and effects of episodic glaciations.

The fossil fuels that lie within these seafloor plateaus may prove to be the greatest threat to this large marine ecosystem. The presence of hydrocarbons indicates that within the past two million years, these shallows were bio-rich tropical landscapes connected to mainland southeast Asia. As the Pleistocene ice age matured, continental glaciers in the upper latitudes grew by drawing water from the oceans and transforming it into snow and ice. This caused sea levels to fall and seafloors to emerge. The repetitive cycle of life, death, and decay of terrestrial and marine life supplied the carbon; erosion, transport, and deposition of sediment locked it beneath the seafloor.

Climate change may impact fishing in the area; as the temperatures of the waters change, and species of fish move in and out of the waters, the nations that rely upon commercial fishing in the South China Sea may become increasingly protective of their areas, vying for whatever fish remain. Not only can climate change impact the marine life living in the South China Sea, but it also may have social and political impacts upon the area, depending on the physical outcomes upon the natural systems here.

To stanch these threats, the United Nations Environment Programme (UNEP) and the Global Environment Facility, along with seven regional nations, have been involved in the South China Sea Project since 2006. Comprehensive scientific surveys on several transboundary ecosystems and subsystems have been completed; these will provide a baseline from which to gauge future changes to this large marine ecosystem.

National governments along with a consortium of international universities are working together to give local people the skills to survey and manage ecosystems; promote public awareness of environmental issues; and understand legal and regulatory issues concerning restoration, preservation, and conservation of local, transboundary environments.

The hope is that increasing cooperation among the people of the South China Sea will lead to sustainable development policies and practices, so that certain geopolitical tensions will no longer threaten to snap the string of cultural and natural pearls adorning the sea.

Ken Whalen

Further Reading

Leinbach, Thomas R. and Richard Ulack. *Southeast Asia: Diversity and Development.* Upper Saddle River, NJ: Prentice Hall, 2000.

Rogers, Will. "Climate Change and the South China Sea." Center for New American Security. http://www.cnas.org/blogs/naturalsecurity/2011/09/climate-change-fisheries-and-south-china-sea.html.

South China Sea Project. "Reversing Environmental Degradation Trends in the South China Sea and Gulf of Thailand." United Nations Environment Programme. http://www.unepscs.org.

Weightman, Barbara A. *Dragons and Tigers: A Geography of South, East, and Southeast Asia, 3rd ed.* Hoboken, NJ: John Wiley & Sons, 2011.

Spain (Northeastern) and France (Southern) Mediterranean Forests

Category: Forest Biomes.
Geographic Location: Western Europe.
Summary: These diverse sclerophyllous forests, dominated by oaks (both deciduous and evergreen), wild olive, and carob have an outstanding environmental history linked to traditional land use.

These northeastern Spanish and southern French forest areas belong to the general context of the western Mediterranean basin, and they include a wide strip located around the coastline. They cover approximately 35,000 square miles (90,000 kilometers) from the Valencia and Catalonia regions in Spain, to the Gulf of Lyon and Provence region in southern France.

Due to the location, general bioclimatic conditions are typically mediterranean, characterized by temperate winters, with average temperatures of about 53 degrees F (12 degrees C), and hot summers, with average temperatures of about 77 degrees F (25 degrees C). Seasonal rainfall is low; the annual average is in the range of 14–30 inches (35–80 centimeters), concentrated during the spring and autumn, with a characteristic summer drought period.

Flora and Fauna

Under these conditions, the main forest communities that occur are: Mediterranean pine forests, Mediterranean oak forests, holm oak forests, and maquis or shrub vegetation. Mediterranean pine forests are present in basal areas, from the coastline up to 2,000 feet (600 meters). They are dominated by thermophilous (heat-loving) species well adapted to low water demand, poor soils, and to regular wildfire occurrence, mainly due to their high resprouting capacity. In this sense, the most extended forest communities are those dominated by Aleppo pine (*Pinus halepensis*) and maritime pine (*Pinus pinaster*).

In more coastal and drier environments, these pine forest stands may occur mixed with *maquis* shrubland; this dense and sclerophyllous (hard-leafed) formation—similar to Californian chaparral—is dominated by evergreen, thermophilous, and spiny shrub plants such as holly oak (*Quercus coccifera*), tree heath (*Erica arborea*), pourret (*Ulex parviflorus*), and Mediterranean dwarf palm (*Chamaerops humilis*), as well as key tree species like wild olive (*Olea europaea* var. *sylvestris*) and carob tree (*Ceratonia siliqua*).

Holm oak forests are perhaps the most outstanding Mediterranean sclerophyllous forest communities, because of their ecological characteristics and their abundance. These woods occur in a wide distribution area, ranging from basal to mid-mountain areas. Usually, these forest stands present a dense tree canopy dominated by holm oak (*Quercus ilex*), and an abundant and diverse understory of evergreen and thermophilous plants, well adapted to Mediterranean conditions (especially to low rainfall and summer drought); these include mastic (*Pistacia lentiscus*), rosemary (*Rosmarinus officinalis*), buckthorn (*Rhamnus alaternus*), and strawberry tree (*Arbutus unedo*).

In certain areas where summer drought is of lower occurrence, cork oak (*Quercus suber*) forest stands thrive. This tree is similar to holm oak, but differs from it mainly in its characteristic outer cork layer, an air-filled bark that protects the tree; perhaps the most famous and widespread use of this natural insulating material is in wine and champagne bottles.

In mid-mountain areas, sub-Mediterranean climate conditions arise with cooler, wetter conditions. In this context, Mediterranean oak forests occur, dominated by downy oak (*Quercus pubescens*) and Portuguese oak (*Quercus faginea*); its understory plants include common broom (*Cytisus scoparius*), wayfaring tree (*Viburnum lantana*), common hawthorn (*Crataegus monogyna*), or European box (*Buxus sempervirens*). These Mediterranean oaks are known as marcescent oaks, also called semi-deciduous oaks, due to their intermediate leaf strategy between evergreen and deciduous. Among the vascular plant population throughout the forests of this ecoregion, there is

A cork oak tree (Quercus suber), showing the outer layer of cork at the bottom of its trunk harvested for use, rises above the rest of a forest in southern France. (Thinkstock)

a 10 to 20 percent rate of endemism, that is, species found nowhere else on Earth. These areas are centers of biodiversity.

Of the vertebrates found in this ecoregion, there are only several large mammal species: the polecat (*Mustela putorius*), otter (*Lutra lutra*), and lynx. There is an abundance of bird species, especially along coastal areas. Coastal areas of this biome support some of Europe's most important littoral wetlands. Commonly seen birds here are egret varieties, duck, tern, and heron. Raptor species include black vulture, osprey, and Elenora's falcon.

This ecoregion has numerous areas that are internationally important for breeding and wintering birds. There are many fish species along the coast; four species are endemic to the Iberian Peninsula, in the Ebro Delta. The delta also sustains mollusks (both freshwater and saltwater), endemic shrimp, amphibians, and reptiles.

Human Impact

Western Mediterranean forests have a long and intense environmental history linked to traditional human land use. For centuries, holm oak has been widely harvested for firewood and charcoal extraction, as well as for timber used in construction and for making implements such as handles, cart wheels, and other tools. Tannins obtained from bark distillation are used in textile and leather work, and acorns feed livestock. This species became the most appreciated and used tree in this part of Europe.

Due to the long, historic use of these forests in addition to grazing and traditional agricultural practices, the forest surface in the western Mediterranean basin reached maximum levels of deforestation during the beginning of the 20th century. Since the 1940s and 1950s, traditional land uses in the western Mediterranean region have been profoundly changed due to demographic growth, abandonment of traditional land uses, mining of fossil fuels, rock quarrying, increased water consumption, increased industrialization, and urban sprawl. Because of deforestation and overgrazing, much of this more open land is now extremely susceptible to brushfire and forest fires.

This condition, compounded with the effects of climate change, is adding significant threats to the health of this biome. Global warming has sped the arrival of spring—by as much as two weeks—in many parts of this region; along with sharply increasing average air temperatures, this is resulting in major shifts in seed germination timing, groundwater availability, and fire regimes. Some migratory bird species have forgone their winter trip to Africa, opting to remain in Spain. Certain other terrestrial species—both plant and animal—must seek cooler habitats at higher elevations. At the same time, the Ebro Delta is experiencing impacts of sea-level rise, including saltwater intrusion, soil erosion, changes in sediment deposition, and related habitat disruption.

F. Javier Gómez

Further Reading

Barbero, M., G. Bonin, R. Loisel, and P. Quézel. "Changes and Disturbances of Forest Ecosystems Caused by Human Activities in the Western Part of the Mediterranean Basin." *Vegetatio* 87 (1990).

Blondel, J. Aronson. *Biology and Wildlife of the Mediterranean Region*. Oxford, UK: Oxford University Press, 1999.

McNeill, J. M. *The Mountains of the Mediterranean World*. New York: Cambridge University Press, 1992.

Quézel, P., F. Médail, R. Loisel and M. Barbero. "Biodiversity and Conservation of Forest Species in the Mediterranean Basin." *Unasylva* 50, no. 197 (February 1999).

Spratly Islands

Category: Marine and Oceanic Biomes.
Geographic Location: South China Sea.
Summary: This marine environment, including more than 750 reefs, islets, atolls, cays, and islands, has a rich ecosystem, but is mired in a tug-of-war between several nations.

The Spratly Islands are located in the central region of the South China Sea. In Vietnam, they are known as the Truong Sa Islands, in the Philippines as the Kalaya'an Island Group, and in China as the Nansha Archipelago. Ownership of the islands is disputed by China, Vietnam, Malaysia, the Philippines, Brunei (Darussalam), and Taiwan. Also, it is important to note that Japan receives 90 percent of its oil shipped through this area. Each of these nations claim some or all of the islands, along with their adjacent marine territories and resources. Despite disputed ownership, the Spratly Islands and surrounding waters have remained prolific marine ecosystems, with many unique reefs, fish, and other marine species.

Geography and Climate

The reefs of the Spratly Islands are spread over 155,000 square miles (400,000 square kilometers) in a region measuring about 500 miles (800 kilometers) from north to south and 560 miles (900 kilometers) from east to west. There are eight low sandy islands, 26 reefs, 21 shoals, and 10 submerged banks, with a total land area above water of less than 2 square miles (5 square kilometers). This is because most of the reefs in the Spratly Islands are submerged at high tide. Atolls here have high species diversity, including many fish, seabirds, and turtles. Ongoing industrial development and additional surveys suggest that there are significant reserves of oil and natural gas beneath the ocean floor here.

The Spratly Islands have a southern tropical climate, with an average annual temperature of 81 degrees F (27 degrees C). Summer is from May to August, with an average temperature of about 86 degrees F (30 degrees C). Winter is not much cooler, with an average temperature of 77 degrees F (25 degrees C). The islands have a seven-month dry season and a five-month rainy season. Southeast monsoon winds blow from March to April, with southwest monsoon winds from May to November. Few of the islands have substantial freshwater resources.

Flora and Fauna

The Spratly Islands encompass several hundred coral reefs in one the world's most diverse seas. The marine environment is a breeding ground for sea turtles, birds, marine mammals, and tuna. It is thought that the larval form of marine invertebrates in this area may supply other marine ecosystems throughout the South China Sea. In the future, the Spratly Islands may play a crucial role in seeding and restoring some of the over-harvested marine populations throughout the South China Sea.

Much is known about the marine environment, but less has been recorded about the terrestrial environment here. Little vegetation grows naturally on these islands, which are subject to intense monsoons. The few larger land surfaces have tropical and scrub forests and grasses. Several islands were in the 1930s covered with shrubs, mangroves, coconut, and pineapple. Papaya, banana, and palm may have been cultivated on some of the islands in more distant history. A few islands have been developed as small tourist resorts; soil has been added and trees planted. Very few humans occupy this marine region on a regular basis, however, with the exception of several military establishments.

Marine turtles and many species of seabirds visit the islands. The green turtle and hawksbill turtle are still found here. Seabirds use the islands for nesting, breeding and overwintering. These avians include streaked shearwater, brown booby, red-footed booby, great crested tern, and white tern.

Human Impact

The terrestrial and marine ecosystems in the Spratly Islands are under stress from increasing human activities. Military groups that occupy the islands on behalf of different countries shoot turtles and seabirds, and raid birds' nests for eggs and young. Harvesting of rare medicinal plants and cutting of timber are additional threats. The marine environment is under stress from overfishing and the use exploitative methods of harvesting fish and invertebrates, such as the use of bottom trawling, explosives, or poison.

The danger of rising sea levels due to global climate change threatens to permanently submerge many of these landforms. More immediate threats to the region appear to be from the strategic importance of location, with seven nations in persistent conflict over the area eying the value of shipping lanes, fisheries, and ocean bed mineral and hydrocarbon resources.

An international marine park has been suggested for portions of the Spratly Islands. If accepted by the countries claiming ownership, this designation could safeguard fish, birds, turtles, and other fauna. The ecological benefits to the biome could also extend to commercial interests, by helping to ensure a steady supply of young fish and invertebrates for regional fisheries.

Magdalena Ariadne Kim Muir

Further Reading

Gyo Koo, Min. *Island Disputes and Maritime Regime Building in East Asia: Between a Rock and a Hard Place.* Dordrecht, Netherlands: Springer Science and Business Media, 2010.

Hong, Nong. *United Nations Convention on the Law of the Sea (UNCLOS) and Ocean Dispute Settlement: Law and Politics in the South China Sea.* New York: Routledge, 2012.

Lally, Mike. "Spratly Islands Strategic Importance and Rising Sea Levels." *Inventory of Conflict and Environment Case Studies* 226 (December 2010).

United States Department of Energy: Energy Information Administration. "South China Sea Energy Data, Statistics, and Analysis—Oil, Gas, Electricity, Coal." March 11, 2008. http://www.eia.gov/emeu/cabs/South_China_Sea/pdf.pdf.

Srebarna Lake

Category: Inland Aquatic Biomes.
Geographic Location: Eastern Europe.
Summary: Srebarna Lake is the only significant Danubian wetland in Bulgaria; it shelters a significant percentage of the endangered Dalmatian pelican.

At roughly 2.5 square miles (6.5 square kilometers), including wetlands and the lake, Srebarna is the largest river lake in Bulgaria. Surrounding and protecting the freshwater lake from intrusion is the Srebarna Nature Reserve, a World Heritage Site. The lake and reserve are located approximately 100 miles (160 kilometers) from the Black Sea to the southeast and 85 miles (140 kilometers) from Bucharest, Romania, to the northwest. The combined ecosystem incorporates former farmland, a belt of forest plantations along the river, three islands in the Danube River, and the water between the island and riverbank. Vegetation here is that of the Ukraine-Kazakh biotic province. Adjacent to Srebarna is the Pelikanite, another nearly 2-square-mile (5-square-kilometer) enclave of protected pelican environment. The main purpose of the reserve is wildfowl protection for half of the avifauna in Bulgaria; this area is on the Western Palearctic bird migratory flyway.

Srebarna Lake is quite close to the Danube River, which provides seasonal flooding of the lake. Surrounding hills serve as a natural boundary as well as a perfect vantage point for observing the waterfowl and other birds that nest, breed, or stopover here. The elevation is only 32–43 feet (10–13 meters). Temperature averages are 28 degrees F (minus 2 degrees C) in January and 73 degrees F (23 degrees C) in July. Average rainfall is 20 inches (50 centimeters), with summers increasingly drier over the past 25 years.

Flora and Fauna

Srebarna is the only significant protected tract of Danube wetlands in northeastern Bulgaria. Wetland habitats here include standing and temporary open water with submerged vegetation, reed beds, swamp, seasonal marsh, river, hay meadows, and poplar stands. The wetland is of the sort that was once common across much of coastal Bulgaria, and it supports increasingly at-risk plant and animal species.

There are more than 130 plant species present in the Srebarna Nature Reserve; approximately one dozen are rare or endangered. Two-thirds of the area is covered in reeds, with varying amounts of lesser reed-mace, water lilies, willows, and osier. There are floating reed islands that serve as breeding and resting places for migratory birds. At the northern end of the lake, reedbeds gradually yield to wet meadows. The northwestern end of the lake, as well as the area along the Danube, includes riverine forest belts that include single old white willow trees.

Srebarna Lake is a significant breeding and wintering ground for a large number of bird species. Ninety-nine species of birds breed here, and some 80 migratory species from other parts of Europe and Africa winter here. Of the more than 200 bird species that have been identified in immediate lake area, nine are globally threatened, and 78 at risk in Europe. Twenty-four bird species that breed in Srebarna are endangered or rare.

The Dalmatian Pelican population is thought to be the only one left in Bulgaria. Globally threatened pygmy cormorants, ferruginous duck, white-tailed eagle, and corn crake found here are among the largest populations remaining worldwide. Other interesting species found in the lake biome include terns, egrets, herons, ibis, and the white spoonbill.

There are approximately 40 mammalian species supported by this wetland. Included are 18 rodents, four carnivores, three ungulates, and seven mustelids (weasels). Threatened mammals include otters, two types of polecat, and a wild cat. Invasive species include muskrats, raccoon dogs, and jackals, as well as wild boar, roe deer, red deer, and hares in large enough numbers to be hunted in nearby areas. Fish species number 18, including six endangered in Bulgaria. Reptiles total 15 species, amphibians 12.

Human Impact

Until creation of a flood-control dike for marshland drainage in 1948, the lake enjoyed annual periods of high Danube water. The dike stopped the annual inflow, causing the lake to deteriorate as vegetational succession accelerated. Organic mud filled the lake, causing it to become shallower, decreasing open water surface, and diminishing fish and bird populations and variety. Underground springs and runoff from the hills helped, but did not replace the Danube.

The two Iron Gates Dams built in Romania, with construction beginning in 1972 and completed in 1984, damaged the lake environment further, as did the increased pollution from neighboring farmlands. Drought between 1982 and 1994 decreased average water depth of the lake; nitrogen and phosphorus from agricultural runoff cre-

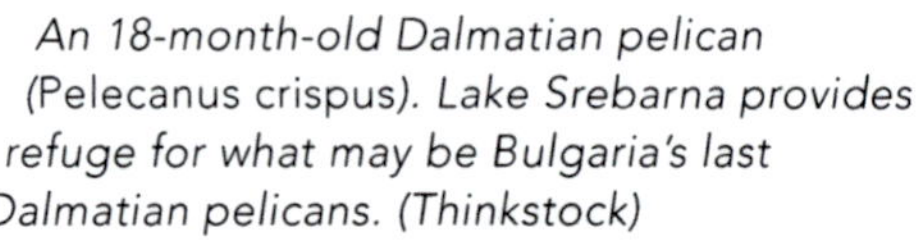

An 18-month-old Dalmatian pelican (Pelecanus crispus). Lake Srebarna provides refuge for what may be Bulgaria's last Dalmatian pelicans. (Thinkstock)

ated a hypereutrophic condition in which the over-abundant nutrients supported the growth of unhealthy algal blooms, leading to lower oxygen content in the water, and, ultimately, to lower biodiversity. The turning of the lake into a marsh adversely affected phytoplankton, fish, and birds.

Reconnection of the lake to the river began in 1979. By 1994, two canals allowed renewed river water flow, and all agricultural and residential activity around the lake was halted. In 1999, the mean depth was reestablished at 7 feet (2 meters), with a maximum depth of 9 feet (3 meters). The current lake contains much of its former open water areas and reedbeds. It is still surrounded by marsh, with low hills and farms beyond that. The 1994 canal slows the deterioration—but continues to restrict flow from the lake to the river. Lake maintenance requires two-way flows.

While the dense reeds are a barrier around the lake, as well as the foundation for reed islands on which birds nest, thickening of this vegetation has also caused its own set of problems. The extension of these reedbeds has attracted predators such as fox, jackal, wild boar, and wild cat that prey on the nests and on the 100-year-old Dalmatian pelican colony. This colony is the gem of all the avian life here, with breeding pair counts ranging from 29 to 127 between 1950 and 1980. The Srebarna Nature Reserve was established as a United Nations Educational, Scientific, and Cultural Organization (UNESCO) World Heritage Site in 1977.

During recent decades, the Srebarna Lake ecosystem has undergone significant changes toward hypereutrophication and degradation because of the disrupted natural connection with the Danube River water, due to both natural and man-made factors. These include the Iron Gates Dams, the dike built in 1948, the drought during 1984–93, and the new canal built between the lake and the Danube in 1994. These phenomena have led to mainly negative changes in the lake and wetland ecosystems, which depend largely on annual flooding from the Danube.

The latest restoration programs, such as canal dredging, have provided successful interventions that have led to a partial recovery of the wetland's functioning. However, the Lake Srebarna biome now faces additional pressures from global warming. Unlike some far-upstream segments that may see average flow diminish as glaciers retreat and rainfall patterns shift, this lower section of the Danube River is expected to experience surface level rises in coming years. Flooding of Lake Srebarna lasting beyond historic seasonal cycles is therefore an anticipated threat. However, great uncertainties in such projections remain; those who wish to help conserve this biome must also plan for the opposite: a climate change outcome where the Danube River flows into the lake are far reduced, leading to a diminishing of the depth, extent, and habitat richness of Lake Srebarna.

John H. Barnhill

Further Reading

Michev, T. *Biodiversity of the Srebarna Biosphere Reserve*. Sofia, Bulgaria: Pensoft Publishers,1998.

Nikolova, Mariyana, Rumiana Vatseva, and Valentin Nikolov. *GIS Assessment of Global Change Impacts on the Dynamics of the Srebarna Lake Ecosystem.* Sofia: Bulgarian Academy of Sciences, Institute of Geography 2010.

Riley, Laura and William Riley. *Nature's Strongholds: The World's Great Wildlife Reserves*. Princeton, NJ: Princeton University Press, 2005.

UNESCO World Heritage Centre. "Srebarna Nature Reserve." 2012. http://whc.unesco.org/en/list/219.

Sri Lanka Rainforests

Category: Forest Biomes.
Geographic Location: Asia.
Summary: This biome is characterized by high endemism, much of which is still unknown, but is critically threatened by direct and indirect human interactions.

Sri Lanka, a large insular landmass off the coast of India, exhibits a rainforest biome in its southwestern portion. These forests exist along an elevation gradient reaching up to more than 8,200 feet (2,500 meters). The stable climate conditions and monsoon rains allow for productive forest environments. Likewise, the isolation from mainland Asia has led to a huge amount of endemism (species found nowhere else) across all taxonomic levels. More new species will undoubtedly be discovered as further exploration and sampling take place.

Sri Lanka comprises about 25,500 square miles (66,000 square kilometers) on the continental shelf of India. Overall, its rainforest ecosystems encompass nearly 15,000 square miles (39,000 square kilometers)—or over half of the southwestern portion of the island. Despite the past geologic land connections between Sri Lanka and India, these rainforests have been isolated from the rainforests of mainland India due to drier, warmer conditions in the areas between them, creating a high degree of endemism on the island.

The two land masses are currently separated by the Palk Straight. While Sri Lanka's species closely resemble those of the Western Ghats (mountain range in western India), studies have concluded that they are actually quite distinct. Likewise, Sri Lanka and the Indian subcontinent are unique biogeographically because of their Gondwana ancestry, which contrasts with the rest of Asia.

Lowland and montane rainforest ecosystems extend along elevation gradients here. The lowland rainforest falls below 3,300 feet (1,000 meters). Climate in the lowland rainforests remains fairly constant, at 81–86 degrees F (27–30 degrees C) and 80–85 percent relative humidity. Deep valleys radiate away from the central mountains and contain the major rivers of the region.

Montane Sri Lanka rainforests lie above 3,300 feet (1,000 meters) in the Central Massif and the Knuckles Mountains. Overall, there is a trend of a roughly 1-degree F (0.5 degrees C) decrease in air temperature for every 330-foot (100-meter) rise in elevation, making the highlands much cooler. Ground frost is evident frequently during nights from December to February. Likewise, cloud forests characterize many of the high peaks.

Both elevation divisions receive 100–200 inches (250–500 centimeters) of annual rainfall, with most occurring during the May-to-September southwestern monsoon season. The northeastern monsoon also brings rainfall from December to March.

Biodiversity

The biodiversity and endemism of the Sri Lanka Rainforests biome are staggering. Almost all the endemic flora and fauna of the island nation are found in these lowland and montane rainforests. More than 34 percent of the island's endemic trees, shrubs, and herbs are located in these ecosystems. The lowland rainforests are characterized by two floral communities: the *Dipterocarpus*-dominated community and the *Mesua-Shorea* community.

Virgin forests of this ecoregion have four strata: a main canopy at 100–131 feet (30–40 meters), a subcanopy at 50–100 feet (15–30 meters), an understory at 15–50 feet (5–15 meters), and a sparse shrub layer. Trees of the emergent layer may reach above the main canopy to 150 feet (45 meters).

Generally, montane rainforests host more endemic flora and fauna than lowland rainforests. Among fauna, there are at least 15 endemic or near-endemic mammals; 25 endemic or near-endemic birds; and a mostly unknown but large number of endemic reptiles, fish, and invertebrates, many species of which have not yet been identified.

Few large megafauna call this ecosystem home because of the large areas of habitat they need for survival, but the montane rainforests support a population of the Sri Lankan leopard (*Panthera pardus kotiya*). The Asian elephant, previously found in the montane forests, is now locally extinct, although there are reports of a few seen in the lowlands. There are 13 endemic and near-endemic species of small mammals here, including shrews, hares, civets, squirrels, and bats.

There are many birds in Sri Lanka, including 20 that are near-endemic, including the Sri Lanka junglefowl (the national bird of Sri Lanka), dull-blue flycatcher, Sri Lanka bush-warbler, Sri Lanka hanging parrot, and the yellow-eared bulbul. Reptiles in Sri Lanka show more endemism than birds and mammals. Frog and lizard species are among those fauna still being newly discovered, along with fish and crab.

Human Impact

Unfortunately, direct and indirect human consequences threaten this biome, with a majority of the rainforest area already lost. Protected areas have been established, but protection and conservation plans are lacking in this rising nation. During the past 200 years, the Sri Lanka rainforests have been cleared for large-scale agriculture under the influence of the British colonizing forces. Human settlements and the development of tea, rubber, and coconut plantations, as well as rice paddies, have encroached on the rainforests. Likewise, exotic trees like Eucalyptus, Cupressus, and Acacia have been planted for construction and firewood. Furthermore, the Mahaweli River Project uses this major river for irrigation and hydroelectric production.

These developments combined have devastated these ecosystems, leaving the majority of lands, habitat, and biodiversity lost or severely affected. Only about 8 percent of the lowland rainforests are not affected; this region contains about 55 percent of Sri Lanka's human population. Nine protected areas have been established throughout the biome and offer the only real refuge. Unfortunately, adequate enforcement measures and conservation plans are largely absent. The projected effects of climate change on the island are still largely undetermined, but the impact on an already-compromised ecoregion could hardly be positive.

There are several reasons for the conservation shortfalls on Sri Lanka. First, Sri Lanka is emerging from a long period of British colonialism followed by extreme internal conflict, with a civil war between the government and the separatist Liberation Tigers of Tamil Eelam only recently ending. The gross domestic product of $106.5 billion is currently ranked 69th in the world but is growing by 9.1 percent annually, ranking it among the top 10 in the world.

Therefore, even as it rockets ahead economically, this fledgling nation needs the support and guidance of international conservation groups, such as the International Union for Conservation of Nature (IUCN) and the World Wildlife Fund (WWF). Continued growth and stability will ideally allow Sri Lanka to protect the huge amount of biodiversity and endemism that exists in its severely threatened rainforests.

Daren C. Card

Further Reading

Bossuyt, Franky, et al. "Local Endemism Within the Western Ghats-Sri Lanka Biodiversity Hotspot." *Science* 306, no. 5695 (October 15, 2004).

Erdelen, Walter. "Forest Ecosystems and Nature Conservation in Sri Lanka." *Biological Conservation* 43, no. 2 (1988).

U.S. Central Intelligence Agency. "The World Factbook 2009: Sri Lanka." https://www.cia.gov/library/publications/the-world-factbook/geos/ce.html.

St. Lawrence River

Category: Inland Aquatic Biomes.
Geographic Location: North America.
Summary: This major river is a complex ecosystem with diverse habitats; it supports a variety of endemic species—as well as 60 million people.

The St. Lawrence River (locally known as the *Fleuve Saint-Laurent*) is located in northeastern North America, forming part of the border between the United States and Canada. It generally runs northeast from the outlet of Lake Ontario into the Gulf of St. Lawrence and the North Atlantic Ocean. The river is the main outlet of the five North American Great Lakes (Ontario, Erie, Huron, Michigan, and Superior), and several tributaries flow into the main stem as well. On an annual basis, 70 percent of the St. Lawrence River discharge at Quebec City, Canada, originates from the Great Lakes via Lake Ontario, 20 percent from the Ottawa River, and 10 percent from smaller tributaries.

Geography and Surrounds

The watershed of the St. Lawrence is 617,763 square miles (1.6 million square kilometers), of which 55 percent is forests (boreal and temperate deciduous), approximately 22 percent is urban area, 20 percent agricultural lands, and 2 percent other types of land cover. Two major physiographic divisions influence the water chemistry of the river: the Precambrian Shield (north and west of the river) and the Appalachian Highlands (south and east). The Precambrian Shield is relatively homogeneous and dominated by silicate rocks; the Appalachian Highlands is more complex, but Paleozoic rock dominates.

More than 60 million people live in the watershed of the river. Several cities (Chicago, Illinois; Cleveland, Ohio; and Detroit, Michigan, in the United States; and Toronto, Ontario; and Montreal, Quebec, in Canada) are among the largest built-up areas along the shores of the Great Lakes or the St. Lawrence. Especially in the fluvial section, natural physiography has been altered by human intervention, such as construction of hydroelectric works, creation of the St. Lawrence Seaway locks and canals, dredging of shipping channels, intensive shore modification, creation of the Expo 67 (held in 1967) artificial islands, and ongoing fill work.

Although it is highly disturbed, the St. Lawrence is still a relatively complex natural river with diverse habitats. It contains three fluvial lakes (Saint-François, Saint-Pierre, and Saint-Louis), a few island chains (such as the Hochelaga Archipelago), and several isolated large islands; these last include Anticosti Island at 3,059 square miles (7,923 square kilometers). The shores of the fluvial lakes form large wetlands, the most notable being Lake Saint-Pierre, a Ramsar World Heritage Site. Wetlands here—excluding aquatic plant communities—cover at least 83,027 acres (33,600 hectares); this is augmented by 21,315 acres (8,626 hectares) of low marsh, 46,016 acres (18,622 hectares) of high marsh, and 15,884 acres (6,428 hectares) of swamps.

The St. Lawrence can be seen as four separate sections: the fluvial, covering more than 149 miles (240 kilometers) from Cornwall, Ontario, to the outlet of Lake Saint-Pierre, Quebec; fluvial estuary, encompassing more than 99 miles (160 kilometers) from the eastern tip of Lake Saint-Pierre to the eastern tip of Île d'Orléans; the upper estuary and Saguenay River, totaling 93 miles (150 kilometers) from the eastern tip of Île d'Orléans to the mouth of the Saguenay River; and the lower estuary and Gulf of St. Lawrence, a semi-enclosed sea that together are more than 143 miles (230 kilometers) long. Salinity gradually increases from Quebec City at the eastern tip of Île d'Orléans to the outlet of the Saguenay River.

Vegetation

The four sections of the river have different shoreline plant zonation patterns. In the upper parts of the river, such as the fluvial section, the strong seasonal fluctuations in water levels is the main driver at the shoreline. Along the riverbank is a silver maple (*Acer saccharinum)* forest inland, with a willow stand at the shore, alongside wet meadows and/or marsh, and beds of aquatic vegetation.

Downstream, the other sections are more influenced by tides and water chemistry. Shoreline vegetation in the freshwater estuary (Grondines-Montmagny) is characterized by a decrease in diversity and the lesser extent of riparian forests, which are often reduced to a narrow band composed of tree willows, species particularly adapted to the wide water-level fluctuations.

In the saltwater estuary, cordgrass (*Spartina* spp.) and sedge (*Carex)* salt marshes clearly dominate. The riparian vegetation of the Gulf has a similar structure to that of the estuary, but sand ryegrass (*Elymus mollis*) and beachgrass (*Ammophila breviligulata*) communities are more widespread.

Biodiversity

For the phytoplankton, seasonal environmental changes are an important control on community composition: *Diatoms* (a major group of algae) and *Cryptophyceae* (freshwater algae) are abundant year round, but *Chlorophyceae* become more plentiful in the summer. Inside the river, the fluvial lakes are major habitats for the plankton. The dense beds of aquatic vegetation in the fluvial lakes are highly productive compared with those in other sections of the river.

For benthic (near-bottom-dwelling) invertebrates, the main drivers of community composition are depth, substrate size, and organic content of the water. Dominant groups of zooplankton inside the freshwater areas are rotifers (wheel animals), cyclopoid copepods (T-shaped, shrimp-like bodies), and small cladocera (water fleas). In the estuary section, one particularly productive habitat for the zooplankton is the confluence—itself a narrow freshwater fjord—of the St. Lawrence and Saguenay rivers, where saltwater upwells in the form of nutrient-rich deep-water flows from the Laurentian channel arise to stimulate plankton growth and support krill populations.

This abundance of krill attracts approximately 21 species of cetaceans and pinnipeds, or seals, to the estuary section. One the most emblematic species of the St. Lawrence is the beluga whale (*Delphinapterus leucas*), which is more commonly found in the Arctic. The beluga is the only whale that stays inside the estuary all year; all other species spend only a few summer months there. This beluga community, estimated at 500 to 1,000 individuals locally, is listed as locally endangered due to overfishing and water pollution.

Fish diversity is relatively low compared with other great fluvial ecosystems, with 87 species of freshwater fish and 18 types of diadromous fish inhabiting the river. Fish endemism (those species found nowhere else) is also relatively low. Only the copper redhorse (*Moxostoma hubbsi*) and pygmy smelt (*Osmerus spectrum*) are true endemic species restricted to the river. Other fish species found in the St. Lawrence are the spring cisco (*Coregonus* spp.), chain pickerel (*Esox niger*), redfin pickerel (*Esox americanus americanus*), bridle shiner (*Notropis bifrenatus*), cutlip minnow (*Exoglossum maxillingua*), and eastern silvery minnow (*Hybognathus regius*).

Relatively few fish species are hunted on a commercial basis in the fluvial ecosystem. Among the species locally important to Native Americans of this region is the American eel (*Anguilla rostrata*). Like a number of other species, the eel population has declined, affected by river modifications such as hydroelectric dams and bridges that have disturbed their long migrations from the Atlantic Ocean to the Great Lakes.

Herpetofauna (reptile life) biodiversity is relatively low due to the northern nature of the river, although there are relatively abundant reptiles near Montreal, on the shores of the islands of the Hochelaga Archipelago, and in the wetlands of the fluvial lakes. In Canada, this section of the river is a national hot spot for herpetofauna biodiversity with several species of turtles and snakes. However, most of these species are considered endangered or at risk because of the human pressure exerted on their habitats.

Many important bird colonies are located on the shores of the river or use the St. Lawrence as a feeding ground. A colony of great blue heron (*Ardea herodias*) is located on Grande Île (Berthier-Sorel Archipelago, Lake Saint-Pierre), with more than 1,000 breeding pairs, perhaps the largest breeding colony for this species in the world. In the gulf, Bonaventure Island on the Gaspe Peninsula is home to the largest breeding outpost of northern gannets (*Morus bassanus*) in the world, with tens

A large colony of northern gannets (Morus bassanus) crowd a rock at Cape St. Mary's Ecological Reserve in Newfoundland. The largest breeding ground in the world for these birds is in the Gulf of St. Lawrence. (Wikimedia/Benutzer)

of thousands of breeding pairs. The St. Lawrence also is located on an important migration route for many bird species reproducing in the Arctic during summer. The St. Lawrence River wetlands provide resting and feeding areas for more than 1 million snow geese (*Chen caerulescens*) during their spring and fall migrations.

Environmental Threats

Regarding invasive species, the St. Lawrence is similar to other large rivers in human-modified environments. Several introduced species threaten the biodiversity and the functioning of the different subecosystems of the river. More than 163 aquatic species have been introduced into the Great Lakes during the past 200 years, and at least 85 are found inside the St. Lawrence, including the zebra mussel (*Dreissena polymorpha*), black spotted goby (*Neogobius melanostomus*), and Chinese mitten crab (*Eriocheir sinensis*). Examples from the plant kingdom include Eurasian watermilfoil (*Myriophyllum spicatum),* common frogbit (*Hydrocharis morsus-ranae),* water chestnut (*Trapa natans),* reed canary grass (*Phalaris arundinaceae),* and common reed (*Phragmites australis* ssp. *australis).*

Climate change poses some challenges to the river. If the water temperature rises too much and precipitation fluctuates too broadly, the river's flow could be altered by changes to the aquatic chemistry balance of the St. Lawrence, an ecosystem already struggling with hypereutrophic (over-abundance of nutrient flow) conditions, low-oxygen zones, areas of high turbidity, and other challenges typical of developed rivers. Higher water levels and more intense storms could increase shoreline erosion and instream sedimentation and turbidity, damaging the balance of flora habitats at the foundation of the food web.

Etienne Paradis

Further Reading

DesGranges, Jean-Luc. "Biodiversity Portrait of the St. Lawrence." Environment Canada. http://www.qc.ec.gc.ca/faune/biodiv/en/table_contents.html.

Environment Canada. "St. Lawrence River." http://ec.gc.ca/stl/default.asp?Lang=En&n=F46CF5F8-1.

Thorp, James H., Gary A. Lamberti, and Andrew F. Casper. "St. Lawrence River." In *Field Guide to Rivers of North America,* edited by Arthur C. Benke and Colbert E. Cushing. San Diego, CA: Academic Press, 2009.

Stromlo Plantation Forest

Category: Forest Biomes.
Geographic Location: Australia.
Summary: The nearly 100-year-old Stromlo plantation forest was heavily damaged in the widespread brushfires of 2003 and is not slated to be replanted.

The Stromlo Plantation Forest biome was an anthropogenic monocultural ecosystem consisting almost entirely of exotic Monterey pine (*Pinus radiata*) trees planted on degraded land on the outskirts of the new capital city of Canberra, Australia, in 1915. Canberra was established in a sparsely populated agricultural district—the Limestone Plains—in the first decade of the 20th century. Mount Stromlo, nearby, was the site of the earliest afforestation planting associated with the new capital.

The fortunes of Stromlo Forest waxed and waned in response to changing public policy and periodic bushfires for 90 years. In 2003, a fast-moving and intense bushfire driven by a convection column consumed nearly all of the forest. Today, the legacy of the Stromlo plantation is reflected in a large arboretum that has been established on part of the site; this now includes the Canberra International Arboretum and the Gardens-Mount Stromlo Forest Park. The rest of the property has been earmarked for urban development, recreation, and conservation.

The range of average temperatures in Mount Stromlo Forest Park is 76–79 degrees F (24–26 degrees C) during the summer months of December through February, and 51–54 degrees F (10–12 degrees C) during the winter. Annual rainfall is approximately 38 inches (973 millimeters).

History

Europeans first visited the Limestone Plains district in the 1820s, and pastoral development proceeded quickly. The rapid uptake of land for pasture was facilitated by the dominant pre-European ecosystem, yellow box-Blakely's red gum (*Eucalyptus melliodora-E. blakelyi*) grassy woodland. When Europeans arrived in Australia, the ecosystem was dominated by large, widely spaced eucalypts with an understory of grasses and forbs that were palatable for sheep and cattle. Agricultural development largely consisted of displacing the indigenous inhabitants, building homesteads, and expanding the supply of water for the sheep flocks.

By 1911, when the Australian Capital Territory (ACT) was established, the nonindigenous human population was 1,714, while the sheep population was 224,764. The high stocking rates of sheep and feral rabbits, combined with the removal of the native trees for timber and firewood, resulted in a highly degraded landscape. Afforestation was an obvious tool for protecting soil, reducing the dust problems, suppressing the rabbits, and creating a supply of softwood timber for the projected increase in the human population.

To meet the land management issues that were affecting the site of the new capital, the Commonwealth Afforestation Branch was established in 1913 under an energetic officer-in-charge, Thomas Weston. One of Weston's tasks was to find out which species would grow well in the Canberra region for the functions of timber production, ornament, climate amelioration, and food. Planting on Mount Stromlo proceeded rapidly, probably encouraged by the establishment of an astronomical observatory on the summit in 1911. The building housing the Oddie telescope (the first telescope to be installed at Mount Stromlo) was the first Commonwealth building in Canberra, and the need for night skies clear of dust probably provided the impetus to set the surrounding land aside for forestry.

Flora

The first trees were planted on Mount Stromlo in 1915, and over the following 10 years, more than 20 different species were trialed, including pines (*Pinus* spp.), poplars (*Populus* spp.), cedars (*Cedrus* spp.), cypress (*Cupressus* spp.), a Babylon willow (*Salix babylonica*), and a range of Australian natives such as eucalypts, wattles (*Acacia* spp.), and silky oak (*Grevillea robusta*). By 1924, more than half the plantation area, 1,307 acres (529 hectares), in the ACT was on Mount Stromlo; this became known as Stromlo Forest.

The trials showed that Monterey pine (*Pinus radiata*) was the most productive species on the site. Of the 982,200 trees planted there in 1915–25, 932,000 were of this species. At this time, the *Pinus radiata* seedlings were spaced in a 5.9-foot-by-5.9-foot (1.8-meter-by-1.8-meter) grid. All planting was preceded by rabbit control, site cleanup, and fencing.

Due to the lack of site-preparation machinery, explosives were sometimes used to prepare the ground, but even so, dry seasons in 1918, 1919, and 1921 caused heavy losses that had to be made up in the following years. Despite the setbacks, a 1925 report to Parliament recommended a total pine-forest estate of 20,016 acres (8,100 hectares), which established the broad objectives of forest policy in the ACT. These were commercial wood production, catchment protection, better water quality, prevention of soil erosion, and improvements to amenities.

When Weston retired in 1926, the Afforestation Branch became the Forestry Branch. The site for the Australian School of Forestry (now known as Westbourne Woods) had been selected near his arboretum.

The success of forestry operations at Mount Stromlo led to a gradual expansion of the program into other areas of the ACT. Canberra's water-supply catchment in the Lower Cotter Valley was planted with pines in an effort to reduce soil erosion, and in the 1930s, worker relief programs planted eroded grazing land. A further large expansion occurred in the 1950s, when government policy called for a total pine-forest estate of 40,031 acres (16,200 hectares). Some of this was achieved by clearing native forest in the mountainous region on the western side of the ACT. In 1967, another government committee recommended that the area of pine forest be maintained at 40,031 acres (16,200 hectares) and that action to initiate a major forest industry be taken.

During the 1970s, silvicultural practices were modernized, with plantations pruned and thinned to produce both saw and peeler logs for plywood. Much of the unthinned plantation, including areas that had self-seeded following major bushfires in the 1950s, was harvested and replanted under the new regime.

Wood from ACT forests was known for its higher density and strength resulting from the relatively slow growth of trees in the dry conditions. The slow growth, however, produced a relatively low internal rate of return. This was partly compensated by the proximity of the wood processors; Stromlo Forest was 4 miles (6 kilometers) from the center of Canberra. It also was close to the area's population center, and the community was pressing for recreation opportunities that required creative management of the forest. The 1970s ushered in a multiple-use policy for the forest estate and the end of converting native forest to pines to accommodate altered community perceptions of the practice.

Biodiversity

The Stromlo-Canberra region is home to a rich assortment of birds, animals and fish. Among the native bird species are red-rumped parrot, red-browed firetail, white-eared honeyeater, and red wattlebird; among those introduced are the common starling and common myna.

Introduced fish species have pushed out the native ones from most of the ACT rivers; these exotics include carp, brown and rainbow trout, redfin perch, mosquitofish, and dojo loach. Native fish species include the Murray cod and golden perch, two-spined blackfish, trout cod, silver perch, Macquarie perch, and mountain galaxias.

Terrestrial animals native to the area include Gould's wattled bat and little forest bat. The dingo, extensively persecuted for years, still survives in the ACT and is interbreeding with feral dogs, threatening the genome. The population of the eastern grey kangaroo is robust in the grasslands here, the swamp wallaby is common in the ranges, and common brushtail possum is common in bushland, as well as in urban areas. Koalas do not live naturally in the ACT any more, but likely did so prior to the 1939 bushfires that decimated their principal food source.

A destroyed section of forest after the 2003 Australian bushfires. The fires destroyed almost all of the Stromlo Forest. While much of the land will now be developed, an arboretum has been established as well. (Thinkstock)

Environment

In 2003, a large, fast-moving bushfire that had ignited in the Brindabella Mountains to the west of Canberra consumed 24,711 acres (10,000 hectares) of ACT Forests pine plantations, including virtually all of Stromlo Forest, the historic Mount Stromlo Observatory, ACT Forests headquarters, and 500 homes in the nearby suburbs. Following the fire, a review of the forestry operations concluded that the Stromlo plantations should not be replanted, and ACT Forests was absorbed into the ACT Parks and Conservation Service.

A large mountain-bike facility was established on the slopes of

Mount Stromlo, and replantings have been mostly of Australian native trees and shrubs. Another part of the former forest was set aside for the International Arboretum, featuring 100 forests composed of rare and significant tree species from Australia and around the world. Much of the rest of Stromlo Forest has been allocated for urban development, and the ACT Fire Management Unit headquarters at Stromlo Depot sits amid the new urban infrastructure. The original ecosystem, yellow box grassy woodland, is listed as nationally threatened, along with several of the species associated with it.

Climate change in the forests poses the threat of increased insect outbreaks and wildfires that could become more widespread and harder to manage. The warmer temperatures could increase the physiological stress on the trees that could augment die-off, made worse by increased drought conditions.

Adam Leavesley

Further Reading

Bartlett, A. G. *Community Participation in Restoring Australian Forest Landscapes.* Canberra: Australia Department of Agriculture, Fisheries and Forestry, 2010.

McComb, Brenda C. *Wildlife Habitat Management: Concepts and Applications in Forestry.* Boca Raton, FL: CRC Press, 2007.

Richardson, D. M. *The Ecology and Biogeography of Pinus.* Cambridge, UK: Cambridge University Press, 2000.

Shea, Syd, Peter Kanowski, Diana Gibbs, Allan Gray, and Ross Smith. *ACT Forests Reforestation Business Case.* Vantaa, Finland: JP Management Consulting, 2003.

Suburban Areas

Category: Grassland, Tundra, and Human Biomes.

Geographic Location: Global.

Summary: This novel and synthetic biome is defined by the degree to which human activity has dominated and transformed it—but suburban ecosystems still exhibit a high level of ecological activity.

Suburban ecosystems are a unique type of synthetic biome whose defining characteristic is the significant degree to which human activity has dominated and transformed the landscape. The key ecological pressures include replacement of native vegetation with impervious surface such as roads, concrete, and buildings; degradation of native vegetation and/or replacement with non-native species; fragmentation of natural/green spaces; the corresponding reduction in wildlife habitat; increased pollutant loads; and changes to the hydrologic regime due to land-cover change, for example, increasing flooding during storms.

Suburban areas are not always ecological wastelands, however. Suburban ecosystems can contain novel collections of biodiversity because of introduced species, and increasing trends in ecosystem restoration are placing an emphasis on restoring ecological functions within the built environment. A defining characteristic of a suburban area is that resources, materials, and energy need to be imported from outside the urban area; therefore, suburban areas do not fit the classical definition of an ecosystem. Natural ecosystems are considered to be relatively self-contained systems, where energy and nutrients are cycled within the boundaries of the ecosystem.

To address these unique aspects, suburban ecosystems need to be examined using three different lenses/perspectives: the suburban area as an ecosystem; ecosystems within suburban areas; and the suburban area as an entity within a larger, regional ecological network.

Characteristics of Suburban Ecosystems

The ecological definition of what constitutes a suburban area can vary, depending on the context. A commonly accepted rule-of-thumb definition is that an area is urbanized if it consists of "residential land at densities greater than one dwelling per acre, consisting of housing, commercial and public institutions, railyards, truckyards, and highways." In this context, both city and suburb are considered to be urban, at the opposite end of the scale from rural areas and/or wilderness.

The ecological differences between urban and suburban areas are relative and are a matter of degree, related primarily to the density and scale of urbanization. There can be outlying areas of a major city that feel more suburban than the inner-core areas of an adjacent suburb; therefore, the delineation from an ecological perspective between city and suburb is not hard and fast. A city and its suburbs can be viewed as lying at different points along a gradient, with the core of the major city at one end, the suburbs in the middle, and farmland and untouched wilderness at the other end of the gradient.

In general, suburban areas are considered to be mainly residential areas, with an abundance of single-family or low-density multifamily housing, where the residential properties often contain yards. This discussion focuses on suburban areas at this medium-density area of the scale.

One of the unambiguous features that definitively distinguishes an urban/suburban ecosystem from its surroundings is the high level of energy use, typically generated from fossil-fuel combustion, that is required to construct and maintain the suburban infrastructure. The energy level in an urban or suburban environment is typically at least an order of magnitude greater than in other ecosystems. While energy use can be relatively complex to measure, it is a helpful metric, as it can help differentiate urban ecosystems from other ecosystems that are also human-dominated but nonurban (for example, intensively managed agricultural land).

Terms and Concepts in Suburban Ecology

Some important terms and concepts used when discussing the ecology of the suburban landscape include:

- *City:* A relatively large or important municipality. Typical size ranges for what is considered to be a city can range from 250,000 to 10 million population. The defining characteristic of a suburb is that it is in proximity to a large city.
- *Urban region:* The area of active interactions between a city and its surroundings. The outer boundary of an urban region is determined by a drop in the rate of flow and movements of people, materials, resources, etc., as one proceeds outward from the city.
- *Metropolitan or metro area:* The nearly continuously built, or all-built, area of the city and adjoining suburbs. From the eye of a satellite, a metro area is prominent as a visible object.
- *Built area:* Land with continuous closely spaced buildings, as on small properties or plots.
- *Suburbs:* Mainly residential municipalities, such as towns, close to a city. A suburb may be entirely within, partially within, or altogether outside a metropolitan area.
- *Peri-urban area:* The area on both sides of a metro-area border, where built and unbuilt areas intermix.
- *Urban-region ring:* The area outside the metropolitan area and inside the urban-region boundary. This ring is a mosaic of greenspace (unbuilt) land types interwoven with built systems and relatively small built areas. Major highways, railroads, and power-line corridors are the prominent built systems that criss-cross the urban-region ring. Urban regions have a city-center nucleus.
- *Greenspace:* Unbuilt area in an urban region—areas without continuous closely spaced buildings—also called open spaces. Greenspace can range from tiny city parks to extensive woodland landscapes, and can range from rounded spots to linear greenways and river corridors.
- *Habitat:* A relatively distinct area, and its physical and biological conditions, where an organism, population, or group of species mainly lives.
- *Urbanization:* The combination of densification and outward spread of people and built areas.
- *Densification:* Increase in density of people and building units, for example, developing/

building on greenspace, or changing from low- to high-rise apartment buildings.
- *Sprawl:* The process of distributing built structures in an unsatisfactory or awkward, spread-out (rather than compact) manner or pattern.

Suburban Area as an Ecosystem

There are multiple perspectives and scales at which the ecology of suburban areas can be examined. One approach is to consider the suburb as its own ecosystem, and study the town's urban metabolism as though it were a discrete organism with its own metabolic processes. The study of urban metabolism examines a town from a holistic perspective as a consumer and digester of resources, and a creator of waste products. When viewing the suburb in this manner, the total inflows and outflows from the town can be quantified, which gives planners tools for ensuring the future availability of resources needed for the town to sustain itself.

Urban metabolism studies typically quantify the inputs, outputs, stocks, and flows of energy, water, nutrients, materials, and wastes. Factors that influence the metabolism of cities and towns include urban density (sprawled low-density cities have more intensive transportation energy requirements per person than compact dense cities), climate, technology, local policies and programs (for example, recycling programs), and the use of vegetation. Overall, trends in per-capita metabolism of urban areas have increased over the past 50 years.

Suburban areas can be examined in terms of the natural environments located within them. In urban and suburban areas, there are at least seven common types of ecosystems that can be considered natural (green or blue, rather than concrete or steel):

- *Street trees:* Stand-alone trees, often surrounded by paved ground.
- *Lawns and parks:* Managed greenspace with a mixture of grass, larger trees, and other plants, including areas such as playgrounds and golf courses.
- *Urban forests:* Less-managed areas, with a more dense tree-stand than parks.
- *Cultivated land and gardens:* Used for growing various food items.
- *Wetlands:* Various types of marshes and swamps.
- *Lakes and sea:* Open water areas.
- *Streams:* Flowing water.

Other areas within the city, such as dumps, abandoned backyards, and alleyways, may also contain significant populations of plants and animals.

Ecosystems in suburban areas provide many ecological services that have important social and economic value. These services can include air-quality regulation (filtering/detoxification of pollutants and generation of oxygen); local climate regulation (reduction of urban heat island effect, shade contributing to decrease in heating- and cooling-related energy use); noise reduction; stormwater drainage and aquifer recharge; sewage treatment; and recreational, cultural, and aesthetic values.

While the ecosystems in suburban areas can be considered natural (as humans are animals that affect their environment, just like any other creatures), they are also novel and relatively new to the planet, as the archaeological evidence suggests that major cities and their suburbs have been in existence for only the past 5,000 years. Therefore, suburban ecosystems are still in a relatively early stage of development. While each suburban area has its own unique aspects, there are still many attributes that can be found in most suburbs.

Suburbs have variables that influence the size of natural habitat patches and the dynamics of plant and animal interactions. These variables are consistently present, but differ according to local factors: buildings, pavement, scattered green areas, street trees, gardens, and house plants, as well as thousands of people and vehicles moving about daily.

Some species experience population explosions after their habitat becomes urbanized, such as the bluegill fish species in ponds, or some pioneer tree species that thrive in vacant lots after the surrounding area has been cleared. Also, many types of roadside plants thrive in urban and suburban areas due to the abundance of roads that did not exist before urbanization. Many bird species thrive in suburban environments, their populations increasing upon urbanization at the same time that the prior native species decline. In the United States, for example, starlings and house sparrows are abundant in every city, but these birds have been on the North American continent only since the early 1900s.

Some species accompany people wherever they go and colonize a region due to being imported by humans during the suburbanization process. Species such as dogs and cats are introduced intentionally, while many other species are inadvertently introduced, such as roaches; rats; mice; beetles that live in stored grain; and insects that arrive on fruits, vegetables, and imported flower pots. Urbanization has also been known to spread invasive pests and diseases that have a devastating effect in their new host environment because the native organisms do not have immunity; examples include Dutch elm disease, chestnut blight, and the Asian longhorn beetle that has attacked tree populations in suburbs across the United States.

In addition to invaders and pests, urbanization can have positive, if novel, ecological effects. Many suburban areas can be classified as forest. In many places, there can be a surprisingly high level of biodiversity in suburban areas.

In North America, common wildlife species found in suburban areas include tree frogs, snails, butterflies, turtles, raccoons, fireflies, moths, newts, muskrats, spiders, rabbits, skunks, honeybees, crickets, wasps, beetles, bats, moles, woodpeckers, dragonflies, foxes, sunfish, toads, squirrels, snakes, woodchucks, praying mantids, crayfish, hawks, mice, opossums, hydra, lizards, shrews, flies, otters, worms, freshwater sponges, fairy shrimp, mayflies, ants, and lacewings.

While in many ways there is a relatively high degree of ecosystem functioning occurring in even heavily built suburban landscapes, suburbanization does present pressures that affect urban ecosystem function. Net primary productivity (NPP) is a key measure of ecological functioning, as it determines the amount of sunlight energy captured and fixed by photosynthesis, making the energy available to drive biological processes throughout the ecosystem. Urban and suburban areas typically have a significantly lower NPP than their natural counterparts.

This neighborhood on the west side of Des Moines, Iowa, was in the process of being transformed from farmland (in the foreground) to tracts of suburban houses. (USDA Natural Resources Conservation Service)

Suburban areas do contribute to biodiversity in novel ways, but urbanization typically fragments native habitats to an extensive degree, and fragmentation of natural habitat patches is one of the best-known effects of human activities on natural biodiversity. In general, native biodiversity is increasingly lost the further one moves from a rural fringe area to the urban core; native vegetation generally has very little chance of surviving once the building density exceeds 1 unit per 1 acre (0.4 hectare). Urban exploiter species typically dominate the urban core; urban adopters dominate suburban areas; and urban avoiders dominate the peri-urban fringe.

Natural materials and nutrient cycles are also affected by urbanization, including nutrient cycling, soil erosion, hydrological flow, and the runoff of pollutants from suburban areas. Urbanization also greatly changes the natural ecological disturbance regime, which alters natural succession and introduces biogeographic barriers (roads, canals, and so on) that decrease the patch size of natural vegetation and introduce chronic stresses such as noise and light.

Larger Ecosystems

Suburban ecosystems can also be viewed as nodes in a network of regional and global ecosystems. The growing discipline of ecological footprint analysis determines the extent and area of natural ecosystems, located outside the actual suburban area, that are exploited to support the suburb. Suburb dwellers are often the sole macro-consumers of vast areas of cropland, pasture, and forest outside the immediate area. The ecosystems providing a majority of the biophysical life support for suburb dwellers are rural and other nonurban ecosystems.

In the Canadian metropolis of Vancouver, British Columbia, the ecological footprint—the total land area needed to provide the goods and services consumed by Vancouver city and suburb residents—is estimated to be about 11,476 square miles (29,722 square kilometers), which corresponds to about 300 times the size of the actual surface area of the metro area.

Trends in Urban Ecosystem Restoration

While the issues contributing to the unsustainability of many suburban environments are becoming increasingly evident, there are also encouraging and accelerating trends toward sustainability and ecological restoration in suburban areas. The trend toward integrated urban planning has accelerated, with contemporary planning approaches explicitly addressing the contributions that healthy ecosystems make to the economy and quality of life. In many areas, there is a growing focus on green infrastructure, which seeks to restore natural vegetation to improve stormwater handling and reduce flooding.

There is also a growing community gardening movement that encourages food cultivation in suburban areas, providing environmental, economic, and social benefits. Many towns are encouraging the installation of green roofs (vegetated rather than hard-surface), which helps mitigate stormwater flows, reduces the urban heat island effect, sequesters carbon, and encourages biodiversity. And a growing local food consumption, or *locavore,* movement encourages farmers' markets, which help restore the connections between suburban areas and the rural ecosystems they depend on.

Eric Landen

Further Reading

Alberti, Marina. "The Effects of Urban Patterns on Ecosystem Function." *International Regional Science Review* 28, no. 2 (2005).

Bolund, Per, et al. "Ecosystem Services in Urban Areas." *Ecological Economics* 29 (1999).

Forman, Richard T. T. *Urban Regions—Ecology and Planning Beyond the City.* Cambridge, UK: Cambridge University Press, 2008.

Garber, Steven D. *The Urban Naturalist.* Hoboken, NJ: John Wiley and Sons, 1987.

Headstrom, Richard. *Suburban Wildlife.* Hoboken, NJ: Phalarope Books, 1984.

Pickett, S. T. A., et al. "Advancing Urban Ecological Studies: Frameworks, Concepts, and Results From the Baltimore Ecosystem Study." *Austral Ecology* 31 (2006).

Rees, William. "Understanding Urban Ecosystems: An Ecological Economics Perspective." In Alan Berkowitz, Charles Nilon, and Karen Hollweg, eds., *Understanding Urban Ecosystems.* New York: Springer-Verlag, 2003.

Sudanian Savanna

Category: Grassland, Tundra, and Human Biomes
Geographic Location: Sub-Saharan Africa.

Summary: A critically endangered mosaic of tropical dry forest, savanna, and grassland, this biome provides sustenance for a growing human population and is home to endangered animals such as the African wild dog.

The Sudanian savanna is a highly diverse ecosystem that extends across the African continent from the Ethiopian Highlands in the east to the North Atlantic Ocean in the west. This tropical biome is divided into two sub-segments, which are separated by the Cameroon Highlands in west-central Africa: the West Sudanian savanna, which runs from the Atlantic to eastern Nigeria; and the East Sudanian savanna that stretches from the Cameroon Highlands to the Ethiopian Highlands. During the dry season of December to February, local temperatures are generally 68–78 degrees F (20–25 degrees C); during the tropical or wet season, the temperature is 78–86 degrees F (22–30 degrees C).

Biodiversity

The Sudanian savanna is composed primarily of a mix of grasslands and woodlands. This mix features large deciduous trees such as *Combretum, Terminalia* and *Acacia*; long elephant grass (*Pennisetum purpureum*), and shrubs such as *Combretum* and *Terminalia*. The ecosystem is vast, stretching some 3,850 miles (6,200 kilometers) east to west, but it has been heavily fragmented and degraded. It is one of the World Wildlife Fund's Global 200 priority ecoregions for conservation, and in 2002 was designated as "critically endangered."

The Sudanian savanna is home to a number of endangered animals, the populations of which have been significantly reduced by over-hunting and habitat destruction. The last population of black rhinoceros (*Diceros bicornis*) surviving within the Sudanian Savanna biome in Cameroon was declared extinct in 2011. The last confirmed sightings of the only remaining population of northern white rhinoceros (*Ceratotherium simum*) within this region—in the Democratic Republic of the Congo—were reported in 2007, although unconfirmed sightings occur from time to time in southern Chad.

Endangered and vulnerable species still occurring within the Sudanian savanna, according to the International Union for Conservation of Nature (IUCN) 2011 Red List of Threatened Species, include the Egyptian vulture (*Neophron percnopterus*), African elephant (*Loxodonta africana*), cheetah (*Acinonyx jubatus*), leopard (*Panthera pardus*), lion (*Panthera leo*), and the African wild dog (*Lycaon pictus*). Antelope species remain relatively stable in the region, with some exceptions such as the giant eland (*Taurotragus derbianus*) in Sudan and the roan antelope (*Hippotragus equinus*) in Sudan and Chad.

Among the birds found here are the East African crowned crane, ostrich, Kori bustard, and lilac breasted roller.

Environmental Impact

Residents of the Sudanian savanna depend heavily on their environment for fuel, food, medicine, income, and feed for livestock; these functions place great stress that is only mounting. The most acute threats to the biome are from the logging, charcoal production, farming including overgrazing by livestock, and wildfire. These factors have been particularly pronounced in the West Sudanian savanna, which is more densely populated. In part, these threats have been attributed to the local populations' traditional lifestyle of seasonal cultivation and herding. However, factors affecting the behavior of the agro-pastoral population—such as rainfall, primary productivity, and the numbers of grazing animals—vary significantly. Therefore, it is difficult to attribute ecosystem change to any single factor.

In the East Sudanian savanna is a relatively undisturbed habitat outside the main protected areas. Approximately 7 percent of the West Sudanian savanna and 18 percent of the East Sudanian savanna consist of protected areas and national parks. Enforcement on the ground is often lacking and many parks struggle with the impact of civil unrest, political instability, and lack of resources. Due to very poor infrastructure in the region, wildlife-based tourism as a source of income is underdeveloped, with the exception of sport hunting in the Central African Republic.

The pressure on environmental resources exerted by the growing human population is amplified by the threat of climatic desiccation in the Sudanian Savanna biome. Recent years have shown higher than average rainfall, with some of the heaviest rains and extensive flooding occurring in 2007 and 2010 across sub-Saharan Africa. Rainfall across the African continent is highly variable in space and time, but the general long-term trends in rainfall pattern point toward extreme drought following periods of increased precipitation in the Sudanian savanna—in other words, greater extremes.

Apart from potentially negative consequences for migratory wildlife, these global-warming-driven changes limit the ability of local habitats to recover from over-use. This is likely to result in increasing resource scarcity, from the direct effects of drought as well as indirect consequences such as soil erosion and leaching of nutrients.

To adapt to the changing environmental conditions and increasingly large-scale mechanized agricultural practices, water management strategies have changed in recent decades, with national and international programs diverting rivers and building dams. Although some of these developments are positive, providing water for irrigation and improving the efficiency of agriculture, other initiatives have resulted in severely damaged natural wetlands and flood plains. This has important implications for migratory wildlife, especially birds and fish, which often depend on these wetlands in the dry season—as well as for the human population, whose dependence on fishing and seasonal farming is tightly linked to the seasonal floods.

Franziska I. Schrodt

Further Reading

East, Rod. *African Antelope Database 1998.* Gland, Switzerland and Cambridge, UK: International Union for Conservation of Natural Resources/SSC Antelope Specialist Group, 1999.

International Business Publications (IBP). *Sudan Ecology and Nature Protection Handbook.* Washington, DC: International Business Publications, 2011.

Olson, David M. and Eric Dinerstein. "The Global 200: Priority Ecoregions For Global Conservation." *Annals of the Missouri Botanical Garden* 89 (2002).

Shorrocks, Bryan. *The Biology of African Savannahs (Biology of Habitats Series).* New York: Oxford University Press, 2007.

Wint, William and David Bourn. *Livestock and Land-Use Surveys in Sub-Saharan Africa.* Oxford, UK: Oxfam Professional, 1994.

Sudd Wetlands

Category: Inland Aquatic Biomes.
Geographic Location: Africa.
Summary: This hydrologically dynamic ecosystem is a vital resource for many animal species as well as domesticated fauna; its location near arid, war-torn areas has disrupted conservation efforts.

The Sudd Wetlands biome of South Sudan, found in the lower reaches of Bahr-El-Jabal segment of the White Nile River, is the one of largest tropical wetlands in the world, and was designated as a Ramsar Wetland of International Importance in 2006. *Sudd* is derived from the Arabic word *sadd,* meaning "blockage of river channels, or obstruction to navigation." The wetlands ranges 310 miles (500 kilometers) from south to north, and 124 miles (200 kilometers) east to west.

Hydrology and Climate

The average expanse of wetland or marshland here is 11,600–15,450 square miles (30,000–40,000 square kilometers), but this mushrooms to as vast an area as 50,000 square miles (130,000 square kilometers) during times of major flooding. The area contains thick grassland and vegetation over heavy clay soils that prevent absorption of surface water. Sandy earth is found 98 feet (30 meters) below ground, evidence that groundwater does not readily percolate to the surface of this biome. In general, Lake Victoria and its influence on the White Nile are the key sources of inundation here.

The mean annual temperature in the Sudd is 91 degrees F (33 degrees C) during the hot season, and 64 degrees F (18 degrees C) during the cold season. Annual rainfall ranges from 23 inches (600 millimeters) in the north to 39 inches (1,000 millimeters) in the extreme south of the wetland. The relative humidity is 23 percent during the dry season and a steamy 88 percent during the wet season. Because of hot, humid conditions, more than half of the inflowing water from the Nile system is lost through evapotranspiration across the permanent and seasonal floodplains. To avoid evaporation losses and increase the amount of water discharged at the outlet of the Sudd for agricultural and municipal uses, a planned 223-mile (360-kilometer) canal—the Jonglei Canal—is being constructed to bypass the swamps and carry some of the river's waters directly to the main channel.

The nearby segment of the Nile River, called alternately the Upper White Nile or Albert Nile locally, is the extreme north outlet of Lake Albert. Its waters run north to the town of Nimule, South Sudan, where it is called Bahr-El-Jabal. The river's meandering path here carves various channels and lagoons during the dry season and expands broadly over the partly flooded grasslands during the wet season.

Part of the wetlands fall within a sprawling system of mudflats in the arid Sahelian region of Africa. Essentially, the Sudd wetlands form as a function of the river channels exceeding their carrying capacity and periodically overflowing their banks. The Sudd wetlands are in a sense composed of interconnected river channels, associated with huge flood plains. Closer to the rivers are the permanent swamp areas. The more substantial part of the Sudd manifests as seasonal swamps, mainly created by Nile flooding but sometimes directly from heavy rainfall events when isolated ponds overflow.

A small village of thatched huts beside a creek in the Sudd wetlands of south Sudan. The Sudd wetlands are one of the largest tropical wetlands in the world, and are home to over 400 species of birds and 100 different mammal species, as well as the Suddia plant genus, which is found nowhere else. (Thinkstock)

Biodiversity

Ecologically, the Sudd Wetlands comprise various ecosystems: open water and submerged vegetation, floating fringe, seasonally inundated woodland, rain- and river-fed grasslands, and floodplain scrubland. The swamps and floodplains, in particular, support a rich biota with more than 400 bird and 100 mammal species. Migratory birds make favored stopovers here, while many wetland birds inhabit the extensive floodplains, including shoebills, great white pelicans, and black crowned cranes.

The wetlands are a haven for an estimated 1.2 million herd animals, including such migrating mammals as antelopes—among them the endangered Nile lechwe, reedbuck, tiang, and the world's largest population of white-eared kob.

The Sudd also is a habitat for many species of freshwater mammals such as hippopotamus, and reptiles such as the crocodile, as well as amphibians. As a giant filter that controls and normalizes water quality and functions as an enormous sponge to stabilize regional water flow, the Sudd is the major source of water—and abundant grazing plants—for domestic livestock as well, especially Nilotic cattle.

The deep open water areas of the biome are mainly surrounded by a permanent swamp zone consisting of *Cyperus papyrus, Vossia cuspidata,* and *Typha* spp., which are important habitats for shoebill stork. This zone, in turn, tends to be surrounded by seasonally flooded grasslands consisting of *Echinochloa stagnina, E. pyramidalis,* and *Oryza longistaminata,* as well as *Hyparrhenia ruffa* at the edge of the wetland. The *Cyperus papyrus,* which is threatened elsewhere by pollution and flood control, flourishes in the pristine Sudd wetlands.

Human Interaction

The Sudd is the home of *Suddia,* a plant genus known only to exist in this region, and which is thus endemic. Trees, shrubs, and both perennial and annual grasses provide various ecological services in the area, and also play a vital subsistence role for nomadic herdsmen, small farmers, and urban dwellers.

The wetlands provide substantial socioecological values for the region's Dinka, Nuer and Shilluk tribal communities. The Dinka and Nuer tribes depend on the annual floods and rain to regenerate floodplain grasses that feed their cattle herds; the indigenous communities also use the wetlands and scrublands for firewood, mud, and other construction material. Through this intimate relationship with the biome, these indigenous peoples have come to take pride in ownership and have developed custodial vigor in efforts to preserve the area's balanced but dynamic natural resources.

Climate change has already arrived in the Sudd. The regular seasonal patterns have been somewhat disrupted, with the dry periods lasting longer than they did formerly, and the rainy seasons more often producing heavier, damaging flood events. The seasonal changes affect crop growth, holding back some farm types while expanding the grazing areas, and may contribute to extinction of plant species; these factors could easily combine to make the area more vulnerable to the less-predictable weather.

The local growth and in-migration of the human population is meanwhile putting new strain on the water supply system. This is presenting tough challenges in a region already wracked by more than a decade of war. These socio-economic developments are forestalling a coordinated system of protection and conservation of the Sudd wetlands.

Suman Singh
Muraree Lal Meena

Further Reading

Dumont, H. J., ed. *The Nile: Origin, Environments, Limnology and Human Use (Monographiae Biological).* Dordrecht, Netherlands: Springer Science & Business Media B.V., 2009.

Postel, S. *The World Watch Environment Alert Series.* New York: Norton, 1992.

Stanton, E. A. "The Great Marshes of the White Nile." *Journal of the Royal African Society* 2 (1903).

Wetlands International. *Waterbird Population Estimates.* 3rd ed. Wageningen, Netherlands: Wetlands International, 2002.

Suncheon Bay Wetlands

Category: Marine and Oceanic Biomes.
Geographic Location: South Korea.
Summary: This protected natural area in South Korea serves as a habitat for migratory birds and marine life. An ecological treasure, Suncheon Bay is one of the largest tidal flats in the world.

Suncheon Bay is a coastal wetland between the Yeosu and Goheung peninsulas in the southwestern Korean province of Jeollanam-do, near the city of Suncheon. The ocean here, which South Korea officially calls South Sea, is part of the northernmost extent of the East China Sea. The climate of the Suncheon Bay Wetlands biome is temperate to warm, rarely below 32 degrees F (0 degrees C) in the winter and frequently sweltering in the summer, with humidity of 65–80 percent and 60 inches (1,524 millimeters) of rainfall every year. The bay, which primarily consists of tidelands, sand bars, salt marsh, the central stream of the Dongcheon River, and fields of reeds—some in distinctive circular forms—is a lively habitat for migratory birds.

The waters are largely unspoiled and serve as a spawning ground for many fish and shellfish species, all within the protected areas of Suncheon Ecological Park. There is some commercial fishing, particularly of shad, mullet, and octopus.

The Suncheon Bay wetlands comprises about 29 square miles (75 square kilometers), including one of the largest tidal flats in the world at about 5,488 acres (2,221 hectares), or about one-third of the total area. It is a major eco-tourist draw, attracting millions of visitors each year. The ecological park here has exhibits that introduce visitors to various habitats and guide them through the wetlands on a set of raised pathways designed—as the park buildings were—to minimize disturbance to the ecosystem.

Biodiversity

The wetlands have at least 116 species of plants in 36 families. Common reed (*Phragmites australis*), Korean starwort (*Aster koraiensis*), and starwort (*A. tripolium*) are hallmark types. Around the periphery of the reed beds are such salt-tolerant flora as cockspur (*Echinochloa crus-galli*) and flax-leaved fleabane (*Erigeron bonariensis*). Along the banks of the brackish and freshwater segments of the river are seablite (*Suaeda glauca*), zoysia grass (*Zoysia sinica*), Korean clover (*Kummerowia stipulacea*), and snoutbean (*Rhynchosia volubilis*).

The extensive reed beds are wintering sites for numerous migratory birds. Notable avians include the hooded crane (*Grus monacha*), spoonbill (*Platalea leucorodia*), Baikal teal (*Anas formosa*), whooper swan (*Cygnus cygnus*), and Chinese egret (*Egretta eulophotes*). The hooded crane breeds primarily in Siberia and Mongolia, and spends its winters in South Korea and adjacent areas of China. The current population is believed to be about 9,500 worldwide.

Interestingly, Korea lacks endemic (found nowhere else) bird species, and has poor diversity in dry-land avian species. However, sites like Suncheon Bay attract a plethora of migratory waterbirds. The Chinese egret, black-faced spoonbill (*Platalea minor*), and Saunders's gull (*Larus saundersi*) are all threatened species that thrive here.

There are as many as 1,000 Saunders's gulls in the wetlands, as well as large numbers of spotted greenshanks, grey-tailed tattlers (*Tringa brevipes*), great scaups (*Aythya marila mariloides*), and eastern taiga bean geese (*Anser fabalis middendoffi*). One of the most populous bird species here is the common shelduck, with more than 15,000 individuals in the Suncheon Bay wetlands.

Common mammals in the area include otter (*Lutra lutra*), raccoon (*Nyctereutes procyonoides*), and weasel (*Mustela sibirica coreana*). The wetlands also provide spawning ground for mollusks, crustaceans, fish, and amphibians.

The Environment

The wetlands and the migratory birds that depend on them face significant threats in the future, as an increasingly industrial Korea, while committed to green principles in other spheres, has consistently opted to put wetlands at risk. Some of the country's wetlands conservation plans have focused on other aspects, neglecting the use of wetlands by migratory birds and failing to preserve estuarine habitats that provide critical shelter, feeding, and rearing areas for waterbirds.

However, the green footprint of Suncheon Bay, widely viewed as a local treasure, has been successfully set aside in a plan designed to preserve the wetlands and help blunt the effects of climate change. Wetlands, such as Suncheon Bay, are thought to collectively absorb about 40 percent of carbon emissions worldwide, and thus are important for slowing the pace of rising global temperatures.

Additionally, on a local ecological service level, the salty swamp here helps in water filtration and purification. Removal and sequestration of heavy metals and certain toxic organics from the water has been documented, for instance. The stems and leaves of the Bay's phragmites reed community tend to block seaweed growth that might otherwise hinder the tidal cycles here; the reeds also help transfer oxygen into the water, keeping the balance ideal for a wide range of species.

Bill Kte'pi

Further Reading

Jun, Chang Pyo, Sangheon Yi, and Seong Joo Lee. "Palynological Implication of Holocene Vegetation and Environment in Pyeongtaek Wetland." *Quaternary International* 227, no. 1 (2010).

Kamala-Kannan, Seralathan and Kui Jae Lee. "Metal Tolerance and Antibiotic Resistance of Bacillus Species Isolated from Suncheon Bay Sediments." *Biotechnology* 7, no. 1 (2008).

Moores, N., S. K. Kim, S.-B. Park, and S. Tobai. *Yellow Sea Ecoregion: Reconaissance Report on Identification of Important Wetland and Marine Areas for Biodiversity Conservation.* Tokyo: World Wildlife Fund-Japan and Wetlands International China Programme, 2001.

Nam, Jungho, Jongseong Ryu, David Fluharty, Chul-hwan Koh, Karen Dyson, and Won Keon Chang. "Designation Processes for Marine Protected Areas in the Coastal Wetlands of South Korea." *Ocean and Coastal Management* 53, no. 11 (2010).

Sundarbans Wetlands

Category: Marine and Oceanic Biomes.
Geographic Location: Asia.
Summary: Characterized by one of the world's largest tidal mangrove forests, the Sundarbans wetlands is a biodiversity hot spot under threat from sea-level rise.

The Sundarbans—*beautiful forests* in Bengali—is among the world's largest contiguous tidal mangrove forests. The biome covers more than 2,510 square miles (6,500 square kilometers), split between Bangladesh (62 percent) and India (38 percent). The Sundarbans is located on the vast Ganges-Brahmaputra-Meghna alluvial delta on the Bay of Bengal, a lush and extensive low-lying coastal area.

The climate here is humid tropical, with monsoon effects. Average temperatures are in the range of 70–86 degrees F (21–30 degrees C). Rainfall totals some 65–79 inches (1,640–2,000 millimeters) yearly. Most precipitation occurs in the May-to-October monsoon season, with heavier rain tending to fall in the eastern areas of the biome. Major storms often bring massive inundation, erosion-causing storm surges, and tidal waves.

Biodiversity

The Sundarbans wetlands form a coastal biodiversity hot spot, supporting 40 mammal species, with such hallmark species as the Ganges river dolphin (*Platanista gangetica gangetica*); 35 reptile species, including the saltwater crocodile (*Crocodylus porosus*); almost 300 species of birds, many migratory; 400 fish species; and almost 30 species of mangrove trees.

The mangrove types of sundari (*Heritiera fomes*) and gewa (*Excoecaria agallocha*) are the dominant tree species here. The former is an exceptionally good hardwood used in furniture, flooring, and boatbuilding, and it is a strong erosion-preventer in its tidal environment, while the latter is used for more mundane commercial functions such as making paper pulp and charcoal—but it also has a growing number of medicinal uses.

*A Bengal tiger (*Panthera tigris*) photographed wading through deep water in the wild in Bangladesh. The Bengal tigers found in the Sundarbans wetlands are thought to be the only ones in the world to have adapted to living in a mangrove forest. (Thinkstock)*

The vegetation community of the Sundarbans Wetlands biome also includes *Avicennia* spp., *Sonneratia apetala, Bruguiera gymnorrhiza, Aegiceras corniculatum,* and *Rhizophora mucronata.* The golpata palm *Nypa fruticans* is distributed throughout the ecosystem; its dense root system, underground stem, trunkless structure, and moderate salt-tolerance are extremely well-adapted to this wetland habitat. Dozens of types of grass, reed, and sedge are found here, as well.

The Sundarbans is perhaps most renowned for its population of Bengal tigers (*Panthera tigris*), thought to be the only tiger population in the world adapted to living within mangrove forests. The presence of large fauna and increasing human population densities in the Sundarbans mean that human-tiger conflicts are an important social issue. An average of 57 people per year were killed by tiger attacks between 1975 and 1984. One step to help both the tiger and the people has been the establishment of the Sundarbans National Park and Tiger Reserve, a major conservation area that comprises some 1,000 square miles (2,585 square kilometers) in the Indian portion of the biome.

Another mammal that benefits from the sanctuary of the reserve is the muntjac or barking deer (*Muntiacus muntjak*). Reptiles sheltering here include the northern river terrapin (*Batagur baska*), olive ridley sea turtle (*Lepidochelys olivacea*), water monitor (*Varanus salvator*), saltwater crocodile (*Crocodylus porosus*), and king cobra (*Ophiophagus hannah*). Amphibians found either here or in the Bangladesh portion of the Sundarbans include the tree frog (*Polypedates maculatus*), skipper frog (*Euphlyctis cyanophylctis*), and green frog (*E. hexadactylus*). The Sundarbans wetlands is rife with such crustaceans as fiddler crab (*Uca* spp.), shrimp (*Penaeus monodon, Metapenaeus monoceros*), and mangrove mud crab (*Scylla serrata*). Hundreds of spider species have also been identified here.

Birds flock here by the millions. Nine species of kingfisher, multiple types of sand piper, many varieties of heron, stork, tern, curlew, and at least 20 raptor species, including the sea eagle, rank among the most abundant.

Human Impacts

Tens of millions of people live in and not far upstream from the Sundarbans wetlands. Among the leading activities that most impinge upon the biota here are fishing, aquaculture, logging, and agriculture. Despite great exposure to the life-threatening dangers of major cyclones, many people live and work here—sometimes by their actions undermining the very protections against storm surge and flooding that the wetlands and their mangrove forests provide. It is far from a static situation: Cyclone Sidr in 2007 detrimentally affected 30 percent of the Sundarbans, while Cyclone Aila in 2009 destroyed up to 15 percent of the mangrove forest.

Sea-level rise is expected to significantly threaten the flora and fauna of the Sundarbans during this century. Most of the Sundarbans is less than 3 feet (1 meter) above sea level; a rise of 11 inches (280 millimeters)—a conservative estimate of sea-level rise by 2100—is predicted to result in nonviable tiger populations in the Bangladeshi portion of the Sundarbans, as habitat is drowned and fragments of the small remaining mangrove patches will be unable to support breeding pairs. Sea-level rise will also affect coastal human populations. By 2020, it is estimated that the Sundarbans will lose 15 percent of its habitable area.

Conservation Efforts

The ecological and socioeconomic functions of the Sundarbans, and the significant threats this ecosystem faces, mean that protecting this ecosystem is important. As of 2007, about one-fourth of the biome was designated under some form of protection, such as a wildlife sanctuary or national park.

The Sundarbans as a whole is listed as a United Nations Educational, Scientific and Cultural Organization (UNESCO) World Natural Heritage Site. Remote sensing via satellite shows that, while the various forms of protective status has held down overall habitat loss in the Sundarbans over the past quarter-century, there has still been significant land-cover conversion, such as from mangrove to open water and barren land and mud flats.

Stepped-up protection measures include augmented patrols and enforcement, more strict permit requirements for the extraction of various products, and restricted extraction at identified key breeding grounds of certain animals. An Integrated Resource Management Plan (IRMP) was prepared in 1998 to sustainably manage the extraction of timber and nontimber products here. However, there is uncertainty as to whether the IRMP has been successful.

Daniel A. Friess
Lee Wei Kit

Further Reading

Banerjee, Anuradha. *Environment, Population and Human Settlements of Sundarban Delta.* New Delhi, India: Ashok Kumar Mittal, 1998.

Giri, Chandra, Bruce Pengra, Zhiliang Zhu, Ashbindu Singh, and Larry Tieszen. "Monitoring Mangrove Forest Dynamics of the Sundarbans in Bangladesh and India Using Multi-Temporal Satellite Data From 1973 to 2000." *Estuarine, Coastal and Shelf Science* 73 (2007).

Iftekhar, M. S. and P. Saenger. "Vegetation Dynamics in the Bangladesh Sundarbans Mangroves: A Review of Forest Inventories." *Wetland Ecology and Management* 16 (2008).

Loucks, Colby, Shannon Barber-Meyer, M. D. A. Hossain, Adam Barlow, and Ruhul M. Chowdhury. "Sea-Level Rise and Tigers: Predicted Impacts to Bangladesh's Sundarban Mangroves." *Climatic Change* 98 (2010).

Superior, Lake

Category: Inland Aquatic Biomes.
Geographic Location: North America.
Summary: The largest freshwater lake on Earth by surface area and the third-largest by volume, this is an ecosystem altered by overfishing, invasive species, and industrial activities.

Lake Superior exceeds all other lakes in the world by surface area, has the third-greatest volume of water for a freshwater lake, and contains the deepest point in the United States. While its formation during the last glaciation was geologically recent, on its shores lie some of the most ancient rocks on Earth. An inland ecological system that has been known by many names and has seen human activities come and go, Lake Superior has undergone troubling changes in water quality and aquatic species.

Geography, Hydrology, Climate

With a surface area of 31,820 square miles (82,410 square kilometers), Lake Superior is the largest expanse of freshwater in the world. Containing 2,900 cubic miles (12,100 cubic kilometers) of water, it is the third-largest lake by volume, and contains more water than all of the other Great

Lakes combined. In surface size, it spreads 350 miles (563 kilometers) in length and 160 miles (257 kilometers) in width, and is bounded by 2,726 miles (4,387 kilometers) of shoreline in Minnesota, Wisconsin, and Michigan in the United States, as well as the Canadian province of Ontario along most of its northern waters.

So large is this lake that the summer sun sets about 35 minutes later on its western shore than its eastern shore. Its maximum depth of 733 feet (223 meters) below sea level represents the deepest point of the contiguous continental United States. Its surface averages 600 feet (183 meters) above sea level.

Like all of the Great Lakes, Lake Superior formed when massive glaciers hollowed out its basin during the Laurentian, or Wisconsin, glaciation 12,000–7,000 years ago. When the climate warmed and the glaciers receded, their meltwater filled the basins. As the westernmost lake in the chain, Superior accepts water from more than 200 rivers. Water leaves the lake via the St. Mary's River to Lake Huron.

Rock on the northern shore of Lake Superior is some of the oldest found on Earth. In particular, the north shore granites formed from magma in the Precambrian Period (between 4.5 billion and 540 million years ago).

Due to its large size, Lake Superior affects regional weather in several ways. Its waters hold the heat of summer and the cold of winter longer than the surrounding land, moderating temperatures as seasons change. Overall, the climate around Lake Superior is more maritime than inland temperate, and resembles Nova Scotia more than Ontario or the surrounding states.

The most pronounced impact is lake effect snow, when westerly winds pick up moisture as they blow over the lake; upon reaching the colder eastern shore, heavy snow falls. Lake effect snows extend 20 to 30 miles inland, primarily on the Ontario shore southeast of the town of Marathon, and from the transboundary city of Sault Ste. Marie to the Wisconsin-Michigan border, that is, nearly the entire south shore. Average annual snowfall in Michigan's Keweenaw Peninsula exceeds 200 inches in places.

The average annual water temperature is about 40 degrees F (4 C). The lake seldom freezes over completely in winter; the last time was recorded in 1979. Recent metrics indicate several changes in the lake. Summer surface water temperature has increased by 4.5 degrees F (2.5 degrees C) since 1979, whereas summer temperature on the surrounding land has increased 2.7 degrees F (1.5 degrees C) during the same period. While the lake water seems to be building up heat faster than the land, it may also be evaporating faster. In the summer of 2007, monthly historic surface-level lows were set: August at 0.67 feet (0.20 meters) below norm, and September at 0.58 feet (0.18 meters) below the norm, which is 601 feet (183 meters).

These data are part of the evidence that global warming impacts are affecting the Lake Superior biome in ways that could—especially as they accelerate—severely threaten habitat along shorelines and associated wetlands, in the surface water layer, and in the deepwater zones.

Biota

Lake Superior is an oligotrophic lake, meaning that it has a low nutrient budget because the surrounding soils release few nutrients. Its deep, cold waters are less productive than those of the other Great Lakes; still, it is home to a variety of species of fish and other organisms. The food web is based on the primary producers, the phytoplantkton, here mainly green algae, blue-green algae, diatoms, and flagellates. These are consumed by a variety of filter-feeding animals, including zooplankton, shrimp, snails, and clams, which in turn are consumed by fish.

Over 80 species of fish have been recorded in the lake. These include key native species such as muskellunge (*Esox masquinongy*), pumpkinseed (*Lepomis gibbosus*), bloater (*Coregonus hoyi*), northern pike (*Esox lucius*), lake sturgeon (*Acipenser fulvescens*), longnose sucker (*Catostomus catostomus*), brook trout (*Salvelinus fontinalis*), and burbot (*Lota lota*).

Introduced fish in Lake Superior include freshwater drum (*Aplodinotus grunniens*), rainbow trout (*Oncorhynchus mykiss*), Atlantic salmon (*Salmo salar*), rainbow trout (*Oncorhynchus *mykiss*), and brown trout (*Salmo trutta*)—as well as some particularly damaging invasive species, such as sea lamprey (*Petromyzon marinus*), round goby (*Neogobius melanostomu*), white perch (*Morone americana*), ruffe (*Gymnocephalus cernuus*), and various Asian carp species, such as the common carp, *Cyprinus carpio*.

This rich fish community forms the basis of a historically important commercial fishery, based especially on lake whitefish (*Coregonus clupeaformis*), yellow perch (*Perca flavescens*), rainbow smelt (*Osmerus mordax*), and cisco or lake herring (*Coregonus artedi*). The first records, from 1879, of commercial lake whitefish indicate 2,356 thousand pounds (1,069 kilograms) of production. Catches have fluctuated over the years, but remain high, with 2,481 thousand pounds (1,125 kilograms) reported from 2006, for example. In 1941, commercial catches of lake herring peaked at almost 41,000 pounds (18,597 kilograms). Production declined over the next 20 years, reaching a low of 356 pounds (164 kilograms) in 2006.

Fish populations have declined due to overfishing and parasitism by the nonnative sea lamprey, which attaches its mouth to other fishes and scrapes away the flesh with a sharp tongue and teeth. Fish victims often die from fluid loss and infection after several attacks.

Threatened plants in the Lake Superior biome include pitcher's thistle (*Cirsium pitcheri*), which grows in the sandy beach and dune habitat at various points along the shore; Fassett's locoweed (*Oxytropis campestris* var. *chartacea*), found between seasonal water-level highs and lows; American hart's-tongue fern (*Asplenium scolopendrium* var. *americana*), a fern that prefers to grow upon stone surfaces; dwarf lake iris (*Iris lacustris*), a denizen of gravel, moist sand, and tree-shaded areas; and Houghton's goldenrod (*Solidago hoghtonii*), which prefers moist environs such as limestone crevices and the swales between dunes.

*A sea lamprey (*Petromyzon marinus*). This invasive parasitic species frequently kills native fish in Lake Superior. (Thinkstock)*

Human Interaction

Unlike the other Great Lakes, Lake Superior does not have large cities or a dense population along its shore. The world's farthest-inland port, Duluth (population 86,265 in 2010), sits on the far western point of the lake. Thunder Bay, Ontario, is the largest city on Lake Superior, with a population of approximately 125,000.

The Lake Superior region is rich in minerals and wildlife, and many of today's towns and cities grew from mining settlements or fur trading posts. Iron, copper, silver, gold, nickel, and uranium are all mined in the area and transported via the lake to global markets. Iron, in particular, has been a major enterprise here. Between 1875 and the 1950s, iron mining dominated industry on much of the north shore of Lake Superior, and contended with grain shipments for top commodity shipped on its waters. By 1896, one harbor was shipping more than 2.2 million tons (2 million metric tons) of iron annually. This led to the expansion of the fleet of lake freighters that ply Great Lakes waters, the St. Lawrence River, and thus the Atlantic Ocean. Although mining techniques and locations have changed, it still comprises a significant portion of the industry and shipping traffic in the Lake Superior region—and contributes to pollution.

The French, and later other Europeans, entered the fur trade here in the 17th century. The eventual depletion of widely abundant populations of furbearing animals caused many settlers to leave the shores of Lake Superior; some companies re-established farther west. Lake Superior today attracts significant tourism, often taking the form of hunting and fishing participants. Isle Royale, Minnesota, for example, which itself contains several lakes, boasts moose and wolf populations that bespeak its relatively intact habitat. Isle Royale is a U.S. National Park; Madeline Island and the Apostle Islands in Wisconsin; and Grand Island, a National Recreation Area in Michigan, are other favorite destinations of boaters.

To combat pollution, the United States and Canada in 1991 formed the Binational Program to Restore and Protect the Lake Superior Basin, a body committed to administering Lake Superior as a demonstration area to prohibit any point source discharge of any toxic, persistent substance. Zero discharge procedures and compliance are monitored by three groups: the Lake Superior Task Force, made up of senior managers from government and environmental bodies; the Lake Superior Work Group, providing technical direction and policy guidance; and the Lake Superior Binational Forum, where representatives from the public and industry regularly air the issues and examine solutions.

The Binational Program has since taken up related issues that affect the biome, such as a comprehensive attack on the deleterious impacts of invasive plant and animal species in Lake Superior. The group has initiatives for monitoring and prevention, ship water ballast testing, mitigation and restoration measures for shoreline habitats, and outreach to such entities as permitting agencies, fishing outfitters and marinas, nurseries, and garden centers.

Susan M. Moegenburg

Further Reading

Ostlie, Wayne. *Great Lakes Ecological Assessment—Threatened and Endangered Plants.* Minneapolis, MN: The Nature Conservancy, 1990.

Risjord, Norman K. *Shining Big-Sea Waters: The Story of Lake Superior.* St. Paul: Minnesota Historical Society Press, 2008.

Thomas, Amy, Sue Greenwood, Roger Eberhardt, Elizabeth LaPlante, Nancy Stadler-Salt, et al. *Lake Superior Aquatic Invasive Species Complete Prevention Plan.* Chicago: Lake Superior Binational Program, 2009.

Swan River (Australia) Estuary

Category: Marine and Oceanic Biomes.
Geographic Location: Western Australia.

Summary: The site of Perth, the capital and largest city of Western Australia, this estuary is the focus of considerable ecological concern because of the cumulative impacts of European settlement.

The Swan River Estuary drains a 49,000-square-mile (126,000-square-kilometer) catchment. Its major tributaries, out of a total of 29, are the Avon River to the north, the Canning to the south, and the Helena to the east. The Avon River contributes the majority of the freshwater flow. The estuary biome is mainly of mediterranean climate, with hot, dry summers and mild, moist winters.

Soils of the Swan coastal plain are generally sandy and low in fertility, with the exception of pockets of more fertile soils around the wetlands that dot the plain and along the river banks. These areas became the focus of early European settlement. The plain has since lost approximately 80 percent of its wetlands to development; the estuary is an urban wetland.

Flora and Fauna

The original vegetation that once fringed the Swan River's waters was unique and diverse, and included samphire flats (*Milyu* is the Aboriginal name for samphire), *Juncus kraussii* sedgelands, forests of paperbark (*Melaleuca rhaphiophylla*) and flooded gum (*Eucalyptus rudis*), and stands of swamp oak (*Casuarina obesa*). It provided habitat for many species of waterbirds and land birds, including migratory wading birds. This vegetation also protected the estuary margins, reducing erosion and filtering nutrients and pollutants flowing into the river.

These areas now encompass mudflats, seagrass beds, and intertidal vegetation. Although little of it remains, natural vegetation along the margins of the Swan River Estuary biome provides many different habitats for a host of animals. Plant species vary significantly with soil-types and other physical considerations. Vegetation in the preserved area includes saltmarshes, samphire flats, sedgebanks, and areas of woodland and shrubland.

Principal among upper-story species is *Eucalyptus rudis*; while *Melaleuca rhaphiophylla* and

Melaleuca cuticularis with *Casuarina obesa* form a middle-story, and *Juncus kraussii* is the main species of sedge. The Swan Estuary suffers extensively from weed infestations, such as black flag (*Ferraria crispa*) in Milyu; Brazilian pepper tree (*Schinus terebinthifolius*), bulrush (*Typha orientalis*), and kikuyu grass (*Pennisetum clandestinum*) in Alfred Cove and Victorian tea tree (*Leptospermum laevigatum*) in Pelican Point.

Perhaps the most important wildlife in the estuary are the birds, the most significant of these being the migratory wading birds. Among these are the tiny red-necked stint, pelican, swan, ibis, egret, red-capped robin, white fronted chat, pallid cuckoo, sacred kingfisher, and osprey. There is a small population of variegated fairy wren. There are 33 wading bird species that have been identified here. The migratory species come from as far as Mongolia and Arctic Siberia.

The fauna in Swan River Estuary has not yet been extensively surveyed. The mammals here are mainly small; southern brown bandicoot (*Isoodon obesulus*) and common brushtail possum (*Trichosurus vulpecula*) may still be present, but fox, rabbit, and feral cat populations seem to have contributed to the demise of these species.

River dolphin use the adjacent waters, and a variety of fish species—including the yellow-eye mullet (*Aldrichella forsteri*), Perth herring (*Nematalosa vlaminghi*), and cobbler (*Cnidoglanis macrocephalis*)—and prawns use the shallows as both a juvenile nursery and feeding ground. Other aquatic fauna include mollusks, jellyfish, polychaetes (marine annelid worms), and crustaceans.

Human Impact

The Swan River Estuary is part of the customary lands of the Nyungah people, which provided them with both physical and spiritual sustenance. While the traditional inhabitants actively managed the river and its catchment, they did not make substantial physical changes to its hydrology or morphology. This ecoregion became the focus of early European settlement beginning in the early 1600s; initial growth was slow, but the importation of convict labour in the 1850s saw the construction of bridges, drainage and land reclamation, and changes to streambank morphology. Of particular note was the commencement of reclamation of marsh and intertidal land directly in front of the city of Perth.

The next major leap came with the gold boom, which led to two major hydrological alterations in the estuary. The first was the creation of Fremantle Harbor. The estuary had been blocked by a bar at the mouth, and there were extensive mudflats further upstream. The bar was removed in 1896; this also had the effect of increasing salinity in the lower reaches. The second was the construction of Mundaring Weir and the associated water pipeline to Kalgoorlie, which exported quantities of the Swan-Canning catchment's water to the desert.

The advent of the railway and automobile decreased the significance of the river for transport, while at the same time increasing its popularity for recreation, including fishing. The river's banks were also used for industry and for waste disposal. For example, the East Perth power station drew cooling water for its operations, and the Attadale foreshore area became a filled-in rubbish dump. A decline in bird numbers was noted in the late 1890s. At the same time, recreational fishing began to be regulated—but it would take until the mid-1950s for an understanding of the ecological role of river shallows to emerge.

Community concern over the health of the estuary has been steadily growing, leading to the proclamation of the Swan River Conservation Act in 1958, and the establishment of the Swan River Conservation Board, which later became the Swan River Trust. The formation of an Environmental Protection Agency in the United States in 1970 corresponded with the establishment of similar organizations in Australia. In the 1980s, the river's indigenous significance hit the headlines with an ongoing dispute over the proposed redevelopment of the Swan Brewery site along Mount's Bay Road.

During the 1990s, algal blooms began to become more noticeable in the estuary. In January 2000, a severe bloom forced the closure of the river to all fishing and recreational uses for 12 days. Public concern lead the government to establish the Western Australia Estuarine Research Foundation in 1994 to set baseline data on water quality and on different species for management and decision-making. The Swan Estuary Marine Park, established in 1999, protects three biologically important areas of Perth's Swan River: Alfred Cove, 494 acres (200 hectares) adjacent to Melville; Pelican Point, a 111-acre (45-hectare) area in Crawley; and Milyu, 235 acres (95 hectares) adjacent to the Como foreshore in south Perth.

Western Australia's marine areas are globally significant; its coastal waters are considered to be among the world's least disturbed. Climate change predictions suggest that all of southwestern Western Australia, however, will become hotter and drier, which has substantial implications for Perth water supply, and for the ecological health of the estuary. Imbalances in this biome could readily impact the offshore ecology, as well as spread inland.

Kylie Carman-Brown

Further Reading

Department of Conservation and Land Management. *Management Plan: Swan Estuary Marine Park and Adjacent Nature Reserves: 1999–2009.* Perth, Australia: Marine Parks and Reserves Authority and National Parks and Nature Conservation Authority, 1999.

Graham-Taylor, Sue, "A Missing History: Towards a River History of the Swan." *Studies in Western Australian History* 27 (2011).

Hamilton, David P. and Jeffrey V. Turner. "Integrating Research and Management for an Urban Estuarine System: The Swan-Canning Estuary, Western Australia." *Hydrological Processes* 15 (2001).

Taylor, William. "Rivers Too Cross: River Beautification and Settlement in Western Australia." *National Identities* 5, no. 1 (2003).

Syrian Desert

Category: Desert Biomes.
Geographic Location: Africa.

Summary: Traversing parts of four countries, the Syrian Desert is an ancient land where only a few very hardy plant and animal species survive—and they are threatened by development and climate change.

Also called the Syro-Arabian Desert, the Syrian Desert is a dry wasteland of steppe and true desert situated in southwestern Asia between the coast of the eastern Mediterranean Sea and the Orontes Valley of the Euphrates River. The Syrian Desert accounts for roughly one-third of the area of Syria. Covering a total area of some 193,000 square miles (more than 500,000 square kilometers), this desert also encompasses the Nafud Desert of northern Saudi Arabia, and parts of eastern Jordan, southeastern Syria, and western Iraq.

Most of the land, which receives less than 5 inches (13 centimeters) of rain annually on average, is covered with basalt from the lava flows deposited there by the Jabal al-Druze raised volcanic field in southern Syria. These lava flows isolated the Syrian Desert from the populated areas of the Levant and Mesopotamia for much of its recorded history; they are often a barrier to life getting a foothold here. The northern area of the desert is characterized by dry river channels, known as wadis, that encompass areas ranging from 93 to 186 square miles (150 to 300 square kilometers). Physical conditions in the Syrian Desert can be harsh, with temperatures frequently rising to more than 100 degrees F (38 degrees C) in the day, and becoming quite cool during autumn and winter nights.

Biodiversity

Drought-resistant plants such as the date palm and the wild olive can survive the harsh climate of the desert. In areas where there is some moisture, herbaceous and dwarf shrub sage brush (*Artemisia herba-alba*) communities tend to dominate in deeper, non-saline soils, and often occur in association with such grasses as *Poa bulbosa* where disturbed by grazing. Tamarisk, Euphrates poplar, and reeds are found along or near the Euphrates River.

Animals endemic (found nowhere else) to the Syrian Desert include several species of gazelle, oryx, jerboa, viper, and chameleon, as well as the Syrian hamster. Large mammals can be found in areas where human settlements are less dense.

Predators such as wolves, Ruppell's sand fox, caracals, and wildcats are occasionally encountered. Badgers (*Meles meles*) can be found in more vegetated areas, and wild boar (*Sus scrofa*) can be found in reed thickets and semi-desert terrain. Loss of habitat has led to animals such as the striped hyena, honey badger, and jackal becoming extirpated in the Syrian Desert biome.

There is a remarkable variety of bird species found in this desert region, especially in the river valley habitats. Included are the houbara bustard, great bustard, and little bustard; the lesser kestrel; the regionally threatened Eurasian griffon vulture; and the lanner falcon. The Euphrates Valley is a major migration route for waterbirds, providing a narrow corridor between the wetlands of southern and central Turkey and the vast wetlands of Mesopotamia in central and southern Iraq. Greater flamingo, the pygmy cormorant, and marbled teal are known to breed here.

The valley is a very important migration route for terrestrial birds such as the turtle dove; the pin-tailed sandgrouse is common on the adjacent plains and visits the river in large numbers to drink. Also of note, the northern bald ibis was considered extinct in Syria for more than 70 years, but was rediscovered on a remote cliff of the Syrian desert in April 2002.

A herder watches over his flock at the side of a highway in the Syrian Desert. The desert covers one-third of the entire country of Syria. (World Bank)

Human Impact

Within the steppes of the Syrian Desert, degradation and desertification have broadened, threatening endemic animal and plant life, and exacerbating the poverty of the people who live there. The southern section of the Syrian Desert is home to Bedouin nomads who are known for the Arabian horses they breed. In more prosperous times, the Bedouins served as guides for travelers and tourists, but now their way of life is threatened. Many have been forced, by economic and political factors, into permanent settlements where they live in black-goat tents and eke out a living selling trinkets to the few tourists who pass by.

The few animals that can be bred in the desert include camels, cattle, goats, and sheep hearty enough to survive the harsh climate. In ancient times, scattered oases were used as staging posts, but in modern times, a major highway runs from Damascus to Baghdad, and a modern railway system facilitates travel throughout the area. The largest oases are in highly populated areas of Damascus and Palmyra (Tadmor). During the gas shortage of the 1970s, concentrated efforts were made to find oil in the area. After those efforts proved successful, oil companies moved in and began laying pipelines that now criss-cross some northern stretches of the desert, destroying ecosystems and increasing possibilities of major oil spills. Other environmental problems include overgrazing, desertification, mining of phosphates, hunting of wildlife, and damage done by off-road vehicles.

Climate change will have an impact on this ecoregion, which includes both key urban centers, wide areas of grain farms, and rather large portions that are sparsely inhabited by humans. Recent years of drought across much of the Syrian Desert may be tied to global warming; it has certainly strained the lives of farmers in the region. Plants and wildlife are affected in some areas where the groundwater availability has dwindled, first due to irrigation schemes that diverted the scant flows away, and second due to the settling drought. The biota that has evolved to hug riverine habitats is also under stress, as higher temperatures and lower moisture has pushed their survivability traits to their limits.

Elizabeth Rholetter Purdy

Further Reading

Boland, Mary. "Bedouin's Way of Life Dying among the Ruins of Palmyra." *Irish Times,* October 18, 2010.

Moffet, Barbara. "New Hope for a Rare Bird in the Syrian Desert." March 18, 2011. http://newswatch.nationalgeographic.com/2011/03/18/northern_bald_ibis_syria.

Talhouk, Salma N. and Maya Abboud. "Impact of Climate Change: Vulnerability and Adaptation—Ecosystems and Biodiversity." In *Arab Environment—Impact of Climate Change on the Arab Countries,* edited by Mostafa K. Tolba and Najib W. Saab. Beirut, Lebanon: Arab Forum for Environment and Development (AFED), 2009.

United Nations Environment Programme. "Cross-Desert Rivers of Non-Desert Sources." http://www.unep.org/geo/gdoutlook/047.asp.

Tahoe, Lake

Category: Freshwater Lake Biomes.
Geographic Location: North America.
Summary: Strikingly beautiful and a moist habitat oasis amid the high and dry peaks of the Sierra Nevada Mountains, Lake Tahoe must contend with overdevelopment by humans and an altered forestscape.

The Lake Tahoe basin, encompassing over 500 square miles (804 square kilometers) astride the California-Nevada border high in the Sierra Nevada Mountains, fills a geological trough, or graben, created by faults from ancient tectonic movements, later reshaped by glaciation. Perched at an altitude over 6,000 feet (1,828 meters), this freshwater lake is the 11th deepest on Earth, reaching a depth of 1,645 feet (501 meters). The 192-square-mile (309-square-kilometer) lake occupies montane and sub-alpine zones, and stretches in a rough oval 12 miles wide (19 kilometers) wide and 22 miles (35 kilometers) long.

Winter storms feeding the lake's copious water supply also limit the growing season to less than 120 days, depending on the alternating El Nino-La Nina cycles that bring moisture off the Pacific Ocean mainly in the form of snowfall. Annual precipitation averages range from 30 inches (78 centimeters) falling as snow on the dry eastern slopes to over 70 inches (178 centimeters) on the western slopes.

In the midst of a desert, the lake's water is a reliable reservoir, and its encircling forests an inland refuge of biological diversity. The mountain range's extent and elevation create a rain shadow effect, giving rise to distinctly different vegetation communities along lines of longitude as well as changes in altitude.

Biodiversity

Sugar pine, once abundant throughout the basin, was extensively logged in the 19th century. The granite-derived, well-drained soils today support Ponderosa pine, lodge pole pine, and incense cedar; while Jeffery pine and piñon pine, and fire-dependent broadleaf-sclerophyll forests, or chaparral, is widely found in the rain shadow along the more arid east shore and south-facing slopes. Great Basin shrub species are also found in the drier terrains below the snowline.

Once-extensive wetlands, especially along the south shore and adjacent to some bays, still provide favorable conditions for sedges, cattails, and wildflowers that filter the runoff and nourish whitefish, trout, and birds, in addition to willows and black cottonwood trees. Thick stands of pine trees lined the lake shores in the 1850s, prior to being clear-cut.

Because of current human-monitored fire suppression, forests in the Lake Tahoe biome commonly contain abundant stands of mountain hemlock, Sierra juniper, quaking aspen and big-leaf maple. At higher altitudes, the red fir and white fir forests dominate below the treeline, above which alpine wildflowers and lichens predominate in the often frigid reaches of passes and old glacial tarns.

Grizzly bears, wolverines, and peregrine falcons, once residents of the region, are extirpated, although the Forest Service currently lists 275 other species common to the basin. Important to this assemblage of wildlife and fisheries was the role of beavers—before the fur trade—in sequestering water in ponds. Beaver dams had the effect of regularizing stream flow from snowmelt through the dry season; and as a keystone species, the beavers maintained ponds, attracting many other plants and animals dependent on the higher water levels.

In areas where seasonal fires rejuvenated fire-dependent species, the black-backed woodpecker, indicative of post-burn timbered lands, thrived in greater numbers than today. As land-use practices shifted due to a striking rise in human population, wildlife retreated. Introduced species have reduced the numbers and range of natives. For example, the Sierra Nevada yellow-legged frog is being restored to lakes in the Desolation Wilderness by removing introduced brook trout and rainbow trout from those stocked waters. The Desolation Wilderness is among the oldest reserved lands, since 1899 extending protection to 63,960 acres (25,883 hectares) of sub-alpine and alpine forest that grew since the initial logging after the California Gold Rush of 1849.

Human Impact

Railroad building and the discovery of silver in the Comstock lode of 1859 led to the loss of virtually all forested landscape on the Nevada, or eastern, side of the lake—and provoked the California Assembly to pass legislation in 1883 to appoint a forestry commission and sparked a movement for the protection of ample water to promote and sustain agricultural land uses in the foothills and valleys of the lower slopes.

A 1976 study of the land-use cover of different types within the basin distinguished five vegetation associations that were discernibly reduced after 1900. Findings revealed that three-quarters of the marshes, half of the mountain meadows, over one-third of the streamside-riparian zones, 15 percent of forests, and 5 percent of shrub lands here had been replaced by suburban expansion.

More recent studies reveal that since World War II, growth had incorporated a 2,000 percent increase in developed land in the southern lake basin. This region had been the more extensive and geographically diverse lakeshore landscape.

Although human influence on the landscape ecology, air quality, and water resources of the Lake Tahoe biome date to the silver mining rush of 1859, the dominant and persistent role of human-induced changes in the watershed changed markedly after 1945. These impacts peaked in the 1940–87 period, in terms of the most extensively affected terrains; this was due to an abrupt rise in the amount of impervious paved surface areas that accompanied residential, commercial, and municipal building.

Nearly 50,000 permanent residents crowd the steep slopes lining south and north shores of the lake, and largely account for the degradation of the area's waters, forests, wetlands, fisheries and wildlife. The fragmentation of biological communities associated with landscape development has disproportionately affected keystone species, and thus diminished the biological diversity of the entire basin. That is because some identifiably important vegetation associations, such as meadows and wetlands, sustain a greater variety and density of wildlife when intact.

Before 1900, a more diverse array of fire-tolerant forests and dependent meadows sustained cattle and sheep here, but shepherding successively destroyed native grass cover and by 1899 there was extensive deterioration of pasturelands. With the failure of an effort to protect the lake as a National Park, in 1917 the State of California banned further

The human population in the area around Lake Tahoe increased by almost 50,000 in the second half of the 20th century. The resulting suburban expansion contributed to the loss of an estimated three-quarters of the area marshes, over one-third of the streamside-riparian zones, 15 percent of the forests, and 5 percent of shrub lands by 1976. (Thinkstock)

fishing in Tahoe, so depleted were the native and introduced stocks from fish hatcheries.

After the construction of the transcontinental railroad over the Sierra summit to Nevada in 1869, tourism began to become a viable industry in addition to timber, mining, and fisheries.

The regeneration of the basin's forests was shaped through management, and much of the policy after 1940 centered on strategies aimed at reducing perceived threats from the natural fire cycle. Fire suppression had the ironic counter-influence of both a loss in tree diversity and increased densities—so that when periodic drought cycles transpired, the loss of timber could be severe, as happened in the late 1980s and 1990s. Like so many drastically altered landscapes and water bodies, these extant natural features comprising the Tahoe region are actually artifacts of social desires and economic choices made long ago, but whose consequences currently linger; the past is persisting and ever intruding into all effective means and each collaborative step to restore the lake water's clarity, scenic beauty, and allure of the mountain air.

In addition to locally and regionally generated direct and indirect pollution effects, climate change impacts have been noted in Lake Tahoe, specifically a noticeable long-term warming trend. Having coped with extractive industries, population growth, unhealthy levels of pollution, and 200,000 tourists annually, and accommodating suburban sprawl beside an enigmatic wilderness, the regional planning safeguards created collaboratively by interests in this lake ecosystem's development now contend with an appreciably murkier and warmer lake.

The terrain now harbors a measurably hotter climate with which its native and introduced species will have to cope. A cradle of biotic health and beauty amidst the granite monotony of a water-scarce Sierra Nevada Mountain Range, Lake Tahoe remains an alpine recreational magnet in an inevitably warming, forested watershed.

Joseph V. Siry

Further Reading

Goin, Peter. *Stopping Time: A Rephotographic Survey of Lake Tahoe.* Albuquerque: University of New Mexico Press, 1992.

Raumann, C.G. and M.E. Cablk. "Change in the Forested and Developed Landscape of the Lake Tahoe Basin, California and Nevada, USA, 1940–2007." *Forest Ecology and Management* 255 (2008).

Strong, Douglas H. *Tahoe: From Timber Barons to Ecologists.* Lincoln: University of Nebraska Press, 1999.

Takla Makan Desert

Category: Desert Biomes.
Geographic Location: Asia.
Summary: Takla Makan in northwestern China is the world's second-largest shifting-sand desert. Despite its dry surface, the desert holds precious resources such as groundwater and oil.

The Takla Makan Desert is located in the Xinjiang Uyghur Autonomous Region, in the northwestern area of the People's Republic of China. Classified as a hyperarid region, it includes xeric (drought-tolerant) shrublands at its edges.

The Tarim Basin is bordered by the Tian Shan Mountains to the north, the Kunlun and Altun Mountains to the south, and the Pamirs to the west. In the center of the Tarim Basin extends the Takla Makan Desert, covering an area of about 129,344 square miles (335,000 square kilometers). While elevations of 3,937 feet (1,200 meters) and even 4,921 feet (1,500 meters) above sea level can be reached, the lowest point is 505 feet (154 meters) below sea level. Based on the classification protocols in use by China, the region is demarked as desert steppe, shrubby and rocky desert with sand dunes, saline soil, and sparsely vegetated areas. As a part of Xinjiang, the Takla Makan is earthquake-prone.

Alluvial deposits from glacial watercourses provide the basic stratum, which is covered with a thick layer of sand. More than 100 mummies, some 4,000 years old, and Buddhist artifacts have been found in the region, proving early human habitation. Unusual, complex wind conditions in the basin form a variety of eolian, crescent-shaped sand dunes that make up 85 percent of the desert area. Larger sand-dune chains average 98–492 feet (30–150 meters) in height and span 787–1,640 feet (240–500 meters) in width, with a distance between the dune chains of 0.6–3 miles (1–5 kilometers). Some are 984 feet (300 meters) high, sculpted by erosive removal of sand, called denudational shapes. Depressions consist of gravel, sand, silt, or clay brought here by water, called depositional shapes. Saltmarsh and saltflats can also be found among the dunes.

Climate

Especially in the spring when the surface sand heats up, ascending air currents develop, and due to the wind-tunnel effect of the surrounding mountains, hurricane-force dust storms blow from the northeast. This is the season of the Kara Buran, the black sandstorm. Year round, constant winds keep the sand moving. Each year the dunes shift about 492 feet (150 meters), which endangers the continuing existence of oases here. The Takla Makan is considered to be China's warmest desert. Average temperatures are 77 degrees F (25 degrees C) in summer and 14 degrees F (minus 10 degrees C) in winter, with peaks of 115 degrees F (46 degrees C) and lows of minus 4 degrees F (minus 20 degrees C).

Located in an extreme inland position, the desert lacks any balancing influence from warm maritime air. Before air masses from the oceans reach central Asia, they lose their humidity, resulting in cold nights even during summertime, and extremely scant precipitation, below 1.5 inches (38 millimeters) per year. Some areas receive less than 0.5 inch (13 millimeters) of precipitation per year. The Tarim Basin is an internal drainage, or endorheic, basin. Perennial freshwater springs are fed by snowmelt from the Kunlun and Tian Shan mountains. Some rivers penetrate about 62–124 miles (100–200 kilometers) into the desert, gradually drying up in the sands. The Tarim River flows along the edge of the desert in an east-west direction.

Biodiversity

The Takla Makan Desert is mostly uninhabited; it is also almost entirely devoid of vegetation and fauna. Only a few, very specialized species live here, such as the Asiatic wild ass. In the Lop Nur Basin, east of the now-dry Lop Nur Lake, live the last 500 wild bactrian camels. Bactrian camels store water in their humps and can survive for months without drinking. The Arjin Shan Nature Reserve, at some 5,840 square miles (15,125 square kilometers), has been established in this area to conserve habitat for this species, but wild camels as well as Asiatic wild asses are still declining. Around oases, people cultivate fruit trees and other types of crops. Native plants include tamarisk, nitre bushes, and reeds. *Alhagi sparsifolia* is a leguminous perennial desert plant able to absorb atmospheric nitrogen. As a fodder plant, it provides protein to livestock in the desert region. Its deep, extensive root structure helps prevent sand erosion.

All plants that grow here tolerate storms, extreme temperatures, and salt. Turanga poplar, oleaster, camel thorn, members of the *Zygophyllaceae* (caltrop) family, and saltworts grow along the edges of the desert, near river valleys. Each species developed a unique survival strategy. In low basins, the shallow groundwater lies only 10–15 feet (3–5 meters) below the surface; some plants' root systems can tap this.

Human Activity and Environmental Threats

The desert's rich natural oil reserves have been exploited since the 1950s. The Chinese government has encouraged settlement in the region, founding the city of Korla. Many trees were planted to reduce the encroachment of the sand. Today, the desert is the headquarters of Tarim Oilfield Co., a unit of PetroChina.

To transport the oil, three desert roads were built from 1995 to 2007, including the world's longest desert road. The present population consists largely of Turkic Uyghur and recently settled Han people. Nonetheless, the Takla Makan ecoregion remains relatively intact. (The consequences of some nuclear tests in Lop Nor are not reported.)

An increasing population uses more water for crop irrigation than is available. Lakes and rivers have dried out through irrigation, threatening the existing oases. Furthermore, the yields drop after a few years. Fields turn into desert rather than reverting to grassland. Oversize herds overgraze the land, further intensifying soil depletion, and making natural regeneration virtually impossible. Climate change impacts may be felt in increasing sandstorms throughout the area, as well as other potential changes including warmer temperatures and even scarcer precipitation.

Manja Leyk

Further Reading

Arndt, Stephan, et al. "Contrasting Patterns of Leaf Solute Accumulation and Salt Adaptation in Four Phreatophytic Desert Plants in a Hyperarid Desert with Saline Groundwater." *Journal of Arid Environments* 10 (2004).

Eyre, S. Robert, ed. *World Vegetation Types.* New York: Columbia University Press, 1971.

Warner, Thomas. *Desert Meteorology*. Cambridge, UK: Cambridge University Press, 2004.

Talamancan Montane Forests

Category: Forest Biomes.

Geographic Location: Central America.

Summary: One of the largest undisturbed forest areas in Mesoamerica, and home to many endemic species, these forests are a vital crossroad for species and ecosystems between South and North America.

Located between Costa Rica and Panama, the Talamancan montane forests represent some of the most intact ecosystems in Mesoamerica. Occurring at 2,461–9,843 feet (750–3,000 meters), this biome includes the largest number of endemic (found nowhere else) species for both countries and one of the richest for the entire region. The area remains mostly undisturbed due to its effective protection under national and international categories, but mostly due to its isolation and inaccessibility.

The region is considered to be the main stepping stone for temperate species from the south and north in the isthmus, a refuge from the Pleistocene, and an independent biogeographic region. The isolation that maintains the area in a good conservation status also represents a unique opportunity for speciation, retaining an important proportion of endemic (found nowhere else) species.

The Talamancan Montane Forests biome is composed of a large intact and continuous forest patch from central Costa Rica south to western Panama. It is almost entirely covered by protected areas, including national parks, indigenous reserves, and an international park shared by the two countries. Due to its biological uniqueness, the area has been designated as a Biosphere Reserve and a Human Heritage Site, and is listed in several global prioritization categories, including Endemic Bird Areas, Key Bird Areas, Key Biodiversity Areas, and Centers of Plant Diversity.

The region includes the highest peaks in both Costa Rica and Panama, such as Cerro Chirripó at 12,467 feet (3,800 meters), Kamuk, and Fabrega. The ecoregion ranges up to 9,843 feet (3,000 meters), where vegetation, solar radiation, and extremely low temperatures create a unique habitat zone of grasslands and marshes known as páramos, a formation typical of the Andes but present in the northernmost range in the Talamancan Mountains, albeit without the characteristic South American *Espeletia* spp.

The typical climate is temperate, with high precipitation and two marked seasons, rainy and dry. However, even during the dry season, the region is very humid and retains high atmospheric moisture levels. Precipitation ranges from 98 to 256 inches (248 to 650 centimeters), occurring as rainfall and cloud drip. This last effect also creates a unique environment that produced important epiphyte richness in most of the elevation range. Above 9,843 feet (3,000 meters), temperature falls dramatically, so plant communities become limited to those species that are tolerant of frost.

Vegetation

Talamancan includes three main ecosystems: the seasonal rainforests on the Caribbean slopes, the seasonally dry but mostly evergreen forests on the Pacific slopes, and the perpetually dripping cloud forests on the higher elevations. The typical ecosystem found in the Talamancan forests is dominated by mixed stands of oak trees (*Quercus* spp.) and other vegetation formations. In the entire region, the ecosystems are dominated by *Quercus costaricensis* and *Q. copeyensis*, while *Magnolia, Drymis,* and *Weinmannia* are important elements. However, the northern portion includes areas dominated by species of the *Lauraceae* family.

In general, vegetation types occur in three elevational

*The brilliantly colored resplendent quetzal (*Pharomachrus mocinno*) is an iconic bird species of Costa Rica and the Talamancan Montane Forests. (Wikimedia/Matt MacGillivray)*

belts, creating unique species associations related to slope, temperature, and precipitation. These belts are the premontane belt at 1,969–4,921 feet (600–1,500 meters), the lower montane belt at 4,265–8,202 feet (1,300–2,500 meters), and the montane belt at 8,202–10,499 feet (2,500–3,200 meters), the latter forming a slow gradual transition to páramo.

Biodiversity

For most of its fauna and flora groups, Talamancan is one of the areas of highest richness and endemism in Central America. Its unique biogeographical features, isolation, and location make it an important area for speciation processes. As the main crossroad for southern and northern biota during all the important biodiversity exchanges between continents, the Talamancan Mountains are a unique mixture of elements from North and South America.

La Amistad International Park, shared by both countries, alone contains some 10,000 vascular and 4,000 nonvascular plant species, including several hundred endemic plant species. The range as a whole is estimated to retain nearly 90 percent of all Costa Rica's plant species.

More than half the avifauna of the highlands of Costa Rica and western Panama is endemic to the region. Endemism among amphibians is high; even in recent expeditions, new amphibian species, especially salamanders, have been described. Also, the area is considered to be the last refuge for various threatened amphibian species that disappeared from several other regions due to climate change and diseases, but have been rediscovered in this biome.

In terms of mammals, the area includes at least 16 species considered to be regional endemics, and 14 rodent, two shrew, one rabbit, one primate, and one bat species. The area also retains the highest number of threatened mammal species, globally and nationally, in Costa Rica. It is one of the only viable habitats for various larger mammal species such as jaguar, puma, tapir, and red brocket deer.

The region's endemic mammal species include Dice's cottontail (*Sylvilagus dicei*) and some pocket gophers (*Orthogeomys* spp.). Among the representative mammal fauna of Talamancan are the Baird's tapir (*Tapirus bairdii*), jaguar (*Panthera onca*), and puma (*Puma concolor*). The small spotted cat varieties (*Leopardus* spp.) are well-represented here, especially the oncilla (*L. tigrinus*), limited to the high portions and reaching its northern range limit in Talamanca. Amphibians and reptiles in Talamancan are often unique, and many species are endemic; salamanders are especially rich.

Bird diversity is high in this biome, which includes three Important Bird Areas for Costa Rica. Among the most representative are the magnificent resplendent quetzal (*Pharomachrus mocinno*), the vulnerable three-wattled bellbird (*Procnias tricarunculatus*), bare-necked umbrellabird (*Cephalopterus glabricollis*), the near-threatened black guan (*Chamaepetes unicolor*), and the harpy eagle (*Harpia harpyja*).

Environmental Threats

The region retains almost 75 percent of its original cover, but as in most tropical biomes, deforestation, hunting, and extractive activities such as coal production threaten the ecoregion. Most of the region is officially protected, requiring law enforcement and active management of protected areas. Climate change impacts on this sensitive area may be especially severe; cloud forest habitats are especially sensitive to changes in the balance of humidity, temperature, precipitation, and air pressure regimes.

José F. González-Maya
Jan Schipper
Annelie Hoepker

Further Reading

González-Maya, J. F., B. Finegan, J. Schipper, and F. Casanoves. *Densidad absoluta y conservación de jaguares en Talamancan, Costa Rica.* San José, Costa Rica: Nature Conservancy, 2008.

González-Maya, J. F., J. Schipper, and K. Rojas-Jimenez. "Elevational Distribution and Abundance of Baird's Tapir (*Tapirus bairdii*) at Different Protection Areas in Talamancan Region of Costa Rica." *Tapir Conservation* 18, no. 1 (2009).

Halffter, G. "Biogeography of the Montane Entomofauna of Mexico and Central America." *Annual Review of Entomology* 32 (1987).

Kappelle, M. *Los Bosques de Roble (Quercus) de la Cordillera de Talamancan, Costa Rica.* Costa Rica: University of Amsterdam/Instituto Nacional de Biodiversidad, 1996.

Pounds, J. A., M. P. Fogden, and J. H. Campbell. "Biological Response to Climate Change on a Tropical Mountain." *Nature* 398, no. 6728 (1999).

Reid, F. *Field Guide to the Mammals of Central America and Southeast Mexico.* Oxford, UK: Oxford University Press, 2009.

Stiles, F. G. and A. F. Skutch. *A Guide to the Birds of Costa Rica.* Ithaca, NY: Cornell University Press, 1989.

Tamaulipan Matorral

Category: Desert Biomes.
Geographic Location: North America.
Summary: On the border between the United States and Mexico, the Tamaulipan matorral is a diversity hot spot for cacti and succulents, but it has been considerably diminished by human activities.

The Tamaulipan Matorral ecoregion occurs at the lower elevations of the eastern slopes of the Sierra Madre Oriental mountain range. It is found in Mexico in the northeastern states of Nuevo León, Coahuila, and Tamaulipas, and in the United States in the southern part of Texas. This ecoregion is characterized by desert scrub vegetation; it is rich in thorny shrubs, cacti, and succulents. The Tamaulipan matorral is an important habitat for mammals and reptiles as well as resident and migratory birds. The area of intact Tamaulipan matorral is much reduced from its original distribution, and human activities are likely to result in further losses.

The Sierra Madre Oriental was formed by uplift, faulting, and erosion that began approximately 23 million years ago. The coastal plain along the Gulf of Mexico lies to the east of the range, and the Mexican plateau lies to the west of the Sierra Madre Oriental. The Sierra Madre Oriental extends 621 miles (1,000 kilometers) from the Mexican state of Coahuila south through Nuevo León, Tamaulipas, south to where it joins the volcanic belt that crosses central Mexico. Abrupt mountains are present in one part of the ecoregion, but most of the landscape is made up of undulating mountains, hills, valleys, and plateaus. Sedimentary rocks of marine origin characterize the ecoregion.

This ecoregion is located where the nearctic and neotropical ecozones meet. The Tamaulipan Matorral biome experiences desert-like conditions. Annual precipitation levels are below 39 inches (99 centimeters) per year, and the geology of the region further compounds these xeric (dry) conditions in that the small quantities of rain that do fall are quickly soaked up by the limestone substrate. The lowest temperatures are generally in December, and severe frosts can occur. In contrast to the climate of this ecoregion, the forests farther south along Sierra Madre Oriental on the coastal plain of Veracruz have a tropical, moist climate.

Vegetation

The vegetation of the Tamaulipan matorral is predominately made up of woody shrubs, small trees, cacti, and succulents that are adapted to withstand the ecoregion's xeric environment and temperature extremes. Plant species that are common in this ecoregion include cacti such as living rock cactus (*Ariocarpus* spp.), star cactus (*Astrophytum asterias*), Christmas cactus (*Cylindropuntia leptocaulis*), pincushion cactus (*Mammillaria* spp.), and paddle cactus (*Opuntia* spp.), as well as spiny shrubs including acacia, mountain mahogany (*Cercocarpus* spp.), leadtree (*Leucaena* spp.), catclaw mimosa (*Mimosa aculeaticarpa* var. *biuncifera*), and mesquite (*Prosopis* spp.).

Other xerophytic plants are *Salvia ballotaeflora*, leatherstem (*Jatropha dioica*), cenizo (*Leucophyllum texanum*), and Spanish dagger (*Yucca* spp.). The composition of the plant communities along the slopes of the Sierra Madre Oriental changes with elevation. Higher up the slopes, the matorral vegetation gives way to pine forests. A total 165 plant species have been reported for the ecoregion. This ecoregion is a center of diversity for cacti and succulents. Species endemic here (not found elsewhere) include *Agave victoria-reginae* and *Astrophytum caput-medusae.*

Fauna

The Tamaulipan Matorral biome is home to mammal species such as the Mexican prairie dog (*Cynomys mexicanus*), Saussure's shrew (*Sorex saussurei*), yellow-faced pocket gopher (*Pappogeomys castanops*), Allen's squirrel (*Sciurus alleni*), Mexican ground squirrel (*Spermophilus mexicanus*), collared peccary (*Pecari tajacu*), and coyote (*Canis latrans*).

The matorral is an important biological corridor, refuge, nesting, and feeding ground for birds. More than 170 avian species belonging to 42 families have been observed in this ecoregion, about half of which are residents; the other half consist of migratory species. Resident species include the burrow owl (*Athene cunicularia*), hooded oriole (*Icterus cucullates*), eastern meadowlark (*Sturnella magna*), long-billed thrasher (*Toxostoma longirostre*), hooded yellowthroat (*Geothlypis nelsoni*), blue bunting (*Cyanocompsa parellina*), and olive sparrow (*Arremonops rufivirgatus*).

Threats and Conservation

Land clearing for agriculture, cattle grazing, logging, and human settlement have resulted in significant habitat losses. Irrigation associated with intensive farming has led to the buildup of salts in the soil. The development of the manufacturing zone along the United States-Mexico border has attracted more people to the region, thereby escalating the rate of habitat destruction.

Both deliberate and accidental fires have destroyed wide swaths of matorral. All in all, 90 percent of the Tamaulipan matorral has been lost in Texas, and only 30 percent of this ecoregion remains intact in Mexico. Global warming is a likely culprit for disruption of some habitat here, as mounting changes in temperature and elevation zones exert pressure on species to move, adapt, or die.

Around 60 of the plant and vertebrate species that live in the Tamaulipan matorral are threatened, endangered, or vulnerable to extinction. Illegal collection of rare endemic cacti such as star cacti and living rock cacti is a major issue for many species. Introduced diseases such as blight (*Phytophtora infestans*) and insect pests including *Cerambycid* beetles have been observed to cause mortality to some rare cacti, such as the star cactus.

Given that this ecoregion is a hot spot for biodiversity and an important habitat for birds, preserving the little remaining Tamaulipan matorral is a conservation priority. Protected areas include the El Cielo Biosphere Reserve in Tamaulipas, Cumbres de Monterrey National Park in Nuevo Leon, the Bajo Río San Juan and Las Lajas national irrigation districts in Nuevo Leon, and the Laguna Madre and Río Bravo delta protected areas for flora and fauna in Tamaulipas.

MELANIE BATEMAN

Further Reading

Goldman, E. A. and R. T. Moore. "The Biotic Provinces of Mexico." *Journal of Mammalogy* 26, no. 4 (1946).

Miller, B., G. Ceballos, and R. Reading. "Prairie Dog and Biotic Diversity." *Conservation Biology* 3 (1994).

Oldfield, S. *Cactus and Succulent Plants—Status Survey and Conservation Action Plan.* Cambridge, United Kingdom: International Union for Conservation of Nature, 1997.

Tanami Desert

Category: Desert Biomes.
Geographic Location: Australia.
Summary: This remote desert region is valued for its biological diversity, its vast new wildlife preserve, and its management by Aboriginal Australians.

Australia is the world's driest continent after Antarctica; arid zones cover 65 percent to 75 percent of its landmass. Situated between Darwin in the north and Alice Springs in the south, and located in the Northern and Western Australia Territories, the Tanami Desert is one of the nation's five major desert regions. Annual precipitation ranges from 8 to 16 inches (200 to 400 millimeters). Together with the neighboring Sandy Desert, Gibson Desert, and Central Ranges xeric (drought-tolerant) scrubland, the combined extent of ecoregion is vast—nearly 500,000 square miles (1.3 million square kilometers) of arid biomes.

Arid regions of Australia tend to have extensive sand dune formations. Underlying much of the Tanami bioregion is the Tanami dune field. The associated surface communities include sand plains, dune fields, and extensive hummock grassland habitats. Other landforms include alluvial plains, rocky hills and rises, clay pans, and freshwater and saline lakes. This habitat diversity contributes to high biological diversity, and the desert serves as a refuge for a variety of threatened species.

In addition to grasslands, which are often dominated by spinifex grasses in the genus *Triodia*, the region contains sparse shrublands, like those containing saltbush (*Atriplex* spp.), and open woodlands with many *Acacia* and *Eucaplyptus* species.

Aboriginal Land Use

Land conditions can be strongly affected by livestock grazing and by fire, a process long present in the Tanami, and often associated with traditional Aboriginal practices. Aborigines use fire for both practical and symbolic reasons, and refer to it as "a tool with many uses." Among these uses are the protection of preferred plant species from larger fires, the encouragement of other valuable plant species that require fire, and the improvement of habitat for hunted wildlife species.

Much of the Tanami region is Aboriginal land that was reclaimed following the introduction of the Aboriginal Land Rights Act of 1976. These traditional landowners, most of whom are Warlpiri-speaking, were granted freehold ownership by the Australian government in 1982. The Australian government is currently working with indigenous groups to understand their fire-setting practices and to help them use this tool to manage the Tanami Desert landscape.

Biodiversity

The Australia government considers the Tanami Desert to be one of the country's most important landscapes for the protection of rare and endangered species. Mammalian species of interest include marsupials, such as the long-tailed planigale (*Planigale ingrami*) and bilby (*Macrotis lagotis*); and nonmarsupials, such as the western chestnut mouse (*Pseudomys nanus*) and little native mouse (*P. delicatulus*). The region has lost some small and medium-size mammals still found elsewhere in Australia, including the western quoll (*Dasyurus geoffroii*). Bird species of significance include the gray falcon (*Falco hypoluecos*) and the freckled duck (*Stictonetta naevosa*).

Despite its desert status, the region also houses a variety of aquatic habitats ranging from isolated desert water holes to larger lakes. Lake Gregory in the Western Australia Territory is a permanent water source and is considered to be a freshwater lake at times of high rainfall when it is full. In the Northern Territory, Lake Surprise, at the end of the Lander River, is the largest body of water in the Tanami Desert when its level is high. It is considered to be a critical resource for birds during times of drought.

Lake Gregory has been designated an Important Bird Area by Birdlife International because it supports more than 1 percent of the world's population of at least six species. Approximately 100,000 to 600,000 individual waterbirds may be found there during many seasons. Elsewhere in the region, the Davenport and Murchison ranges contain waterholes that are critical resources for birds in times of drought. These catchments also hold species of fish of biogeographical significance because of their isolation from other river systems.

Human Settlement

Because of its remote location in north-central Australia, the Tanami was not fully explored by scientists until the 20th century. However, Aboriginal occupation and management of the region are thought to extend back as far as 40,000 years, and continues into the present. Human population density is low, probably less than one person per 0.4 square mile (1 square kilometer). Modern land uses are grazing, mining, and various indigenous practices, including the harvesting of native plants and animals. Tourism has also become increasingly important.

Climate change threatens to alter this area significantly. The entire Australian Outback is receiving more annual precipitation than ever before, which could alter the size and scope of this desert habitat.

Environmental Threats

The most significant land-management issues in this region are land degradation due to grazing, fire, and the effects of nonnative species on the native flora and fauna. The Tanami Desert is of great national and international importance because of its high biological diversity value.

Site conditions vary considerably throughout the region. Where significant land-management issues occur, they mostly stem from one of three factors: grazing, fire, or invasive species. Grazing has its strongest negative effects near bodies of water, where livestock tend to congregate. Some grassland communities have also been degraded by grazing.

In combination with grazing, fire has negatively affected native vegetation. Fire is a natural and regular feature of this landscape; however, changes in traditional burning practices have resulted in a shift from frequent burning and small average fire sizes to less frequent burning that results in larger and more intense fires. This can inflict serious, long-lasting damage on habitats.

Several exotic vertebrate and plant pest species cause significant problems for native species and for human land use in the Tanami Desert. Non-native rabbits and foxes, as well as feral cats, cause problems for native flora and fauna as competitors, grazers, or predators of native species. At least 21 weedy plant species have been recorded here, including buffel grass (*Cenchrus ciliaris*), an invasive that is native to North Africa and the Middle East. This species has shown the capacity to outcompete native grasses and to increase the frequency of wildfires.

Despite these problems, the Australian Land Disturbance Database lists most of the Tanami as having high "biophysical naturalness and wilderness quality." In 2012, the Australian government declared an immense southern swath of the Tanami Desert—nearly 25 million acres (10 million hectares)—its newest, conservation zone. Easily the largest such area set aside in Australia, this desert-and-savanna reserve was established to help the stabilization and recovery of such endangered species as the bilby and the great desert skink, a burrowing lizard. Aboriginal rangers will manage this new sanctuary, dubbed the Southern Tanami Indigenous Protected Area.

John Mull

Further Reading

Kelly, K. *Tanami—On Foot Across Australia's Desert Heart.* Sydney, Australia: Pan Macmillan, 2003 .

Morton, S. R., J. Short, and R. D. Barker. *Refugia for Biological Diversity in Arid and Semiarid Australia, Biodiversity Series Paper No. 4.* Canberra, Australia: Department of Environment and Heritages, 2004 .

Newell, Janet. "The Role of the Reintroduction of Greater Bilbies (*Macrotis Lagotis*) and Burrowing Bettongs (*Bettongia Lesueur*) in the Ecological Restoration of an Arid Ecosystem: Foraging Diggings, Diet, and Soil Seed Banks." *Research Theses of the School of Earth and Environmental Sciences, University of Adelaide* (2009).

Tanganyika, Lake

Category: Inland Aquatic Biomes.
Geographic Location: Africa.

Summary: Lake Tanganyika is the world's second-deepest freshwater lake, and supports an amazing variety of fish species.

Lake Tanganyika is the second-deepest lake in the world, at 4,823 feet (1,470 meters), after Lake Baikal in Siberia. It is also the second-largest freshwater lake in the world by volume. It is located within the East African Great Rift Valley and divided among four countries: Burundi, Democratic Republic of the Congo (DRC), Tanzania, and Zambia. Most of the lake is within the DRC (45 percent) and Tanzania (41 percent). The water of the lake flows into the Congo River system and ultimately into the Atlantic Ocean.

Lake Tanganyika is narrow; it extends about 404 miles (650 kilometers) in length, by 31 miles (50 kilometers) wide, on average. It covers 12,703 square miles (32,900 square kilometers), with a shoreline of 1,136 miles (1,828 kilometers), a mean depth of 1,870 feet (570 meters), and a maximum depth of 4,823 feet (1,470 meters) in the northern basin. It holds an estimated 4,534 cubic miles (18,900 cubic kilometers) of water.

The lake has an average surface temperature of 77 degrees F (25 degrees C) and a pH averaging 8.4, certainly a bit on the alkaline side. Studies show that beneath the 1,640 feet (500 meters) of water are about 14,764 feet (4,500 meters) of sediment lying over the rock floor.

Setting and Hydrology

The lake catchment is approximately 89,190 square miles (231,000 square kilometers), with the Rusizi River entering from Lake Kivu in the north and the Malagarasi River entering from the east side of the lake, as well as numerous smaller rivers and streams.

The lake is surrounded by mountainous areas. The eastern side has poorly developed coastal plains; on the western coast, the steep side walls of the Great Rift Valley reach 6,562 feet (2,000 meters) in relative height from the shoreline. Its sole effluent river, the Lukuga, starts from the middle part of the western coast and flows westward to join the Lualaba, a tributary of the Congo River, which flows on to the South Atlantic Ocean.

The water balance of the lake is largely determined by rainfall and evaporation. A total of 31–47 inches (78–119 centimeters) per year of rain falls in the vicinity of the lake, while evaporation averages 60 inches (152 centimeters) per year. The many rivers and streams that enter the lake, including its major tributaries, the Rusizi and the Malagarasi, play small roles in the lake's water balance. The flushing time of water in the lake (lake volume divided by river outflow) is up to 7,000 years because the rivers have such a small influence over the lake.

Due to the lake's enormous depth and location within the tropics, there is little or no mixing of the shallower and deeper waters, which qualifies the lake as meromictic: The noncirculating hypolimnion (bottom layer) does not mix with the circulating upper layer (epilimnion). The mixed oxygenated layer extends to about 164–820 feet (50–250 meters), influenced by the seasons. The distribution of aquatic life is limited to this depth, as beyond this depth the environment is anaerobic, and accumulated particulate matter makes the water much denser. Permanent stratification keeps much of the nutrients in the noncirculating hypolimnion.

Biodiversity

The lake has some of the richest lake fauna on Earth, making it an important biological resource for the study of speciation in evolution. The lake's great depth is thought to have given its fauna an evolutionary advantage. Over its long history, estimated to be 9 million to 12 million years, Lake Tanganyika must have served as a refuge for aquatic organisms during extremely dry periods when other water bodies desiccated.

The split into two or three separate basins during low lake levels seems to have had important effects on the evolution and distribution of the ichthyodiversity, and facilitated allopatric speciation. The age of the lake has also facilitated further differentiation of the fish fauna compared with the fauna of lakes Malawi and Victoria. The Lake Tanganyika cichlids, for example, are considered to have evolved from eight ancestral lineages, more than in Lake Malawi or Victoria.

The lake's biodiversity exhibits high levels of endemism (species found exclusively here) within several taxonomic groups at the species and genus levels. It holds at least 250 species of cichlid fish and 150 noncichlid species, most of which live along the shoreline to a depth of approximately 591 feet (180 meters).

The fish fauna of Lake Tanganyika have strong affinities with the Congo basin, and the two systems are still connected hydrologically. In the late Miocene-early Pliocene, the Congo basin contained a large internal lake that covered the Cuvette Centrale.

It is believed that Lake Tanganyika's fauna originated in this ancient environment, although the details and timeline are not complete. The 23 fish families of Lake Tanganyika are all present in the Congo basin fauna.

Human Activities

It is estimated that 25–40 percent of the protein in the diet of the approximately 1 million people living around the lake comes from lake fish. Fishery products, especially the Tanganyika sardine (*Stolothrissa tanganikae*), are vital for the local economy. Currently, around 100,000 people are directly involved in the fisheries, operating from almost 800 sites. The lake is also vital to the estimated 10 million people living in the greater basin area.

Lake Tanganyika fish are exported throughout eastern Africa. Commercial fishing began in the mid-1950s and has had an extremely heavy effect on the pelagic fish species. In 1995, the total catch was around 198,416 tons (180,000 metric tons). Former industrial fisheries, which boomed in the 1980s, have subsequently collapsed.

The lake also serves as a transport gateway to the larger East Africa region. Regular ship lines connect Kigoma, Tanzania; Kalemie, Zaire; and other coastal towns as an essential part of the inland traffic system of the region. Two ferries carry passengers and cargo along the eastern shore of the lake: the *MV Liemba,* which runs between Kigoma and Mpulungu, and the *MV Mwongozo,* which runs between Kigoma and Bujumbura.

Agriculture, livestock, and the processing of their related products, as well as mining (tin, copper, and coal, mainly) are still the main industries in the drainage basin of Lake Tanganyika.

Environmental Threats

Some of the major environmental threats are deforestation, resulting from agricultural expansion, and overgrazing due to an influx of livestock, mostly from Sukuma-land in Tanzania. These activities have led to increased erosion, causing more flooding in coastal areas and the siltation of the lake. There are also destructive fishing practices; the use of beach seines is rampant, especially in the inshore areas of the lake.

Artisanal, small-scale gold mining is another challenge to the lake ecosystem, particularly the use of mercury for binding the gold. In many cases, mercury residues escape to pollute the rivers used by the miners to process their gold, not to mention the neurotoxic effect of mercury on the miners themselves. Climate change impacts upon Lake Tanganyika have been studied by researchers. Their findings indicate that the lake's waters are warming, a trend which may also have significant impact upon the fish species living there.

Okeyo Benards

Further Reading

Coulter, G. W. *Lake Tanganyika and Its Life.* Oxford, UK: Oxford University Press, 1991.

Lowe-McConnell, Rosemary H. "Fish Faunas of the African Great Lakes: Origins, Diversity and Vulnerability." *Conservation Biology* 7 (1993).

Patterson, G. and J. Makin. *The State of Biodiversity in Lake Tanganyika: A Literature Review.* Chatham, UK: Natural Resources Institute, 1998.

Roberts, T. R. "Geographical Distribution of African Freshwater Fishes." *Zoological Journal of the Linnean Society* 57 (1975).

Salzburger, W., A. Meyer, et al. "Phylogeny of the Lake Tanganyika Cichlid Species Flock and Its Relationship to the Central and East African Haplochromine Cichlid Fish Faunas." *Systematic Biology* 51, no. 1 (2002).

Tasman Sea

Category: Marine and Oceanic Biomes.
Geographic Location: South Pacific.
Summary: A biodiverse sea between Australia and New Zealand, the Tasman Sea is home to numerous unique marine species.

Nestled in the southwestern portion of the Pacific Ocean, the Tasman Sea is a saline body spanning the distance between the landmasses of Australia to the west and New Zealand to the east. The Tasman Sea is approximately 1,200 miles (1,931 kilometers) from west to east, and 1,700 miles (2,736 kilometers) north to south.

The Maori name for the Tasman Sea is *Te Tai-o-Rehua,* or *the great sea of Rehua,* the divine being who is the son of the sky god and earth goddess associated with immortality and various stars. The European name pays homage to Abel Tasman, the Dutch explorer who was the first European to discover New Zealand and the island of Tasmania, formerly called Van Diemen's Land. Tasman's voyage, in 1642-43, mapped substantial portions of the region; in the following century, British explorer Captain James Cook's first voyage extensively navigated the Tasman Sea, producing a complete and substantially correct map of the New Zealand coastline.

Australians and New Zealanders commonly refer to the Tasman Sea as the Ditch, especially in the phrase *crossing the Ditch*, referring to passage across the sea in either direction.

Like other Australian waters, the Tasman Sea is generally nutrient-poor, and its ecosystem relies on processes such as upwelling to bring nutrients to the food chain. The mid-ocean ridge of the Tasman Sea was formed about 85 million years ago. Much later, Australia became drier, and Antarctica cooler when the island of Tasmania separated from Antarctica and changed the ocean currents around the major landmasses. As the rift valley sank, sediment accumulated, forming the Otway and Strzelecki ranges in southeastern Australia. The mid-ocean ridge is considerably closer to Australia than New Zealand, having formed essentially at the midway point between their continental margins.

Where the Otway ranges meet the Tasman Sea is Dinosaur Cove, a minor bay of seafront cliffs with fossil-bearing strata dating back over 1 million years. Many of the fossils show the kinds of dinosaurs that lived in the Tasman Sea region, including the taxa called the polar dinosaurs of Australia, which some paleontologists believe may have been warm-blooded.

Biodiversity

The seamounts of the Tasman Sea are extremely biodiverse, home to hundreds of species of fish, including the orange roughy, which can live for over a century. In 2004, the deep-sea research ship *Tangaroa* found more than 100 previously undiscovered species of fish during a four-week expedition of the Tasman Sea, as well as fossilized teeth of the megalodon, the prehistoric shark that was twice the size of the modern-day great white shark.

Seamounts, undersea mountains which are isolated from many disturbances, are good homes for coral reefs, which are both long-lived and slow-growing. One of the southernmost platform reefs in the world, Middleton Reef is home to the protected species *Epinephelus daemeli,* one of the fish known as black cod. Declines were noted in the early 1900s, and began to decline severely no later than the 1950s. The cod is vulnerable because of commercial interest in the fish and resulting overfishing, as well as its slow speed, large size, and territorial use of inshore habitats, making it an easy catch for spear fishers. Furthermore, some of the black cod populations in the Tasman Sea are believed to be non-breeding populations as a result of drifting larvae, slowing the replenishment of populations.

On the other side of a pass some 25 miles across is another platform coral reef, Elizabeth Reef. At low tide, most of the reef flat is above water; at high tide only one cay (Elizabeth Island) remains, just barely above sea level. Australia's National Heritage Trust manages the Elizabeth and Middleton Reefs Marine National Park. In addition to black cod and a variety of other species, Elizabeth is home to a Galapagos shark (*Carcharhinus galapagensis*) nursery, large numbers of sea cucumbers (*Holothuria whitmaei*), and a recently growing number of crown-of-thorns starfish (*Acanthaster planci*).

The crown-of-thorns starfish is the second-largest starfish in the world, a solitary animal covered in venomous spines. Though it sometimes preys on brittle stars, the crown-of-thorns is principally a corallivore, feeding on the polyps of reef coral by climbing the reef, extruding its stomach onto the structure, and liquefying coral polyps in order to absorb nutrients from them. Its spines, which exude a strong neurotoxin, protect it from shrimp, mollusks, and large reef fish who attempt to prey on it.

The crown-of-thorns is best known as a threat to the coral reef ecosystem of the Great Barrier Reef elsewhere in Australian waters. A single crown-of-thorns consumes 65 square feet (6 square meters) of living coral reef surface each year, and the other factors that jeopardize a reef ecosystem exacerbate the effects of the crown-of-thorns's predation. It may also promote the spread of coral diseases. So far, the crown-of-thorns is not seen in Elizabeth Reef in the same numbers as in the Great Barrier Reef, but the population has shown noticeable growth since 2003.

A crown-of-thorns starfish sits atop coral, on which it feeds; one starfish can kill 65 square feet (6 square meters) of coral surface per year. Their numbers have increased in the Tasman Sea since 2003. (Thinkstock)

Middleton and Elizabeth Reef are part of the same underwater plateau, the Lord Howe Rise, consisting of about 932,057 square miles (1.5 million square kilometers) of the Tasman Sea. The central-east area of the rise is known as the Lord Howe platform, and includes a seamount capped by Lord Howe Island and Ball's Pyramid. Lord Howe Island is a crescent-shaped volcanic remnant about 6 miles (10 kilometers) long and 1 mile (1.6 kilometers) wide at its widest point, with a population of 348 in the 2006 census.

The island is a distinct terrestrial ecoregion and more than half of its plant species are endemic (found only here) to the island; many locals are involved in the Kentia palm industry, revolving around the endemic genus of *Arecaceae Howea*. The critically endangered creeping vine *Calystegia affinis* is native only to Lord Howe Island and nearby Norfolk Island. The island is also known for its banyans, pandanus trees, the spring-flowering bush orchid, and the glowing mushrooms (*mycena chlorophanos* and *omphalotus nidiformis*, both endemic to the island) that crop up in the palm forests after heavy rain.

Ball's Pyramid is an eroded remnant of a shield volcano, consisting of an uninhabited volcanic stack. The most notable life on the Pyramid is a population of Lord Howe Island stick insects living among the growth of *Melaleuca howeana* shrubs that grow out of crevices in the rocks.

The central and southern Tasman Sea experiences a spring (September to December) bloom, with the maximum chlorophyll biomass occurring in early October. The surface of the sea includes a rich assortment of microscopic flora and fauna, including crustacea, pteropoda, heteropoda, radiolarians, dinoflagellates, foraminifers, polycaeta, chaeotgnatha, urochorda, vorticellids, and polychlads. Seabirds include the endangered kiwi species *Apteryx haastii*, *A. rowi*, *A. owenii*, and *A.*

australis along the New Zealand coast, as well as the brown booby, brown tern, black-naped tern, bridled tern, crested tern, roseate tern, sooty tern, black noddy, red-tailed tropicbird, silver gull, wedge-tailed shearwater, and eastern reef egret.

Marine macrofauna include the the humpback whale, the spinner dolphin, the dumbo octopus (which navigates with the help of two flaps, giving it the appearance of the Dumbo cartoon character), and the Pacific spookfish, which uses a long snout to detect the electrical impulses of hiding prey. Along the Fjordland coast, the Fjordland Crested Penguin nests.

Warming Threats

Climate change could affect the coral reefs with some intensity. As sea temperatures rise, the coral reefs of the Tasman Sea could be put at risk; those of Lord Howe Island are considered to be the most vulnerable. Furthermore, changes to ocean chemistry as the waters become acidified by absorption of increased greenhouse gas would hurt the shellfish of the Tasman Sea, making it more difficult for them to grow their shells, as well as the coral reefs.

Changes in water temperature have already been observed, as the East Australian Current reaches further south than before, bringing subtropical species into temperate waters. The long-spined sea urchin, previously found only off the shores of New South Wales, has recently been introduced to the western Tasman Sea. The European shore crab, introduced to Tasmania in the 1990s, has spread to the entire eastern coast, and other new species arrivals have impacted the ecosystem of Tasmania's eastern coast.

Over the next 50 years, the temperature of the Tasman Sea is expected to keep rising incrementally. Phytoplankton are expected to push further south, and algal blooms may increase, while the phytoplankton populations of the temperate region will shrink. Temperature changes can also affect the sex of the embryos of turtles and other species, biasing the sex ratio toward females, which may lead to long-term fertility problems in some populations and overpopulation in others.

Bill Kte'pi

Further Reading

Ayling, Tony and Geoffrey Cox. *Collins Guide to the Sea Fishes of New Zealand.* Auckland, New Zealand: William Collins Publishers, 1982.

International Hydrographic Organization. *Limits of Oceans and Seas, Special Publication No. 23.* Monte Carlo, Monaco: International Hydrographic Organization, 1953.

Longhurst, Alan R. *Ecological Geography of the Sea.* New York: Elsevier Science and Technology Books, 2006.

Tengger Desert

Category: Desert Biomes.
Geographic Location: Asia.
Summary: This large desert has been partially revegetated in an effort to stabilize its sand dunes and prevent land degradation.

The fourth-largest desert in China, the Tengger Desert covers about 15,444 square miles (40,000 square kilometers), most of which lies in the Inner Mongolia autonomous region and the Shapotou District of Zhongwei, Ningxia Province. The southern region of the Tengger Desert is sometimes called the Shapotou Desert. Shapotou is the site of the Shapotou Desert Experimental Research Station (SDERS), on the banks of the Yellow River on the south edge of the Tengger's dune sea, where research focuses principally on the use of microbial mats and grasses to stabilize the dunes.

The SDERS's work in the past has led to former sand dunes being used to grow fruit and vine crops, to help with China's recurring food shortages. The trans-Asian Baotou-Lanzhou, or Baolan, railway also depends on the stabilization of sand dunes in the Tengger, in order to prevent its repeated burial. The Baolan railway is nearly 621 miles (1,000 kilometers) long, connecting Inner Mongolia to China's Fansu province. It is the first Chinese railway built through deserts, and an important east-west connection, operating since the 1950s thanks to the work of the SDERS.

The Tengger Desert is home to singing sand dunes. Shear stress, caused by wind passing along the dunes, or by people walking or riding camels over the dunes, results in sounds of either high or low frequency; the alternation between the two notes sounds like tuneless singing.

In 2010, China began a far-reaching project to plant a greenbelt between two deserts for the first time in Chinese history; the goal is to prevent the spread of the Tengger and Badain Jaran deserts into one another. The project, with an expected completion date of 2015 or 2016, consists of planting vegetation between the two deserts along a 126-mile (202-kilometer) stretch ranging from 3 to 9 miles (5 to 15 kilometers) wide. The greenbelt replaces a bush forest that once stood between the two deserts, which has been reduced to half its size since 1960 by land degradation, caused principally by human activity, particularly deforestation to create pasture land.

The greenbelt will be fenced after the project has finished rolling and leveling sand dunes, packing and consolidating the sand, paving it with clay, and planting a variety of drought-resistant shrubs. Fencing will prevent livestock from eating the shrubbery. Drought has caused serious forest damage in China over many years, and recently climate change has become an additional possible factor aggravating this trend. The construction of the greenbelt is one of several efforts by the Chinese government to protect forest ecosystems, along with cuts in timber production and bans on forest hunting.

Vegetation

Much of the vegetation in the greenbelt will be the same as or similar to the vegetation used to stabilize dunes for crop growing or to keep the Baolan railway working. Revegetation is key to stabilizing the dunes, because of the positive correlation between the fractal dimension of soil particle size and the clay content of the shallow soil profile in the desert shrub ecosystem; the longer a dune is stabilized, the greater the soil clay content is. Revegetating develops soil structure better and increases that clay content. The sand-binding vegetation used by the SDERS to stabilize the dunes around the railway includes *Caragana korshinskii, Hedysarum scoparium,* and *Artemisia ordosica,* all dwarf desert shrubs existing in an ecosystem with a microbiotic soil crust cover atop the dunes. The stabilization efforts result in decreased soil particle size and increased subsoil layer thickness, microbiotic crust thickness, and volumetric soil moisture.

The richness of species in revegetated areas has been found to correlate positively with soil alkalinity balance and the concentration of carbon and nitrogen in the soil. Such revegetation has also restored cryptogamic diversity to the desert. Cryptogams are plants that propagate by spores, such as algaes, mosses, and lichens. In the Tengger Desert's case, no lichens have been observed, but 24 species of algae and five mosses have been established since the SDERS's revegetation efforts began.

Fauna

Animals in the region include the antelope, wild horse, wild camel, and ostrich, as well as the endangered golden-haired monkey or golden snub-nosed monkey (*Rhinopithecus roxellana*), which is endemic (found only here) to the Tengger Desert and surrounding forests. It feeds primarily on lichen and has been found in colder temperatures (such as the Tengger in winter) than any other nonhuman primate species. The monkey is organized in packs ranging from small social groups of fewer than a dozen individuals to bands of hundreds; the social organization is obscure and complex. The planting of the green belt may assist with mitigating the monkey's endangered status in the face of widespread habitat loss.

Bill Kte'pi

Further Reading

Li, Xin-Rong, Hong-Lang Xiao, Jing-Guang Zhang, and Xin-Ping Wang. "Long-Term Ecosystem Effects of Sand-Binding Vegetation in the Tengger Desert, Northern China." *Restoration Ecology* 12, no. 3 (2004).

Li, Xin-Rong, H.-Y. Zhou, X.-P. Wang, Y.-G. Zhu, and P. J. O'Conner. "The Effects of Sand Stabilization and Revegetation on Cryptogam Diversity and Soil Fertility in the Tengger Desert, Northern China." *Plant and Soil* 251, no. 2 (2003).

Wang, Xin-Ping, Xin-Rong Li, Hong-Lang Xiao, and Yan-Xia Pan. "Evolutionary Characteristics of the Artificially Revegetated Shrub Ecosystem in the Tengger Desert, Northern China." *Ecological Research* 21, no. 3 (2006).

Tepuis Forests

Category: Forest Biomes.
Geographic Location: South America.
Summary: The dramatic sandstone mesas that tower over the lowland forests and savannas of southern Venezuela and its neighbors served as the inspiration for Sir Arthur Conan Doyle's *The Lost World.*

In southern Venezuela, and to a lesser extent in western Guyana, northern Brazil, and eastern Colombia, more than 100 dramatic table mountains called *tepuis* rise 3,281–9,843 feet (1,000–3,000 meters) above the lowland forests and savannas. These mountains generally have flat tops and nearly vertical walls interspersed with talus slopes. Many of the tepuis are virtually inaccessible by foot. The word *tepui* comes from the Pemón people and means *house of the gods.*

While the tepuis used to form a contiguous landmass, they have long since been separated from one another. The environment on the summits of the tepuis can be very different from that of the lowlands. Due to the combination of extended isolation and climate differences, the tepuis summits have truly unique flora and fauna.

Reports of expeditions to this remote region in the 19th century served as the inspiration for Sir Arthur Conan Doyle's novel *The Lost World.* In spite of such occasional publicity, its remoteness has helped maintain this ecoregion relatively intact.

The Guayana Shield, on which the tepuis rest, consists of a rock basement that began to form 1.8 billion years ago. Most of this rock basement was overlain with sediment, which was subsequently compressed and cemented together to form a thick layer of quartzitic and sandstone rocks. Uplifting broke this sandstone layer apart, and erosion resulted in the separation and isolation of the tepuis.

Angel Falls, shown here, drops from a massive tepui in Venezuela. The waterfall's 3,212-foot (979-meter) drop is the longest in the world. (Thinkstock)

Tepui soils are generally sandy and poor in nutrients. Weathering has carved deep canyons, gorges, and sinkholes in the tepui summits. The tepuis are some of the oldest geological formations in all of South America. Pico da Neblina in Brazil, at 9,888 feet (3,014 meters), is the tallest tepui. Angel Falls, dropping from the broadest, 270-square-mile (700-square-kilometer) Auyantepui in Venezuela, is the world's tallest waterfall at 3,212 feet (979 meters).

Because the ecoregion is located just north of the Equator, variations in climate are mainly due to the effects of the trade winds and differences in elevation. Rainfall is distributed evenly though the year, 79–94 inches (2,000–2,400 millimeters) on average. The average annual temperature is approximately 75 degrees F (24 degrees C) but can vary considerably with elevation. At the tops of the tallest tepuis, temperatures can drop to 32 degrees F (0 degrees C). The summits also differ from the lowlands in that they experience intense light and strong winds.

Zoned Biota

The vegetation of the tepuis changes with elevation, and both evergreen rainforests and cloud forest are found upon the tepuis. Generally, the plant communities found at the base, along the talus slopes, on the cliff walls, and on the summit are each distinct. Lowland species associated with the highland savanna or evergreen rainforest that surround the tepuis can be found at elevations below 1,640 feet (500 meters).

Cloud forest occurs on the talus slopes above 1,640 feet (500 meters) elevation. The bare sandstone of the cliff walls is home to plants that are adapted to grow in crevices, such as bromeliads (the pineapple family). The summits are home to a mixture of elfin forest, scrub, savanna, and bog adapted to withstand the intense environment.

The tepuis have exceptionally high plant diversity; more than 2,000 species of plants are known here. Approximately one-third of the species, and nearly 80 genera are endemic (found nowhere else) to the ecoregion. The highest concentrations of endemic species are found on the summits. Some summit endemics occur on most or all mountains, while others are highly localized, occurring on a single summit. Many carnivorous plants are found among the tepuis, possibly because the soil is so nutrient-poor.

The tepuis are home to almost 200 species of mammals. Many species' natural distribution is confined to the Tepuis Forests ecoregion or the wider region of Amazonia. Nine species of primates and five species of cats live in the Tepuis Forests biome.

More than 600 species of birds have been observed here, including 41 endemic bird species. Likewise, more than 150 species of reptiles and amphibians have been recorded for this ecoregion, including venomous snakes such as fer-de-lances, coral snakes, and bushmasters.

Effects of Human Activity

The lowlands surrounding the tepuis are sparsely populated, and the tepuis themselves are difficult to access. As a consequence, much of this ecoregion has remained relatively untouched. However, anthropogenic disturbances are on the rise. These disturbances include the setting of accidental and deliberate fires, logging, over-collecting of flora and fauna, and illegal mining for gold, diamonds, and bauxite.

Tourism is also having an effect, as visiting hikers litter, trample vegetation, and strip woody plants for firewood. Plants on the summits of tepuis are often adapted to the climate and the poor soils tend to support slow-growing types, making natural recovery from disturbance a protracted process. Climate change, too, could have a significant effect on the flora and fauna of this ecoregion for similar reasons.

Conservation Efforts

Steps have been taken at national and international levels to conserve the Tepuis Forests biome, but the status of the conservation efforts varies by country. In Venezuela, all the tepuis above 2,625 feet (800 meters) are protected as natural monuments, national parks, or biosphere reserves. Canaima National Park is a World Heritage Site. Commercial logging is prohibited in the Venezuelan state of Amazonas.

By contrast, legislation to establish a National Protected Areas System in Guyana was passed only in 2011. Even where protected areas have been established, they are often just "paper parks," lacking infrastructure or funds. The vast sizes of the protected areas and their remoteness make enforcement of conservation regulations difficult.

Melanie Bateman

Further Reading

Berry, P. E., B. K. Holst, and K. Yatskievych, eds. *Flora of the Venezuelan Guayana*. St. Louis, Missouri: Botanical Garden and Timber Press, 2003.

Hollowell, T. and R. P. Reynolds. "Checklist of the Terrestrial Vertebrates of the Guiana Shield." *Bulletin of the Biological Society of Washington* 13 (2005).

Zimmer, Carl. "It's Not So Lonely at the Top: Tepui Ecosystems Thrive Up High." *New York Times*, May 7, 2012.

Terai-Duar Savanna and Grasslands

Category: Grasslands, Tundra, and Human Biomes.

Geographic Location: Asia.

Summary: Poverty and international tensions are preventing conservation efforts across this haven for rare species and the tallest grass type in the world.

The Terai-Duar Savanna Grassland biome is a narrow belt of marshy grasslands, savannas, open woodlands, and forests stretching south of the Himalayan foothills. It covers parts of Nepal, India, Bhutan, and Bangladesh, and extends 21,437 square miles (34,500 square kilometers) from the Brahmaputra River in the east to the Yamuna River in the west.

Biodiversity

The Terai-Duar savanna grassland is a globally significant ecoregion for its rich biodiversity. The main features are its highly productive alluvial tall grassland, particularly for one of the tallest grasses of the world, *Saccharum*, which attains heights of 9–13 feet (3–4 meters). The biome is also unique due to its nine distinct plant assemblages and eight succession phases.

This biome is a habitat of globally threatened species such as tiger (*Panthera tigris*), Asian elephant (*Elephas maximus*), greater one-horned rhinoceros (*Rhinoceros unicornis*), and swamp deer (*Cervus duvauceli*). Endangered medium-sized mammals such as the hispid hare (*Caprolagus hispidus*) and pygmy hog (*Porcula salvania*) are found here, as well.

Threats and Conservation

Due to anthropogenic disturbances, this unique ecoregion is highly threatened. The grassland and forest areas have in large degree been converted to agricultural use. The Terai-Duar savanna grassland mostly remains only within protected areas. The protected system includes: Koshi Tappu Wildlife Reserve, Chitwan National Park, Parsa Wildlife Reserve, Bardia National Park, and Shukla Phanta Wildlife Reserve, all in Nepal; and Manas, Dudhwa, Katarniaghat, Mahananda, Buxa, and Garumara Wildlife Reserves and National Parks in India.

The degradation of the Terai-Duar savanna grassland is causing great loss to biological diversity. The major causes of the loss include habitat alteration, over-harvesting;,species and disease introduction, pollution, and climate change. Associated issues include poverty, exploitation of the environment by the wealthy, real estate development and infrastructure underdevelopment, rural over-population—and shrinking rural populations in some areas, traditional slash-and-burn agriculture, and modern commercial agricultural production. Many of these pressures lead to habitat damage and biodiversity loss.

While irrigation projects and water diversion systems are degrading and undermining the unprotected areas of grassland habitat here, even protected sections of the Terai-Duar savanna grassland are facing the problems of poaching and some overgrazing.

Government authorities of each country are combatting each of these stressors, in different ways and to varying degrees. In addition, there are several programs managed by international organizations. India and Nepal are serious about cooperating in managing this very important ecosystem and the protected areas; they have joined in various joint projects initiated by the World Wildlife Fund (WWF) such as Sacred Himalayan Landscape and the Terai Arc Landscape Project. However, this ecoregion has weak conservation mechanisms overall. The generally weak conservation output is due to the scarcity of resources, and to various types of civil and political conflicts in the region. With this backdrop, the conflict in natural resource utilization is seen as a normal and day-to-day problem across this ecoregion. For better management, more collaborative efforts are needed.

MEDANI P. BHANDARI

Further Reading

Johnsingh A. J. T., K. Ramesh, Q. Qureshi, et al. *Conservation Status of Tiger and Associated Species in the Terai Arc Landscape, India*. Dehra Dun: Wildlife Institute of India, 2004.

MacKinnon, J. *Protected Areas Systems Review of the Indo-Malayan Realm*. Cambridge, UK: World Bank, 1997.

Olson, David M. and Eric Dinerstein. "The Global 200: A Representation Approach to Conserving the Earth's Most Biologically Valuable Ecoregions." *Conservation Biology* 12, no. 3 (1998).

Rodgers, W. A. and H. S. Panwar. *Planning Wildlife Protected Areas Network in India, Vols. 1–2*. Dehra Dun, India: Department of Environment, Forests, and Wildlife, 1988.

Wikramanayake, E. D., C. Carpenter, H. Strand, and M. Mcknight, eds. *Ecoregion-Based Conservation in the Eastern Himalaya, Identifying Important Areas for Biodiversity Conservation*. Washington DC: WWF and Center for Integrated Mountain Development, 2004.

Texas Blackland Prairies

Category: Grassland, Tundra, and Human Biomes.

Geographic Location: North America.

Summary: The Texas Blackland Prairies are the surviving portions of a once-vast tallgrass prairie and home to a variety of grassland ecosystems.

The Texas Blackland Prairies are a temperate grassland ecoregion running from the Red River in north Texas to San Antonio in south-central Texas, an area of about 20,000 square miles (51,800 square kilometers). The area consists of a main belt of 17,000 square miles (44,030 square kilometers) and two islands of tallgrass prairie grasslands. Tallgrass prairie is an ecoregion dominated by tall grasses, as opposed to the shortgrass prairies characteristic of the western Great Plains, with less than 10 percent tree cover. A similar grass-dominated ecoregion with 10–49 percent tree cover is the savanna.

Tree seedlings and invasive species are regularly eliminated in the prairie by periodic wildfires often set by lightning in the dry season. The fires contribute to the accumulation of nutrient-rich loess soil and organic matter, leading to deep and good fertility.

Tallgrass prairie once covered a much larger portion of North America, including much of the Midwest and the Canadian prairies, but is now perhaps the most endangered large ecosystem in North America. More than 99 percent of the original North American tallgrass prairie has been converted to farmland. Most of the surviving tallgrass prairies owe their survival to rocky hill country that made plowing difficult, as in Kansas; to cattle ranchers who preserved the prairies for grazing, as in Oklahoma; or to nature reserves. In Texas, less than 1 percent of the prairie remains.

Precipitation ranges from 30 to 45 inches (76 to 114 centimeters) a year, mainly in the wet spring, and temperature is mild and temperate, averaging 63–70 degrees F (17–21 degrees C).

The two prairie islands of this biome are the Fayette Prairie and the San Antonio Prairie, which are surrounded by the Piney Woods in the northeast and the east central Texas forests in all other directions. The main belt is divided into four narrow areas aligned north to south: the Eagle Ford Prairie, the White Rock Cuesta, the Taylor Black Prairie, and the Eastern Marginal Prairie. Each of these areas has a characteristic soil type—and that goes a long way toward determining its plant and animal community makeup.

Eagle Ford and Taylor Black have soils that are mainly vertisol clays—soils with a large amount of montmorillonite, an expansive clay that forms deep cracks in dry seasons. These cracks can be big enough to injure grazing livestock that lose their footing. In Texas, vertisol clay soils are commonly called black gumbo and are classified as Ustert vertisols, a category in American soil taxonomy meaning that the cracks are open for at least 90 cumulative days a year.

Vertisol regions are marked by hog wallows, small lakes usually a few feet (meters) across and 1 foot (0.3 meter) or so deep that appear after a rainfall, that is, mud puddles big enough for a pig to roll around in.

The White Rock Cuesta soils are mollisols, similar to vertisols but with less clay content and without the crack-forming characteristics. Mollisols are calcareous and rich in loess, and similar to the soils of the Great Plains and the Argentinian pampas.

The Eastern Marginal Prairie and the San Antonio Prairie are predominantly composed of alfisols, while the Fayette Prairie is composed of both alfisols and vertisols. Alfisols are a soil with abundant aluminum, iron, calcium, magnesium, and potassium, as well as clay-enriched subsoil. Alfisol regions are easily identified by the presence of Mima mounds (named for the Mima Prairie in Washington state), naturally occurring domelike or flattened mounds roughly 1–6 feet (0.3–2 meters) high and 5–150 feet (1.5–46 meters) in diameter, covered in a blanket of tallgrass. There are varying theories as to how Mima mounds are formed, with the leading schools arguing for wind-based origins or the tunneling activities of gophers. It is likely that they can be formed by several factors, either independently or in conjunction.

*Invasive fire ants (*Solenopsis invicta *Buren) have destroyed 99 percent of the native ant population in the Texas Blackland Prairies biome. (Thinkstock)*

Vegetation

On Eagle Ford, Taylor Black, the Eastern Marginal, and the prairie islands, the dominant grasses are little bluestem (*Schizachyrium scoparium*) and indiangrass (*Sorghastrum nutans*), both of which regrow well after fires. Big bluestem (*Andropogon gerardii*), though found on the vertisol regions, is dominant only on White Rock Cuesta. Mixed switchgrass (*Panicum virgatum*) and gamagrass (*Tripsacum dactyloides*) prairies are found in bottomlands throughout the ecoregion, and near hog wallows in the vertisol uplands.

The Fayette Prairie is home to mixed prairies of little bluestem and brownseed paspalum (*Paspalum plicatulum*), while in the north, where the soil is acidic, mixes of Silveanus dropseed (*Sporobolus silveanus*) and mead's sedge (*Carx meadii*) are found. The White Rock Cuesta is home to seep muhly (*Muhlenbergia reverchonii*), hairy grama (*Bouteloua hirsuta*), and species of *Dalea* such as prairie clover. The Eastern Marginal is home to the prairie blazing star (*Liatris pycnostachya*), a tall aster with fluffy purple flowers, and the large-flowered tickseed (*Coreopsis grandiflora*), a yellow flower resembling a daisy.

Invasive plant species that have threatened the indigenous flora include bastard cabbage (*Rapistrum rugosum*), giant reed (*Arundo donax*), Johnson grass (*Sorghum halepense*), Chinese tallow tree (*Triadica sebifera*), King Ranch bluestem (*Bothriochloa ischaemum*), field blindweed (*Convolvus arvensis*), bermudagrass (*Cynodon dactylon*), chinaberry tree (*Melia azedarach*), heavenly bamboo (*Nandina domestica*), and Chinese privet (*Ligustrum sinense*).

Steps are being taken to encourage planting of native grasses in the area, which are thought to mitigate some of the effects of climate change. Native prairie plants are useful to sequester carbon and provide a high-energy biofuel source.

Both are incentives for landowners to replant areas with prairie grasses, native wildflowers, and other native plant species.

Fauna

The faunal species that has had the most effect in this biome as an invasive type is the fire ant (*Solenopsis invicta Buren*), the arrival of which resulted in the decimation of 99 percent of the native ant population and a 66 percent decline in ant-species richness.

Bison and pronghorn antelope were once prominent here, but neither species has lived here in wild herds since the 19th century; bison are raised in small numbers as livestock. Today, there are about 500 species of wildlife, including 327 bird species, 15 of which are threatened. There are seven threatened species of reptiles in the Texas Blacklands Prairies biome.

Other reptile species include the southern prairie lizard (*Sceloporus undulatus consobrinus*); northern prairie lizard (*S. u. garmani*); the southern prairie skink (*Plestiodon septentrionalis obtusirostris*); Texas horned lizard or horny toad (*Phrynosoma cornutum*), which has horns made of true bone that extend from its cranium and feeds on harvester ants and termites; Louisiana pine snake (*Pituophis ruthveni*), a nonvenomous constrictor that feeds mainly on the Baird's pocket gopher (*Geomys breviceps*), a rodent that may be responsible for the Mima mounds and thus can be found near them; and the timber rattlesnake (*Crotalus horridus*), which usually hibernates with copperheads (*Agkistrodon contortix*).

The Arctic bird Smith's longspur (*Calcarius pictus*) spends its winters in the blackland prairies. Other common bird species include the yellow warbler (*Dendroica petechia*), cardinal (*Cardinalidae*), turkey vulture (*Cathartes aura*), and hairy woodpecker (*Picoides villosus*). Mammals in the ecoregion include the coyote (*Canis latrans*), northern pygmy mouse (*Baiomys taylori*), fulvous harvest mouse (*Reithrodontomys fulvescens*), collared peccary (*Pecari tajacu*), and ringtailed cat (*Bassariscus astutus*).

Bill Kte'pi

Further Reading

Brennan, Leonard Alfred. *Texas Quails: Ecology and Management.* College Station: Texas A & M University Press, 2006.

Erlickman, Howard J. *Camino del Norte: How a Series of Watering Holes, Fords, and Dirt Trails Evolved into Interstate 35.* College Station: Texas A&M University Press, 2006.

Peacock, Evan and Timothy Schauwecker. *Blackland Prairies of the Gulf Coastal Plain: Nature, Culture, and Sustainability.* Birmingham: University of Alabama Press, 2003.

Ricketts, Taylor H., Eric Dinerstein, David M. Olson, Colby J. Loucks, et al. *Terrestrial Ecoregions of North America: A Conservation Assessment.* Washington, DC: Island Press, 1999.

Thames River

Category: Inland Aquatic Biomes.
Geographic Location: Europe.
Summary: The Thames has great historic and ecological significance to London. Its slow, decades-long recovery from infrastructure damage in the 20th century may now be jeopardized by climate change.

At 215 miles (346 kilometers), the River Thames is the longest river in England and the second-longest in the United Kingdom. It begins in Gloucestershire and flows eastward through Oxford, Reading, Windsor, and central London before reaching the Thames Estuary on the North Sea. The river has both seawater and freshwater stretches, is fed by dozens of tributaries, and has a catchment area covering much of southeastern England. Much of the Thames' fame comes from its presence in London and its effect on London's layout, geography, and local ecosystem.

The river has also been key to trade, as the riverside Port of London, once the world's largest port, has been a center of international trade since the city's founding in the 1st century C.E. Major port industries on the Thames have included iron-working, brass and bronze casting, paper milling, arms manufacturing, submarine communication cables, and timber. For most of British history, London was the nation's shipbuilding center, and the volume of shipping in the 19th century supported 33 dry docks devoted to ship repair.

Today, sugar refining, vehicle manufacturing, and edible oil processing remain major port industries. In London, the river's tide has a rise and fall of 23 feet (7 meters), with high tide reaching Teddington Lock.

Many gas works and power stations are located along the Thames and its canals and tributaries, the most prominent of which are the Beckton and East Greenwich gas works and the Brimsdown, Hackney, West Ham, Kingston, Fulham, Lots Road, Wandsworth, Battersea, Bankside, Stepney, Deptford, Greenwich, Blackwall Point, Brunswich Wharf, Wollwich, Barking, Belvedere, Littlebrook, West Thurrock, Northfleet, Tilbury, and Grain power stations.

Catchment, Flow, and Tides

Thames Head, the source of the river, is 1 mile (1.6 kilometers) north of the Gloucestershire village of Kembile, in the Cotswolds. The Thames River Basin District covers 6,229 square miles (16,133 square kilometers). The western part of the catchment is primarily rural, but the eastern and northern parts include heavily urbanized areas.

A total 38 main tributaries feed the Thames, including brooks, rivers, and canals, covering 3,842 square miles (9,951 square kilometers). Tributary rivers include the Churn, Leach, Cole, Ray, Coln, Windrush, Evenlode, Cherwell, Ock, Thame, Pang, Kennet, Loddon, Colne, Wey, and Mole, as well as the manmade Longford River. A complex of three locks and a weir, Teddington Lock was built in the early 19th century at Ham, in London's western suburbs. Downstream of the lock, the portion of the Thames known as the Tideway is tidal and is governed by the Port of London Authority.

Upstream, the river is governed by the Environment Agency. An obelisk shortly below the lock marks the boundary between these navigation authorities. When spring leads to especially high tides, the headwater can rise above Teddington, reversing the river flow for a short period; at such times, tidal effects can be observed much farther upstream than usual.

The human-made Jubilee River, completed in 2002, is a hydraulic channel 7 miles (11 kilometers) long and 148 feet (45 meters) wide, built to alleviate flooding by taking overflow from the Thames via the east bank upstream of Boulter's Lock near the town of Maidenhead, and channeling it to the northeast bank downstream from Eton.

The Teddington Lock is about 55 miles (89 kilometers) upstream from the Thames Estuary. That 55-mile (89-kilometer) stretch, the Tideway, is subject to the North Sea's tidal activity. The main tributaries on the Tideway are the rivers Brent, Lea, Roding, Darent, Wandle, Effra, Ingrebourne, Fleet, Westbourne, and Ravensbourne. When it was founded as the new capital of Roman Britain at a natural hub for trade, London was established on Ludgate Hill and Cornhill, both of which were elevated enough to be out of reach of spring tides. Tide tables today are based on the tide at London Bridge. The London waters are brackish, a mixture of the freshwater upstream and the saltwater of the North Sea.

Engineering and Reconfiguration

The Thames contains nearly 100 islands. In several places, the islands are created by the river's splitting into multiple streams. In Oxford, for example, the Thames splits into Seacourt Stream, Castle Mill Stream, Bulstake Stream, and several smaller runs, creating Fiddler's Island, Osney, and other small islets. Upstream of London proper is Thorney Island, named for the brambles that posed challenges for the monks who settled there to establish Westminster Abbey. The Palace of Westminster, commonly referred to now as the Houses of Parliament, is also located on Thorney Island, on the north bank of the Thames.

The development of London and the construction of embankments destroyed the marshes that once surrounded the greater London stretch of the Thames; much the same soon happened in the other major cities by which it passes. Thorney Island, as one example, is no longer identifiable as an island to the casual observer, as the river has been completely embanked. The estuary was narrowed to about one-third of its natural width, so it is much deeper and faster-flowing to accommodate the same amount of water.

Nineteenth-century public health reforms aimed at improving the health and hygiene of the city backfired; the requirement of flushing public streets and the Thames' tributaries to sluice away the waste simply deposited that waste and sewage in the Thames, and therefore into London's drinking water. Disease outbreaks followed, including a succession of cholera outbreaks. The effect on the Thames ecosystem was significant, with pathogens introduced into the food chain, dissolved oxygen levels reduced, and the balance of microflora and microfauna upset.

The water quality of the estuary became so bad that numerous written records from the time attest to the stench, and in 1858, called The Year of the Great Stink, Parliament, housed on Thorney Island on the banks of the river, was unable to convene for the simple reason that the smell of the river under the summer sun was overwhelmingly nauseating. Extensive engineering efforts followed in the 1860s, creating interceptor sewers to divert discharges to outfalls, which succeeded in alleviating the situation in London but simply created a new problem in the mid-estuary area around Barking, where mudbanks of waste lined the river, and a fishery with more than 1,000 workers was destroyed. When two ships collided in 1878, three-quarters of the 800 people on board died without reaching shore; it was commonly accepted that the water was so polluted and stank so badly that swimming was too difficult.

Eventually, ponds were constructed to collect solid waste from discharges, a solution that was retained until the end of the 20th century. The water quality of the estuary gradually improved, and fish populations returned, even becoming plentiful. When London's population increased again after World War I, the new sewage system was able to maintain water quality with no troubles, though the bombing during World War II damaged the infrastructure enough to temporarily set things back to 19th-century levels. The poor state of the postwar British economy made a full repair of infrastructure impossible until the 1960s.

Biodiversity

The Thames supports more than 120 species of fish and marine mammals, including populations of dolphins, seals, and whales in the Tideway, and the Thames Estuary is commonly considered to be the cleanest metropolitan estuary in the world. The ecology of the estuary was deeply affected by the 19th century, not only because of the construction of the Teddington Lock, but also because of the Industrial Revolution, the rising prominence of the United Kingdom, and the growing urbanization of the country, factors that resulted in London's growing to become the largest city in the world. The population reached 4.7 million people by the end of the 19th century.

Well before pollution became a problem, the series of weirs and locks installed to enable boats to navigate upstream to Reading and points beyond created obstacles for the habits of migratory fish. The Thames salmon population declined to near zero by 1820, because adults had too much trouble reaching their spawning grounds upstream.

Freshwater eels have long been plentiful in the Thames and are a traditional British food, from the jellied eels particularly common in East London (and once one of the cheapest forms of animal protein available to Eastenders) to elvers, young eels prized as a delicacy. Eels travel through the river for up to 20 years before journeying across the Atlantic to spawn in the Sargasso Sea; their larvae return to complete the cycle.

Environmental changes, however, have made this once-plentiful creature a rarity. A 2010 study suggested that in only five years, the eel population of the Thames was more than decimated, reduced to 2 percent of its 2005 levels. A total of 1,500 eels were counted in catch-and-release traps in 2005; only 50 were counted in 2010. After the significant damage to the Thames ecosystem in the past, eels were one of the first species reintroduced to the Thames Estuary after the 1960s, and the rapid collapse of local eel populations could be a signal of a sharp reduction in the overall health of the ecosystem.

Birds are plentiful along the Thames, including both sea and shorebirds. Cormorants, black-headed gulls, and herring gulls are the most common. The mute swan and the rare black swan are both traditional birds in the area.

Plants found along the banks of the Thames include yellow flag iris, marsh marigolds, and purple loosestrife. The Lodden lily and snake's-head fritillary are both rare plants that bloom only on flooded areas during the spring season.

Environmental Threats

One environmental problem that the river faces is discharge of raw sewage during heavy rains as sanitary sewers overflow. This can be a serious problem for the fish, plankton, microflora and microfauna living in the river, as well as the animals that depend on it for food or water, including Londoners, who receive two-thirds of their drinking water from the Thames. The resulting decrease of dissolved oxygen can have consequences on Thames life far beyond the obvious pathogenic issues associated with untreated sewage.

Since the 1960s, the Thames ecosystem has been the subject of frequent study and serious interest, in tandem with studies of its water quality. The number of species found in the estuary has steadily increased, including seahorses, goldfish, stingrays, and angler fish. The level of suspended solids in the estuary has remained essentially constant since the early 1990s, after a gradual drop since 1977, indicating that the present level is most likely the natural muddiness inevitable with an estuary, rather than an artificial elevation caused by proximate human habitation.

Heavy metals have decreased in the same period, and pesticide levels have dropped steadily since record-keeping began in 1988. On the other hand, dissolved oxygen, a key factor in measuring the health of an aquatic ecosystem, has repeatedly decreased and increased since 1977, and in the 21st century, it is at very low levels similar to those of the 1970s, when the ecosystem was still being replenished.

The estuary is getting warmer, mainly because of climate change, and the increase has been rapid. Warmer waters mean faster bacterial breakdown, reducing oxygen levels in the water when organic material is present. The combination of warm water and major storm events, which can overflow sewer systems, leads to significant increases in pathogens such as *Escherichia coli* in the summer, and this has caused numerous mass fish kills as pathogen levels exceed the threshold that Thames fish populations can tolerate. Climate change also contributes to drought-like conditions in southeastern England, reducing the amount of freshwater contributed to the Thames, which over time will increase the brackishness of the Tideway.

Bill Kte'pi

Further Reading

Attrill, M. J., ed. *A Rehabilitated Estuarine Ecosystem. The Thames Estuary: Environment and Ecology.* Dordrecht, Netherlands: Kluwer Academic Publishers, 1998.

Hall, Jenny and Ralph Merrifield. *Roman London.* London: HMSO Publications, 1986.

National Trails. "Wildlife on the Thames Path." Natural England, 2012. http://www.nationaltrail.co.uk/thamespath/text.asp?PageId=30.

Thar Desert

Category: Desert Biomes.
Geographic Location: Asia.
Summary: The Thar Desert, also called the Great Indian Desert, is one of the smaller deserts in the world, but it is the largest in India, exhibiting a wide variety of habitats and biodiversity.

Deserts are territories that generally receive less than 5 inches (120 millimeters) of rainfall, averaged over more than 30 years. The Thar, also called the Great Indian Desert, includes the arid portions of western India, eastern Pakistan, and southern Afghanistan. It covers an area of about 172,202 square miles (446,000 square kilometers). The Thar Desert is not one of the largest deserts in the world, but it exhibits a wide range of habitats and biodiversity. It is the most thickly populated desert in the world, with an average density of 83 people per 0.4 square mile (1 square kilometer), whereas in other deserts of the world, the average is only seven people per 0.4 square mile (1 square kilometer). The Thar is considered to be an important desert in terms of its location where Palaearctic, Oriental, and Saharan elements of biodiversity are found.

Climate and Water Resources

The Thar area has a tropical desert climate. April, May, and June are the hottest months. The average maximum and minimum temperatures during this period are 106 degrees F (41 degrees C) and 75 degrees F (24 degrees C), respectively, while December, January, and February are the comparatively coldest months, with average maximum and minimum temperatures of 82 degrees F (28 degrees C) and 48 degrees F (9 degrees C).

Rainfall varies from year to year. Most of the rain falls in the monsoon months from June and September, whereas winter rains are insignificant. The Thar Desert of Afghanistan receives annual rainfall of 13 inches (324 millimeters), and some groundwater comes from the Helmand River flowing through this region. The water resources of the Pakistan desert include water of the Indus River and its tributaries, the Kabu, Jhelum, Chenab, Ravi, Beas, and Sutlej. Local rainfall and usable groundwater from the aquifers underlying the plains are the other available water resources. The average annual rainfall in this region is 10 inches (250 millimeters).

The Thar is a transition zone between major wind belts. Midlatitude cyclones produce moderate amounts of winter precipitation in the northern and western portions, while the eastern portion receives its rainfall from the monsoon circulation that dominates the subcontinent in the summer. The monsoon movement of moist air terminates in western India, resulting in light and irregular rainfall in the Thar. Summers are hot and winters warm throughout the area.

For the Thar as a whole, the entire desert consists of level to gently sloping plains broken by some dunes and low, barren hills, interspersed with sandy and medium- and fine-textured depressions or river terraces and floodplains. Stony and gravelly soils are confined to the slopes of the mountains on the south and east, and the plateaus in the northwest.

Biodiversity

The biodiversity of the region is strongly influenced by soil conditions and water availability, with communities varying distinctively among sand, gravel, and rock areas. Despite its comparatively small area, the Thar Desert has high avian diversity thanks to its location on the crossroads of the Palaearctic and Oriental biogeographic regions. As the Thar Desert is not isolated, avian endemism (species found only here) is low. To the west, it is connected through the Sind Plains with the Persian and then the Arabian Deserts; to the northeast, to the Gangetic plains; and to the east,

A camel feeding on a shrub in the Thar Desert in India. The inhabitants of this most populous desert in the world depend on camels (Camelus dromedarius) for transportation and call them the "ships of the desert." (Wikimedia/Dan Searle)

it joins the semiarid biogeographic zone. In the south, it merges with the Rann of Kutch Desert.

Most species of birds of the Thar are widely distributed; a total 250 to 300 species have been reported here overall. This variation is mainly due to the fact that some authors include Kutch, parts of Saurashtra, and the western side of the Aravalli Mountains in the Thar Desert, while others have more a restrictive definition of the desert that includes only nine districts of western Rajasthan and Kutch in Gujarat.

Tremendous changes in the avifaunal structure of the Thar Desert are taking place due to the Indira Gandhi Nahar Project (IGNP), and species never seen here previously are now regularly found near this canal area. However, the project is wreaking havoc with the desert ecosystem by changing the crop pattern and traditional grazing regimes. Also, the desert is being colonized by new people who do not have the same conservation value system that the desert people have always had.

With the canal providing easy availability of water everywhere, unsustainable livestock grazing is taking place, and the famous Sewan grasslands, which have survived for hundreds of years with low grazing pressure, are now under tremendous pressure. These grasslands are the major habitat of the highly endangered great Indian bustard (*Ardeotis nigriceps*), the winter migrant (*Houbara*), and the Macqueen's bustard (*Chlamydotis macqueeni*).

Other important desert species are the cream-colored courser (*Cursorius cursor*); greater hoopoe-lark (*Alaemon alaudipes*); and various species of sandgrouse, raptors, wheatears, larks, pipits, and munias. In the Rann of Kutch of Gujarat, both greater flamingoes (*Phoenicopterus roseus*) and lesser flamingoes (*P. minor*) breed when conditions are suitable. These nesting colonies come under increasing pressure due to tourist disturbance, and a large number of nests have been reported to have been destroyed. As the sites of the nesting colonies shift, it is difficult to protect them.

Vegetation

Ecologically, the vegetation of the major part of the region falls in the categories of thorn forest type or scrub forest type. Vegetation in this region is sparse, but a surprisingly large number of plant species exist and have immense economic value. Some of the important plants yielding fiber for cottage industries are *Acaccia jacquemontii*, *Leptadenia pyrotechnica*, *Saccharum bengalense*, *S. munja*, *Calotropis procera*, *Acacia senegal*, and *Acacia nilotica*. Natural dyes are extracted from *Butea monosperma* and *Lawsonia alba*. *Citrillus colocynthis* and *C. lanatus* yield nonedible oil for the washing-soap industry.

Famine-food plants that provide grain include *Cenchrus biflorus*, *Panicum turgidum*, and *P. antidotale*. Plants of medicinal value are *Plantago ovate* and *Commiphora wightii*. In the extreme west of the region, there is hardly any tree to meet the tired eyes of a traveler except the king of desert trees, the khejri (*Prosopis cineraria*), which grows only near wells.

Wildlife

The Great Indian Desert is fairly rich in fauna. About 38 species of fish occur in the perennial lakes in the desert. In the larger lakes, *Crocodilus palustris* is fairly abundant, but now its numbers are declining, partly due to the drying of lakes during droughts and partly due to human persecution.

More than 50 mammalian species inhabit the Thar Desert, including large carnivorous flying bats and tiny rodents. The panther (*Panthera tigris*) is usually associated with hilly regions, but its numbers are also dwindling. Rodents constitute one of the largest animal groups in the region. Among insect pests, the desert locust (*Schistocerca gregaria*) is the most pernicious destroyer of vegetation.

The inhabiting populations of the region have domesticated the camel (*Camelus dromedarius*), upon which they are widely dependent for purposes including transportation. The camels in the region are also called the "ships of the desert" because they are the only mode of transportation in the Thar, where human feet and car tires sink into the soft dunes.

Human, animal, and plant desert dwellers are generally well adapted to face the problem of scarcity of water, food, and shelter. An excellent equilibrium has been maintained between the resources available in this area and their use, so that the ecological balance has not been seriously disturbed. But the desert ecosystem is a fragile one, so it has to be handled very carefully in a tender fashion and with great mercy. Slight carelessness, poor planning, and climate change may spoil it to the point of no return.

MONIKA VASHISTHA

Further Reading

Baqri, Q. H. and P. L. Kankane. "Deserts: Thar." In *Ecosystems of India,* edited by J. R. B. Alfred. Kolkata, India: ENVIS-Zoological Survey of India, 2001.

Champion, H. G. and S. K. Seth. *A Revised Survey of the Forest Types of India.* Delhi: Manager of Publications, University of Michigan, 1968.

Gupta, R. and I. Prakash. *Environmental Analysis of the Thar Desert.* Dehradun, India: English Book Depot, 1975.

Mathur, C. M. "Forest Types of Rajasthan." *Indian Forester* 86 (1960).

Planning Commission, Government of India. "Report of the Task Force on Grasslands and Deserts." http://planningcommission.nic.in/aboutus/committee/wrkgrp11/tf11_grass.pdf.

Tewari, D. N. *Desert Ecosystem.* Dehradun, India: International Book Distributors, 1994.

Ward, D. *The Biology of Deserts.* Oxford, UK: Oxford University Press, 2009.

Whitford, W. G. *Ecology of Desert Systems.* Waltham, MA: Academic Press, 2002.

Tian Shan Desert

Category: Desert Biomes
Geographic Location: Asia.
Summary: This ecosystem contains a diverse flora—many rare, endangered, and endemic species—and represents a special mixture of species from steppe, desert, and mountain areas.

In the southern range of the Eurasian steppe belt, west of the Junggarian Basin, a special ecosystem can be found along the northern and northwestern slopes and foothills of the Tian Shan Mountains. This area, which features the Tian Shan Desert at its core, is situated in the vicinity of the lost cities of the Silk Road of ancient times and represents a high diversity of plant species and communities, many of them endemic to the area, that is, found nowhere else on Earth. The flora of the region is highly diverse, with more than 2,000 species. The fauna, too, is valuable, with many rare and endangered species. Overgrazing and hunting threaten this unique ecosystem, but the biggest danger is the exploitation of coal, oil, copper and iron deposits.

The Eurasian steppe-belt is a wide zone of the once endless grasslands stretching from the Carpathian Basin in central Europe to the Amur River Basin that flows to the Pacific Ocean. It appeared when the Eurasian Mountain System and the Himalaya Mountains emerged to form their rain shadow across the center of the Asian continent.

The steppe narrows at two points, dividing it into three major parts. The Tian Shan Desert and Mountains area extends in the east-west direction as much as 1,553 miles (2,500 kilometers) across central Asia. An extensive mountain system comprises part of the basin-and-range topography of northwestern China.

The western part of the Tian Shan breaks down into a complex series of ridges and lake basins that extend westward toward the steppes of central Asia, away from the steep and dry, basin-and-range topography of northwestern China. Thus, this area is exposed to moist Arctic air from western Siberia. Increased precipitation in the northern and northwestern part of the Tian Shan promotes vegetation at low elevations that is more moisture-friendly than in the southern and southeastern part of the range, where mountains rise from the deserts and semideserts of the Taklimakan and Junggar Basin.

Flora

Most of the area of the Tian Shan Desert ecosystem is covered by raised foothill plains and disjunct loess foothills. Because of somewhat increased precipitation and more erratic topography yielding more diverse habitats, the western foothills of the Tian Shan are richer in plant species than other parts of the range. The flora is varied, rich, and includes many endemics.

The lower areas of the foothill plains are occupied by low vegetation called *savannoides*. As the altitude increases, high grasses begin to dominate the plant communities. In low areas of the western Tian Shan region, colorful tall herbs are important members of the plant communities. The tall umbellates appear higher in loess foothills.

River basins support a shrub and meadow savanna with poplars and large sand dunes. Grasses along the Ili River in eastern Kazakhstan include feathergrass and fescue. Dominant shrubs are *Artemisia* species, a drought-tolerant type that is very similar to the sagebrush of the North American Great Basin. Lower foothills support semi-desert vegetation dominated by salt-tolerant shrubs such as tamarisks together with *Artemisia* steppe.

The higher valleys provide moist places sheltered from cold, dry winter winds but exposed to moist air from the west. Here the remains of ancient deciduous forest can be found, including Sievers' apple, Ansu apricot, and maple tree species. These trees represent the remnants of broadleaf temperate forests that flourished in this area during the Tertiary, but were nearly extirpated during the Pleistocene glaciations of the past 2 million years. In this altitude, grass diversity is higher than in the eastern desert foothills of the Tian Shan range, and includes an endemic grass species; wild tulips, some of which are endemic; and desert candle.

Fauna

The diversity of vertebrates in the Tian Shan Desert biome includes representative species from the integrated steppe, desert, and mountain zones. This mixed fauna can be characterized with red fox, corsac, jackal, wolf, brown bear, ground squirrels, marmots, pikas, hamsters, voles, mole vole, gerbils, and jerboas.

The ecosystems of Tian Shan and arid steppes also include such rare ungulates as the Asian wild ass, or kulan; goitered gazelle; saiga antelope; Kizylkum wild sheep; Tian Shan argali; and Karatau wild sheep. The kulan has been recently introduced in this ecoregion to the Kapchagai preserve from Turkmenistan, where the only remaining wild population lives in the Badghyz Reserve.

Among the rare carnivores here are hyena, dhole (Asian wild dog), steppe and marbled polecat, and Pallas's cat. The most spectacular large predator, the Turanian tiger, lived in this ecosystem in the 19th century, but was hunted to extinction about 100 years ago. Rare birds include saker falcon, Barbary falcon, short-toed eagle and steppe eagle, Egyptian and black vulture, white-headed duck, Jankowski's bunting, and bustard. Notable reptiles include the chalcid skink, sunwatcher toadhead agama, Central Asian tortoise, gray monitor and the Central Asian cobra.

Conservation and Threats

There are numerous protected areas in this ecosystem. This mosaic of habitats which occupies the foothills is characterized by intensive agricultural activity. Almost all areas suitable for field crops are plowed, and the grazing pressure on pastures is very high. Cotton cultivation in the foothill territories is particularly intense.

Throughout the ecosystem, the diversity and density of ungulates, predators, and birds of prey have been seriously affected by poaching and improperly managed hunting tourism. Overgrazing, oil and mineral extraction, and poaching are the major threats to this ecosystem. Some climate scientists agree that the most arid areas of the Tian Shan Desert may be expanding as global warming effects increase. This could offset some of the efforts to preserve species in the grassland and shrub areas at the fringes, but the hilly habitats are thought likely to provide more stable environments for many species.

Attila Németh

Further Reading

Bragina, T. M. and O. B. Pereladova. *Biodiversity Conversation of Kazakhstan: Analysis of Recent Situation and Project Portfolio*. Alma Ata, Kazakhstan: World Wildlife Fund, 1997.

Lioubimtseva, Elena. *Climate Change in Arid Environments: Revisiting the Past to Understand the Future.* Allendale, MI: Grand Valley State University, 2004.

National Aeronautics and Space Administration. "NASA Studies Life's Limits in China's Extreme Deserts." http://www.nasa.gov/centers/ames/multimedia/images/2007/extremedesert.html.

Tibetan Plateau Alpine Steppe, Central

Category: Grassland, Tundra, and Human Biomes.
Geographic Location: Asia.
Summary: This cold, inhospitable plateau has prevented significant human incursion from upsetting the balance of the local ecosystem.

The Tibetan Plateau is a vast elevated plateau that includes most of the Qinghai province of Tibet and China, as well as parts of Jammu and Kashmir in India. It is sometimes called the Roof of the World, because it is the highest and largest plateau, at almost 1 million square miles (2.6 million square kilometers), surrounded by enormous mountain ranges, including the Himalayas to the south.

An area of grasslands and shrublands about the size of Texas, the alpine steppes of central Tibet includes a large part of western Tibet, the southern part of the Changtang plateau, and high enclosed basins extending northeast to Qinghai Lake. Plant cover is about 20 percent. The vegetation is too sparse and the climate is too cold to support human-directed pastoralism, so the ecosystem of roaming herds of ungulates, the plants they graze on, and the predators that hunt them is less disturbed by human incursions than many places elsewhere in the world.

Like the rest of the plateau, the climate of the alpine steppe is affected by seasonal monsoons. Climate change may not only affect the plateau, but also in time completely change the character of the alpine steppe. Because the plateau includes the world's third-largest store of ice, warming temperatures that accelerate melting will make the steppe more hospitable to agriculture and grazing, and the resulting introduction of new plant and animal species could destroy many of the indigenous species in the ecosystem through grazing, predation, and competition.

At the moment, though, the climate is predominantly cold. Even the summer months do not exceed 50 degrees F (10 degrees C), and the mean annual temperature is just below 32 degrees F (0 degrees C). Portions of the steppe more closely resemble semiarid desert.

Vegetation

The dominant vegetation of the steppe consists of purple feather grass (*Stipa purpurea*), a grass with feathery flowering spikes, and cushion plants and alpine forbs such as *Arenaria bryophylla, Thylacospermum caespitosum, Saussurea,* and *Leontopodium.* In the colder, drier regions, *Kobresia pygmaea* is more common; this dense, turf-forming sedge grows alongside thick mats of cushion forbs such as *Ceratoides compacta.*

The higher the elevation, the more likely *Carex* sedges are to replace *Stipa* feathergrasses. Less-arid elevated areas support dwarf shrubs such as *Ajania* and *Potentilla fruticosa,* as well as herbaceous legumes such including *Thermopsis lupinodides* and locoweed (species in the *Oxytropis* and *Astragalus* genera), so named because they are generally toxic to grazing animals. Where the steppes transition into the Himalayan alpine shrub and meadows ecosystem, more herbaceous plants can be found, primarily species of *Aster, Anemone, Jurinea, Gentiana, Delphinium,* and *Pedicularis,* as well as shrubs of rhododendrons.

Fauna

The Changtang plateau includes the Changtang Cold Desert Wildlife Sanctuary, in an area of deep gorges and broad plateaus, and including 11 lakes and 10 marshes. It is one of the few habitats of the kiang (*Equus kiang*), the largest of the wild asses,

Black-Necked Cranes

The Changtang Cold Desert Wildlife Sanctuary provides habitat for the nearly endemic black-necked crane (*Grus nigricollis*), which can be found in China, Vietnam, Bhutan, and India but has the majority of its population (about 10,000 individuals worldwide) concentrated on the Tibetan plateau.

This crane, which is revered by many Buddhist sects, forages in small groups, often with one crane watching for predators. Foraging takes most of the day, and a group may walk several miles (kilometers) a day while foraging, feeding on roots, tubers, earthworms, insects, and frogs. The concentration of their numbers in the Tibetan plateau is due to the destruction of their habitats in other countries, in regions more hospitable to human activity. The black-necked crane is classified as vulnerable and is legally protected in China, India, and Bhutan.

Two black-necked cranes in a wetland area. Only about 10,000 of the cranes remain in the wild, with the majority living on the Tibetan Plateau. (Thinkstock)

which is endemic (found nowhere else) to the Tibetan plateau. It has a shoulder height of nearly 5 feet (1.5 meters) and weighs up to 0.5 ton (0.45 metric ton).

There are three subspecies of the kiang, all found in the ecoregion and the northern frontier of Nepal: the western kiang (*Equus kiang kiang*), eastern kiang (*E. kiang holdereri*), and southern kiang (*E. kiang polyodon*). The kiang is an herbivore, feeding on feathergrass and digging for *Oxytropis* roots in the winter. Most of their water comes from the plants they eat and winter snows. They sometimes gather in herds on a temporary basis, but older males are nearly always solitary. They are preyed upon by wolves.

The steppes are also home to the bearded vulture (*Gypaetus barbatus*), one of the rarest raptors. Mated pairs of vultures hunt, breed in, and defend huge territories in which the presence of other vultures is not tolerated. It is the only known animal with a diet consisting primarily of bone; it drops bones from a large height to break them, feeding on the marrow first and then on small pieces of bone. This adaptation allows the bird to feed on carcasses already picked over by predators or other scavengers.

Other animals in the region include the endangered Tibetan antelope (*Pantholops hodgsonii*), the fine underfur of which has made it a target for poachers, and which saw its population halved in the last two decades; one of its predators, the snow leopard (*Panthera uncia*), which also hunts partridges, hares, and Tibetan snowcocks; the wild yak (*Bros grunniens*); the white-lipped deer (*Cervus albirostris*); the wolf (*Canis lupus*); the threatened Himalayan brown bear (*Ursus arctos*); and the argali (*Ovis ammon*), a wild sheep that grazes on the sparse grasses at elevations of 9,843–16,404 feet (3,000–5,000 meters).

Bill Kte'pi

Further Reading

Chang, D. H. S. "The Vegetation Zonation of the Tibetan Plateau." *Mountain Research and Development* 1, no. 1 (1981).

Schaller, G. B. *Wildlife of the Tibetan Steppe*. Chicago, IL: University of Chicago Press, 1998.

Zheng, Du, Qingsong Zhang, and Shaohong Wu. *Mountain Geoecology and Sustainable Development of the Tibetan Plateau*. New York: Springer, 2000.

Tigris-Euphrates Alluvial Salt Marsh

Category: Inland Aquatic Biomes.
Geographic Location: Middle East.
Summary: This unique marshland, with many dramatic seasonal features, is home to many endemic species, a crucial fish nursery, and a vital migratory bird area—and is still struggling to recover from extreme anthropogenic damage.

The Tigris-Euphrates alluvial salt marsh, known as *Ahwar* in Arabic, is located in the southern reaches of the Fertile Crescent of Mesopotamia, or current day Iraq. The marshes begin where these two major rivers meet to the north of Basrah, the second largest city in Iraq, and drain into the northwest Persian Gulf via the Shatt al-Arab estuary. As of 1970, the marshes were estimated to cover approximately 12,427 square miles (20,000 square kilometers) in southern Iraq and the province of Arabistan in western Iran.

This saltmarsh biome stretches across a complex system of marshes and lakes, connected by waterways and canals that provide habitat for a wide variety of fish and wildfowl, and shelter nursery grounds for commercially important marine fishes and shrimp. This system of marshes can be divided into three main sections: the southern Al Hammar Marshes, the Central Marshes, and the northeastern Al Hawizeh Marshes.

What follows are descriptions of the marshes prior to the destruction and diversion of the marsh waters in 1991. From 1991 to 2003, the marshes lost more than 70 percent of their integrity. Since 2003, the natural flow has been restored and further restoration efforts are currently underway.

Al Hammar Marshes

The Al Hammar Marshes are situated to the south of the Euphrates, extending from the southern Iraqi town of Al Nasiriyah in the west, to the area north of Basrah on the Shatt al-Arab in the east. The Al Hammar Marshes are bordered to the south by a sand dune belt of the Southern Desert. Estimates of this marsh area range from 1,739 square miles (2,800 square kilometers) of contiguous permanent marsh and lake, to a total area of over 2,796 square miles (4,500 square kilometers) during periods of seasonal inundation.

Al Hammar Lake, which dominates these marshes, is the largest water body in the lower Euphrates. It is approximately 75 miles (120 kilometers) long and 16 miles (25 kilometers) wide. The lake is brackish because of its proximity to the Persian Gulf, as well as eutrophic and shallow. The maximum depth of Al Hammar Lake is just 5 feet (1.8 meters) on average, and about 9 feet (3 meters) at high tide.

The Al Hammar Marsh complex has one of the most important waterfowl areas in the Middle East, both in terms of numbers and species diversity. The vast and dense reed beds provide ideal habitat for breeding populations, while the ecotonal mudflats support shorebird feeding grounds.

Central Marshes

The Central Marshes are at the heart of the Mesopotamian wetlands ecosystem, bounded by the Tigris River to the east and the Euphrates River in the south. This section of the marsh system receives water from Tigris tributaries, as well as the Euphrates along its southern limit. The Central Marshes cover an area of about 1,150 square miles (3,000 square kilometers); during flood periods, this may expand to over 2,485 square miles (4,000 square kilometers).

This freshwater marsh complex is densely covered in tall reed beds interspersed with several large lakes. Along the marshes' northern fringes, dense networks of tributary deltas are the site of extensive rice cultivation. The Central Marshes are considered to be a highly important breeding, staging, and wintering area for large populations of waterfowl species. Endemic (found only here) sub-species of the smooth-coated otter have been reported in the Central Marshes region.

Al Hawizeh Marshes

The Al Hawizeh Marshes are to the east of the Tigris River, straddling the Iran–Iraq border. The Iranian section of the marshes is known as Hawr Al Azim. In the west, they are largely fed by two main tributaries of the Tigris River. During spring flooding, the Tigris may directly overflow into the marshes. Another important influx comes from the Karkheh River in the east.

Extending for about 49 miles (80 kilometers) from north to south and 18 miles (30 kilometers) from east to west, the marshes cover an approximate area of at least 1,150 square miles (3,000 square kilometers). During periods of inundation, the area expands to over 3,106 square miles (5,000 square kilometers). The northern and central parts of the marshes are permanent, but toward the lower southern sections they become increasingly seasonal.

Flora and Fauna

Natural wetland vegetation typically covers the bulk of the marshes. Common reed (*Pharagmites communis)* dominates the core of the permanent marshes, gradually yielding to reed mace (*Typha augustata*) in the ephemeral seasonal zone. Deeper, permanent lakes support rich submerged aquatic vegetation typified by species such as hornwort (*Ceratophyllum demersum*), eelgrass (*Vallisneria* spp.), and pondweed (*Potamogeton lucens*).

The Mesopotamian marshes are located on the inter-continental flyway of migratory birds, and constitute a key wintering and staging area for waterfowl traveling between breeding grounds in western Siberia to wintering quarters in the Caspian region, Middle East, and northeast Africa. Known as the West Siberian-Caspian-Nile Flyway, it represents a major waterfowl migratory route in western Asia. Two-thirds of west Asia's wintering wildfowl, estimated at several million, are believed to reside in the marshes of Al Hammar and Al Hawizeh.

Particularly dependent on the marshlands are the Dalmatian pelican, pygmy cormorant, marbled teal, and an endemic subspecies of the little grebe (*Tachybaptus ruficollis iraquensis*). The goliath heron, sacred ibis, and African darter, whose world population has been steadily falling, are also known to breed in the marshes.

Furthermore, the marshes have been singled out as one of the 11 non-marine wetland areas in the world with Endemic Bird Area status. The marshes support almost the entire global population of two species, the Basrah reed warbler and Iraq babbler.

The marshes support a rich variety of fish species. Many of the fish are of economic and scientific importance. It is estimated that 60 percent of Iraq's inland fish catch is caught in the marshes. Fish from the carp family *Cyprinidae* are dominant in the marshlands. They are of special scientific interest because of their importance in the study of evolution. At least one barbel species, the gunther (*Barbus sharpeyi*), known locally as bunni, is endemic to the marshands and is of high commercial value.

A number of species are known to spawn mainly in the marshes: the endemic giant catfish (*Silurus glanis* and *S. triostegus*); the Hilsa shad (*Tenualosa ilisha*); and a pomphret (*Pampus argenteus*). A wide range of marine fishes migrate upstream via the Shatt al-Arab. Of major commercial importance is the seasonal migration of penaeid shrimp (*Metapenaeus affinis*) between the Persian Gulf and nursery grounds in the marshlands.

Indigenous Population

The Mesopotamian marshlands are considered home to the Ma'adan, or Marsh Arabs, who have occupied this region for the past 5,000 years, and are descendants of the ancient Sumerians and Babylonians. From study of early settlement of the Tigris-Euphrates basin, there is evidence of water management projects dating back over six millennia. Throughout this period, the power base has consistently been constructed on the wealth generated by irrigated agriculture. Following the early civilization at Ur, and later after the rise of the Ottoman Empire, lack of proper land management led to the degradation of arable lands and encroachment of the western desert onto the Mesopotamian agricultural fields.

It was only in the late 19th and early 20th centuries that major irrigation development began to appear once more in the lower part of the Tigris-Euphrates watershed. During this whole period, the system of water management in place was one that attempted to minimize the risks to crop growth. Another characteristic feature of this era was that only a small proportion of the total water in the river was being utilized for human activity.

The vast majority of the water flowed unused into the Persian Gulf.

Threats

The entire area suffers from desertification and poor soils. Sadaam Hussein's government initiated projects that drained various marshlands and streams, reducing or eliminating marshlands. It remains to be seen whether any or all of this can be reversed, or if these habitats have been lost forever.

While many other regions are studying the potential impacts of climate change, the governments in this area have been slow to acknowledge and respond to the threats. Many governments, already overwhelmed by political turmoil and humanitarian crises, have put climate change studies low on their list of priorities. Nevertheless, in an area where potable water and food sources are critical, any changes to the climate may have a devastating impact on the region, increasing the frequency of droughts, creating food shortages, and reversing efforts to help bring the Tigris-Euphrates Salt Marsh biome back to full recovery.

Nasseer Idrisi

Further Reading

BBC News. "Iraq's Marshes in Doubt." August 30, 2006. http://news.bbc.co.uk/2/hi/science/nature/5295044.stm.

Hussain, Najah A., et al. "Structure and Ecological Indices of Fish Assemblages in the Recently Restored Al-Hammare Marsh, Southern Iraq." BioRisk 3 (2009).

McNab, Christine, et al. *United Nations Integrated Water Task Force for Iraq. Managing Change in the Marshlands: Iraq's Critical Challenge.* New York: United Nations Integrated Water Task Force for Iraq, 2011.

Tigris River

Category: Inland Aquatic Biomes.
Geographic Location: Asia.

Summary: Long a fertile provider to natural species and human cities alike, the Tigris River has been damaged and many of its diverse habitats disrupted.

The Tigris River bounds Mesopotamia from the east; the Euphrates River sets the western boundary. The name *Mesopotamia*, or *between the rivers*, speaks to the great fertile lands between these great flows that gave rise to many ancient civilizations.

The Tigris originates in the Taurus Mountains of eastern Turkey and flows southward through Iraq, passing through Baghdad, and joins the Euphrates near the southern Iraqi marshlands, then flows through the Shatt al-Arab estuary to drain into the northern reaches of the Persian Gulf.

From its origin in tiny Lake Hazar to the gulf, the Tigris is about 1,180 miles (1,900 kilometers). About 248 miles (400 kilometers) of its length runs through Turkey; the next 27 miles (44 kilometers) sets the border between Syria and Iraq; and the remaining approximately 900 miles (1,450 kilometers) runs through Iraq. The Tigris catchment exceeds 230,839 square miles (371,500 square kilometers). From one-third to one-half of the drainage into the Tigris River originates in Turkey, with other source water streaming down from the Zagros Mountains of Iran.

The Tigris system experiences major spring flooding, which has been in recent times heavily controlled by the erection of dams in Turkey, Iran, and Iraq. These controls, added by the mid-20th century, have dramatically changed the hydrology of lower Mesopotamia, especially the dynamics in the southern marshes and alluvial plains. Large dams mark a fundamental departure in the course and focus of the basin's historical riverine development. They induced a major shift from ancient downstream diversion activities by barrages and irrigation canals in the lowlands of southern Iraq in ancient times, to modern water storage and hydroelectric projects in the upper basin of the Tigris and its associated tributaries.

As the Tigris flows southward beyond the headwaters, it nears the third-largest city in Iraq, Mosul, where the largest dam in Iraq was constructed and completed in 1984 to provide hydroelectric power.

Thatched shelters beside the Tigris River where it flows through the district of Hasankeyf, an ancient town in southeastern Turkey. (Wikimedia/Htkava)

Flora and Fauna

The range of aquatic plants in the Tigris includes submerged varieties such as pondweed (*Potamogeton lucens*) and eelgrass (*Vallisneria* sp.), emergent types such as reed mace (*Typha domingensis*), and the iconic papyrus (*Cyperus* spp.).

Water lilies, duckweed, hornwort, and stonewort are also found in abundance, interspersed with broad swaths of common reed (*Pharagmites communis*).

In the southern reaches of the Tigris, and particularly around the marshlands north of the Shatt al-Arab estuary, many stands of date palm (*Phoenix dactylifera*) have been grown, both naturally and by cultivation, since antiquity.

Among Tigris River animals that are endemic, or found nowhere else on Earth, two are birds: the Iraq babbler (*Turdoides altirostris*) and the Basra reed warbler (*Acrocephalus griseldis*). Both find sanctuary among the reeds and ponds of the southern marshlands, which serve as a vital stopover for millions of birds migrating between the three continents of Africa, Asia, and Europe. Among them are pelicans, cormorants, gulls, herons, ducks, and storks.

The marshes are also a vast nursery for shrimp, mollusks, and fish that are key to the ecosystem here, and to the economic livelihood of many Iraqis. At least 60 species of fish dwell in the Tigris River biome, including the endemic catfish *Glyptothorax steindachneri*. The Hilsa shad (*Tenualosa ilisha*) is anadromous, spawning in the waters of the estuary, the marshes, or the Tigris proper, and growing to maturity in the Persian Gulf or the greater Indian Ocean to which it is linked.

Threats

Climate change is exerting an irrepressible rise in temperatures across the Tigris River region, bringing great stress to many habitats. However, numerous species here have evolved in a climate regime that has seen many cycles of drought and flood across the millennia. Still, the current warming occurs against a background marked by unprecedented human-driven impact: construction, damming, water diversion and drain-offs, air pollution, tainted water, war and its dislocations, poaching, ill-considered land-use, and agricultural over-use of fertilizers and pesticides. In this case, the habitat pressures of global warming may push some species to the brink.

NASSEER IDRISI

Further Reading

Geopolicity. *Managing the Tigris-Euphrates Watershed: The Challenges Facing Iraq*. Abu Dhabi, United Arab Emirates: Geopolicity, 2010.

Hamden, M. A., T. Asada, et al. *Vegetation Response to Re-Flooding in the Mesopotamian Wetlands, Southern Iraq*. Madison, WI: Society of Wetland Scientists, 2010.

Trondalan, Jon Martin. *Climate Changes, Water Security and Possible Remedies for the Middle East*. Paris, France: United Nations Educational, Scientific, and Cultural Organization (UNESCO), 2009.

Titicaca, Lake

Category: Inland Aquatic Biomes.
Geographic Location: South America.
Summary: Lake Titicaca is the highest commercially navigable lake in the world.

Lake Titicaca is always described first as the world's highest navigable lake. It is at an elevation of 12,500 feet (3,810 meters). Nestled in the Andes Mountains about halfway along the north-south spine of South America, this prehistoric lake was the cradle of Peru's ancient civilizations and has been inhabited continually for 10,000 years.

The primary economic activities today mirror those from past centuries. Inhabitants fish in the lake's icy waters; exist as subsistence farmers in the poor, rocky soils around the lake; or herd llamas and alpacas on slopes that leave visitors gasping for air. Despite the combined efforts of joint Peruvian and Bolivian legislative bodies, sociocultural improvements lag environmental gains, because of a high level of poverty in the region.

The lake is located at the northern end of the high plateau, or altiplano, region between two ranges of the Andes Mountains. The ring of high peaks surrounding the plateau creates a closed hydrological system referred to as the TDPS system, which encompasses four interacting basins: Lake Titicaca (T), the Desaguadero River (D), Poopó Lake (P), and Coipassa Salt Lake (S). Lake Titicaca is the largest portion of this system, sitting directly on the border between Peru to the west and Bolivia to the east.

Hydrology and Climate

The lake itself is the second-largest in South America, after Lake Maracaibo, spanning 3,200 square miles (8,288 square kilometers). It ranges more than 120 miles (193 kilometers) in a northwest-to-southeast direction and is 50 miles (80 kilometers) across at its widest point. A narrow strait, Tiquina, nearly separates the lake into two bodies of water. The larger portion to the north is called Lake Chucuito in Bolivia and Lake Grande in Peru. The smaller body to the south is known as Lake Huiñaymarca in Bolivia and Lake Pequeño in Peru.

Lake Titicaca is one of fewer than 20 ancient lakes in the world. It is estimated to be 3 million years old and was formed by both tectonic activity and glaciation. The lake currently averages 460–600 feet (140–183 meters) in depth, but the bottom tilts sharply toward the Bolivian side, reaching a maximum recorded depth of 920 feet (280 meters) off Isla Soto in the northeast corner.

However, the lake level fluctuates both seasonally and cyclically over periods of several years. The water level rises during the rainy season (December through March) and recedes through the dry winter months. More than 25 rivers empty into Titicaca, but only one small river, the Desaguadero, drains the lake, at the southern end. A total of 95 percent of the water in the system is lost by evaporation from the fierce winds and strong sun of the arid altiplano. The high rate of evaporation results in a slight increase in salinity annually.

Titicaca has a cold but tropical climate with moderate seasonal changes. Precipitation varies widely across the region, with more humid Lake Titicaca receiving approximately 55 inches (1,397 millimeters) of rain per year, while the arid Coipasa Salt Lake to the south receives only about 7 inches (178 millimeters). Unequal precipitation levels present challenges to water-distribution efforts throughout the area, but the greater issue is variability from year to year. It is not unusual here for a year of severe drought to be followed by a year of extreme flooding.

Biodiversity

A limited variety of animal populations inhabit Lake Titicaca. Two species of killifish (*Orestias*) and a catfish (*Trichomycterus*) were the principal inhabitants of the lake until the 1930s, when trout and mackerel were deliberately introduced for their high economic value. Populations of some native fish species are now vulnerable due to the competition generated by these and other invasive species.

Relatively few mammals live around the lake. Species that have adapted to the high elevation and extreme temperatures include Andean camelids such as the vicuña, llama, guanaco, alpaca, puma, and Andean fox. Avian life is far more prolific.

More than 60 resident bird species share the grasslands, forests, and wetlands of the Titicaca basin with many migratory species.

The most recognized amphibian is the Lake Titicaca frog, or giant frog. Growing nearly 1 foot (0.3 meter) long, this loose-skinned frog spends much of its life under rocks in the marshy edges of the lake.

Effects of Human Activity

Ruins along the shore and on many of the 41 islands in the lake provide evidence of continuous human occupation dating back to approximately 10,000 B.C.E. The Aymara people living in the Titicaca basin still practice their ancient agricultural methods on stepped terraces that predate the Incan culture. Barley, oats, alfalfa, quinoa, and potatoes, along with other legumes and tubers, are common crops. Yields are often low due to the harsh climate and primitive farming methods.

Simple agriculture, fishing, and herding livestock such as llamas and alpacas are the main economic activities in the Titicaca region. Poverty and its associated problems such as poor nutrition, health issues, illiteracy, lack of sanitation, and a fragile environment combine to make it nearly impossible to improve the lives of the population.

Humans have had a great effect on the delicate environment of the TDPS system. Before 1000 C.E., the high plateau was covered with native forest. Around 1100 C.E., a severe 80-year drought changed the surface cover, and the forest disappeared. After 1500 C.E., inappropriate agricultural practices and imported livestock permanently altered the native vegetation.

In the past century, humans have had little additional effect on the environment, primarily due to the arid environment and lack of vegetation. Natural climate change appears to have impacted the region the most, creating periods of heavy rain and inundations, offset by drought.

Future of the Titicaca Region

Despite the historical significance of this region, its future is uncertain. Irregular precipitation and extreme hydrological events limit governments' ability to plan for economic development. Currently, high levels of pollution from the few urban centers and from mining operations in the region are affecting the watershed. The high level of poverty within the TDPS region is responsible for the lack of education and new technology, resulting in overexploitation and inefficient use of land and water resources.

The surface of Lake Titicaca and the entire watershed are evenly distributed between Peru and Bolivia, ensuring integrated stewardship of the entire hydrologic system. Since the 1950s, the countries have been developing a shared vision for environmental protection and socioeconomic development.

A master plan was created with the cooperation of the European Community from 1991 to 1993. Titled the *Master Plan for Flood Prevention and the Usage of Water Resources of the TDPS System,* this document consists of a framework for future development of the entire biome. It focuses on three key components: creating a system for the sustainable use of natural resources; reestablishing ecological integrity by protecting endangered species, and by mitigating human effects on the system; and promoting human development within the basins. There has been measurable success already on the first two points, but until the issues of rampant poverty are addressed, human development in the Titicaca region will have a low level of success.

JILL M. CHURCH

Further Reading

Binational Autonomous Authority of Lake Titicaca. "Lake Titicaca Basin, Bolivia and Peru." United Nations, 2003. http://www.unesco.org/new/en/natural-sciences/environment/water/wwap/case-studies/latin-america-the-caribbean/lake-titicaca-2003.

Dejoux, Claude and Andre Iltis. *Lake Titicaca: A Synthesis of Limnological Knowledge.* New York: Springer, 1992.

Revollo, Mario Francisco, Maximo Liberman Cruz, and Alberto Lescano Rivero. "Lake Titicaca: Experience and Lessons Learned Brief." February

2006. http//:www.ilec.or.jp/eg/lbmi/pdf/23_Lake_Titicaca_27February2006.pdf.
Stanish, Charles. *Lake Titicaca: Legend, Myth, and Science.* Los Angeles, CA: Cotsen Institute of Archaeology Press, 2011.

Tocantins River

Category: Inland Aquatic Biomes.
Geographic Location: South America.
Summary: One of the largest rivers in South America runs through the Cerrado and Amazon, connecting populations in central Brazil and bringing energy and sustainability to several native populations.

The Tocantins River rises in the state of Goiás in central Brazil and runs northward through different sedimentary basins for 1,553 miles (2,500 kilometers) in the states of Tocantins, Maranhão, and Pará. It is formed by two main tributaries, the Paraná and Maranhão rivers. At the end of its course, the Tocantins flows into the Amazon River near Belém. *Tocantins* means "toucan's beak" in the Tupi indigenous language.

The landscape drained by the river is relatively flat, with flooded plains; the banks of the river sometimes flood, creating white-sand igapó forest. The soils are nutrient-rich in many areas, but other areas have low nutrient availability in the soil.

The basin of the Tocantins is formed by another important river, the Araguaia, and has an average discharge of 388,461 cubic feet (11,000 cubic meters) per second. Rapids and falls are the most common aquatic habitats in the Tocantins, dominating the upper course; they are less common in the middle reaches, but form an important habitat on the lower course, which is now largely inundated by the Tucuruí Reservoir.

Rocky and sandy islands and beaches are usually found in the middle course; clay islands dominate in the lower reaches. Floodplain lakes are less frequent along the Tocantins. The period of rising water is October to April, with high-water peaks in February (upper Tocantins) and March (middle and lower courses). The Tocantins-Araguaia system dries up from May to October, with the lowest water levels in September.

The Tocantins and its tributaries can be classified as clear, nutrient-poor, low-ion, and sediment-load rivers. Recently, heavy anthropogenic interventions are changing this pattern. The lower Tocantins has many marginal lakes and numerous islands; it is influenced by both the annual rise and fall of the main river and by tidal cycles. This part of the river course is a complex and fragile ecosystem.

Floodplain lakes and seasonally inundated forests in tributaries of the Tocantins are key habitats for the maintenance of aquatic food webs.

Human Settlement

The history of human occupation in the region dates back 11,000 to 12,000 years, with the first evidence of human presence in the middle Tocantins. This period corresponds to the beginning of the Holocene. The period of Portuguese colonization started in 1625, when a group of Jesuit missionaries established the first settlement in the middle Tocantins. Later, *Bandeirantes* from São Paulo went through the entire region in search of gold and minerals, and pioneered commercial navigation along the river.

Currently, there are several indigenous tribes in the region belonging to different cultures. The Apinajé and Karajá are the major indigenous people living in the region, along with the Xerente, Javaé, Xambioá, and Krahô. These tribes live in the lands on the left bank of the Tocantins and the right bank of the Araguaia, in the north of Tocantins state.

Captain S. C. Bullock provided a vivid report on one expedition that took place in the spring of 1922 through the Tocantins-Araguaia rivers, describing the geography, climate, landscape, and human diversity. The expedition was mounted to confirm the navigability of the river and to search for minerals with economic value in the vicinity.

Biodiversity

The region of the Tocantins River is biologically rich. The eutrophic soils sustain a rich and diverse biota, with many unique species that are found only in this place in the world, that is, they are endemic here. The forests in the region are generally classified as evergreen tropical rainforest. Additionally, the zoological diversity is remarkable. A total 153 species of mammals have been recorded, of which eight are primates and 21 are rodents. Also noteworthy is bat diversity, with more than 90 species.

Many other eye-catching species can be found in the region, such as spectacled caimans (*Caiman crocodilus*), black caimans (*Melanosuchus niger*), yellow-spotted sideneck turtles (*Podocnemis unifilis*), American manatees (*Trichechus inunguis*), and two species of river dolphins (*Ina geoffroyensis* and *Sotalia fluviatilis*). Bird diversity is particularly high, with 527 known species, including toucans, hawks, and the marvelous scarlet macaw (*Ara macao*).

The Tocantins-Araguaia Rivers harbor more than 300 fish species from 30 different families, most of which are migratory species. Upstream reproductive migration occurs in the early rainy season (October to March), followed by return movements by the end of the high-water season. An interesting pattern is the decrease in species richness of fish from mouth to headwaters.

However, fish diversity in the Tocantins is considered to be low by Amazonian standards. For the sake of comparison, the much larger St. Lawrence River basin in Canada, spanning 1,900 miles (3,058 kilometers), has 67 recorded fish species in 20 families. This fact emphasizes the rich biological diversity along the Tocantins river channel and the long-recognized diversity gradient between tropical and temperate ecosystems.

Effects of Human Activity

Four large dams were constructed along the Tocantins River. The two largest ones are Tucuruí and Serra da Mesa, which together generate more than 9,000 megawatts of energy. With the construction of the Tucurui Dam—the largest dam ever built in a tropical rainforest—the natural flow of Tocantins changed, and the migratory fish movements become dependent on the controlled water flow of the hydroelectric dam. Long-distance migratory species, such as large catfish, were directly affected because their upstream movements were interrupted by the dam.

However, other migratory species whose life cycles are completed downstream from the dam were also adversely affected, probably due to separate reasons, but perhaps stemming from altered sediment loads, different patterns of river bank exposure, and related changes in the flora communities downstream.

Despite helping with the development of the region, these dams have further negative effects. Twenty years after the river flow-closure by the Tucuruí Dam, a substantial reduction in fish-species richness occurred in the mid-Tocantins River. The main factor explaining this reduction may be associated with the lack of transference for fish from downriver to upriver. In addition, there is further evidence that the changes in the river channel modified not only species composition, but also size distribution of fish along the middle Tocantins River. It is estimated that only 20 percent of the species are large after the dam's closure.

There is vivid local fishery activity in the middle Tocantins, both professional and artisanal, with an interesting symbiosis between fishers and dolphins (*Ina geoflrrensis*), which are trained to herd the fish against fences, to be rewarded with some trapped fish. Fisheries on the mid-Tocantins have a seasonal pattern. Almost half of the total capture is concentrated from May to August. This fact could be explained by the low level of the Tocantins River in this period, in parallel with the increase in the density of fish assemblage and the presence of large beaches. Nonetheless, the closure of the Tucuruí Dam and others along the river also negatively influenced this activity.

The large commercial species have decreased in number and are qualitatively reduced in that area of the Tocantins. Although total species richness remained the same, changes in fish community occurred in the long term, such as the dominance of predatory species. Fishers also noticed that the Tucurui Reservoir has led to an unprecedented increase in the abundance, length, and weight of some migratory fish species upstream of the dam. However, immediately after the closure of the reservoir in 1984, catches in the lower Tocantins decreased by 65 percent in the two subsequent years. Climate change may impact the migratory patterns of some fish species as well, and alter the mix in varieties of fish that thrive in the different segments of the river.

The Tocantins-Araguaia region is an agricultural frontier and one of the most deforested regions of legal Amazonia. This region has been ravaged by fire, commercial logging, agriculture, and cattle raising, mainly following the roads. It is estimated that the Tocantins has lost about 50 percent of its forest cover. Although deforestation fuels the expansion of development frontiers, selective slashing of inundated trees in floodplains and riparian forests may pose a direct threat to fisheries. These changes affect the discharge of the river and patterns of flow and depth all along its length.

Conservation Efforts

Currently, there are only five protected areas in the basin. The 24,711-acre (10,000-hectare) biological reserve of Águas Emendadas and Chapada dos Veadeiros National Park protect the headwaters of the Tocantins River. The 1,388,732-acre (562,000-hectare) National Park and Indigenous Reserve of Araguaia preserves the middle Araguaian floodplains, whereas a small biological reserve north of Mocajuba preserves part of the lower Tocantins landscape.

Diogo B. Provete

Further Reading

Bullock, C. S. "Tocantins and Araguaya Rivers, Brazil." *Geographical Journal* 63, no. 5 (1924).

Costa, Marcos Heil, et al. "Effects of Large-Scale Changes in Land Cover on the Discharge of the Tocantins River, Southeastern Amazonia." *Journal of Hydrology* 283 (2003).

Garavello, J. C., et al. "Ichthyofauna, Fish Supply, and Fishermen Activities on the Mid-Tocantins River, Maranhão state, Brazil." *Brazilian Journal of Biology* 70, no. 3 (2010).

Instituto Socioambiental. "Indigenous Peoples in Brazil." http://pib.socioambiental.org/en.

Lickens, Gene E., ed. *River Ecosystem Ecology*. San Diego, CA: Academic Press, 2010.

Ribeiro, Mauro César Lambert de Brito, et al. "Ecological Integrity and Fisheries Ecology of the Araguaia-Tocantins River Basin, Brazil." *Regulated Rivers: Research & Management* 11 (1995).

Silverman, Helaine and William Isbell, eds. *Handbook of South American Archaeology*. New York: Springer, 2008.

Tonlé Sap Wetlands

Category: Inland Aquatic Biomes.
Geographic Location: Asia.
Summary: Characterized by annual climatic changes and hydrological fluctuations, these wetlands form a unique ecosystem that has great economic, social, environmental, and cultural significance.

Located in central Cambodia, the Tonlé Sap, or Great Lake, is the largest freshwater lake in Southeast Asia. The lake's drainage basin includes many major tributaries and covers almost 27,027 square miles (70,000 square kilometers). Its low-lying floodplain extends up to 186 miles (300 kilometers) and spans up to 62 miles (100 kilometers) in width.

The Tonlé Sap and its surrounding wetlands are rich in biodiversity and have been recognized by both local and international organizations as a unique ecosystem with significant economic, social, environmental, and cultural value. In October 1997, Tonlé Sap and parts of its floodplain were nominated as a Biosphere Reserve under the Man and the Biosphere Program of the United Nations Educational, Scientific, and Cultural Organization (UNESCO). This was followed by a Cambodian Royal Decree in April 2001 that officially established the Tonlé Sap Biosphere Reserve. The area is to serve various functions, including fostering sustainable development, providing support for projects and education, and conserving biological diversity.

The Tonlé Sap wetlands are characterized by annual climatic changes and hydrological fluctuations that make this ecosystem a unique ecological

The photograph shows one of the 170 floating villages in Tonlé Sap, the largest freshwater lake in southeast Asia. Approximately 470,000 people live within the inundation zone of the lake and a total of 1 million live in the greater area surrounding the Tonlé Sap wetlands. Eight fish sanctuaries help protect the lake from overfishing. (Thinkstock)

phenomenon. The area is subject to a tropical monsoon climate, with distinct dry and rainy seasons. As a result, the Tonlé Sap expands and contracts seasonally following precipitation patterns.

The hydrological fluctuations also cause the flow of the Tonlé Sap to reverse twice a year. During the dry season, water flows out from the Tonlé Sap to the Mekong River. However, during the rainy season, the flow direction reverses, due to large volumes of rainwater entering the Mekong, and water flows back from the Mekong into Tonlé Sap.

The lake volume can fluctuate significantly, from 0.3 cubic miles (1.3 cubic kilometers) during the dry season to as much as 17 cubic miles (70 cubic kilometers) during the wet season—with its depth increasing from 3 to 33 feet (1 to 10 meters). At the peak of the wet season, the lake is estimated to cover up to 8 percent of the land area of Cambodia, as its surface area expands from 965 square miles (2,500 square kilometers) to 6,178 square miles (16,000 square kilometers).

Biodiversity

The annual flood pulse, nutrient cycling, and high sediment fluxes in the Tonlé Sap wetlands support the survival and proliferation of many species of vegetation and wildlife, making the Tonlé Sap ecosystem one of the most productive inland waters in the world. The area is home to approximately 200 plant species, many of which have adapted to withstand significant water-level fluctuations. Many of these flora species are found exclusively in the Tonlé Sap wetlands, that is, they are endemic to the biome. The vegetation in the area includes aquatic herbaceous communities, seasonally flooded freshwater swamp forests, and dense short-tree shrublands.

The Tonlé Sap Wetlands habitat also supports close to 50 species of mammals and more than 40 species of reptiles and amphibians, including globally threatened species such as the long-tailed macaque, Germain's silver leaf monkey, Siamese crocodile, and yellow-headed temple turtle. The lake and its wetlands also form an important breeding, nesting, and feeding site for birds. About 225 bird species have been observed, including globally threatened species such as the spot-billed pelican and black-headed ibis.

Out of all the freshwater systems of the world, the Tonlé Sap wetlands support the greatest known biodiversity of snails. Other invertebrates, such as crabs, shrimp, and more than 200 species of insects, have also been found in the area.

In terms of biomass and species diversity, fish represent the largest and most important vertebrate faunal group in the Tonlé Sap wetlands. The area yields about 197,865–271,169 tons (179,500–246,000 metric tons) of fish annually, accounting for more than half of all freshwater fish catches in Cambodia and providing food for more than

3 million people. Due to the unique fluctuating hydrological patterns of the Tonlé Sap ecosystem, many of the fish species in the area migrate laterally within the ecosystem itself or longitudinally to the Mekong. Globally threatened fish species, such as the Mekong giant catfish and Jullien's golden carp, have been found in Tonlé Sap.

Human Activity and Conservation Efforts

Tonlé Sap and its wetlands also support a huge human population. The lake has played a significant role in shaping Khmer history and culture. It is believed that Tonlé Sap helped the Khmer Angkor civilization thrive, and many traditions and festivals still revolve around the lake. Currently, an estimated 470,000 people live within the lake's inundation zone, and more than 1 million people live in the immediate surroundings of Tonlé Sap and its wetlands.

About 170 floating villages, ranging from two to 100 households per village, exist on the lake itself. The people living on and in the periphery of the Tonlé Sap depend heavily on its natural resources for survival. The main economic activities of the Tonlé Sap communities include fishing, rice cultivation, and wood collection.

The Tonlé Sap ecosystem has been significantly affected by the large volume of human activity in the area. Overhunting and overfishing of animals have resulted in the extirpation of many species here, such as the Irrawaddy dolphin and the greater flamingo. Significant amounts of vegetation have been destroyed by human activity, such as through the conversion of animal habitats for rice cultivation.

Pollution is also prevalent in the area, especially through excessive noise, use of agrochemicals and petrochemicals, and unregulated release of untreated urban and domestic waste. Climate change impacts in the area may include greater than normal areas of flooding and more prolonged flooding.

Conservation efforts have been undertaken by various organizations to protect the Tonlé Sap ecosystem. Eight designated fish sanctuaries have been established in the reserve, which prohibit fishing and boat entry during the dry season. However, more needs to be done to promote conservation and ensure that human activity in the area does not result in irreversible long-term effects to this unique and important ecosystem.

Janelle Wen Hui Teng

Further Reading

Bunnara, Min and Cornelis van Tuyll. *Case Study: Tonle Sap Biosphere Reserve.* Phnom Penh, Cambodia: MekongInfo, 2000.

Davidson, Peter J. A. *The Biodiversity of the Tonle Sap Biosphere Reserve: 2005 Status Review.* Phnom Penh, Cambodia: Wildlife Conservation Society, Cambodia Program, 2006.

Matsui, Saburo, et al. *Tonle Sap: Experience and Lessons Learned in Brief.* Kasatsu, Japan: International Lake Environment Committee Foundation, 2006.

Torngat Mountain Tundra

Category: Grassland, Tundra, and Human Biomes.
Geographic Location: North America.
Summary: This ancient and mountainous landscape at the northern extremity of the Labrador coast supports a pristine and fragile ecosystem.

The Torngat Mountain Tundra biome occupies northern Labrador and the western slopes of the Torngat Mountains in Quebec, Canada, covering 12,500 square miles (32,375 square kilometers). The region consists of steep mountains, deep U-shaped valleys, and impressive fjords carved by glaciers along the Labrador Sea. The valleys contain Arctic tundra shrubs and riparian willow and alder thickets. At higher elevations, vascular plant vegetation becomes increasingly sparse and transitions into a rocky landscape covered by moss and lichens.

The native Inuits refer to this region as *Torngait,* for "the place of the spirits." This name was derived from the presence of Torngarsoak, who was believed to control the life of the sea animals and took the form of a polar bear. The Torngat Mountain tundra has been home to the Inuit and their predecessors for thousands of years. The incredible and pristine Torngat wilderness is known for its impressive and deep fjords, the highest mountains east of the Rockies, the only tundra-dwelling black bears, and the southernmost denning of polar bears on the North American east coast. In January 2005, a significant part of this region became established and protected as Torngat Mountains National Park.

The Torngat Mountains include a collection of geological features explaining the processes that formed this spectacular landscape over the past 3.9 billion years. The slow repetition of plate tectonic events is clearly seen in the rock record. Major mountain-building events occurred during the Archean period (before 2.5 billion years), early Proterozoic (1.6 billion to 2.5 billion years), and in the late Jurassic to Tertiary (after 200 million years).

These major geologic events have created spectacular mountains that are the highest in eastern continental Canada, with peaks over 5,000 feet (1,524 meters). Throughout the mountains, glaciers have carved deep, U-shape valleys. More than 40 small glaciers exist in the Torngat Mountains and are the only remnants of the last ice age left in eastern continental North America.

The northern Labrador region has a tundra climate comprised of short, cool, moist summers and long, cold winters. The mean temperature is 1 degree F (minus 17 degrees C) in winter and 39 degrees F (4 degrees C) in the summer months. The coastal ice can persist here into July, and permafrost underlies the majority of the landmass in this region. The mean annual precipitation ranges from 16 to 28 inches (400 to 700 millimeters), with higher annual precipitation in the central high-elevation regions.

The photo shows rocks covered with moss and lichens high in the Torngat Mountains in Labrador, Canada, in 2008. (Wikimedia/Gierszep)

Vegetation

The rocky landscape and cold climate support areas of distinct vegetation. Toward the north of the biome, the dwarf shrubs decrease in size. The tundra vegetation includes ferns, flowering plants, mosses, liverworts, wildflowers, Arctic sedge, and patches of Arctic mixed evergreen and deciduous shrubs. If the conditions are moist enough, typically along the edge of rivers, Arctic black spruce and mixed evergreen and deciduous shrubs flourish.

Other shrubs include dwarf birch, willows, and heath species. Bearberry and mountain cranberry provide bears food in the late summer. While in bloom, the Arctic poppy and mountain cranberry, or redberry, dot the landscape in yellow and red patches. Where the conditions become too harsh and vegetation becomes sparse and less diverse, shrubs fade away, and lichens inhabit rock surfaces.

Fauna

The harsh climatic conditions and steep geologic formations keep human visitors to a minimum, yet support a variety of boreal and Arctic species.

The Torngat Mountains are home to mostly small mammals, such as the Arctic ground squirrel, but do provide seasonal habitat for polar bears and caribou. As the seasons change, both boreal and Arctic species migrate. Boreal species seek habitat during the summer months until they reach their northern limit, while Arctic species seek refuge during the winter months.

Globally, this region is known to be the home of the only tundra-dwelling black bears. The Torngat Mountain tundra also provides habitat for the southernmost-denning polar bears on the North American east coast. Arctic hare and Arctic fox also roam this region.

The creation of federal sanctuary areas here protects these animals and several important migratory bird species that use this area for breeding and nesting grounds, including Canada geese, whistling swans, and oldsquaw ducks, among others.

Conservation Efforts

The Torngat Mountain Tundra ecosystem is relatively intact. Portions are now protected through the establishment of Torngat Mountain National Park. This ecosystem is extremely fragile, sensitive to temperature gradients, and vulnerable to climate change. In addition, any human development would drastically interfere with the wildlife in this region.

A severe threat to the Torngat Mountain tundra and tundra regions all over the world is the warming temperatures associated with climate change. The melting of permafrost creates a severe risk for the vegetation and animal communities that inhabit these areas.

In addition, permafrost stores an enormous amount of carbon. When permafrost melts, this carbon is released back into the atmosphere as carbon dioxide and methane, both of which are greenhouse gases. Therefore, protection of the Torngat Mountain Tundra ecosystem is quite important for both the preservation of the natural communities here, and for global climate change mitigation.

Megan Machmuller

Further Reading

Commission for Environmental Cooperation (CEC). *Ecological Regions of North America: Toward a Common Perspective.* Montreal, Canada: CEC, 1997.

Payette, S. "Contrasted Dynamics of Northern Labrador Tree Lines Caused by Climate Change and Migrational Lag." *Ecology* 88, no. 3 (2007).

Ricketts, T. H. *Terrestrial Ecoregions of North America: A Conservation Assessment.* Washington, DC: Island Press, 1999.

Tubbataha Reef

Category: Marine and Oceanic Biomes.
Geographic Location: Asia.
Summary: This World Heritage Site is a rich coral habitat that faces several threats, but has begun to be better managed and protected.

Tubbataha Reef is a coral atoll reef in the Sulu Sea, part of the territory of the Philippines. The reef is protected as part of Tubbataha Reef National Marine Park and is also a United Nations Educational, Scientific, and Cultural Organization (UNESCO) World Heritage Site, since 1993. Protected by the Philippines Department of National Defense, the reef has two atolls, North and South Reefs. At low tide, the atolls are exposed to the air and connected, resulting in the *tubbataha* name which means *a long reef exposed at low tide.* The park is managed and patrolled by rangers year round, and is under the jurisdiction of the municipality of Cagayancillo, 81 miles (130 kilometers) to the north.

Diverse Biota

The special marine sanctuary status of the Tubbataha Reef area exists because the reef harbors some of the highest diversity of marine life found anywhere on the planet. There more than 1,000 species of marine animals overall, with more than 360 species of corals and 600 species of fish. The colorful reefs teem with life and also harbor many big marine animals such as manta rays, sea turtles, 12 species of sharks, and 13 species of dolphins and whales.

Established in 1988, the marine sanctuary had a total area of 128 square miles (332 square kilometers). In 2006, the park boundaries were increased by 200 percent: It is now 374 square miles (968 square kilometers).

Because of the amazing richness of marine life in the region, the park is considered a prime destination for avid divers and underwater photographers who flock to the area during Tubbataha Season, mid-March to mid-June. The usual starting point is Puerto Princessa City, 93 miles (150 kilometers) to the northwest of the marine park; here, the many coral walls, where shallow reefs abruptly drop to great depths, attract divers. Such walls offer an impressive perspective into the massiveness of coral reefs and also offer the opportunity to see many kinds of fish, from smaller types living along the walls—such as parrotfish, Moorish idols, moray eels, anemone fish, damselfish, and butterflyfish—to pelagic, or open ocean, species such as barracudas, giant trevally, and sharks.

The Tubbataha Reef also harbors many globally endangered species, such as the hawksbill sea turtle (*Eretmochelys imbricata*).

The importance of this national marine park has been recently highlighted by marine ecology studies showing that Tubbataha Reef acts as a marine species seed bank for the region, sourcing coral and fish larvae out to the greater Sulu Sea area, and helping maintain coral and fish populations far beyond park boundaries. This is important because the Philippines is a tropical island nation and, like many other tropical island nations, highly dependent on marine resources for food and livelihoods. It is vital environmentally as well, due to this seeding function and the role it is thought to play in the broader western Pacific Ocean region.

Threats and Protection

In 2009, the Tubbataha Reef experienced a crown-of-thorns sea star infestation. The crown-of-thorns sea star (*Acanthaster planci*) is the second-largest sea star in the world, ranging from 10 to 14 inches (250 to 350 millimeters). It has venomous spines and feeds on coral polyps. Outbreaks of such sea stars on coral reefs often result in large-scale wipeout of corals unless there are reef managers controlling the crown-of-thorns populations. The infestation on Tubbataha Reef was notable, but not disastrous for the reef. Marine ecologists and reef managers are investigating what causes these outbreaks to prevent them in the future, and to develop ways to mitigate their effects on the functioning of the coral reef ecosystem.

Tubbataha Reef has been well known to Filipino fishers for many years. Before the 1980s, it was mostly subsistence, artisanal fishermen from the closest municipality who would go during the summertime south to Tubbataha in small wooden sailboats. In the 1980s, access to the reef and fish resources increased with greater use of motorized boats. This fact, in addition to the increased use of destructive fishing practices (such as poison fishing and dynamite blast fishing), eventually led to public outcry in the late 1980s.

As a result, the provincial government of the Palawan region teamed with environmentalists and requested presidential endorsement to protect the reefs in the Tubbataha region. This initiative led, in 1988, to president Corazon Aquino declaring support for the establishment of Tubbataha as a national marine park.

Park enforcement and protection is carried out via a field station based on North Atoll. The station hosts park rangers from the Philippines navy and coast guard, the municipality of Cagayancillo, and the Tubbataha Management Office (TMO). TMO is responsible for day-to-day park operations and is the executive arm of the reef management body, the Tubbataha Protected Area Management Board (TRAMB).

Well-managed marine protected areas (MPAs) have shown that with good practices, fish populations increase both inside and outside the MPAs, leading to higher fish catch in nearby communities, and healthier biomes overall. Fishers across the Philippines have witnessed for themselves such positive outcomes; hence, one of the main goals of Tubbataha Reef and its management team is the preservation and conservation of marine biodiversity, which helps maintain natural resource productivity. In other words, MPAs such as Tubbataha Reef help save reefs without destroying subsistence fishers' livelihoods.

The Tubbataha Reef is also known as a bird sanctuary; an above-the-tides islet at South Atoll harbors large populations of nesting masked red-footed boobies, terns, and frigate birds, with more than 100 species of birds. There is a small monitoring station operated by the Philippine Coast Guard at one of the permanent sandbars, aimed at protecting the large bird nesting areas from intrusions.

LIDA TENEVA

Further Reading

Bos, Arthur R. "Crown-of-Thorns Outbreak at the Tubbataha Reefs UNESCO World Heritage Site." *Zoological Studies 49*, no. 1 (2010).

Palawan Council for Sustainable Development. "Tubbataha Reef National Marine Park." Office of the President of the Philippines. http://pcsd.ph/protected_areas/tubbataha.htm.

Tubbataha Management Office. "Tubbataha Reefs Natural Park." http://www.tubbatahareef.org.

Turkana, Lake

Category: Inland Aquatic Biomes.
Geographic Location: Africa.
Summary: The world's largest desert lake is a source of life and sustenance in a remote and harsh landscape, and is now threatened by climate change and large-scale development of its tributaries.

Lake Turkana is one of Africa's most remote and enigmatic lakes. It is located within a hot, dry desert landscape in the Great Rift Valley in the northwestern part of Kenya, just south of the border with Sudan and Ethiopia. The lake sits in an area of volcanic activity that has shaped its existence over time and contributed to the unique ecology and chemistry of the lake.

Lake Turkana has a surface area of 2,606 square miles (6,750 square kilometers) and a volume of 49 cubic miles (204 cubic kilometers). The maximum depth of Lake Turkana is 358 feet (109 meters), and there is a mean depth of 98 feet (30 meters). The lake level fluctuates, with changes of 3–5 feet (1–1.5 meters) commonly observed. The catchment area for Lake Turkana is large and located a great distance from the lake in the western highlands of Kenya and the southern highlands of Ethiopia, covering an area estimated to be more than 50,193 square miles (130,000 square kilometers).

The main tributary of Lake Turkana is the Omo River, which enters from the north, originating in the Ethiopian highlands. The Omo contributes more than 90 percent of the total water flowing into the lake at present. There are many other streams and seasonal rivers originating from mountains and hills around the lake, but all of these are temporary, or seasonal, flows.

The second-largest river flowing into Lake Turkana is the Turkwel River, which originates from the Cherangani Hills in Kenya. The Turkwel River has been dammed at the Turkwel Gorge about 93 miles (150 kilometers) west of the lake. This dam was intended for hydroelectric power generation, but it still has not filled to the levels that were expected.

Notably, Lake Turkana has no outlet—it is an endorheic, or closed basin—and this means that water is lost from the lake mainly by evaporation. The evaporation rate has been estimated at 92 inches (2,335 millimeters) per year. Water levels of Lake Turkana are, therefore, determined by the balance between the inflow from rivers and groundwater and evaporation from the lake's surface. Evaporation rates are high and compounded by the fact that the lake is exposed to high strong winds from the southeast. These factors all contribute to the fact that the level of Lake Turkana is extremely sensitive to climatic variations, droughts, and human-made interference with the rivers and catchment area.

Lake Turkana is located in one of the planet's hottest and most arid areas. The mean annual rainfall in the region directly surrounding the lake is less than 10 inches (250 millimeters), and years can go by here with no rains. Rainfall is typically unpredictable and can be locally heavy and torrential when it does fall.

Weather data from the town of Lodwar in northern Kenya shows a seasonal pattern, with the lowest temperatures experienced in July and August, and a wide range from 68 to 104 degrees F (20 to 40 C), with a mean daily temperature of 84 degrees F (29 degrees C). In some areas, the daytime temperature can exceed 122 degrees F (50 degrees C), and soil surface temperatures of up to 158 degrees F (70 degrees C) have been recorded.

Despite these harsh conditions, Lake Turkana is a thriving ecosystem and an important global site for the study of evolution. The Lake Turkana region is most famous for fossil discoveries that have provided insights into the origin and evolution of hominids, including our own species, *Homo sapiens sapiens*. Hominids have been present around the lake for several million years, and the fossil record of the area, brought to light by the work of the Leakey family, is extraordinary.

Biodiversity

The current flora and fauna of the Lake Turkana region includes a wide range of species that are both unique to the lake as well as typical of the dryland biome of northern Kenya and the Somali-Maasai center of endemism, which means hosting species found nowhere else on Earth.

Lake Turkana supports a large, thriving local fishery. The two most important commercial species are the Nile perch and Nile tilapia, both of which occur naturally in the lake. The annual fish catch was estimated at 16,535 tons (15,000 metric tons) in 1988, but records are scarce, and it is unclear what the current fish catch trends are today. More than 40 species of fish have been recorded from the lake and its tributaries, including several endemic species.

Large numbers of nile crocodiles make their homes in Lake Turkana and its islands, which are havens for wildlife in one of the world's hottest regions. (Wikimedia/ Derek Ramsey)

The vegetation of the lake includes a community of phytoplankton that provides the basis and primary production for the food web in the lake. Lake Turkana is fringed in areas by vegetation with a submerged plant called *Potomageton*. These areas are important spawning ground for fish and occur in more sheltered bays. The deltas of the Omo, Turkwel, and Kerio rivers have swampy, vegetation-including areas of papyrus and bulrushes. However, today the deltas are rapidly being overtaken by invasive species, including the thorny shrub *Prosopis* and the water hyacinth. The invasive species are having a serious effect on the ecology of the deltas and the local fisheries.

Insect life includes large numbers of dragonflies and damselflies. Some of the species that occur in the lake are more closely related to those in the Mediterranean rather than those of tropical Africa, which further indicates the uniqueness of the biogeography of this desert lake.

Bird life at Lake Turkana is abundant, and it is one of the region's Important Bird Areas. Large numbers of the pink-backed pelican, greater flamingo, spur-winged plover, and little stint occur on the lake. More than 30 paleartic migrant species from Europe and Asia make use of the lake on their journey south through Africa. Several important waterbirds breed on the lake and its islands, including the African skimmer and goliath heron.

Lake Turkana contains three islands: North, Central, and South Islands, which are all national parks. The islands are an important nesting area and sanctuary for Nile crocodiles, of which the lake was estimated to contain 14,000 in 1968. The islands are remote and difficult to protect, and invasion by goats or disturbance by fishers is a major issue for the species on these islands.

Human Settlement

The communities surrounding the lake are increasingly shifting from a nomadic, pastoralist, and artisanal fishing lifestyle to a sedentary lifestyle. There is limited cultivation of millet, sorghum, and other crops in the deltas and areas near the rivers. This area has also been recently a zone of drought and famine, possibly exacerbated by climate change, and remains one of the least-developed parts of East Africa.

Large-scale development projects along the Omo River include a proposed dam and plantations for biofuels. These projects would seriously affect the levels of the lake, given that the Omo River is the only real source for Lake Turkana at present.

DINO J. MARTINS

Further Reading

Beentje, H. *Kenya Trees, Shrubs and Lianas.* Nairobi: National Museums of Kenya, 1994.

Bennun, L. and P. Njoroge. *Important Bird Areas in Kenya.* Nairobi, Kenya: East Africa Natural History Society, 1999.

Hopson, A. J., ed. *Lake Turkana; A Report on the Finding of the Lake Turkana Project 1972–1975, Vols. 1–6.* London: Overseas Development Administration, 1982.

Tyrrhenian-Adriatic Sclerophyllous and Mixed Forests

Category: Forest Biomes.
Geographic Location: Europe.
Summary: These forests, which are extraordinarily adapted to hot, dry Mediterranean summers, are home to diverse flora and fauna, but much has been lost to agriculture and development.

The Tyrrhenian-Adriatic Sclerophyllous and Mixed Forests ecoregion has heterogeneous vegetation in which shrubby plants with evergreen, leathery (*sclerophyllous*) leaves are common. This ecoregion occurs in the lowlands along the coasts of the Adriatic and Tyrrhenian seas, which flank the east and west coasts of the Italian peninsula, respectively. This ecoregion's geologic history and Mediterranean climate have given rise to a wealth of specially-adapted species. The ecoregion is a diversity hot spot, with many locally-adapted endemic (found nowhere else) species. The primary threats to this ecoregion are development for tourism, urbanization, agriculture, water shortages, forest fires, and unsustainable collection of rare plants.

The Tyrrhenian-Adriatic sclerophyllous and mixed forests grow in the coastal areas of mainland Italy and on the islands of Corsica, Dalmatia, Malta, Sardinia, and Sicily. Three major geologic systems are found in this ecoregion. Paleozoic substrates such as granite, schist, micaschist, diorite, and gneiss are found on Corsica and Sardinia. Volcanic rocks such as tufa are associated with the band of volcanoes that runs along the boundary between the Eurasian Plate and the African Plate, which includes active volcanoes such as Etna, Vesuvius, and Stromboli. Mesozoic substrates such as limestone and sandstone are found on the southern Italian peninsula, Sicily, Malta, and the Dalmatian Islands.

The biome is located at a meeting point between the Iberian and North African fauna and flora with the Asian Balkan front; this area marks a transition point between temperate and tropical-arid climates. The Mediterranean summers are hot, while winters are relatively temperate, typically ranging from 41 to 50 degrees F (5 to 10 degrees C). Frosts and snowfalls are rare and, when they do occur, ephemeral. Annual rainfall ranges from 16 to 47 inches (40 to 119 centimeters), most of which falls during the winter months. Summer months are characterized by long periods of drought.

Vegetation

Adaptations such as the sclerophyllous leaves, thick bark, deep roots, and other physiological traits to conserve water help the specialized vegetation of this region resist or tolerate the hot, dry summers. Xeric-adapted species of trees such as oak, pine, ash, and juniper are common, as are shrubs such as olive and heather. Because southern Italy was a refuge during the last ice age, this region served as the source from which many species then recolonized the rest of Europe. At present, this biome is the extreme southern limit of many species with a large European coverage, and it includes several relict species.

Tyrrhenian-Adriatic Sclerophyllous and Mixed Forests biome is noteworthy for its high number of endemic plants, which account for approximately 10 percent of the total flora. The ecoregion is also important for rare plants. Examples of such species on the International Union for Conservation of Nature (IUCN) and national Red Lists of threatened plant species include *Salvia fruticosa, S. brachypodon, Portenschlagiella ramosissima, Phyllitis sagitata, Ornithogalum visianicum, Orchis quadripunctata, Iris adriatica, Geranium dalmaticum, Euphorbia rigida,* and *Dianthus multinervis.*

Fauna

Although there are fewer endemic animal species, faunal diversity is relatively high. Two rare and endemic herbivores, mouflon (*Ovis aries musimon*) and Corsican red deer (*Cervus elaphus corsicanus*), live in Sardinian forests. The forests of this ecoregion are also home to a wide range of birds, including endemic species such as Marmora's warbler (*Silvia sarda*), and threatened raptors such as the griffon vulture (*Gyps fulvus*), Eleonora's falcon (*Falco eleonorae*), lanner falcon (*Falco biarmicus*), levant sparrowhawk (*Accipiter brevipes*), and Bonelli's eagle (*Hieraaetus fasciatus*). Many unique species of amphibians and reptiles are found in this region. The Dalmatian Islands in particular are home to many rare butterfly species.

Ecosystem services provided by the Tyrrhenian-Adriatic sclerophyllous and mixed forests include soil stabilization and water retention, which help prevent erosion, habitat fragmentation, and desertification.

Human Activity

The coastal areas throughout the ecoregion are popular tourist destinations. Products such as pine nuts, cork, carob, culinary herbs, plant-based liquors, and ornamental plants come from this ecoregion. For millennia, much of this biome has been used by humans for pasture and cork production. The traditional management of this landscape was adapted to the adverse environmental conditions imposed by the low-quality soils and the harsh climate.

Environmental Threats

Human activity has escalated, and old management regimes are breaking down, having a significant effect on this ecoregion's long-term viability. While forests here are adapted to withstand fire, many fires due to negligence and arson occur with a frequency and intensity that make recovery difficult.

The majority of forest cover has been lost to land-clearing for roads, railroads, urban settlements, industry, agriculture, and pasture. Invasive species such as the cochineal insect (*Matsococcus feyitaudi*) are having a significant effect on native flora. Potential climate change impacts on the area may push habitat zones to higher elevations, applying pressure to plant and animal species alike. Higher temperatures could exacerbate the already erratic fire regime.

Conservation Efforts

Portions of this ecoregion have been set aside as protected areas. Along the Adriatic coast, protected areas include the Gargano Peninsula national park and the Monte Conero regional park. Protected areas along the Tyrrhenian coast include national parks such as Archipelago Maddalena, Cilento, Vallo di Diano e Alburni, Cinque Terre, Circeo, Gennargentu, and Vesuvio; the United Nations Educational, Scientific, and Cultural Organization (UNESCO) World Heritage Site the Aeolian Islands; and the Maremma regional park.

Melanie Bateman

Further Reading

Davis, George W. and David M. Richardson. "Mediterranean Type Ecosystems." *Ecological Studies* 109 (1995).

Medail, F. and P. Quezel. "Hotspots Analysis for Conservation of Plant Biodiversity in the Mediterranean Basin." *Annals of the Missouri Botanical Garden* 84 (1997).

Pignatti, S. *The Woods of Italy: Synecological and Biodiversity*. Turin, Italy: University of Turin, 1998.

Ural River

Category: Inland Aquatic Biomes.
Geographic Location: Eurasia.
Summary: This river, winding through low-rainfall areas, brings life to many fish species and supports a vital migratory bird stopover where it meets the Caspian Sea.

The Ural River originates in Russia, in the southern reaches of the Ural Mountains—the river and mountains together form the dividing line between Europe and Asia—flows southwest into the neighboring country of Kazakhstan, and at its delta meets the Caspian Sea at the Kazakh harbor city of Atyrau. The total length is roughly 1,500 miles (2,400 kilometers); the catchment area is some 90,000 square miles (233,000 square kilometers).

The river first flows down the eastern slopes of the Urals, from a source on Mount Kruglaya, and south to the Russian city of Orsk, where the first major tributary, the Or, joins in as the Ural turns west to skirt the southern end of the mountains. Reaching Orenburg, the Ural is swollen by the next major tributary, the Sakmara. The third major tributary, the Illek, flows in before the final turn to the south, at the city of Oral.

Near its mouth, the Ural River enters the Caspian Depression, where it waters great mudflats and shoreline reeds. Here, the Ural forms two main distributaries—Yaik and Zolotoy—as the mainstem continues out into the still waters of the Caspian Sea, forming a digitate delta, with distinct branches channeling sediment-loaded waters out at acute and right angles. This delta as recently as 1977 extended 20 miles (32 kilometers) straight out into the sea. The Caspian surface level has risen since then, and the delta in its present-day form reaches about 8 miles (13 kilometers) into the sea.

The Ural River biome in general has a continental climate, with hot summers and cold winters; rainfall is rather low, given its location at the juncture of European dry steppes and central Asian semidesert areas. The river is at least partly frozen every winter; maximum flow occurs in April and May. Water temperature in the Ural River ranges from close to freezing to as high as 77 degrees F (25 degrees C).

Annual precipitation can average as little as 4 inches (100 millimeters), and a bit higher near the delta, at 6.3 inches (160 millimeters). Snowmelt is the most significant source of its direct and tributary flow; therefore, the flow volume of the Ural varies considerably each year. At spring flood, the

Ural typically spreads up to 5 miles (8 kilometers) wide across much of its floodplain, and even double that width near the confluences with its major tributaries. The middle reaches of the Ural River basin are dotted with many small lakes and ponds; their niche habitats depend greatly on whether or not the annual spring flood recharges their waters.

Biodiversity

Flora in the Ural catchment transitions from mountain meadows and scrub to forested foothills, and to grassy steppe along much of its middle length—with treed areas fairly consistent along the river itself in its upper and middle segments, transitioning to willow shrub and finally disappearing as the river approaches the Caspian Depression.

In the uplands, conifer coverage includes Siberian pine, Siberian and Norway spruce, downy birch, and larch. Meadow communities include dropwort, clover, and grasses such as Arctic brome. The dry steppes bring a regime of saltwort, sagebrush, Siberian peashrub, and stands of poplar or willow. It its final stretch, the wetlands are a realm of saltmarsh vegetation and submerged reedbeds.

More than 40 species of fish from at least 10 families inhabit the Ural River and its tributaries. Sturgeon is the characteristic fish of the biome. Beluga sturgeon and starry sturgeon, valued for their roe, or caviar, are very high in commercial value and are managed here through harvest control and hatchery supplementation. The beluga sturgeon catch in the Ural, measured as landings, peaked in the 1960s and has since declined. The sturgeon can make its spawning migrations fairly freely here, as the Ural has not been the site of much dam construction.

A freshly caught sturgeon lies on nets in the Caspian Sea region. Sturgeon are able to complete their spawning migrations relatively easily in the Ural River biome because few dams block the river. (Thinkstock)

The Ural hosts freshwater fish types from family *Cyprinidae,* such as carp, minnows, barbs, and barbels, as well as anadromous types such as sturgeon, salmon, and the occasional black-back shad or Caspian cisco which migrate upstream. Other commercial species include bream, perch, and catfish. A fish that is endemic—meaning found nowhere else—to the Ural and the neighboring Volga River system is the Volga whitefin gudgeon, an indicator species for clear, well-oxygenated water.

The wetlands and delta area, at the northeastern lobe of the Caspian Sea, form a rich feeding, resting, and nesting ground for migratory waterfowl. This vital stop on the Asian Flyway is host to at least 20,000 waterfowl; major flocks of great egret, Dalmatian penguin, great cormorant, and Pallas's gull are perennial migrants among the more than 250 avian species recorded here. The birds are drawn by the alluvial banks and inundated marsh areas that support populations of gastropods, mollusks, and crustaceans.

The firmer land surrounding these wetlands supports modest-sized terrestrial species, including bog turtles, Caspian whipsnake, muskrat, the occasional fox or wild boar, and the endemic marbled polecat.

Threats and Conservation

Upstream parts of the Ural River are destined to be tarnished from time to time by the activities of mining operations extracting gold, nickel, bauxite, and other mineral deposits. Oil and coal production in the greater Volga-Ural region is another escalating environmental concern, joining long-standing worries about nuclear waste dumps and facilities in the southern Ural Mountains area.

Global warming trends, if they reduce snowfall in the Ural Mountains region, will have serious negative effects on the flow levels of the Ural River and its capacity to support various species. There is already evidence that the Urals are warming, and the somewhat reliable cycle of snowmelt is giving way to less-likely rainfall events. Many types of fish are quite sensitive to water temperature for triggering spawning activity; such species might migrate to cooler upstream areas of the Ural River and its tributaries, or they may sustain major population reductions as warming takes hold in this biome.

Because it is largely undammed, has generally healthy fish populations, and is a key migratory-bird wetland, the Ural River biome has attracted some worldwide attention and support on ecological matters. The Ural River Delta has been named an Important Bird Area by BirdLife International. Conservation of the sturgeon in the river is a leading goal of the Ural River Basin Project, an international effort that grew out of a 2007 workshop. The government of Kazakhstan, working with the United Nations and other bodies, in 2011 established the Akzhaiyk Nature Reserve in the delta area. Each of these initiatives is a cause for some optimism.

Medani P. Bhandari

Further Reading

Chaves, H. M. L. and S. Alipaz. "An Integrated Indicator Based on Basin Hydrology, Environment, Life, and Policy: The Watershed Sustainability Index." *Water Resource Management* 21 (2007).

Jewitt, Graham. "Can Integrated Water Resources Management Sustain the Provision of Ecosystem Goods and Services?" *Physics and Chemistry of the Earth, Parts A/B/C* 27, nos. 11–22 (2002).

Lagutov, Viktor, ed. *Rescue of Sturgeon Species in the Ural River Basin.* Berlin, Germany: Springer, 2008.

Urban Areas

Category: Grassland, Tundra, and Human Biomes.
Geographic Location: Global.

Summary: The urban ecosystem is a synthetic biome whose defining characteristic is the significant degree to which human activity has dominated and transformed the landscape.

Urban ecosystems are a unique synthetic biome with defining characteristics influenced by large populations and commensurate human activities. The key ecological pressures include replacement of native vegetation with impervious surfaces such as roads, concrete, and buildings; degradation of native vegetation or replacement with nonnative species; fragmentation of natural or green spaces with a corresponding reduction in wildlife habitat; increased pollutant loads; and changes to the hydrologic regime due to land cover change, such as increased flooding during storms.

Urban areas are not always ecological wastelands, however. They can contain novel collections of biodiversity from native as well as introduced species. Today, there is an increasing emphasis on restoring ecological functions within urban areas.

Because populous urban areas require resources, materials, and energy to be imported from outside the locale, these communities do not fit the classic definition of an ecosystem—which is a relatively self-contained system where energy and nutrients are cycled within its boundaries.

Characteristics

The ecological definition of an urban area can vary depending on the context. A commonly-accepted definition is that an area is urbanized if residential dwellings exceed more than one per acre (0.4 hectare), and if the designated area includes such components as housing, commercial and public institutions, railyards, truckyards, and highways. In this context, both city and suburb are considered urban. The ecological differences between urban and suburban areas come down to a matter of degree, related primarily to the density and scale of urbanization.

One of the unambiguous features that definitively distinguishes an urban ecosystem from its surroundings is the high level of energy use. Typically this is generated from fossil fuel combustion, which is required to construct and maintain the urban infrastructure. The energy level in an urban environment is typically much greater than in other ecosystems. While energy use can be relatively complex to measure, it is a metric that can help differentiate urban ecosystems from other ecosystems that are also human-dominated but nonurban.

More broadly in the United States, the Department of Agriculture uses the terms *developed areas, urban,* and *built-up areas.* They include cities, ethnic villages, and built-up areas larger than 10 acres (4 hectares), as well as industrial sites, railroad yards, cemeteries, airports, golf courses, shooting ranges, and other large developments. Typically, urban areas contain a city, such as New York City, Washington, D.C., or Dallas, Texas, with a population of 250,000 to 10 million people. In the United States, the largest city, in terms of population, is New York City, followed by Los Angeles; Chicago; Washington, D.C.; Boston; and Philadelphia.

Urban regions consist of active interactions between a city and surroundings, with the outer boundary determined by a drop in the rate of flow and movements—people, materials, and resources, etc.—radiating outward from the city center.

Metropolitan areas are nearly continuously built, or completely constructed, and have adjoining suburbs. These areas also contain continuous closely-spaced buildings, mostly on small lots. In contrast, suburbs consist mainly of residential municipalities, such as towns, and may be located entirely or partially within, or altogether outside, a metropolitan (metro) area. Peri-urban areas are positioned on both sides of a metro border, where built-up and open areas intermix.

Urban regions begin with a city center, and also feature a development ring that includes an area outside the metro space but still inside the urban-region boundary. This ring is a mosaic of greenspace, interwoven with developed systems, and relatively small built-up areas. Major highways, railroads, and powerline corridors are the prominent systems that criss-cross this environment.

Urban greenspaces, also known as natural areas or natural habitat, are open, unbuilt areas in an urban region that can range from tiny city parks to extensive woodland landscapes, and they can range from self-contained tracts to linear greenways and public river corridors.

Urban Ecology

Cities can be viewed as ecosystems with a metabolism. Using that analogy, the city can be seen from a holistic perspective as a consumer and digester of resources, and a creator of waste products. Total inflows and outflows can be quantified, which gives urban planners tools for ensuring the future availability of resources needed for the city to sustain itself.

Included in this model are inputs, outputs, stocks, and flows of energy, water, nutrients, materials, and wastes. Factors that influence the metabolism of cities include urban density—sprawled, low-density cities have more intensive transportation energy requirements per person than compact dense cities—climate, technology, local policies, programs such as recycling initiatives, and the use of vegetation. Overall, trends in per capita metabolism of urban areas have generally increased over the past 50 years.

Urban areas can be examined in terms of the ecosystems and natural environments located within cities and suburbs. In cities, there are at least seven common types of urban ecosystems that are considered natural, including:

- street trees—stand-alone trees, often surrounded by paved ground;
- lawns and parks—managed green areas with a mixture of grass, larger trees, and other plants, including areas such as playgrounds and golf courses;
- urban forests—less managed areas with a more dense tree stand than parks;
- cultivated land and gardens—used for growing various food and floral types;
- wetlands—various types of marshes and swamps;
- open water areas—lakes and ponds;
- flowing waters—streams, rivers, and seas.

Ecosystems in urban areas provide many services that have important social and economic value. These services include air quality regulation, filtering, detoxification of pollutants and generation of oxygen; local climate regulation such as reduction of the urban heat island effect; shade that helps decrease cooling-related energy use; noise reduction; stormwater drainage and aquifer recharge; sewage treatment; and recreational, cultural, and aesthetic values. In the case of heating and cooling functions, one major city estimated that a 10-percent increase in tree cover, or about three trees per building lot, would reduce the heating and cooling energy cost per dwelling by $50–$90 per year.

Colonizing Species

While the ecosystems in urban areas can be considered natural, they also are novel and relatively new to the planet. Each urban area has unique aspects, but there are still many attributes that can be found in most urban locations. These include: opportunities for some species that do not exist in rural areas; new food sources; places to live; species to parasitize; buildings to roost upon; warm, polluted, nutrient-rich waters; underground sewers for breeding; and train and/or auto tunnels and pipes to inhabit.

Many variables within cities influence the size of natural habitat patches and the dynamics of plant and animal interactions. These variables are consistently present, but differ according to local factors such as buildings, pavement, scattered green areas, street trees, gardens, and houseplants, as well as thousands of people and vehicles moving about daily.

Some species experience population explosions after their habitat becomes urbanized. This has been the case with the bluegill fish in urban ponds, and some pioneer tree species that thrive in vacant lots after the surrounding area has been cleared. In addition, many types of roadside plants thrive in urban areas near thoroughfares that did not exist before urbanization. Many bird species thrive in urban environments. In some cases, their populations increase upon urbanization at the same time as the area's previous native species decline in the same habitat. In the United States, starlings and house sparrows are abundant in every city. However, both of these birds have only been on the North American continent since the early 1900s.

A number of species accompany people wherever they go and colonize a region, imported by humans during the urbanization process; chief among these creatures are dogs and cats, which are intentionally introduced. Many other species are inadvertently introduced into urban environments, among them roaches, rats, and mice; beetles that live in stored grain; and insects—which can arrive in fruits, vegetables, imported flowerpots, and other conveyances.

Urbanization has been known to spread invasive pests and diseases to devastating effect. This has occurred when a pest encounters a new host environment where the native organisms do not have adequate immunity. Among the examples of this scenario are Dutch elm disease and chestnut blight, which destroyed many trees in the United States.

In addition to invaders and pests, urbanization can also have positive, if novel, ecological effects. Nearly 60 percent of many urban areas could be classified as forest; there is also a surprisingly high level of biodiversity in urban areas. One example of this is the heavily-built inner city region of Cleveland, Ohio, which has more than 400 species of plants identified within its city limits.

Environmental Issues

While in many ways there is a relatively high degree of ecosystem functioning in heavily-built urban landscapes, urbanization does pressure everyday functions, as measured by net primary productivity (NPP). This is a key measure of ecological functioning, as it determines the amount of sunlight energy that is captured and fixed by photosynthesis. This process makes the energy available to drive biological functioning throughout the ecosystem.

Urban areas typically have a significantly lower NPP than their natural counterparts, but these areas do contribute to biodiversity in novel ways, as previously discussed. However, urbanization typically fragments native habitats to an extensive degree, which impacts human activities and natural biodiversity.

In general, native biodiversity is increasingly lost the further populations move from a rural

Flowers cover a green roof in Toronto, Canada. Adding green roofs to new or existing buildings is one measure people in urban areas can take to help address some of the environmental problems in cities. Green roofs may mitigate stormwater flows, reduce the urban-heat-island effect, sequester carbon, and encourage biodiversity. (Wikimedia/Sookie)

fringe to the urban core. Native vegetation generally has very little chance of surviving once the building density exceeds one unit per acre (0.4 hectare). Urban exploiter species typically dominate the urban core, while urban adopters dominate suburban areas, and urban avoiders dominate the peri-urban fringe.

Natural materials and nutrient cycles are affected by urbanization, including nutrient cycling, soil erosion, hydrological flow, and the runoff of pollutants from urban areas. Urbanization also greatly changes the natural ecological disturbance regime, which alters natural succession and introduces biogeographic barriers (e.g., roads, canals) that decrease the patch size of natural vegetation and also introduce chronic stresses such as noise and light.

Urban ecosystems can be viewed as nodes in a network of regional and global environments. The growing discipline of ecological footprint analysis determines the extent that natural ecosystems located outside the actual urban area are exploited to support the urban area. Urban dwellers are often the sole macro-consumers for vast areas of cropland, pasture, and forest outside the urban area. The ecosystems providing a majority of the biophysical life-support for urbanites are rural and other nonurban ecosystems. For example, the ecological footprint—the total land area needed to provide the goods and services consumed—of metro Vancouver, Canada, is estimated to be about 11,495 square miles (29,722 square kilometers), about 300 times the size of the actual surface area of the city.

While the issues that may contribute to the unsustainability of many city environments are becoming increasingly evident, there also are encouraging and accelerating trends toward sustainability and ecological restoration in urban and suburban areas. The trend toward integrated urban planning is accelerating, with the approach explicitly addressing the contributions that healthy ecosystems make to the urban economy and quality of life. The Chicago Plan, initiated at the turn of the 20th century, was an early documented example of accounting for open space in urban planning, and the more recent Chicago Goto2040 project has intensified this focus by comprehensively addressing the natural environment and its contribution to civic well-being.

In many urban areas, there is a growing focus on green infrastructure, which seeks, for example, to restore natural vegetation to improve stormwater handling and reduce flooding. There is a growing community-gardening movement, which encourages food cultivation in urban areas, providing environmental, economic, and social benefits. Many cities are encouraging the installation of green roofs (vegetated rather than hard-surface rooftops), that help mitigate stormwater flows, reduce the urban heat island effect, sequester some carbon, and also encourage biodiversity. A growing local food, or locavore, movement encourages local farmers markets and community-supported agriculture plans, which help restore the connections between urban areas and the rural ecosystems they depend on.

Eric Landen

Further Reading

Alberti, Marina. "The Effects of Urban Patterns on Ecosystem Function." *International Regional Science Review* 28 (2005).

Bolund, Per, et al. "Ecosystem Services in Urban Areas." *Ecological Economics* 29 (1999).

Forman, Richard T. T. *Urban Regions—Ecology and Planning Beyond the City*. Cambridge, United Kingdom: Cambridge University Press, 2008.

Garber, Steven D. *The Urban Naturalist*. New York: John Wiley and Sons, 1987.

Pickett, S. T. A., et al. "Advancing Urban Ecological Studies: Frameworks, Concepts, and Results from the Baltimore Ecosystem Study." *Austral Ecology* 31 (2006).

Piracha, Awais, et al. *Urban Ecosystem Analysis: Identifying Tools and Methods*. New York: United Nations University Institute of Advanced Studies, 2003.

Uruguayan Savannas Pampa Grasslands

Category: Grassland, Tundra, and Human Biomes.
Geographic Location: South America.
Summary: One of the most diverse grassland ecosystems of the world, the Uruguayan Savannas Pampa Grasslands biome harbors a unique mix of native winter and summer plant species.

The landscape of the Uruguayan Savannas Pampa Grasslands biome is shaped by large areas of natural grasslands, mostly under grazing by cattle, interspersed with gallery forests and shrublands. The 289,577-square-mile (750,000-square-kilometer) biome encompasses parts of three countries—southern Brazil, southeastern Paraguay, and northeastern and eastern Argentina—and the whole of Uruguay. The climate is temperate within most of the biome, and subtropical in the northern reaches, with less extreme conditions in comparison with areas of similar latitude, because of the influence of the Atlantic Ocean.

The mean annual rainfall here is 20–51 inches (500–1,300 millimeters) along a gradient that runs roughly from southwest to northeast. Temperature is greatly variable across the region, with means of 50–77 degrees F (10–25 degrees C) and occasional frosts. Summer temperatures are more uniform than winter ones, generally 68–90 degrees F (20–32 degrees C) in the north, 63–86 degrees F (17– 30 degrees C) in most of the areas, and 59–81 degrees F (15–27 degrees C) along the moderating Atlantic coast. Extremely hot summer days can hit 104 degrees F (40 degrees C).

The Uruguayan Savannas Pampa Grasslands biome cradles a rich biodiversity of plants and animals; it is among the most extensive grasslands on Earth. Grasslands serve as a backbone to food production and environmental stability wherever they exist worldwide, providing for meat, milk, wool, and leather production by rangelands. They also contribute to soil formation and global temperature regulation through the sequestration of carbon dioxide, with their plant roots helping to control soil erosion, and their biomass providing vital genetic material for the species that consume them.

Flora and Fauna

The fertile plains here include a variety of grasses such as those in the *Poaceae* family, which dominate and determine most of the landscape. Other grasses include legumes (*Fabaceae*), sunflowers (*Asteraceae*), and sedges (*Cyperaceae*). Pampas grass (*Cortaderia selloana*) is an iconic component of this ecoregion, a genus of some 300 large perennial grasses known variously as feather grass, needle grass, and spear grass; they are widespread. There is no complete account of grassland plant species for the entire biome, but up to 380 species of grass have been recorded for Argentina, and 400 each for southern Brazil and Uruguay.

Although apparently uniform at a first glance, the Uruguayan Savannas Pampa Grasslands are extremely rich and diverse, with 30 or more herbaceous plant species often occurring in a single 11-square-foot (1-square-meter) plot. In addition to their overall variety, grassland ecosystems vary widely in space and time. In particular, species composition and abundance vary in space according to differences in soil, latitude, climate, and human land use.

These grasslands present a unique mix of winter and summer native species. The winter types apply the C_3 photosynthetic cycle, an alternative way that plants in low-carbon-dioxide (CO_2) areas convert the CO_2 into a usable form during photosynthesis. The summer types use the C_4 cycle. The C_3-C_4 ratio varies along a latitudinal gradient within the biome: C_4 species predominate in the northern areas, under subtropical climate, and become less expressive toward the south, whereas C_3 species follow an inverse pattern. The distinct phenological rhythm of C_3 and C_4 species adds to the variation of grassland vegetation over time.

Although forests in this ecoregion are mostly restricted to strips along water courses, larger forest patches can be found in the southern Brazilian and Uruguayan highlands. Forests within the biome are comprised of *Araucaria*, a genus of evergreen coniferous tree; forests in the northern parts are seasonal deciduous; and parkland vegetation is restricted to the western boundaries of the grasslands.

The biome has 450–500 bird species, approximately 60 of which are endemic, or found only here. These include cinereous harrier (*Circus cinereus*), buff-breasted sandpiper (*Tryngites subruficollis*), yellow cardinal (*Gubernatrix cristata*), and saffron-cowled blackbird (*Xanthopsar flavus*). There are more than 100 species of mammals, many of which are threatened with extinction as climate change, anthropologically-driven land-use changes, and invasive species encroachment accelerates. Among the flagship mammals are Pampas deer (*Ozotocerus bezoarticus*), puma (*Puma concolor*), maned wolf (*Chrysocyon brachyurus*), and giant anteater (*Myrmecophaga tridactyla*).

History and the Environment

Fossil records indicate that the Uruguayan Savannas Pampa Grasslands biome was populated by large grazing herbivores in the geological past, although the extent to which these animals influenced the modern ecosystem is under debate. The extinction of these herbivores coincided with an increase in the rate of fire disturbances in grasslands, which may have been related to hunting and land-management practices carried out by indigenous people in past millennia. Widespread plant adaptations, such as underground storage organs and high resprouting ability after disturbance, indicate that the grasslands have a long history of co-evolution with dry climatic conditions, herbivory, and/or fire.

In the 17th century, European settlers introduced cattle to the region, and this influence on local economies continues today. The most expressive and traditional economic activity in the biome is beef-cattle livestock farming, fostered by the natural fit of the ecosystem with this activity. Productivity in such systems is often low, but may be increased threefold when simple management practices such as controlling of stocking rates and forage-offer are followed.

Natural pasture is the main livestock feed in this system, which is important for the conservation of the ecosystem. Seventy percent of the region's natural grassland cover is still preserved in Uruguay, as opposed to 45 percent in southern Brazil and 30 percent in Argentina. Also, livestock farming is culturally important because of its relationship with the gaucho, the typical South American cattle herder—part of a rich folklore and cultural legacy.

Current scientific evidence supports the hypothesis that the grasslands here are natural, ancient ecosystems that once dominated even greater expanses of southern South America. However, there also is evidence that forests have been expanding over these grassland-dominated landscapes for roughly 5,000 years, as the climate has turned increasingly favorable to forest establishment. Grazing and fire seem to be disturbances that keep this process at bay, especially at the northern limits of the biome.

Today, large areas of these savannas are being transformed into plantations of introduced tree species such as pine, eucalyptus, or acacia, threatening the ecology of the region. The introduction of nonnative plants potentially threatens the grasslands by reducing acreage for grazing. Other threats include farmland encroachment, animal poaching, and illegal development.

Climate change poses additional challenges. Increased rainfall in the region has led to increases in pasture productivity by 7 percent in Argentina and Uruguay during the past decade, as well as higher yields of soybean, corn, wheat, and sunflowers. More lush pastures could have positive effects on livestock production. Although they may seem positive, these changes could push humans to clear more of the grasslands for conversion to cropland, or tip the grazing use into overgrazing.

Pedro Maria de Abreu Ferreira
Gerhard Ernst Overbeck

Further Reading

Ferreira, Pedro Maria de Abreu and Ilsi Iob Boldrini. "Potential Reflection of Distinct Ecological Units in Plant Endemism Categories." *Conservation Biology* 25 (2011).

Gibson, David J. *Grasses and Grassland Ecology.* Oxford, UK: Oxford University Press, 2009.

Overbeck, Gerhard Ernst, Sandra Cristina Müller, Alessandra Fidelis, Jörg Pfadenhauer, Valério De Patta Pillar, Carolina Casagrande Blanco, et al. "Brazil's Neglected Biome: The South Brazilian Campos." *Perspectives in Plant Ecology, Evolution and Systematics* 9 (2007).

Pallarés, Olegario Royo, Elbio J. Berretta, and Gerzy Ernesto Maraschin. "The South American Campos Ecosystem." In J. M. Suttie, S. G. Reynolds, and C. Batello, eds., *Grasslands of the World.* New York: Food and Agriculture Organization (FAO) of the United Nations, 2005.

U.S. Hardwood Forests, Central

Category: Forest Biomes.
Geographic Location: North America.
Summary: These forests are biologically rich, even after two centuries of continually expanding habitat fragmentation.

The central hardwood forests of the United States are characterized by temperate broadleaf (angiosperms) and mixed broadleaf-coniferous (angiosperm-gymnosperm) stands. They cover more than 100 million acres (40 million hectares) in all. The boundaries of the ecoregion are rather ambiguous, but the commonly accepted range occurs along transition zones or ecotones, with several other bioregions, with the northern hardwood-conifer forest forming a northern border, the eastern and southern boundaries covered by the southeastern-pine forest, and the tallgrass prairie region bordering the west.

First referenced by ecologist E. Lucy Braun in 1950, the definition of Central Hardwood Forest has developed over time and pertains to the forest communities of deciduous and coniferous trees existing mainly on the floodplains of the Ohio-Mississippi River confluence area in the east-central United States, and extending into the more rugged topographic regions of the Ozark Plateau and some hill-and-valley regions in the western parts of Tennessee and Kentucky.

The region today consists of some of the most diverse floral habitats in North America, and the climate is temperate. The Central U.S. Hardwood Forests biome covers approximately 116,000 square miles (300,000 square kilometers).

Biodiversity

Before European contact, the regions consisted of both open oak savannas and forests dominated by oak and hickory species. Oaks in the intact remnants here include red (*Quercus rubra*), white (*Q. alba*), and black oak (*Q. velutina*); prevalent hickories are shagbark (*Carya ovata*) and bitternut hickory (*C. cordiformis*).

Most of these native forests and historic oak savannas have been disturbed, with more than 95 percent ecosystem alteration in some regions, and great swaths have been replaced by agriculture, urbanized areas, or introduced species. The oak savanna regions, of which only about 0.2 percent remains, are fragmented and heavily interspersed with other native vegetation, led by flowering dogwood (*Cornus florida*), sassafras (*Sassafras albidum*), and hop-hornbeam or ironwood (*Ostrya virginiana*). The ecoregion toward its northern boundary contains patches of open prairie. The riparian zones, bottomlands, and marsh areas feature tulip (*Liriodendron tulipfera*), sweetgum (*Liquidambar styraciflua*), and American elm (*Ulmus americana*) as the main tree types.

Along with native flora, the fauna that exist in hardwood forests in the east-central United States are quite diverse and include eastern gray squirrels, opossums, chipmunks, raccoons, deer, and American black bear among major mammals. An array of bird species includes warblers, vireos, wood thrush, sapsuckers and woodpeckers, tanagers, hawks, eagles, owls, and black-billed cuckoo, among many others. Also present are large populations of moths and butterflies, snakes and tortoises, and salamanders and frogs.

Policies of fire suppression implemented in the 1930s precipitated a major decline event in this biome, altering the composition of flora and fauna. With fire no longer keeping savannas in an open state, closed-canopy forests were allowed to develop in many regions. Shade-tolerant species such as maples and yellow poplar now dominate the formerly shade-intolerant composition of oaks; thus, decreases in oak regeneration and establishment have been noted. Invasive vegetation—such as kudzu, garlic mustard, and privet—threaten forest understory species throughout the Central Hardwood Forests biome.

While most forest stands of this region are predominantly angiosperms, softwood conifers are found throughout. A common conifer across the central hardwoods is the red cedar (*Juniperus virginiana*), which has adapted to disturbed sites and does well in drier conditions. Other common conifers are more often found in xeric, or low-moisture, locations in the Ozark Mountains of southern Missouri, and in some of the hillside zones on the eastern fringes of the biome; these include shortleaf and Virginia pine (*Pinus echinata* and *P. virginiana*).

Threats

Much damage has already been done to this biome. Forest-clearing for pasture, cropland, monoculture plantation forests, urban and suburban development, and transportation infrastructure have combined to destroy the majority of the original habitat area. Native species in the fragmented stands that remain must contend with continued counter-cyclical fire suppression regimes, human overuse for recreation, climate change that is leading to a shift in habitat by latitude and elevation, and the persistent onslaught of invasive species. Insects and disease also exact a toll more readily than in the past, as some parts of the complex web of defenses have vanished along with the flora and fauna that once provided them.

*A hillside hardwood forest in the southeastern United States carpeted with white atamasco lilies (*Zephyranthes atamasca*) in spring. Central hardwood forests like these cover about 100 million acres (40 million hectares) of the United States and are one of the country's most diverse regions for flora. (U.S. Fish and Wildlife Service/Pete Pattavina).*

There are some generally small reserve areas throughout the biome. Among them: Flat Rock Cedar Glades and Barrens, in central Tennessee, with 846 acres (342 hectares) protecting such species as leafy prairie-clover (*Dalea foliosa*), Boykin's milkwort (*Polygala boykinii*), and wavy-leaf purple coneflower (*Echinacea simulata*); Hoosier National Forest, in Indiana, with 202,000 acres (82,000 hectares) protecting oak and hickory stands, along with pileated woodpecker, wild turkey, white-tailed deer, and fox; and Shawnee Hills National Forest, in Illinois, with 280,000 acres (113,000 hectares) protecting cherrybark oak (*Quercus pagoda*), cerulean warbler (*Setophaga cerulea*), northern copperbelly watersnake (*Nerodia erythrogaster neglecta*), and bigclaw crayfish (*Orconectes placidus*).

Alexis S. Reed

Further Reading

Fralish, James S. *The Central Hardwood Forest: Its Boundaries and Physiographic Provinces.* St. Paul, MN: U.S. Forest Service, 2003.

Illinois Endangered Species Protection Board (IESPB). *Checklist of Endangered and Threatened Animals and Plants of Illinois.* Springfield, IL: IESPB, 1999.

Shawnee National Forest. *Biological Evaluation of Regional Forester Sensitive Species.* Washington, DC: Shawnee National Forest, U.S. Forest Service, 2004.

Uvs Nuur, Lake

Category: Inland Aquatic Biomes.
Geographic Location: Asia.
Summary: A salty lake located between desert and tundra, Uvs Nuur is important to many migrating birds and supports some unique fish species.

Lake Uvs Nuur is a highly saline lake in an arid, high-altitude basin located mostly in Mongolia and partly in Russia. The largest lake in Mongolia, at 1,300 square miles (3,350 square kilometers), Uvs Nuur has a surface elevation of 2,490 feet (759 meters). The lake averages a depth of just 20 feet (6 meters). Uvs Nuur is a remnant of an ancient lake that covered 35,500 square miles (92,000 square kilometers) and plunged 2,430 feet (740 meters) deep. Ancient mountain uplift separated the present Uvs Nuur basin from the Khirgis Nuur basin to its south, causing major depletion of these waters. Glacial action, winds, and fluid erosion processes molded the present lake basin. The lake's surface level was more than 130 feet (40 meters) higher in the last ice age, but evaporation has steadily reduced it to the current state, with high groundwater salinity part of the proof.

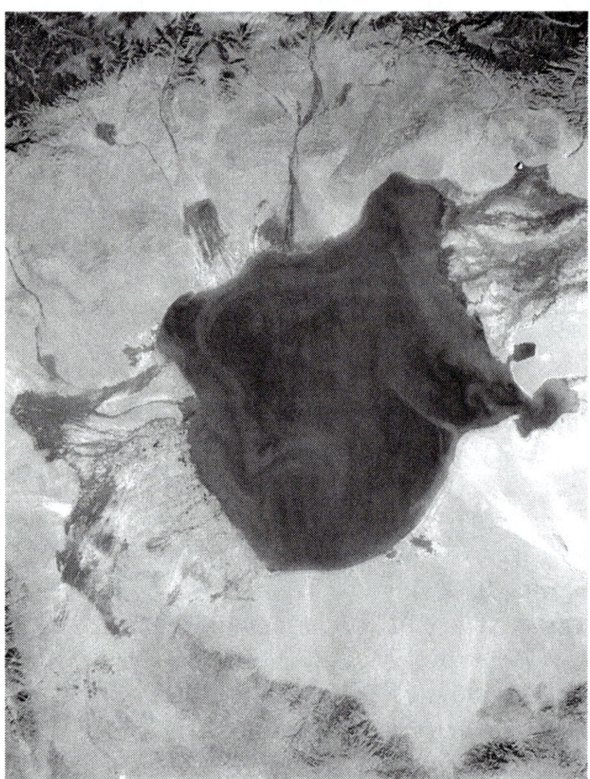

A satellite view of Uvs Nuur Lake in western Mongolia and southern Siberia, Russia. The lake is a remnant of a much larger ancient lake that once extended as much as 35,500 square miles (92,000 square kilometers). (NASA)

The lake is fed by several rivers that originate at the edges of the basin, either in the Altai Mountains to the west or the Khangai Mountains in the east. There is no outlet; the basin is endorheic, meaning draining to the interior. This contributes to the degree of salinity, which is roughly half that of ocean water. The basin spans the geoclimatic boundary between Siberia and central Asia, making for extreme variance in temperatures. From a high of 117 degrees (47 degrees C) in summer, the temperature can reach minus 72 degrees F (minus 58 degrees C) in winter. The water temperature gradient responds to these changes; for instance, in summer a surface temperature of 77 degrees F (25 degrees C) can decrease to 66 degrees F (19 degrees C) at the bottom. Uvs Nuur is ice-covered from October to May, despite its high salinity.

Between 1999 and 2008, water salinity shifted and conditions changed from low algal production and nutrient content, accompanied by clear and oxygenated waters, to grey-colored, somewhat turbid waters with medium levels of nutrients and an intermediate level of productivity.

Flora

Uvs Nuur contains diverse flora communities. Its shallow shore habitats are occupied by species-poor spreads of reed marshes. These are often severely and continuously disturbed by cattle; their grazing and trampling results in the development of annual mudbank communities. The region also includes low reedbeds of mare's tail and spike-rush that develop in shallow water bodies. Aquatic vegetation is primarily represented by fennel pondweed, with water milfoil, stoneworts, and gutweed present. In addition, free-floating duckweed communities appear in several oxbow lakes and pools around the fringe.

At times when the water level drops, short-living stands of dwarf rushes establish themselves on the moist mudbanks, alongside alkali seepweed. Celery-leaved buttercup appears in places, along with oak-leaved goosefoot.

Some salt-tolerant plants grow in shallow depressions that have been cut off from Uvs Nuur, such as glasswort around the bare central part of salt pans with a thick salt crust, followed by stands of alkali seepweed. The transition areas from this groundwater-dependent, halophytic vegetation to the steppe and semidesert are often inhabited by belts consisting of tall and coarse tussocks of Mongolian derris and its associated species.

Fauna

Uvs Nuur is home to 49 species of phytoplankton, 83 types of phytobenthic algae, 45 kinds of aquatic macrophytes, 66 different zooplankton, and 118 species of zoobenthos. The littoral invertebrate community includes scuds, small waterfleas, oar-feet crustaceans, seed shrimp, and insect larvae.

Uvs Nuur is an important habitat for 46 resident waterfowl species, as well as 215 different kinds of birds migrating south from Siberia. At the northernmost area of central Asia, Lake Uvs Nuur and a few smaller neighboring lakes are key nesting sites for many birds, including at least 10 species that overwinter here. Altogether, 20,000–50,000 waterfowl every year utilize the 24-mile (40-kilometer)-wide delta of the Tes-Khem River, a major lake tributary that meanders through an extensive wetland complex. The Uvs Nuur is critical for waterfowl conservation for such endangered species as the white-headed duck and swan goose.

A few dozen mammal species live or pass through the Uvs Nuur biome, including the globally endangered snow leopard.

The lake supports several fish species from genus *Oreoleuciscus,* which is considered near-endemic to the region, meaning it is found scarcely anywhere else in the world. Because of the relative isolation of Uvs Nuur, many of its fish species are endemic or near-endemic. Altai osman, Mongolian grayling, and various species of the ray-finned genus *Triplophysa* also inhabit the lake.

Environmental Issues

As a moist, if salty, oasis in a high basin sited between tundra and desert, Uvs Nuur has not been polluted or developed by industrial or commercial interests. Much of its human interaction takes the shape of pastoral and nomadic activity. The area is generally one of low population density, although livestock herd size in the province has expanded in recent years. There has also been some diversion of water from tributaries, which at times makes fish migration difficult. As environmental protection actions increase in Mongolia, some attention has been paid to Uvs Nuur, which the government classifies as a Strictly Protected Area. Nearby, a national park is located on one of the lake's tributaries. Various international designations have been conferred on the lake and its basin, such as Biosphere Reserve, World Heritage Site, Ramsar Site, and Important Bird Area.

Máximo Florín

Further Reading

Alonso, Miguel. *Limnological Catalog of Mongolian Lakes*. Barcelona, Spain: Endesa, 2011.

Batima, Punsalmaa, Nyamsuren Batnasan, and Bernhard Lehner. *Freshwater Systems of the Great Lakes Basin, Mongolia: Opportunities and Challenges in the Face of Climate Change.* Ulan Bator: Mongolia Programme Office, World Wildlife Federation, 2004.

Dulma, Auyur. "Hydrobiological Outline of the Mongolian Lakes." *Internationale Revue der Gesamten Hydrobiologie und Hydrographie* 64, no. 6 (1979).

Fernández-Giménez, María E. "Spatial and Social Boundaries and the Paradox of Pastoral Land Tenure: A Case Study from Postsocialist Mongolia." *Human Ecology* 30, no. 1 (2002).

Hilbig, Werner. "The Distribution of the Vegetation in the Uvs-Nuur Basin and its Surrounding Mountain Ranges." *Feddes Repertorium* 114, nos. 7–8 (2003).

Shinneman, Avery L. C., James E. Almendinger, Charles E. Umbanhowar, Mark B. Edlund, and Soninkhishig Nergui. "Paleolimnologic Evidence for Recent Eutrophication in the Valley of the Great Lakes (Mongolia)." *Ecosystems* 12 (2009).

Valdivian Temperate Forests

Category: Forest Biomes.
Geographic Location: South America.
Summary: A virtual continental island between the Pacific Ocean and the Andes Mountains, these forests boast high species endemism but face growing threats from human activities and climate change.

The Valdivian Temperate Forests biome is one of the few temperate rainforest areas in South America. Sandwiched between the Pacific Ocean and the Andes Mountains in the southern reaches of Chile and a small area of Argentina, the distribution area constitutes a virtual continental island. Species richness is relatively low, but the rate of species found only here, or endemism, is exceptionally high—about 45 percent for all vertebrates, and up to 90 percent of seed plants.

Phylogenetic uniqueness is also high; for example, there are 32 genera of trees, of which four-fifths are monotypic, meaning the genus comprises just a single species. These factors all indicate long and ancient isolation.

Geography and Climate

Wet temperate forests dominate the narrow strip between the western slopes of the Andes and the Pacific Ocean, from just north of the Chilean capital Santiago to the southern tip of the Taitao Peninsula. The forests are found in two north-to-south running mountain ranges as well as the intermediate valley. The peaks of the Andes are higher toward the north, about 23,000 feet (7,000 meters) higher than the southern range, which peaks at 13,123 feet (4,000 meters). Treelines are the same, at roughly 7,874 feet (2,400 meters). Forests are replaced by montane grasslands and shrublands above the treeline in the north, and by temperate grasslands, shrublands, and savannas further south.

The Andes were extensively glaciated during the Pleistocene Era, although volcanic rock and deposits have since covered most of the glaciated surfaces. The coastal range, Cordillera de la Costa, is a mountain belt rising to 4,265 feet (1,300 meters). It lies parallel to the coast and the Andes, which remained largely unglaciated. It submerges off the shore south of 42 degrees south, but then forms the large Chiloé Island and the Chonos Archipelago. Between the two mountain ranges lies a low structural depression valley, which submerges into the ocean at the same point.

The depression is filled with volcanic ash, erosion and glacial deposits; this fertile valley in the northern reaches of the biome is Chile's agricultural heartland. Its climate results in a mediterranean vegetation type which separates the temperate forests of the eastern and western mountain ranges. These two temperate systems join further south.

The northward flowing Pacific Ocean currents combine with the dominant moist, westerly onshore winds to produce a maritime cool climate overall. This produces higher humidity and precipitation in the coastal range than in the Andes, and on the western slopes than on eastern slopes. In the south, rain falls year-round, with annual precipitation exceeding 236 inches (6,000 millimeters). A shift toward a winter rainfall regime and annual precipitation of about 39 inches (1,000 millimeters) occurs in the north. Coastal upwelling causes year-round coastal fog. Average annual temperatures vary between maxima of 55–70 degrees F (13 to 21 degrees C) and minima of 39–45 degrees F (4 to 7 degrees C) in the north and south, respectively.

Biodiversity

Southern beeches of the genus *Nothofagus* are widespread; this genus is found around the southern Pacific Rim, indicating a common evolutionary history reaching back to Gondwana.

Precipitation and, to a lesser extent, disturbances and latitudinal and altitudinal temperature differences, determine the distribution of vegetation types. The shade-intolerant *Nothofagus* species, which do not regenerate in undisturbed old-growth forests, are characteristic. They are widespread in the Andes due to periodic disturbances including volcanic eruptions and avalanches. Disturbances are less important in the coastal mountains where shade-tolerant trees are abundant.

Five main types of forest ecosystems can be distinguished. Northern deciduous forests with dominant *Nothofagus* species, including rauli and roble, mark the transition to mediterranean forests. Valdivian laurel-leaved forests are typical for the *Nothofagus* gap, and are dominated by broadleaf evergreen tree species, including tepa, ulmo, tiaca, tineo trees, and an understory of myrtle trees and arrayán.

Northern Patagonian forests are dominated by evergreen species such as coihue (*N. dombeyi*). Patagonian Andean forests feature the monkey-puzzle tree, a living fossil and Chile's national tree, and the valuable and threatened alerce. High Andean scrublands with *Nothofagus* dominate nearer the treeline. Magellan's beech (*N. betuloides*) and bogs of *Sphagnum* mosses typify southern evergreen forests.

Endemic bamboo species are characteristic understory species and can form dense, pure thickets. Edible, large-leaved perennial nalca and ferns are widespread. The copihue, Chile's national flower, is a representative pioneer in disturbed areas.

Mammal endemism is relatively low here, but there are five endemic genera, including the *monito del monte,* an arboreal marsupial. Endemism levels for both reptiles and amphibians are high, with many species restricted to very small areas. They constitute important species targeted by the Alliance of Zero Extinction scheme.

About 30 percent of the bird species here are estimated to be endemic. There is just a single species of hummingbird, (*Sephanoides sephaniodes*), but it is vital to the one-fifth of woody

The monkey puzzle tree (Araucaria araucana), which is considered a living fossil, is Chile's national tree and an important species in the Valdivian Temperate Forests. (Wikimedia/Norbert Nagel)

plant genera that depend on its visits to spread pollen between their typically red, tube-like flowers.

Threats and Conservation

Before the arrival of the Spanish colonizers centuries ago, the indigenous Mapuche people cultivated only a few open areas, leaving the forest cover intact. After the colonizers' arrival, heavy logging, burning, land clearance and habitat degradation began. Invasive species and, since the 1970s, ever-increasing pine and eucalyptus plantation forests, have caused significant damage. Today, few primary forests remain, especially in the coastal range.

Chile's economy is one of the fastest-growing in Latin America, exercising strong human pressures on remaining forest fragments through tourism, construction of highways, and forest clearing for power lines, connecting the economic heartland with the hydroelectricity-producing areas in the south. There is a large network of protected areas, but—especially in the north—they are concentrated at middle elevations in the Andes, leaving the coastal range under further pressure. The area with the highest biodiversity (36–41 degrees south) has the lowest percentage of protected areas.

The analysis of historic plant distributions shows substantial variations, and indicates a high sensitivity of temperature regimes. Climatic change and geographic isolation seem to have resulted in a net loss of species over time, which lends credence to current observations that this ecoregion is highly susceptible to global warming.

The Valdivian Temperate Forests ecoregion is part of a Conservation International hot spot. Together with the Juan Fernández Islands it constitutes a World Wildlife Fund Global 200 ecoregion, and also belongs to the Top 100 Ecoregions, a list of those biomes with the highest richness-adjusted endemism of vertebrates in the world.

However, intensive logging and land conversion for forest plantations, agriculture, and economic development severely threaten this biome—in particular by fragmenting habitats and opening the door to invasions by exotic plants and vertebrates.

Stephan M. Funk

Further Reading

Funk, S. M. and J. E. Fa. "Ecoregion Prioritization Suggests an Armoury Not a Silver Bullet for Conservation Planning." *PLoS ONE* 5 (2010).

Smith-Ramirez, C. and J. J. Armesto. "Plant Phenology in a South American Temperate Rainforest, Chiloe, Chile." *Journal of Ecology* 82 (1994).

Veblen, T. T., C. Donoso, F. M. Schlegel, and R. Escobar. "Forest Dynamics in South-Central Chile." *Journal of Biogeography* 8 (1981).

Vanuatu Rainforests

Category: Forest Biomes.
Geographic Location: Pacific Ocean.
Summary: The Vanuatu archipelago features rugged terrain with a patchwork of native and invasive forest cover and some remarkable synergy between fruit bats and trees.

The Vanuatu archipelago in the South Pacific Ocean is comprised of more than 80 islands in a grouping roughly 800 miles (1,300 kilometers) in length. The location is some 1,100 miles (1,800 kilometers) east of Australia and 105 miles (170 kilometers) south of the Solomon Islands. The total land area of the Vanuatu nation is about 4,750 square miles (12,300 square kilometers).

Topography and Climate

The capital, Port Vila, is on the island of Efate. The archipelago is part of the Pacific Ring of Fire, with volcanic activity as the origin of each island, and an active seismic regime. While there are some stretches of flat, sandy beaches, most shorelines are rocky, with water depth dropping quickly. The same steep topography continues inland; there is little bottomland, savanna, or meadow area. Soils tend to be shallow or loose and subject to landslides. Freshwater is not abundant.

Because of the steep nature of the islands, little flat or bottomland is available. Forest tracts are extensive, covering nearly 77 percent of the land area. Mid-height forests, 66–99 feet (20–30 meters) in height, and low forests, 33–66 feet (10–20 meters) in height, comprise approximately 35 percent of the total. The largest island, Espiritu Santo, holds the tallest mountain peak, Mount Tabwemasana, at 6,165 feet (1,879 meters).

Biodiversity

The discovery of extremely valuable sandalwood (*Santalum austrocaledonicum*) growing wild in Vanuatu and neighboring New Caledonia caused land-rush harvesting in the mid-19th century,and nearly 100 percent of the wild plants were removed. Strict regulations adopted in 1987, accompanied by concentrated horticultural efforts, may have saved this species. The tree has now recovered well, but the lingering effects of the clear-cutting and habitat degradation on the other native flora and fauna of the islands has yet to be determined.

Timber harvesting of the remaining and replanted sandalwood and other trees is now more tightly controlled, and currently only two species here are being harvested at or below sustainable levels: whitewood (*Endospermaum medullosum*) and melektree (*Antiaris toxicara*), However, some intact native forests still remain, primarily in steep mountainside locations.

Vanuatu rainforests tend to be shorter in height than similar island forests, with nearly all the trees being shorter than 100 feet (30 meters) in height. This is likely a result of multiple factors, including increased seismic activity and landslide, location of the islands in a zone of frequent cyclones, poor soil formation due to steep terrain and high rainfall levels, and human activities, such as habitat fragmentation and erosion during World War II, when control of the islands was transferred between the Japanese and Allies in several heavily fought engagements.

One notable type of fauna found in the Vanuatu Rainforests biome is the fruit bat, or flying fox. Fruit bats are distributed throughout Old World Tropics regions, including the islands of the southern Indian and Pacific Oceans. Many such islands have their own endemic species (those found nowhere else), often at low densities, which makes them of extreme importance for conservation and research. Vanuatu has five species: four in the genus *Pteropus* and one in the genus *Notopteris*. All are declining and most face risk of extinction.

Fruit bats are extremely valuable to the ecosystem in that they serve as pollinators or seed dispersers for native trees. For example, on the Samoan Islands, between 80 and 100 percent of all tree seeds found on the forest floor are distributed by fruit bats. In Vanuatu, some bat-dependent tree species are of considerable economic importance, such as ebony (*Diospiros* spp.) and mahogany (*Sweitenia* spp.).

Many forest trees have co-evolved with the fruit bats in order to increase the likelihood of pollination or seed dispersal. Some tree species, for instance, produce prodigious quantities of nectar to attract the bats. Other trees bloom only at night to prevent other potential pollinators from capturing the resource. On the island of Guam, where the two endemic species of fruit bats have been extirpated by brown tree snakes (*Boiga irregularis*) and human encroachment, several tree species have stopped blooming.

Fruit bats tend to suffer heavily at the hands of humans. Most islands are protein-poor environments, and given that fruit bats are relatively large and conspicuously roost communally during the day (often on exposed tree branches), they are easy targets for hunters. Fruit bats, although protected by the government in many areas, are sold in markets and served as delicacies in restaurants across many areas of the Pacific. Most species are declining, and two on Vanuatu are listed as endangered.

There is a discrepancy in scientific literature on whether there are four or five species of fruit bat on the Vanuatu Islands chain, and whether or not they are all endemic. For example, the long-tailed fruit bat (*Notopteris macdonaldi*) is listed as a Vanuatu endemic, yet it is also found in Fiji and New Caledonia. In either case, it is considered an endangered species.

Despite being tropical, with heavy vegetation cover and moist conditions, Vanuatu has fewer vertebrate species than other island chains of similar size. Besides the four to five fruit bat species, there are eight species of smaller bats (family *Microchiroptera*) and no other native species

A steep forest lined with coconut palms along the coast of Vanuatu. The archipelago's forests have begun to recover from the loss of nearly all their wild sandalwood through overharvesting, which has since been curbed. (Thinkstock)

of land mammals. The domesticated and feral mammals are typical of many places humans have colonized, and include three species of rat (*Rattus* spp.), mouse (*Mus musculus*), cattle (*Bos taurus*), goats (*Capra hircus*), pigs (*Sus scrofa*), feral cats, and dogs. Of particular note are the feral pigs, which cause massive forest destruction through their foraging on young plants and the creation of wallows they use for cooling and protection from parasites. This habitat destruction, coupled with high rainfall, results in serious erosion problems and resulting drops in water quality.

There are 19 species of lizards in the Vanuatu Rainforests biome, one of which is declining to the point of extinction, and two snakes, the flower pot snake (*Ramphotyphlops braminus*) and the Fiji boa (*Candoia bibroni*). Both snakes are wide-range dispersers and are found throughout the Southern Hemisphere, including most of the South Pacific Islands.

A total of 139 bird species have been recorded on Vanuatu; most are not restricted to the rainforest highlands. Of the species present, nine are endemic. Eight have been introduced by humans and have become established. Twelve species are threatened or endangered on a global basis.

Although found in a variety of locations, rainforest species of interest include the green palm lorikeet (*Charmosyna palmarum*), Vanuatu mountain pigeon (*Ducula bakeri*), peregrine falcon (*Falco peregrinus nesiotes*), Vanuatu flycatcher (*Neolalage banksiana*), and golden whistler (*Pachycephala pectoralis*). Populations of these species vary widely, in part due to the cyclonic nature of the weather. Large population crashes are noted after severe storm activity, yet rapid recoveries tend to occur during intervening years.

Robert C. Whitmore

Further Reading

Bowen, J. 1997. "The Status of the Avifauna of Loru Protected Area, Santo, Vanuatu." *Bird Conservation International* 7 (1997).

Bule, L. and G. Daruhi. 1990. "Status of Sandalwood Resources in Vanuatu." In *Proceedings of the Symposium on Sandalwood in the Pacific,* by L. Hamilton and C. Conrad. Berkeley, CA: Pacific Southwest Research Station, U.S. Forest Service, 1990.

Mourgues, A. *Republic of Vanuatu Environment Profile.* Port Vila, Vanuatu: Government of Vanuatu, 2005.

Vasyugan Swamp

Category: Inland Aquatic Biomes.
Geographic Location: Asia.

Summary: A nearly pristine peat-bog ecosystem, the extensive Vasyugan Swamp is the largest swamp in the Northern Hemisphere.

The Vasyugan Swamp, also known as the Great Vasyugan Mire, is an ecosystem unlike any other on Earth. Geomorphologically unique, the Vasyugan features unique types of swamp massif. Located in the Tomsk region of western Siberia, Russia, the Vasyugan Swamp is the largest swamp system in the Northern Hemisphere and one of the largest swamps in the world, spreading across an area of 20,500 square miles (53,095 square kilometers). The swamp represents about 2 percent of the total area of the planet's peat bogs. From east to west, it extends more than 350 miles (563 kilometers), and from north to south, it extends around 200 miles (322 kilometers).

Formed in the last ice age, the watersheds of the Ob and Irtysh Rivers accumulated sediment, giving rise to the Vasyugan Swamp. It is considered to be a continuous and even accelerating natural phenomenon that has been expanding since it appeared some 10,000 years ago. Most of its growth has happened in more recent times; three-quarters of its current area was developed in the past 500 years.

Located in the central sector of the western Siberian plain in a transitional zone between a subregion of small-leaved forests and the southern taiga, the Vasyugan Swamp sits on the Vasyugan plain. The plain lies within the boundary of four regions of the Russian Federation: Tyumen, Omsk, Tomsk, and Novosibirsk. It extends along the west bank of the Ob River at the confluence of the Ob and Irtysh Rivers.

The swamp has a continental climate with short, hot summers and long, cold winters. Average temperatures range from minus 5 degrees F (minus 21 degrees C) in January to degrees 64 F (18 degrees C) in July. Precipitation is in the range of 17–19 inches (440–480 millimeters) annually.

The primary source of freshwater in the region, with about 100 cubic miles (417 cubic kilometers) of water reserves, the swamp includes almost 800,000 small lakes and is also the source of numerous rivers. It holds massive reserves of peat, estimated at more than 1 billion tons, which reach depths of more than 30 feet (9 meters).

While the climate, topographical relief, and geological structure of the Vasyugan Swamp are typical for temperate-zone swamps, its particular development, distinct lithogenic and biological characteristics, and specific location in the southern portion of the western Siberian plain give the swamp its unique quality. Evenly comprised of bogs, fens, and forested mires, the landscape of this complex ecoregion is perpetually in development. It features springs, peat beds, valleys of rivulets, swamps in transitional phases, and temporary water canals extending from vast waterlogged interfluves.

Its central regions feature swamps and ombrotrophic (watered only by precipitation) mossy pine bogs, known as ryam islands, which rise 20–30 feet (6–9 meters) above the low-lying swamps that line the periphery. The swamp features different types of peat deposits, diverse flora, and distinct forest and swamp landscapes along its northern and southern edges. These differences are caused by varying rates of soil alkalinity and salinity within the mineral bed of the swamp.

Vegetation

The Vasyugan Swamp features a complex mosaic of botanical communities covering the southern taiga forest zones and multiple swamp zones. Vegetation occurs in about 60 percent of this ecoregion, which supports distinct birch-aspen forests, spruce-fir-cedar forests, and dark coniferous forests featuring layers of dwarf pine. There are also low shrubs, such as leatherleaf (*Chamaedaphne calyculata*), marsh Labrador tea (*Ledum palustre*), and bog cranberry (*Oxycoccus microcarpus*), as well as herbs such as cloudberry (*Rubus chamaemorus*) and common sundew (*Drosera rotundifolia*).

The open water area features floating vegetation mats of species such as mud sedge (*Carex limosa*), rannoch-rush (*Scheuchzeria palustris*), white beak-sedge (*Rhynchospora alba*), hare's-tail cottongrass (*Eriophorum vaginatum*), and red cottongrass (*E. russeolum*), as well as several mosses, including Baltic bog-moss (*Sphagnum balticum*), Jensen's sphagnum (*S. jensenii*), and Papillose bog-moss (*S. papillosum*).

On the surface of the interfluvial watershed zone between the Ob and Irtysh Rivers are water bogs that support a variety of flora, such as sedges (family *Carex*), moss (family *Hypnum*), and peat moss (*Sphagnum*), the most dominant of these being brown sphagnum (*Sphagnum fuscum*).

Fauna

The swamp supports a wide variety of insects throughout its many subregions, including butterflies (families *Geometridae* and *Noctuidae*) and beetles (*Cerambycidae, Chrysomelidae, Ipidue, Carabidae*, and *Staphylinidae*). Various suborders of ants, wasps, and bees (*Hymenoptera*) are found in its low-lying grass and bush layers; mayflies (*Ephemeroptera*), stoneflies (*Plecoptera*), and true flies (*Diptera*) along the riverbanks; and dragonflies (*Odanata*) and blood-sucker flies (*Hipoderma bovis, Oestris ovis,* and *Hippoboscidae* spp.) in bogs.

Because the swamp is extremely difficult to access, there is lack of human development in this region; thus, the ecosystem has remained a virtually undisturbed habitat for many native animals. Mammal species include brown bear (*Ursus arctos*), elk (*Cervus canadensis*), Eurasian lynx (*Lynx lynx*), sable (*Martes zibellina*), squirrel (*Sciuridae*), Russian mink (*Mustela lutreola*), Eurasian otter (*Lutra lutra*), and wolverine (*Gulo gulo*). Of the mammals, reindeer (*Rangifer tarandus*) is a rare species.

Bird species include wood grouse (*Tetrao urogallus*), black grouse (*T. tetrix*), willow grouse (*Lagopus lagopus*), hazel grouse (*Tetrastes bonasia*), aquatic warbler (*Acrocephalus paludicola*), and slender-billed curlew (*Numenius tenuirostris*), as well as several vulnerable bird species, including golden eagle (*Aquila chrysaetos*), white-tailed eagle (*Haliaeetus albicilla*), osprey (*Pandion haliaetus*), great grey shrike (*Lanius excubitor*), and peregrine falcon (*Falco peregrinus*).

Amphibian species in the Vasyugan Swamp biome include the Siberian salamander (*Salamandrella keyserlingii*), true frog (*Ranidae*), common toad (*Bufo bufo*), common lizard (*Zootoca vivipara*), and viper (*Viperidae*).

Environmental Threats

Due to the unsuitability of the land in this biome for standard dwellings or cropland development, there has been very little human settlement, logging, or agrarian activity such as herding. There are growing concerns about the state of the endangered species in the Vasyugan Swamp, however, due to the effects of infrastructure development, as well as air, water, and ground pollution by the oil and gas industry, which operates primarily in the western part of the swamp.

Global warming is a serious threat, as it is thought to be heightening evaporation and thereby altering the moisture retention capacities of the peat bogs. Increasing average temperatures are leading to a northward procession of habitat types; this stresses local habitats and compels migration of fauna. Not every species will be able to adapt rapidly enough to keep pace with the warming trend.

The eastern region of the Vasyugan Swamp has been nominated for inclusion on the United Nations Educational, Scientific, and Cultural Organization (UNESCO) World Heritage List.

Reynard Loki

Further Reading

Kulikova, G .G. "History of Formation of Forests of the Vasyugan Swamp During the Holocene." *Moscow University Biological Sciences Bulletin* 34 (1979).

Pisarenko, O. Yu, E. D. Lapshina, and E. Ya Mul'diyarov. "Cenotic Positions and Ecological Amplitudes of Mosses in the Vegetation of the Great Vasyugan Swamp." *Contemporary Problems of Ecology* 4, no. 3 (2010).

United Nations Educational, Scientific, and Cultural Organization (UNESCO). "The Great Vasyugan Mire." World Heritage Centre, 2007. http://whc.unesco.org/en/tentativelists/5114.

Verhoeven, Jos. T. A., Boudewijn Beltman, Roland Bobbink, and Dennis F. Whigham, eds. *Wetlands and Natural Resource Management*. New York: Springer, 2006.

Veracruz Coral Reef System

Category: Marine and Oceanic Biomes.
Geographic Location: Caribbean Sea.
Summary: This interconnected near-shore system of reefs faces severe pressure from human activity.

The Veracruz Coral Reef ecosystem is comprised of approximately 23 reefs that form a submarine range across the continental shelf of the southwestern Gulf of Mexico. The reef clusters are separated into two groups by the mouth of Mexico's Jamapa River, with one cluster to the north of the river, offshore from the port city of Veracruz, and the other to the south near Punta Antón Lizardo. The total surface area is approximately 202 square miles (52,238 hectares).

The reef systems face the classic triple-threat of overfishing, pollution (from sewage, agricultural runoff, and land-use change), and climate change. Management action in the 1990s designated a 200-square-mile (518-square-kilometer) Parque Nacional Sistema Arrecifal Veracruzana (Veracruz Coral Reef System National Park) around this system, which has protected it to some degree. The biome is still under stress from coastal pollution and climate change, and is generally considered to be one of the most threatened coral-reef systems in the greater Caribbean region.

The Veracruz Coral Reef system is located in a portion of the Gulf of Mexico that is relatively isolated from similar reef systems, lacking a major current connection to other reef areas in the Yucatan area and the greater Caribbean. Freshwater input to the system comes largely from the Jamapa, Papaloapan, and La Antigua rivers; it totals, on average, roughly 10 billion gallons (378 billion liters) per year. This is not enough to affect salinity in the reef system, but bears a great deal of sediment and pollution from untreated human sewage, as well as agricultural discharges. Other threats to the system include ship groundings, oil spills, port construction, and continued fishing pressure.

Reefs range from directly adjacent to the shoreline to approximately 12 miles (19 kilometers) from shore. Depth ranges from shallow lagoons less than 6 feet (2 meters) below the surface to relatively deep reef formations at about 150 feet (46 meters) in depth. The maximum depth in the region is approximately 230 feet (70 meters).

In addition to the natural reefs, there are at least seven artificial (human-made) reefs located within the park boundaries, and there may be as many as 350 shipwreck sites. The abundance and diversity of readily accessible near-shore venues within recreational diving limits make the Veracruz Coral Reef system a popular destination for tourists, snorkelers, and divers alike. It also attracts many types of fauna, from invertebrates to fish, eels, and marine mammals.

Nearly 30 species of hard coral and various soft coral types have been documented here, along with coralline algae, turf algae, and macroalgae species. Together, they provide habitat foundation for numerous mollusks, clams, snails, starfish, shrimp, lobsters, sea urchins, and at least 150 species of fish.

Fish species thriving in this reef system range from more than one dozen species of bass and grouper, to snapper, jack, grunt, wrasse, goby, puffer, damselfish, and parrotfish. There are at least four types of moray eel in residence—green, goldentail, spotted, and reticulate—as well as southern stingray, spotted eagle ray, and lesser electric ray.

Threats and Conservation

In 1992, Mexican president Carlos Salinas de Gortari designated the 130,000-acre (52,609-hectare) Veracruz Coral Reef System National Park, one of the largest such parks in Mexico, in an attempt to curb irrational exploitation of marine resources in this area and to protect resources for future generations.

In general, while well-enforced legal protection is effective in reducing point-source threats, it can do little to prevent stressors outside the boundaries of the preserve. The Veracruz system is no exception; while effective legislative and management efforts can control direct effects to the reef, such as fishing, boat traffic, and dredging, the reef still suffers from high nutrient loading (which encourages the growth of smothering algae over corals)—predominantly from untreated sewage, but also from agricultural chemical use. Deforestation of near-shore mangrove ecosystems, which serve as nurseries for many reef fish and also absorb nutrients and trap sediments, has exacerbated this problem.

The relative isolation of this system reduces larval recruitment from other reef systems. This makes it almost wholly dependent on internal recruitment, which has declined dramatically for the above reasons. As a result, the Veracruz Coral Reef System remains one of the most threatened such systems in the greater Caribbean region.

Recently, efforts have been made by park managers, in collaboration with the National Coral Reef Institute, to inventory the park's biological assets and correlate the abundance and disease of fish and coral to freshwater loading and other human effects in the region. Hard coral cover was in the range of 4–38 percent in areas surveyed, although many reef communities showed evidence of coral disease. While the nature and mechanics of coral pathogens are not fully understood, it is well known that stressed corals are more susceptible to disease. Warmer water temperatures, an effect of global warming, tend to be another stressor of corals and can even lead to bleaching events. Climate change also fuels more severe storms, which often wreak havoc below the waves as coral reefs take the brunt of heavy seas.

These ongoing projects are critically important for establishing effective management strategies, as well as understanding the effect of human activity on the region. The Veracruz Coral Reef system is far from unique in the threats which it faces, as coral reef ecosystems throughout the Caribbean and the world face varying degrees of stress from climate change, overfishing, and pollution. While areas such as Veracruz Coral Reef System National Park have proven to be important tools in the management and protection of these fragile ecosystems, they are by no means a magic-bullet solution and must be coupled with social, behavioral, and legislative change on every scale, from local to global.

Fortunately, coral reefs provide a valuable array of ecosystem services, ranging from tourism and fishing to shoreline protection from storms, to providing spawning and nursery habitat for many marine species that migrate to other parts of the sea. Because of this, there is strong incentive to protect these systems to the greatest degree possible. Human scientific and conservation actions of the past few decades offer hope that this unique ecosystem can be protected and can fully recover.

Jason Krumholz

Further Reading

Rangel Avalos, M. A., L. K. B. Jordan, B. K. Walker, D. S. Gilliam, E. Carvajal Hinojosa, and R. E. Spieler. *Fish and Coral Reef Communities of the Parque Nacional Sistema Arrecifal Veracruzano (Veracruz Coral Reef System National Park) Veracruz, Mexico: Preliminary Results.* Puerto Morelos, Mexico: Gulf and Caribbean Fisheries Institute, 2007.

Salas-Perez, J. J. and A. Granados-Barba. "Oceanographic Characterization of the Veracruz Coral Reefs System." *Atmosphera* 21, no. 3 (2008).

Withers, K. and J. W. Tunnell. "Reef Biodiversity." In J. W. Tunnell, E. A. Chavez, and K. Withers, eds., *Coral Reefs of the Southern Gulf of Mexico.* College Station: Texas A&M Press, 2006.

Victoria, Lake (Africa)

Category: Inland Aquatic Biomes.
Geographic Location: Africa.
Summary: Lake Victoria is the second-largest freshwater lake in the world. It supports millions of people directly, but suffers from pollution, overfishing, and invasive species.

Fabled source of the Nile River, Lake Victoria holds a special place in the hearts and minds of many people, both within eastern Africa and beyond. Lake Victoria is an essential source of sustenance for many species in habitats within its waters and around its shores. It is also vital, in terms of both water and food, for tens of millions of people in the Lake Victoria basin and beyond. Like many lakes around the world, Lake Victoria is also exploited, abused, and polluted constantly through careless and intentional human actions every day.

Lake Victoria's surface area covers some 26,641 square miles (69,000 square kilometers), making it the second-largest freshwater lake in the world. The catchment area, where the water comes from countless rivers and streams, stretches over 69,884 square miles (181,000 square kilometers) through five countries: Kenya, Tanzania, Uganda, Rwanda, and Burundi. The average depth of the lake is 131 feet (40 meters), with a maximum depth of 276 feet (84 meters). The volume of water stored is immense, estimated at about 662 cubic miles (2,760 cubic kilometers). This is a significant portion of the entire planet's freshwater resources that are directly available, not frozen in glaciers or polar ice caps or locked deep beneath subterranean rocks.

Lake Victoria itself is shared by three countries, Kenya, Tanzania, and Uganda. The lake is divided unequally, mainly between Tanzania and Uganda, with a small portion (primarily the Winam Gulf) falling within the jurisdiction of Kenya. The immediate basin is home to some 35 million people, most of whom depend on the lake for their water (in cities like Mwanza, Kisumu, and Bukoba) or for a reliable source of protein in the form of fish taken from the lake.

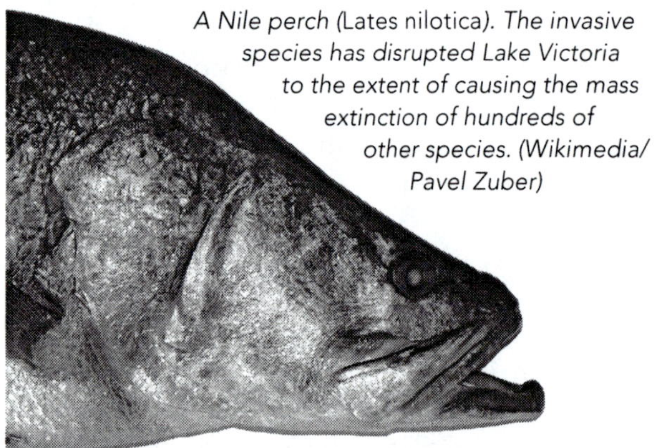

*A Nile perch (*Lates nilotica*). The invasive species has disrupted Lake Victoria to the extent of causing the mass extinction of hundreds of other species. (Wikimedia/ Pavel Zuber)*

Effect of Introduced Species

In the 1950s and early 1960s, the ecology of Lake Victoria was changed forever through the deliberate introduction of the Nile perch. What resulted in Lake Victoria is now a textbook case of mass extinctions resulting from the introduction of an aggressive invasive species to a stable biome. The Nile perch (*Lates nilotica*) was present in the Nile and other aquatic systems further north; it had been kept out of Lake Victoria by the steep waterfalls along the river's course. Therefore, all of the species in the lake had evolved and adapted without ever having to deal with this predator.

The introduction of the Nile perch was done in the northern part of the lake. By 1961, the Nile perch had reached Mwanza Gulf, the southernmost portion of Lake Victoria. Today, it is common throughout the lake, and it is likely to remain a component of the lake's ecology as long as the lake itself exists. The largest recent mass extinctions of vertebrates, including the loss of cichlids and other fish species in Lake Victoria, have occurred in the decades since the introduction of the Nile perch. Of the lake's original 500 or so cichlid species, more than half are thought to be extinct.

The Nile perch has totally disrupted the ecological balance of the lake. Where hundreds upon hundreds of species of fish, invertebrates, reptiles, birds, and mammals once thrived, the lake's freshwater ecology today is dominated by the Nile perch, a single species that seems to have replaced virtually all other piscine predators.

In the open waters, many zooplankton-eating fish have been wiped out, and the lake sardine, popularly called dagaa or omena (*Rastrineobola argentea*), has undergone a population explosion through the elimination of its competitors. Another species whose numbers have risen in the wake of the Nile perch is the freshwater prawn (*Caridina nilotica*). Again, this is because the prawn's population is expanding to occupy the niche hitherto filled by cichlids feeding on detritus along the lake bottom.

Throughout its life, the Nile perch is voracious. Young perch prey mostly on insect larvae, crustaceans, and other invertebrates. Older Nile perch feed exclusively on fish, and will eat anything measuring up to a third of their body length.

Spot-necked otters, mammals that relied on hunting cichlids in clear, shallow waters, have also been affected. In some places, they have managed to change their diet to include the larger fish, but the otters' large family groups that were once seen hunting in the shallows of the lake have declined.

Other invasive species have also made their way into the lake. Several tilapia species, including the Nile tilapia (*Oreochromis niloticus*) and redbreast tilapia (*Tilapia zillii*), were also deliberately introduced by fisheries. These species have also proliferated, and form an important portion of the fish caught and exported from the lake today. Their broader ecological effects on the fish and other biota here remains to be determined.

Still other newcomers that are changing the lake's ecology include the Louisiana crayfish, an invasive freshwater snail, and the water hyacinth. A pretty, flowering plant that bears spikes of attractive lilac-hued flowers, the water hyacinth originates in Central and South America. Fed by the nutrient-enriched waters near towns along the lake, and freed from threats that would have been imposed by grazers and parasites in its native habitat, it has spread across Lake Victoria with impunity.

The leaf stalks of the water hyacinth are spongy and filled with air sacs, enabling it to float on the water. As it grows, it forms thick mats that soon begin to rot in the lower layers. This depletes oxygen in the water below, in addition to blocking off light. The result is that the native plants, particularly in the littoral zone (from the lake-shore edge to the last rooted, submerged plants), are severely disrupted.

Fisheries

The fisheries of Lake Victoria today form a large, dynamic, and thriving economic sector for the region, especially in Kenya, Tanzania, and Uganda. Entire towns and cities, like Mwanza in Tanzania and Kisumu in Kenya, rely on fishing, fish processing, and exporting as the main drivers of their economies.

In the decades following the introduction of the Nile perch and tilapia, the introduced species reproduced rapidly but were not yet important to the lake fisheries. However, by the 1980s, the number of Nile perch exploded and came to dominate the fish catch, along with the Nile tilapia and the lake sardine (dagaa and omena). Ecologists have labeled the population explosion an *irruption,* as when numbers swell rapidly.

By the early 1990s, fish exports from Lake Victoria were 220,462–330,693 tons (200,000–300,000 metric tons) annually. Huge profits have been, and continue to be, made off the lake. The rapid expansion of the fisheries, of course, has come at a high environmental price.

Environmental Threats

Based upon all the evidence to date, the fate of the Lake Victoria biome is gloomy. The fishing boom, now past its peak, is predicted to be relatively short-lived. Recent trends include fish killed by poisoning from commercial synthetic pesticides that are poured into the water at night by unscrupulous fishers. These poisons cause the fish to rise to the surface as they die slowly. The larger tilapia and Nile perch are incapacitated but still alive; they are then pulled from the water to be processed fresh.

The wholesale export of fish and the extinction of cichlids also have some human consequences. Most of the small communities along the lake shore do not have the fuel, motorized boats, or nets needed for catching the large tilapias and Nile perch. This trade is dominated by foreign-owned companies with fleets of trawlers.

Traditionally, women were able to access protein from cichlids year round, and much fishing could be done with simple techniques in the shallows. After the introduction of the Nile perch, this system changed. Today, many poor women trade sex for fish, which, in the light of the AIDS epidemic, is a serious health risk.

Deforestation in the catchment area of the lake has severely reduced the amount of water flowing into the lake. Soil erosion from the loss of forest cover is washing increasing amounts of silt into the lake, which makes the water murky. This disrupts the balance between sunlight penetrating the water and the growth of algae, which is important for many fish and other species. Pollution from towns on the lake shore (Mwanza, Bukoba, and Kisumu), including raw, untreated sewage, also enters the lake daily.

Burning and clearing of the papyrus wetlands that fringe the lake is a growing trend. This deprives the water of a vital filtration system that cleanses and oxygenates it. Sheltered nests for many young fish, birds, and insects are also wiped out through this practice, which is driven by the high population pressure for more agricultural land.

Overly enriching the water with tons of raw sewage and fertilizer runoff is leading in many areas to large pockets of anoxic, or zero-oxygen, water, often called the dead zone. This zone, once limited to the deepest, darkest parts of the lake, is spreading rapidly.

Despite all these ecological problems, Lake Victoria remains one of the most dynamic and productive of the African great lakes. There will be huge challenges in managing it sustainably in an area with a rapidly growing population. Countries sharing the Lake Victoria basin and the Nile River basin are engaged in discussions to better manage this lake, which is important for so many people.

DINO J. MARTINS

Further Reading

Bennun, L. and P. Njoroge. *Important Bird Areas in Kenya*. Nairobi, Kenya: East Africa Natural History Society, 1999.

Martins, D. J. "Differences In Odonata Abundance And Diversity In Pesticide-Fished, Traditionally-Fished And Protected Areas In Lake Victoria, Eastern Africa (Anisoptera)." *Odonatologica* 38, no. 3 (2009).

Pringle, R. M. "The Origins of the Nile Perch in Lake Victoria." *BioScience* 55 (2005).

Vistula Estuary

Category: Marine and Oceanic Biomes.
Geographic Location: Europe.
Summary: A unique coastal marine environment, this estuary hosts rich marine and avian populations, but needs every effort to counteract damage from reckless anthropogenic activities upstream.

Shared by Russia and Poland, the Vistula Estuary is one of the largest transboundary estuaries in the southern Baltic Sea. Sometimes called Vistula Bay or Vistula Gulf, the estuary is unusual in that it takes the form of a mainly freshwater lagoon—an arm-shaped, elongated body of water 57 miles (91 kilometers) long, its width ranging from 5 to 8 miles (8 to 13 kilometers). The estuary covers an area of 330 square miles (855 square kilometers). Approximately 44 percent of the drainage basin area lies in Poland, while 56 percent falls in Russia. The total surface area of the catchment basin of the estuary is 9 square miles (24 square kilometers) within both countries.

While some brackish waters from the Baltic Sea circulate past the extremely long sand bar that separates the two bodies, most of the inflow to the lagoon comes from the Vistula River, the longest river of Poland. The Vistula River originates on the western slopes of the Carpathian Mountains in southeast Poland and flows northwest through the cities of Krakow, Warsaw, and Torum before reaching the lagoon. This is a total length of 677 miles (1,090 kilometers). Also adding freshwater input to the estuary are Nogat and Pregolya, two major eastern streams of the Vistula River delta. They enter the lagoon by the Strait of Baltiysk in Gdańsk Bay.

The sand bar, one of the longest such features in the world, is known as Vistula Spit; about 50 kilometers) long, it varies from .25 to 1.5 miles (0.4 to 2.5 kilometers) wide. There are built-up stretches in the form of small beach towns, some with extensive tree cover, mainly on the lagoon side. Beaches on the sea side are generally kept clear for wildlife, as well as tourism.

The Vistula Estuary is quite shallow, with an average depth of 7 feet (2 meters) and a maximum depth of 16 feet (5 meters). The water volume is estimated to be 0.5 cubic miles (2 cubic kilometers). Exposed to both marine and continental impacts, climate in the estuary is mainly temperate. Summers are cool, with an average temperature of 63 degrees F (17 degrees C), while the winter temperatures average approximately 39 degrees F (4 degrees C). Average annual precipitation ranges from 24 to 33 inches (600 to 850 millimeters).

Biodiversity

The Vistula Estuary is a biologically productive ecosystem with its foundation built upon plankton, in such forms as 34 species of *Rotatoria,*16 of *Cladocera,* and 21 of *Copepoda.* Further up the food chain are 42 fish species. Herring, bream, pikeperch, and eel are among the most abundant and valuable commercial fish species here. Other species include perch, flounder, ruffe, burbot, ziege, roach, and trout.

Marine mammals, mainly seals and porpoises, venture in and out of the estuary, no doubt drawn by both the relative ease of securing prey and the relative safety from larger predators.

Hundreds of thousands of birds on the Scandinavian-Iberian Flyway make the Vistula Estuary a major stop. Some 240 species of birds touch down here, with more than 100 selecting some part of the lagoon area as a nesting zone. The largest European colony of great cormorants has established itself at Katy Rybackie here, numbering over 30,000.

The sandspit is the dominant feature along the north side of the estuary; it features a variety of dune habitats, and even forested swaths on the higher areas that also support small human settlements. There are marshlands near the mouths of the rivers, mostly clustered around the eastern and southeastern reaches of the lagoon. Here dwell reptile and amphibian species that either dodge the preying birds or turn predator and attack nests. Adjacent to the Vistula marshland is the Elblag upland, where mammals such as otter, muskrat, and rodents are found in some abundance.

Threats and Conservation

Polluted waters from upstream farms and industries—only lightly regulated for most of the 20th century—have long plagued this estuary. Phosphorus, oil, nitrogen, mercury, cadmium and zinc were among the most-noted substances causing widespread problems. Things began to turn around in the wake of political changes, following the 1989 collapse of the former Soviet Union. Several years of stagnant economic activity gave the ecosystem something of a breather, but recent economic boom years have once again raised fears that the waterways here will be seen as dumping grounds. However, there is now a better-established environmental point of view in much decision making and administrators recognize the value of intact biomes in attracting tourism revenues, as well as in supporting the longstanding fishing industry here.

Real dangers to the ecological balance include the ongoing warming of the Baltic Sea waters, offshore algal blooms, and the changes in the foodweb that these pressures may foretell. The crash of the cod fishery in the North Sea and Baltic Sea indicate that other species may move into that ecological niche, which may cascade other significant changes that would reach to the Vistula Estuary.

The huge cormorant rookery here is so successful at taking fish from the waters that, combined with commercial fisheries, it is thought to lead to periodic declines in bream and pike perch through overfishing. Even climate change is looked at in terms of its interaction with the cormorants, as the flock is seen to gain breeding-season days as average temperatures rise, meaning the population is on track to grow still more. (On the other hand, earlier successful breeding by the cormorant flock often leads to earlier departure for other fishing grounds. This phenomenon has been recorded as taking place nearly four weeks earlier than normal over the last several decades in various parts of Europe.)

There are various invasive species that the natives have given some ground to, although few are considered especially pernicious. One such nuisance is the Chinese mitten crab, introduced a century ago and still playing havoc with fishing gear.

Quite a few nature reserves dot this biome, and most of them are havens for the vast numbers of migratory birds that stream through twice each year. They range from Vistula Spit Landscape Park, Sea-Holly Dunes, and Vistula Spit Beeches (also on the spit are two bird sanctuary areas, Katy Rybackie and Gull Sandbank) to three other areas favored by both large and small flocks, amounting to well over 100 species in each case. These are Elblag Bay, Nogat Estuary, and Druzno Lake.

There are two year-round, no-fish corridors in the lagoon, established to help ensure that sufficient spawning and fry development activity can take place in support of the commercial fishery here.

Rituparna Bhattacharyya

Further Reading

Miotk-Szpiganowicz, GraŜyna, Joanna Zachowicz, and Szymon Uścinowicz. "Palynological Evidence of Human Activity on the Gulf of Gdansk Coast During the Late Holocene." *Brazilian Journal of Oceanography* 58 (2010).

Naumenko, E. N. "Zooplankton in Different Types of Estuaries (Using Curonian and Vistula Estuaries as an Example)." *Inland Water Biology* 2, no. 1 (2009).

Paturej, Ewa and Marek Kruk. "The Impact of Environmental Factors on Zooplankton Communities in the Vistula Lagoon." *International Journal of Oceanological and Hydrobiological Studies* 40, no. 2 (2011).

Rolbiecki, Leszek and Jerzy Rokicki. "Parasite Fauna of the Eel, *Anguilla anguilla* (Linnaeus, 1758), From the Polish Part of the Vistula Lagoon." *Wiadomooeci Parazytologiczne* 52, no. 2 (2006).

Volga River

Category: Inland Aquatic Biomes.
Geographic Location: Russia.
Summary: Meandering and often slow, the Volga winds southward through the heart of Russia, giving life to a fantastic array of habitats and species.

One of the longest and most ecologically critical rivers in eastern Europe, the Volga River flows more than 2,300 miles (3,700 kilometers) north to south across Russia, starting in low northwestern hills and ending at a huge delta on the Caspian Sea. The elevation of the source of the Volga in the Valdai Plateau is only 738 feet (225 meters) above sea level. It enters the Caspian Sea at a point about 92 feet (28 meters) below global sea level—the Caspian, an inland sea, does not entirely fill the Caspian Depression. The Volga makes an average drop of only 4.3 inches per mile (7.0 centimeters per kilometer); it is decidedly flat. The great old stream flows generally in a meandering way across an often sodden, ancient floodplain.

The Volga River takes tributary flow from an amazing network of more than 150,000 mainly short-run rivers of less than 6 miles (10 kilometers) each, with about 2,600 of them draining directly into the Volga. This occurs across a vast catchment of 540,000 square miles (1.4 million square kilometers). The Volga drains approximately one-third of European Russia.

The climate is generally moderate continental; the average annual temperature varies from 37 degrees F (3 degrees C) in the north to 41 degrees F (5 degrees C) in the south. The mean annual precipitation ranges from 28 inches (706 millimeters) in the north to 7 inches (175 millimeters) in the south. The Volga and its tributaries, on average, remain frozen for about 161 days in the north and about 100 days in the south.

Biodiversity

Among the more celebrated fish species in the Volga are its sturgeon—starry, beluga, and sterlet. Other key species are herring, whitefish, and white-eyed bream.

The delta area alone is a biodiversity hot spot. Its 500 channels and associated mudflats, tidal pools, submerged reedbeds, and marshes provide shelter and habitat for innumerable mollusks and crustaceans—at least 800 known species of aquatic invertebrates—as well as small fish that draw upwards of 250 species of birds to feed, nest, and rest on migration routes. These birds include herons, ibis, swans, terns, ducks, and such iconic species as great white egret and the Dalmatian pelican. The delta sprawls across 10,512 square miles (27,224 square kilometers).

The middle and upland segments of the Volga River wash through many types of landscapes, mainly low-lying but with occasional higher topography. The Samarskaya region is one of the

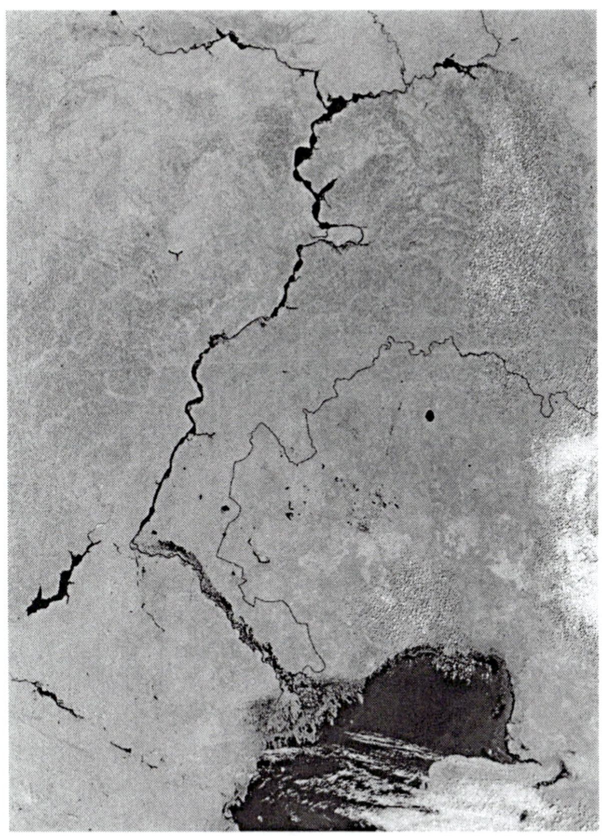

This satellite photograph shows the Volga River flowing into the Caspian Sea, at bottom. The vast delta, with some 500 channels, is visible where the river meets the sea. (NASA)

latter types. Here, the river winds through forested low mountains. Mixed deciduous-conifer stands of oak, pine, aspen, and lime trees are found on the ridges and slopes, transitioning to riverine zones of white willow, black alder, black poplar, and smooth-leaved elm. This region also hosts the endemic (found only here) Zhigulevsky spurge and Zhigulevsky mist-grass. Wildflowers and herbs are also found in great abundance in these forests and wooded steppe areas.

Elk, hare, squirrel, marten, badger, fox, and wild boar are among the major mammals in evidence. The woodland and wetland avians include teal, goldeneye and mallard duck, as well as wood, hazel, and black grouse.

On the flat savanna areas, plant cover varies based largely on latitude. Moisture is at a premium in many of these zones, with the prevailing continental climate tending toward a semiarid regime here. The vegetation is often desert sagebrush habitat, featuring *Artemisia* spp. and *Stipa* spp. Antelope, jerboa, hare, and hedgehog share the land with reptiles such as the sand boa and glass-snake, and birds including the demoiselle crane and lark.

Threats and Conservation

Agricultural runoff has led to nutrient overload conditions, especially in downstream sections and the Volga Delta. This has led to more frequent episodes of blue-green algal blooms and depleted oxygen levels, which can severely set back a wide range of plant and animal species in any aquatic habitat. Damming and water diversion has been a challenge to the river's health throughout much of its human history. Continuing plans for such development threaten fish migration, recovery of aquatic vegetation communities in areas with damaged banks and dredged bottoms, loss of water clarity, altered water temperature and resultant disruption of spawning cycles, and other impacts.

Global warming is affecting habitats along the entire range of the Volga River. In its northern reaches, forest composition is under duress, as longer summers and widespread fire damage in recent years is impressing a northward march by tree species that require colder regimes. In the middle stretches, water quality has suffered as tributary flow dwindles during near-drought conditions. Around the delta, saltwater inundation from the rising Caspian Sea is putting pressure on estuarine habitats.

Preservation efforts for the Volga River biome are diverse. One program monitors aquatic and terrestrial mammal populations along the Oka River, at the Okski Nature Reserve. Such keystone species as Russian desman, a vulnerable semi-aquatic mammal that preys on small amphibians, freshwater crustaceans, and insects, are closely studied here as indicators of overall habitat health. The Russian desman was long prized for its fur, and poachers now must contend with legal measures to protect the species.

In the lower Volga Delta, the Astrakhansky reserve protects about 260 square miles (67,000 hectares) of wetland habitat. Channeled stands of reeds, river meadows, marginal willow forests, and areas of floating vegetation such as white water-lily combine for a full mini-biome that hosts fauna from carp, herring, freshwater turtles, and lake frogs to mute swan, white-tailed eagle, and wild boar.

Rituparna Bhattacharyya

Further Reading

Hamline Univeristy. "Rivers of Life—The Volga River." Center for Global Environmental Education. http://cgee.hamline.edu/rivers/Resources/river_profiles/Volga.html.

Kolomyts, Erland G., et al. *Forests of Volga River Basin Under Global Warming.* Hauppauge, NY: Nova Science Publishers, 2012.

Schletterer, Martin and Leopold Füreder. "The River Volga Headwaters: Inventory, Biodiversity and Conservation." *River Systems* 19, no. 1 (2010).

Volta River

Category: Inland Aquatic Biomes.
Geographic Location: Africa.
Summary: A complex system of waterways and a giant artificial lake connect a wide range of habitats in a watershed shared by six countries that sometimes have conflicting interests.

Deriving its name from the Portuguese word for *meander,* the Volta River is the major river system for the western African nation of Ghana, and the areas surrounding it are home to some 70 percent of the Ghanaian population. Its catchment area also draws upon regions of the countries of Mali, Burkina Faso, Côte d'Ivoire, Benin, and Togo. Because of the length of the Volta, the land habitats are varied, ranging from bush to rainforest and from wooded gorges to grassy plains. The Volta is fed by the Black Volta, the White Volta, the Red Volta, and the Oti River. Both the Red and White Volta Rivers are dry from January until the rains begin in May.

In Burkina Faso, where the Black and White Volta Rivers originate, the Black Volta merges with the Sourou River. This entire area is surrounded by extensive floodplains. Flowing mostly in a southward direction for 1,000 miles (1,609 kilometers), the Volta River ultimately merges into the Gulf of Guinea in the equatorial Atlantic Ocean. The Volta River is made up of numerous headwaters and streams that tend to be intermittent in the upper reaches, evolving into permanent bodies of water in the lower reaches.

The catchment as a whole comprises some 154,400 square miles (400,000 square kilometers). About 39 percent of the basin—including massive Lake Volta, the engineered result of major dam-building—lies within Ghana's borders; some 46 percent lies within Burkina Faso, and smaller sections are found in Togo (6 percent), Benin (4 percent), Mali (3 percent), and Côte d'Ivoire (2 percent).

The climate of the Volta River Basin is subhumid to semiarid, an aspect of its location within the West African Savanna Zone. Because of topography and shifting climate patterns, rainfall varies greatly, with the southeastern area averaging more than three times the rainfall of the extreme northern area. Within the Volta River basin as a whole, however, some 70 percent of rain falls between July and September. The average rainfall overall in Ghana tends to be about 29 inches (237 millimeters). There is almost no rain between November and February. The year-round temperature range is 69–79 degrees F (20–26 degrees C) around the region.

Biodiversity

Great expanses of savanna border the major streams within the Volta River basin. These vary from savanna grassland in the north, to savanna woodland in the center, and finally yield to some swaths of rainforest in the southwest. Conditions within the river are similar to those within the Niger River—which arcs across the northern reaches of the Volta catchment—and it is believed that the Black Volta may have originally been a tributary of the Niger.

At least 145 species of fish live in the waters of the Volta River, including mormyrids, mochokid catfish, and characins. Most of these species are insectivores, substrate feeders, or fish predators. They have adapted to seasonal flooding by spawning and feeding upriver as the floods begin. Many native species have disappeared because of habitat degradation and damming of waters along the river. Only nine fish and a single crab species are endemic (found nowhere else) to the Volta. The river is also home to 40 species of aquatic reptiles and 25 aquatic mollusks. Mammals within this ecoregion include the marsh mongoose, African clawless otter, spot-necked otter, hippopotamus, and West African manatee, the last of which is considered vulnerable.

One of the major wetland areas within the Volta River basin is Muni-Pomadze, which lies within the Muni Lagoon watershed near the river's mouth. It is located some 35 miles (56 kilometers) from the capital of Ghana, Accra, on the Atlantic coast. Recognized as one of the most important coastal wetlands in the world, this wetland has both local and global significance. It is a major breeding and nesting site for migratory birds, and is home to a host of waterbirds, insects, and terrestrial vertebrates. Blackchin tilapia is a key fish species of the lagoon.

People from all over the world come to Muni-Pomadze to attend the Aboakyer Festival, an ecotourist attraction. For locals, the festival is an occasion for traditional hunting trips. In earlier times, the woods were full of bushbucks, leopards, antelopes, lions, mongooses, and a variety of other animals. Many of those animals, particularly lions, have dwindled in number in response to hunting and environmental degradation.

Vegetation within the wetland includes grassland, thicket islands, and savanna trees. Studies have shown that degradation of the wetland is occurring as the result of wood being cut down indiscriminately for fuel, frequent bushfires, hunting of wildlife, and agriculture.

Human Activity

Population has been rapidly increasing in the countries of the Volta River Basin, surpassing 16 million around the year 2000 and projected to double by the late 21st century. The countries sharing the basin generally have high poverty rates and economies that remain undeveloped. There is widespread use of unsustainable agricultural methods, which result in significant environmental degradation.

Because of competing demands on the resources of the Volta River, much of the area's environment has become vulnerable to a variety of influences. Biodiversity is decreasing, fishing and groundwater resources are being depleted, increased flooding is a major concern, and pollution—agricultural, industrial, and municipal—within the river is on the rise.

In the mid-1960s, the 912-megawatt Akosombo Dam was built on the Volta River in partnership with a foreign company, creating Lake Volta, the largest human-made lake in the world by surface area. From the dam, Lake Volta spreads northward for 250 miles (402 kilometers); its total surface area is 3,275 square miles (8,485 square kilometers). The lake is maintained chiefly for its ability to generate hydroelectric power; up to 80 percent of Ghana's power has been derived from the lake.

Lake Volta is also a major transportation and trade source. More than half of the labor force in Ghana is employed in agriculture and fishing. Consequently, farmers use the lake to irrigate their crops, and fishers depend on its abundant resources for survival. As the land of the Volta River Basin stretches away inland from the coast, it becomes more arid, resulting in shorter growing seasons and more erratic rainfalls—making the lake's irrigation capacity all the more valuable.

Overpumping of the river's waters has been a major issue. In 2006, the riparian nations established the Volta Basin Authority as a means of managing the watershed. However, cooperative administration remains illusive. Relations have been particularly tense between Ghana and Burkina Faso, with Ghana demanding greater hydroelectric capabilities and Burkina Faso insisting on more water for irrigation. Both countries have constructed additional dams and power facilities, but have not been scrupulous about communicating intentions, for example, opening floodgates without notice in some cases.

Some efforts are being made to combine profit-making activities with environmental responsibility in the region. In 2009, for example, officials in Ghana began working with a Canadian timber firm on a sustainable form of artisanal logging that does not involve cutting down healthy trees. Instead, laborers use sonar to locate trees that have been submerged, and then recover them for timber use. Many such trees have been at the bottom of the Volta River since the building of Akosombo Dam in the 1960s. At that time, 100 species of trees including ebony, teak, and mahogany were submerged. Today, those recovered trees may be worth thousands, or even millions, of dollars.

ELIZABETH RHOLETTER PURDY

Further Reading

Gao, Yongzuan and Amy Margolies. "Transboundary Water Governance in the Volta River Basin." Center for Development Research of Bonn University. www.zef.de/fileadmin/webfiles/downloads/press/transboundary_water_management_volta.pdf.

Gyasi, Edwin A. and Juha I. Uitto, eds. *Environment, Biodiversity, and Agricultural Change in West Africa: Perspectives from Ghana.* New York: United Nations University Press, 1997.

Ntow, William Joseph. "Pesticide Residues in Volta Lake, Ghana." *Lakes and Reservoirs: Research and Management* 10, no. 4 (2005).

Wuven, A. M. and D. K. Attuquayefio. "The Impact of Human Activities on Biodiversity Conservation in a Coastal Wetland in Ghana." *West Journal of Applied Ecology* 9 (2006).

Wadden Sea

Category: Marine and Oceanic Biomes.
Geographic Location: Europe.
Summary: Crucial to the migration of nearly 12 million birds and host to thousands of other native species, this ecosystem has been upset by increased ecotourism and the effects of aquaculture here.

Situated in the southeastern North Sea, the Wadden Sea is one of the largest and most naturally thriving intertidal zones in the world. The word *wad* in Dutch translates to "mud flat," the defining characteristic of this area. This intertidal zone ranges from the banks of the Netherlands, Germany, and Denmark to the northwestern Frisian Islands (also known as the Wadden Islands), an archipelago of barrier islands that separates the Wadden from the North Sea. The Wadden Sea covers a total area of approximately 3,900 square miles (10,100 square kilometers), and has about 310 miles (500 kilometers) of coastline.

The chain of Frisian Islands, of which there are approximately 50, was formed some 7,000 years ago, making this isolated sea one of the youngest bodies of water in the world. Even as recently as 1,000 years ago, the barrier islands underwent significant erosion that detached them from the mainland, truly making them a group of islands and defining the Wadden Sea from the North Sea.

Furthermore, the landscape of the sea is constantly changing. The turbid sea and wind of both the North and Wadden Seas actively remove and deposit sediment from sandbars and islands over many years, causing the layout of the islands and water channels to vary over short time scales. The general movement of the islands is from west to east.

The climate in the region is mainly temperate and marine, with cool summers and mild winters. The average temperature ranges from 34 to 41 degrees F (1 to 5 degrees C) in January and from 55 to 72 degrees F (13 to 22 degrees C) in July. It rains throughout the year, and extreme cold is rare.

Biodiversity

Because of daily tide cycles and easy access to its mudflats, the Wadden Sea is an important research site for studying organismal adaptation to changes in salinity, populations of benthic micro- and macroorganisms, and shorebird behavior. It is estimated that more than 10,000 species rely on this ecoregion for all or part of their life cycles. At the center of the ecosystem are the microorganisms, which are critical to nutrient cycling and largely dominated by mudflat bacteria. Despite a troubled history of aquaculture, mollusks continue to play an important commercial and ecological role in the sea; more than 27,000 tons (24,500 metric tons) of blue mussels (*Mytilus edulis*) are harvested annually for commercial purposes. However, harvests are closely monitored because of the reliance of multiple shorebird species on the blue mussel.

The Wadden Sea has been named one of the top 100 birding sites in the world. In addition to the permanent shorebird populations here, the mudflats are a critical habitat for more than 50 species of migratory birds. This makes the area one of the largest migratory bird habitats in western Europe. The region is a crucial element of the East Atlantic Flyway; an estimated 12 million birds migrate to the Wadden Sea every year.

An incredible array of bird species can be found on the mudflats and dunes of the coast and islands, including the pied avocet (*Recurvirostra avosetta*), osprey (*Pandion haliaetus*), barnacle goose (*Branta leucopsis*), sandwich tern (*Thalasseus sandvicensis*), and bar-tailed godwit (*Limosa lapponica*). Other birds of note include the spoonbill, eider, ringed plover, oystercatcher, and the peregrine falcon.

The rich diversity of the islands and marshes also makes the Wadden Sea an attractive migratory location. More than 900 species of plants, 300 mosses, 650 fungi, and 350 lichens are found in and around the sea and the islands. Some areas of the sea boast an abundance of 2.2 pounds (1 kilogram) of shorebird food per 11 square feet (1 square meter).

This ecosystem is also home to large marine mammals, including the harbor seal, the gray seal, and the harbor porpoise. These animals can avoid predators in the shallow, warm water of the sea, and the seals use the sandbanks for nursing their young. Additionally, the strong water currents and daily tides in the region provide the Wadden Sea approximately 20 times more nutrients than the adjacent North Sea, allowing this ecoregion to sustain its incredible diversity.

Human Impact

Like many waterways, the Wadden Sea has been significantly affected by chemical and agricultural runoff from the mainlands and increased turbidity from mainland diking. In the 1960s, shorebirds that relied on the shallow waters of this area for fishing experienced population declines when the water became too polluted from chemical runoff. In the mid-21st century, the sea also became used for aquaculture, which resulted in a continuous disturbance of the sea floor and increased turbidity. Besides pollution, recent studies suggest that climate change has influenced and will continue to influence the migration of birds to the Wadden Sea. Some scholars suggest that the shoreline is one of the most human-modified environments in the world.

The Wadden Sea became a tourist destination for many Europeans in the 19th century, when several of the islands opened resorts. Since then, more than 1.8 million visitors per year have had the opportunity to hike on the mudflats during low tide to observe shorebirds and invertebrates. Because of the increased boating and mudflat hiking in the region, the Learning Center of the Wadden Sea (*Schutzstation Wattenmeer*), operated by the German government, has established various conservation areas (Schutzzone) throughout the zone and has initiated research to ensure the protection of critical mudflat species.

Many of the islands continue to be occupied by residents today. Hallig Hooge, the second-largest halligen (German island), has a population of 83 residents living on 10 terpen, which are artificially formed mounds where homes and businesses have been built. These mounds are similar to compact neighborhoods. Some mounds host smaller structures, such as the local school and church, whereas the hallig nature center, movie theater, gift shop, grocery store, and various cafes are located on a larger mound.

These terpen are essential for hallig living, because the smaller islands experience annual floods that would otherwise destroy homes and businesses. Additionally, some of the homes and businesses on the terpen are remarkably sustainable, using geothermal and solar energy for their main sources of heat and electricity. Because the low-lying areas of the island are used for research, tourism, and cattle rearing, residents do have to visit the mainland for specialty items or bulk food.

The Wadden Sea is a remarkably diverse habitat, and is considered to be one of the last pristine intertidal zones in the world. The International Union for Conservation of Nature and Natural Resources (IUCN), an international conservation organization, has characterized the sea as a Category II national park. This recognizes the immense value of this area for its contribution to ecosystem processes and educational, scientific, and cultural enrichment.

In June 2009, the Dutch and German portions of the sea were placed on the World Heritage Site List by the United Nations Educational, Scientific, and Cultural Organization (UNESCO). The list denotes places or objects of cultural or natural significance around the world, furthering protection for the region. Since its induction in 2009, the area's perceived threat for ecological harm has remained at zero.

In 2005, an annual Wadden Sea Day was established as an opportunity for researchers, educators, and conservationists to meet and discuss important ecological trends in the region. One recent Wadden Sea Day focused on the food web of the sea, particularly the importance of fish, a currently understudied group of animals in the area. In 1997, Germany, Denmark, and the Netherlands signed the Trilateral Wadden Sea Plan to ensure shared management and protection of the sea by the three countries. Following this plan in 2002, they developed the Wadden Sea Forum to oversee the conservation of the sea and monitor the protection status of the region.

CATHERINE G. FONTANA

A view of the Wadden Sea's mudflats, one of the largest intertidal zones in the world, which support over 10,000 species during some period of their life cycles. (Thinkstock)

Further Reading

De Jonge, V. N., K. Essink, and R. Boddeke. "The Dutch Wadden Sea: A Changed Ecosystem." *Hydrobiologia* 265, nos. 1–3 (1993).

United Nations Educational, Scientific, and Cultural Organization (UNESCO). "The Wadden Sea." World Heritage Centre. http://whc.unesco.org/en/list/1314.

Van Beusekom, Ruud, Nick Langley, and Manon Tentij, eds. *The Wadden Sea: A Vision for the Conservation of a Natural Heritage.* Cambridge, UK: BirdLife International, 2009.

Vollmer, M., M. Guldberg, M. Maluck, D. Marrewijk, and G. Schlicksbier. "Landscape and Cultural Heritage in the Wadden Sea Region: Project Report." *Wadden Sea Ecosystem* 12 (2001).

Wakatobi Archipelago Coral Reefs

Category: Marine and Oceanic Biomes.
Geographic Location: Asia.

Summary: This biome hosts a large degree of reef biodiversity with local and global ecological and economic importance, and is supported by a huge effort for its conservation.

The Wakatobi Archipelago coral reefs are located off the southeast tip of Sulawesi, Indonesia, the world's eleventh-largest island. The archipelago's name is actually an acronym given for the four main Tukanbesi islands: Wangi-wangi, Kaledupa, Tomia, and Binongko. These small islands, located between the Banda and Flores Seas, house huge coral-reef biodiversity. Hundreds of species of corals, fish, and marine invertebrates call these reefs home and make it one of the most biologically impressive locations on Earth.

The region has been designated as a marine national park—the Wakatobi National Park—and local and international efforts have been influential in conserving this biome. Since 2005, the park has been listed as a tentative United Nations Educational, Scientific, and Cultural Organization (UNESCO) World Heritage Site.

The Wakatobi Archipelago has 25 groups of coral reefs, including fringing reefs, barrier reefs, and atolls, totaling about 375 miles (600 kilometers) in circumference. Numerous other small islands also exist in the archipelago. These islands are all composed of coral that was deposited in previous environments over millions of years. The underwater explorer and conservationist Jacques Cousteau called this area 'underwater nirvana,' and Wakatobi is now one of the most popular diving locations in the world.

Biodiversity

The species richness of the Wakatobi coral reefs is staggering. There are about 850 known coral species globally, and these reefs possess about 750 of these species. A survey conducted in the Wakatobi National Park in 2003 identified 396 species of coral belonging to 68 genera and 15 families. These include *Acropora formosa, A. hyacinthus, Psammocora profundasafla, Pavona cactus, Leptoseris yabei, Fungia molucensis, Lobophyllia robusta, Merulina ampliata, Platygyra versifora, Euphyllia glabrescens, Tubastraea frondes, Stylophora pistillata, Sarcophyton throchelliophorum,* and *Sinularia* spp.

More than 900 fish species can be found on these reefs, and hundreds of additional species of marine invertebrates are evident. Fish species include the peacock grouper (*Cephalopholus argus*), bluespine unicornfish (*Naso unicornis*), titan triggerfish (*Balistoides viridescens*), two-spot snapper (*Lutjanus biguttatus*), spotted rabbitfish (*Siganus guttatus*), ovalspot butterflyfish (*Chaetodon specullum*), beaked coralfish (*Chelmon rostratus*), and longfin bannerfish (*Heniochus acuminatus*), to name just a few.

The reefs support many seabirds, including the brown booby (*Sula leucogaster plotus*), Malaysian plover (*Charadrius peronii*), and common kingfisher (*Alcedo atthis*). Marine turtle species, including the hawksbill seaturtle (*Eretmochelys imbricata*), loggerhead seaturtle (*Caretta caretta*), and Pacific ridley seaturtle (*Lepidochelys olivacea*), also inhabit the reef's surroundings.

Humans derive many resources from the Wakatobi reefs. Many fish and marine invertebrates are captured for food (for private as well as commercial consumption), while others are exported as ornamental species that are found in fish tanks worldwide. There are about 93 species of ornamental and commercially valuable fish. Unfortunately, overfishing, population growth, tourism, and global climate change are threatening these reefs to some degree.

*The Wakatobi coral reefs are home to several marine turtle species, including the hawksbill seaturtle (*Eretmochelys imbricata*), shown here; the loggerhead seaturtle (*Caretta caretta*); and the Pacific ridley seaturtle (*Lepidochelys olivacea*). (Thinkstock)*

Human Impact

Some fishing practices have had detrimental effects on reef health. First, certain species, especially large, pelagic fish, are exploited preferentially over smaller, benthic (deepwater) fish.

This causes trophic cascades that disrupt the overall biome food web. Additionally, overfishing has become a major problem as human populations continue to grow on the various islands throughout Wakatobi. This can be exacerbated by the method of fishing, with blast fishing and cyanide fishing having harmful effects on the environment.

Corals, the basis of the biome, are also negatively affected in many ways. Water quality can have a profound impact, as human-induced terrestrial runoff causes turbid water, which blocks the sunlight needed by the symbiotic zooxanthellae that exist within the coral polyps. These single-celled organisms are essential for providing nutrients to reef-building coral.

Likewise, the loss of grazing fish can cause algae levels to grow, also stealing valuable sunlight. Finally, changes in water chemistry due to climate change could be detrimental to many coral species. Coral bleaching is a consequence of these scenarios and has been observed on Wakatobi reefs. Fortunately, no widespread bleaching events have been observed here yet.

Because of the high ecological and economic importance of the coral reefs, more than 5,000 square miles (13,000 square kilometers) of islands, reefs, and surrounding waters were established as Wakatobi National Park (WNP) in 1996. This immense park is one of a growing number of marine parks worldwide, and is the third-largest in Indonesia.

As part of a rapid ecological assessment, a management plan was put into place to help protect the important coral reefs. The objectives of this plan are to improve management in the park in ways that can be supported by most stakeholders, to initiate monitoring to assess where further management is needed and how effective management has been, and to garner support by increasing stakeholders' understanding and appreciation of the WNP and its management.

International groups have also been influential in Wakatobi coral reef conservation. The Nature Conservancy and the World Wildlife Fund of Indonesia helped Wakatobi adopt a zoning system approved by all forms of government and by the communities living within the WNP. These groups also worked directly with the WNP's local communities to make them more aware of their effects on the fisheries and the environment. Establishing no-take areas helped populations rebound, benefiting the local communities and the overall biome. Efforts have also been made to provide training in marine protected area management and to encourage sustainable tourism in the region.

It is believed that global climate change is already impacting the archipelago. Coral bleaching events may have occurred because of elevated sea surface temperatures in the area. Bleaching is caused when higher than normal sea temperatures make light toxic to the coral's food-producing symbiotic algae, the zooxanthellae. Under these conditions the zooxanthellae are expelled from the coral polyps, which then turn white. Bleaching causes corals to starve and is a temporary state; if thermal stress reverses, corals can return to their normal condition—but if the stress persists, corals can die in great numbers.

One of the management strategies to address this issue is to identify reef sites which may be more resilient to changing water temperatures, and include them in marine protected areas. It is thought that reducing direct human threats in these critical areas, such as destructive fishing and overfishing, will increase the chance that coral reefs will survive from climate change impacts.

The combined efforts of the Indonesian government, local stakeholders, and international conservation organizations provide hope that the Wakatobi coral reefs will persist into the future.

Daren C. Card

Further Reading

Clifton, Julian, Richard K. F. Unsworth, and David J. Smith., eds. *Marine Research and Conservation in the Coral Triangle: The Wakatobi National Park.* Hauppauge, NY: Science Publishers, 2010.

Haapkyla, J., R. K. F. Unsworth, A. S. Seymour, J. Melbourne-Thomas, M. Flavell, B. L. Willis, and D. J. Smith. "The Spatio-Temporal Coral Disease Dynamics in the Wakatobi Marine National Park, South-East Sulawesi, Indonesia." *Diseases of Aquatic Organisms* 87, nos. 1–2 (2009).

Wilson, J. R., R. L. Ardiwijaya, and R. Prasetia. "A Study of the Impact of the 2010 Coral Bleaching Event on Coral Communities in Wakatobi National Park." *The Nature Conservancy, Indo-Pacific Division, Indonesia,* 7 (2012).

Wasatch and Uinta Montane Forests

Category: Forest Biomes.
Geographic Location: North America.
Summary: The coniferous montane forests of the Wasatch and Uinta mountain ranges are valuable resources threatened by wildfires, beetle infestation, and invasive plant species.

At the eastern edge of the Great Basin in the western United States, the Wasatch and Uinta Mountain ranges begin their dramatic ascent to elevations over 12,000 feet (3,600 meters). Beneath their peaks lies a deep green conifer forest, part of a unique ecoregion that stretches from the far corner of southwestern Wyoming to the Colorado plateau in southern Utah. The Wasatch and Uinta Montane Forests biome includes both the Wasatch Mountains, which form the backdrop for the Ogden, Salt Lake City, and Provo, Utah metropolitan areas, and the Uintas, the highest east-west oriented range in the contiguous United States.

Though inaccessibility and conservation efforts have preserved pockets of undisturbed habitat throughout the region, most of the area has been impacted over the years by logging, mining, livestock grazing, and recreational use. As a consequence, forest plant communities have been compromised, resulting in degradation of plant and animal habitat.

One hundred miles east of Salt Lake City, the treeless peaks of the Uinta Mountains rise 11,000–13,500 feet (3,400–4,100 meters) into rarefied air. Consisting of quartzite, shale, and slate, these broad, flat monoliths are the products of 70 million years of geological uplift and glaciation.

The Wasatch and Uintas are bordered by the Great Basin Desert to the west, and because they are also in the rain shadow of the Sierra Nevada range—much further to the west—they are more arid than the rest of the Rockies. They receive less than 20 inches (50 centimeters) of precipitation per year. The higher peaks, however, receive a great deal of dry snow.

Biodiversity

Different species of conifers grow depending on varying soil types—in this case, either limestone or quartzite. Douglas fir thrives in limestone soils, while regions with quartzite support forests of lodgepole pine. Altitude is also a factor. Ponderosa pine and lodgepole pine have a range of about 8,000–9,000 feet (2,440–2,900 meters), while Engelmann spruce is only found over 10,400 feet (3,170 meters).

The Wasatch Range is generally steeper, more rugged, and not quite as dry as the Uintas; however, these mountains are not as high in elevation. Similar to the Uinta mountains, the canyons and passes of the Wasatch support coniferous forests of ponderosa pine, Douglas fir, subalpine fir, Engelmann spruce, and trembling aspen. Gambel oak also grows over a wide area.

Wildlife is abundant and diverse in the Wasatch and Uinta montane forests. Large mammals include elk, mule deer, pronghorn, mountain goat, bighorn sheep, fox, badger, coyote, black bear, and mountain lion. Ground squirrel, woodchuck, marmot, hare, jackrabbit, and other small mammals species can be found here as well. Beaver and otters inhabit areas near streams and rivers.

Bald and golden eagles, the great horned owl, falcon, kestrel, and more than 100 additional bird species inhabit these mountain ranges. There are also at least seven amphibian and 14 reptilian species that have been identified.

Human Impact

Native people first lived in and around the Wasatch and Uinta Mountains around 12,000 years ago. The indigenous Ute, or Uintah, people lived mainly in the lower hills and plains, but hunted and gathered food in the forested uplands in warmer seasons.

By the early 1800s, European explorers and trappers were crossing the plains to the mountains of Wyoming and Utah. Fur trading companies held gatherings just north of the Uinta Mountains, where trappers could trade beaver and fox pelts for supplies and gold. Individual explorers and small groups of traders made little impact on the surrounding ecosystem; however, this would soon change.

In 1846–47, Mormon emigrants traveled to the Salt Lake Valley on the Mormon Trail, which led them through the forested canyons of the Wasatch. In 1849, gold was discovered in California, and a steady stream of settlers passed through Utah to find their fortunes in the west. When the gold rush ended, some miners returned to the mountains of Utah. Construction of silver and gold mines in Little Cottonwood Canyon and Park City required vast quantities of timber from the Wasatch and Uinta mountains.

When the transcontinental railroad reached Utah in 1868–69, demand for lumber grew yet again. Loggers cleared the north slope of the Unitas for railroad ties, leaving nothing but acres (hectares) of stumps. As branch railroad tracks and roads spread their web over northern Utah, more of the Wasatch and Uinta forests disappeared. Grasslands replaced trees, and ranchers began grazing their cattle and sheep on the logged slopes.

The snow that falls in the mountains of Utah is some of the driest in the country—perfect for skiing, snowboarding, and snowmobiling. By 1940, ski enthusiasts began clearing the forests for ski runs, but the now well-known ski resorts did not impact forest ecology until the 1970s. Today, there are 13 multi-sport resorts in Utah, most of them within an hour of Salt Lake City.

There are several factors in the ongoing decline of forest health in the Wasatch and Uinta montane forests. Past logging and grazing practices have contributed to forests that are extremely dense, with large areas of woody debris that provide volatile fuel for wildfires. The density of the forest stresses individual trees, causing them to compete for soil and water. Global warming may also be adding stress, as warmer conditions accelerate evapotranspiration and open the door to in insect assault. As a result, the forest becomes susceptible to bark beetle infestation. Invasive species of weeds are also a problem. These factors combine to reduce biodiversity, alter wildlife habitats, and damage critical watersheds.

More people live adjacent to and enjoy the forests of the Wasatch and Uinta mountains than ever before. Increased development, especially adjacent to ski resorts, impacts the remaining fragmented habitat areas. Proactive forest management that embraces the use of conservation practices and treatments, such as controlled burning and thinning, can improve and preserve the health of this essential forest ecoregion for future generations.

Deborah Foss

Further Reading

Elias, Scott. *Rocky Mountains*. Washington, DC: Smithsonian Institution Press, 2001.

Stokes, William Lee. *Geology of Utah*. Salt Lake City: Utah Museum of Natural History, 1986.

Surhone, Lambert M., Mariam T. Tennoe (ed.), and Susan F. Henssonow (ed.). *Wasatch and Uinta Montane Forests*. Saarbrucken, Germany: Betascript Publishing, 2010.

Veranth, John. *Hiking the Wasatch*. Salt Lake City, UT: Wasatch Publishers, 1991.

Wasur Wetlands

Category: Inland Aquatic Biomes.
Geographic Location: Asia.
Summary: The largest wetland in Papua, Indonesia, is an important habitat of unique biodiversity, but is threatened by human activity and invasive species.

Wasur National Park stretches across more than 1,550 square miles (4,000 square kilometers) of savanna, swamp forest, coastal forest, bamboo forest, and areas of sago swamp forest. These wetlands are located along the south coast of West Papua, Indonesia, adjacent to the border of Papua, New Guinea. The island of New Guinea lies directly northeast of Australia across the Arafura Sea, approximately 100 miles (160 kilometers) from Queensland at the closest point.

Climate and Ecosystems

This tropical region is influenced by a monsoon climate with two seasons: the wet season from roughly December to May, and the dry from June to November. However, New Guinea has one of the most variable climates on Earth in terms of precipitation. Temperatures in the Wasur Wetlands biome range from 72 to 86 degrees F (22 to 30 degrees C). Annual rainfall is about 94 inches (240 centimeters). Most of the Wasur National Park is a natural swamp area during rainy season, and becomes more savanna-like during the dry season.

The area is geographically flat in the coastal area near Arafura Sea, rising only to about 300 feet (90 meters) above sea level to the north. There are six main ecosystems identified in Wasur: (1) seasonal swamp brackish water, (2) permanent fresh water swamp, (3) coastal plain and freshwater, (4) inland fresh water, (5) coastal brackish water, and (6) inland brackish water.

Biodiversity

The dominant plants include mangroves; genus *Terminalia,* with its magnolia-type flowering shrubs and trees; and myrtles of the genus *Melaleuca.* Low swamp grasslands show an abundance of *Pseudoraphis spinescens,* a spiny mudgrass forming low swards in the dry season, and adapted to wetlands by growing taller during rainy season.

The dominant tree species are mostly *Melaleuca cajuputi, M. viridiflora,* and *M. leucadendron*; these are followed by *Acacia auriculiformis, Eucalyptus* spp., *Alstonia actinophilla, Dillenia alata, Nauclea orientalis.* In the coastal forest, *Avicennia leucaliptifolia, Barringtonia tetraptera, Bruguiera* spp., and *Rhizophora* spp. can be found.

Wasur National Park provides habitat for more than 300 bird species, of which at least 80 are endemic, meaning found nowhere else on Earth. Some of endangered species here include southern crowned pigeon, New Guinea harpy eagle, dusky pademelon, black-necked stork, and little curlew. Some of the more common species are Pesquet's parrot (*Psittrichus fulgidus*), southern cassowary (*Casuarius casuarius sclaterii*), blue crowned pigeon (*Goura cristata*), greater bird of paradise (*Paradisea apoda novaeguineae*), king bird of paradise (*Cicinnurus regius rex*), and red bird of paradise (*Paradisea rubra*).

Also utilizing the area for habitat are endemic birds species such as the fly river grassbird and grey-crowned munia. The migrant birds from Australia come to Wasur from the wet season to the early period of dry season.

The Wasur wetlands are an important habitat of rare mammal species such as dusky wallaby *Thylogale brunii*), agile wallaby (*Macropus agilis*), and other marsupials.

A Leichhardt tree (Nauclea orientalis), one of the dominant tree species in the Wasur wetlands, in bloom. Wasur National Park, along with three adjacent wildlife management areas, is part of the longest continuous protected area in Indonesia. (Wikimedia/Tony Rodd)

There are at least 26 species of reptiles here, including the New Guinea crocodile (*Crocodylus novaeguineae*) and saltwater crocodile (*Crocodylus porosus*); eight types of lizards, including three types of monitor lizard (*Varanus spp.*); four types of turtle including the pig-nosed turtle (*Carettochelys insculpta*); eight types of snakes (genera *Condoidae, Liasis,* and *Pythonidae*); and one type of chameleon (*Calotus jutatas*). At least three species of frogs have been identified: tree frog (*Hylla crueelea*), Irian tree frog (*Litoria infrafrenata*), and green frog (*Rona macrodon*).

There are also termites that build unique nests, up to 15 feet (5 meters) in height, made from mud and grass litter. The coastal area provides habitat for various species of fish, shrimp, mollusks, and other aquatic life that help sustain the food web for land ecosystems.

Human Impact

In Wasur National Park there are four tribes of indigenous peoples—Kanume, Marind, Maori, and Yei. These tribes hunt and fish for sustenance; they consume mainly sago, sweet potato, deer, bandicoot, wild boar, and wallaby. Their local wisdom is to use the natural resources sustainably through customary regulation in a mode they call *sasi.* It is believed that sasi provides an opportunity for animals to breed and may thus save many species from extinction.

Rawa Biru Lake, located in Wasur National Park, is an important watershed area supplying the Merauke district. Sedimentation was found to be one of the factors leading to water quality decline in Rawa Biru recently, stemming from forest degradation in the surrounding area. Illegal logging is a problem.

Natural flooded swamp ecosystems here are threatened by large scale changes to scrub and woodland, as well as invasive plant species. Climate change may also impact the coast in the forms of sea-level rise and more intense and frequent storm surges; both effects can undermine habitat integrity and drive out native species.

Wasur National Park was recognized as a Wetland of International Importance under the Ramsar Convention in 2006. Together with three other adjacent wildlife management areas, Wasur park is now part of a larger protected area in the Trans-Fly ecoregion. Named for the Fly River, this coastal region of grasslands, savannas, wetlands, and monsoon forests creates the longest continuous protected area in the country.

Subekti Rahayu
Degi Harja

Further Reading

Bartolo, R. E., M. Bowe, N. Stronach, and G. J. E. Hill. "Landscape Change and the Threat to Wetland Biodiversity in Wasur National Park, West Papua (Irian Jaya)." University of the Sunshine Coast, Queensland, Australia. http://research.usc.edu.au/vital/access/manager/Repository/usc:1030.

BirdLife International. "Endemic Bird Area Factsheet: Trans-Fly." http://www.birdlife.org/datazone/ebafactsheet.php?id=179.

Marshall, J. A. and B. M. Beehler, eds. *The Ecology of Papua.* Singapore: Periplus Editions, 2006.

Water Management

Category: Grassland, Tundra, and Human Biomes.
Geographic Location: Global.
Summary: Water management is the process of developing, distributing, and managing water resources with the goal of optimizing the use of water resources for balanced needs and reducing negative effects on the environment. Growing human populations are leading to water scarcity and increased contamination of water resources worldwide.

Water management is the process of developing, distributing, and managing water resources. Water resources are essential to agricultural, industrial, household, recreational, and environmental activities. Almost all these water uses by humans require sources of freshwater. Freshwater is defined as having little dissolved salts or other dissolved solids. Though freshwater is a renewable resource, the world's supply of readily available, clean freshwater is decreasing and is being put under growing pressure as populations increase globally.

The goal of water management is optimizing the use of water resources and reducing negative effects on the environment. In an ideal world, water management balances the competing water demands of human populations for sanitation, drinking, manufacturing, leisure, and agriculture.

Water and Ecosystems

Water resources are an important component of ecosystems. Many biomes are defined by their rainfall and water levels. Tropical rainforest ecology, for example, is highly dependent on its namesake year-round high levels of precipitation and humidity. Water also plays an integral role in ecosystem services. Ecosystem services are benefits that humans derive from natural systems, including a wide range of material resources and processes.

These services fall into four main categories: provisioning, which includes the production of food and water; regulating, which includes the control of climate and disease; supporting, which includes nutrient cycling and crop pollination; and cultural, which includes spiritual and recreational benefits. Water resource use can affect each of these categories of ecosystem services.

As human populations grow, so do pressures on ecosystems and water resources. More than half the world's wetlands were lost during the 20th century as a result of human activity and improper water resource management. Surface and groundwater systems are being heavily drawn upon and contaminated. Ocean ecosystems are also being contaminated and overfished. Due to improper water resource management and climate change, deserts are the only ecosystems increasing worldwide, and desertification is a significant global environmental problem.

Water Use

Uses of water can be consumptive, where water afterward is not immediately available for other uses, or nonconsumptive (also called renewable), where water can be put to additional use after it is used. Consumptive uses of water include activities that allow seepage into the ground, evaporation into the air, or incorporation into a product (such as agricultural crops). Nonconsumptive uses include treated sewage that is returned as surface water. The main categories of human water use are agricultural, urban, industrial, household, recreational, and environmental.

Agriculture is the biggest consumer of freshwater resources, using about 70 percent of available water resources, and 15–30 percent of that water is used for irrigation in an unsustainable manner, drawing more water than an area can provide in the long term. An estimated 793 gallons (3,000 liters) of water is needed to produce enough food to feed one person daily, compared with the 0.5–1.3 gallons (2–5 liters) that one person drinks daily.

The amount of water required to produce food for each person is increasing as people tend toward eating more meat and produce, which are more water-intensive to grow than grains. Agriculture can cause water pollution through use of fertilizers, pesticides, and livestock manure that is then washed into rivers by rainfall.

Half of the world's people live in towns and cities, and this proportion is expected to reach two-thirds by 2050. Agricultural areas surrounding urban centers must compete for water from the same local supply as the city. Traditional water sources, both ground and surface, are being polluted by urban wastewater. Cities offer the best choice to sell produce, so many farmers are located near urban areas and use polluted waters to irrigate their crops. These polluted waters, depending on the city treatment processes, can pose health hazards when consumed in produce. Urban wastewater is a mixture of pollutants from kitchens, toilets, and rainwater runoff, and contains excessive nutrients and salts, pathogens, heavy metals, trace antibiotics, endocrine-disrupting chemicals, and estrogens.

Developing countries have the least amount of water treatment. Pathogens of greatest concern include bacteria, viruses, and parasitic worms in

World Water Facts

Water is an essential resource for all life. A total 97 percent of the water on planet Earth is saline, or saltwater, found in the oceans. Of the 3 percent that is freshwater, 69 percent is frozen in ice caps and glaciers, about 30 percent is groundwater, 0.3 percent is fresh surface water, and the rest is fixed within such various forms as water vapor.

Of the fresh surface water, about 2 percent is in rivers, 11 percent in swamps, and 87 percent in lakes. Of the liquid ground and surface water, about 20 percent is in remote, inaccessible areas, and much seasonal rainfall cannot be easily used. In sum, less than 1 percent of the world's freshwater is suitable for use by humans at present.

Only half of all people worldwide have access to piped water in their homes; one-third have access to some public improved water source, such as a public well or stand pipe. One-fifth have no access to improved sources of water whatsoever. Further, two-fifths of all people globally have no improved sanitation facilities. One reason is that water supply and sanitation projects are very expensive. Specialized infrastructure is needed in the form of pipe networks, pumping stations, and water-treatment works. Added to that cost are the significant ongoing costs of maintenance, personnel, energy, chemicals, and more.

the water supply. These pathogens can directly affect the health of farmers who come into contact with them, as well as indirectly affect consumers who ingest affected produce.

Contaminated water and produce can carry a variety of pathogens, such as salmonella, cholera, and diarrhea. Diarrhea kills 1.1 million people annually and is the second-most-common cause of infant death.

Industry uses about 20 percent of available water resources for such functions as running hydroelectric dams, thermoelectric power plants, refineries, and manufacturing plants. While industry requires large supplies of water, it is not as water-consumptive as agriculture, but it can cause a large amount of pollution by discharging wastewater that is tainted with manufacturing chemicals. In addition, plants that use water from rivers and lakes as coolant and return the warmed water directly to the water body can cause damage to ecosystems through temperature changes.

Households are estimated to use 8 percent of the world's available water. This includes water for drinking, cooking, sanitation, bathing, and gardening. It is estimated that—excluding gardening—one person requires around 13 gallons (50 liters) of water per day for household uses. Drinking water, also called potable water, must be of high quality such that it can be consumed without causing immediate or long-term harm. In most developed countries, water supplies piped into houses are all drinking-water quality, despite the fact that only a small percentage of that water is used for drinking.

Recreational water use makes up perhaps the smallest percentage of water resource use. An example of this type of water management is when reservoirs are used for recreation. Often, water levels are kept higher in reservoirs to allow for recreational activities such as boating, which, while nonconsumptive in nature, might prevent that water from being used by agriculture downstream. Sometimes, the release of water from reservoirs is timed to enhance recreation, such as whitewater rafting downstream of the reservoir. Golf courses use large amounts of water to irrigate their lawns, and also apply large amounts of fertilizers and pesticides that can contaminate water supplies through runoff.

Human use of water resources to augment the environment makes up a very small percentage of overall available water use. Such activities include restoring natural or artificial wetlands and lakes, creating fish ladders to allow the movement of fish around dams, and releasing water from reservoirs to help fish spawn or restore river flow regimes. Again, these uses are mostly nonconsumptive, but may reduce water availability for other needs.

Threats to the Water Supply

There are many difficulties with proper water management, including the fact that sources of water often cross national boundaries. However, water that spans international boundaries has been shown by the International Union for the Conservation of Nature (IUCN) to be a source for cooperation over water issues rather than a reason for conflict. In addition, many nonconsumptive water uses are hard to give financial value to, and are generally difficult to manage. These nonconsumptive uses could include protecting water bodies as habitats for rare species or for ecosystem services, as well as protecting ancient groundwater reserves. Climate change may also affect the availability of freshwater by changing precipitation cycles, melting ice caps, and raising ocean levels.

Other challenges facing proper water management include water scarcity and water contamination. To improve water availability, there is a need to improve data, treasure the environment, reform governance, revitalize agricultural water use so that farmers can increase productivity to meet growing needs for food, control urban and industrial water use, and empower underrepresented people (such as women) in water management.

A woman tending to an irrigated field of peppers. Agriculture uses as much as 70 percent of all available freshwater resources, with 15–30 percent of that water being used in an unsustainable manner. (Thinkstock)

Water Scarcity

Water scarcity is becoming an increasing concern as human populations grow exponentially, and as water use for industry, development, and agriculture for food and emerging biofuels increase. A total of 1.2 billion (20 percent) of the world's people live in areas of physical water scarcity where there is not enough water to meet demands. A total of 1.6 billion people live in areas of economic water scarcity where there is a lack of investment in water infrastructure or insufficient human capacity to supply the demand for water.

Rainwater harvesting is one potential solution for water scarcity. Rainwater is collected with rooftop harvesting systems, water barrels, or water gardens. Harvested rainwater can then be stored for later use. Rainwater can be used as it is for activities like gardening, or can be treated for consumption as drinking water.

Water recycling, using treated wastewater for additional uses, is another solution to water scarcity. Recycled water is used in a wide range of activities, such as agriculture, landscaping, public parks, golf-course irrigation, cooling water for power plants and oil refineries, processing water for mills, watering plants, toilet flushing, dust control, construction activities, concrete mixing, artificial lakes, and replenishing or recharging groundwater basins. Recycling water is especially useful in water-scarcity areas, but is also used to save cost and energy.

Another type of recycled water is called gray water. Gray water is reusable wastewater from residential, commercial, and industrial bathroom sinks, bathtub shower drains, and laundry-equipment drains. Gray water is generally reused on-site and requires the use of nontoxic, low-sodium soaps to protect vegetation.

Water Treatment

Water treatment is the process of making water more acceptable for its desired use. Water is often treated for use in industry, drinking water, and medicine, as well as to return it to the environment without adverse effect. Treating water generally involves either removing contaminants or reducing them to a satisfactory level.

As stated earlier, water contamination can be harmful to humans through exposure to pathogens, toxins, and chemicals, directly and through uptake into produce. In addition, contaminated water can harm ecosystems and wildlife health. Actions to reduce or remove contaminants in water have the potential to save lives and to improve ecosystems and livelihoods in affected areas.

The multiple-barrier approach to preventing produce contamination involves analyzing the food production process from growth to consumption, looking for where a barrier to contamination might be possible. Such barriers could be safer irrigation practices, on-farm wastewater treatment, actions to kill pathogens, or the implementation of crop washing after harvest in markets and restaurants before sale and consumption.

Treating water to make it safe for drinking can solve both the problems of water scarcity and contamination by making otherwise unusable water safe and available. There are many processes involved in treating drinking water. Solids are separated out using physical means such as settling and filtration, chemical means such as disinfection and coagulation, and biological means such as aerated lagoons, activated sludge, and slow sand filters.

Water is generally pre-chlorinated to control algae and bacteria growth; aerated to remove dissolved iron and manganese; coagulated, sedimented, and filtered to remove solids and particles; and disinfected to remove bacteria. This process varies from location to location based on local water quality, needs, and regulations. To ensure quality for drinking water, its treatment, transport, and distribution are generally highly regulated.

Another type of water treatment involves desalinization, also called desalination, which is the removal of dissolved salts and other minerals from water to make it safe for drinking or irrigation. Desalinization is expensive both in terms of energy and infrastructure, and therefore is considered only in areas where freshwater sources are otherwise limited. Australia, for example, receives little rainfall and relies on desalinization to augment its recycled water supplies.

There are two main types of desalinization, membrane and thermo. Membrane desalinization uses membranes to filter salt out of saline water through reverse osmosis. This uses less energy than the alternative but more widely-used process of thermo or vacuum distillation. Vacuum distillation boils the water to separate it from the salt. To reduce the energy needed to boil the water, vacuum distillation is performed at below-atmospheric pressures, which lowers the boiling point of the water, saving energy and, therefore, money.

Desalinization can have negative environmental effects by killing marine creatures caught in ocean inflow water, and increased temperature and salinity in outflow water returned to the ocean. On the positive side, desalinization sometimes produces table salt as a byproduct.

Sewage treatment has the goal of returning water that was used for household or industrial purposes back to the natural environment without causing adverse effects to the ecosystem. This generally involves removing human waste and chemicals from water that is piped to treatment plants. The process of removing these contaminants is similar to the one used to treat drinking water and involves multiple physical, chemical, and biological steps.

Sewage treatment generally results in the production of a liquid effluent that is safe to dispose into the environment, as well as a solid waste sludge. This sludge is generally treated further and discarded, or is sometimes reused as an agricultural fertilizer. With new advanced technology, it is even possible to reuse wastewater safely as drinking water—but Singapore is the only country to date to adopt this practice on a production scale.

In areas where sewage is not properly treated, or when sewage pipes leak and sewage is released before it reaches treatment, there can be negative effects on the environment. Increased nutrient inputs from organic waste like feces can cause eutrophication in water bodies, leading to algal blooms, low oxygen conditions, and fish die-offs. Contaminants from wastewater can have a diverse range of effects on the ecosystem. Endocrine-disrupting chemicals, such as those found in the form of synthetic estrogens in many pharmaceuticals, can cause reproductive deformities in vertebrates and have even been found to cause intersex individuals among fish and frogs.

Hannah Bement

Further Reading

Chartres, C. and S. Varma. *Out of Water: From Abundance to Scarcity and How to Solve the World's Water Problems.* Upper Saddle River, NJ: FT Press, 2010.

Gould, John and Erik Nissen-Peterson. *Rainwater Catchment Systems.* London: Intermediate Technology Publications, 1999.

Molden, D. *Water for Food, Water for Life: A Comprehensive Assessment of Water Management in Agriculture.* London: Earthscan/IWMI, 2007.

Pickford, John. *Developing World Water.* London: Grosvenor Press International, 1987.

Vickers, Amy. *Water Use and Conservation.* Amherst, MA: Water Plow Press, 2002.

White Sea

Category: Marine and Oceanic Biomes.
Geographic Location: Russia.
Summary: The White Sea is an Arctic water body of international importance that faces environmental threats, with few ecosystem management plans in place.

The White Sea is located at the northwest corner of Russia, just touching the Arctic Circle parallel of latitude; it is a southern inlet of the Barents Sea. The White Sea is almost entirely surrounded by land, with the Kola Peninsula to its north, Karelia to the west, and Arkhangelsk Oblast to the south and east. The strait connecting to the Barents Sea runs along the Kanin Peninsula to the northeast.

Geography and Climate

There are four main bays or gulfs in the sea. In the south is Onega Bay; in the southeast is Dvina Bay, which holds the major port of Arkhangelsk. Kandalaksha Gulf is in the western part of the White Sea and is the deepest part of the sea, reaching 1,115 feet (340 meters). Opposite the Kola Peninsula, on the east side of the strait, is Mezen Bay. The main rivers draining into these bays or gulfs are the Onega, Northern Dvina, Mezen, and Kuloy. Other significant rivers that flow into the White Sea are the Vyg, Niva, Umba, Varzuga, and Ponoy.

The White Sea climate varies between polar and moderate continental. The winter is long, severe, and often unpredictable. The mean temperature in February is about 5 degrees F (minus 15 degrees C) and may fall as low as minus 22 degrees F (minus 40 degrees C). The ice on the surface of the sea may be up to 5 feet (1.5 meters) thick. At times, warm air from the Atlantic raises the air temperature to a "comfortable" 43 degrees F (6 degrees C). The temperature of the water in winter is about 28–30 degrees F (minus 1 to minus 2 degrees C). The salinity of the White Sea is 27.5–28 parts per thousand, lower than the mean salinity of the Arctic Ocean.

The White Sea is affected somewhat by the Atlantic Ocean, as its hydrology reflects processes linked with the warm Gulf Current. Together with its generally Arctic character, the Atlantic and freshwater links define many of the species in the White Sea. Most recently, nutrients, pollutants, sedimentation, and invasive species issues have recurred, because of many disturbances across the inland watersheds.

Biodiversity

Since 1930, a large part of the coastline of the Kandalaksha Bay and about 100 small islands in the

White Sea have been declared protected areas. This was mainly to protect the eider ducks during nesting season and while the ducklings were growing up. During this time, hunting, fishing, and even visiting of these areas has been forbidden or strongly regulated.

The White Sea provides marine habitat for more than 700 species of invertebrates, about 60 fish species and five marine mammal species. The fishing industry is relatively small, mostly targeting harp and ringed seals (hunted mainly by indigenous peoples), herring, saffron cod, European smelt, Atlantic cod, and Atlantic salmon. There is also a small but developing seaweed industry.

Human Impact

The White Sea has been used by the indigenous population for thousands of years, and has played a role in western civilizations for more than 1,000 years. The sea was known to the Novgorod peoples since at least the 11th century, and probably to the Vikings before them. It was used to link Russian trading cities with northern and central European cities.

One of the earliest settlements near the sea was established in the late 14th century in Kholmogory on the Northern Dvina River. Trade in furs, grains, and fish continued through the centuries between Russia, Denmark, England, and Netherlands. Of note is the White Sea Canal that opened in 1933, which—along with the existing natural waterways, Lake Vygozero, Lake Onega, Lake Ladoga, Svir River, and the Neva River—connects the White Sea to the Baltic Sea. The canal was built almost entirely with forced manual labor.

Global climate change impacts all of the Arctic regions, with habitat-altering changes recorded in sea-ice cover, sea level, air and water temperatures, and precipitation patterns. Strategic military and extraction-industry interests present other concerns in the White Sea, such as waste dumping, ongoing for over 70 years. Arctic shipping routes, oil and gas production, commercial fisheries, and inland watershed misuse have left strong traces of pollution in this otherwise robust and somewhat intact environment.

There is a small amount of tourism here in the form of deep-sea diving, but little threat to the environment stems from this. A Russian management regime has dominated the White Sea for centuries; more recently, a Norwegian and European Union agreement has been provisionally established. In either case, an efficient and sustainable management regime has not yet been implemented in a manner or scale sufficient to aid full recovery of the White Sea biome.

Global relevance of conservation practices in the White Sea has not been realized on an international level yet. There is some controversy whether, and what type of, human-made changes have already occurred in the White Sea region. Some Russian sources report little or no changes for local regions in the White Sea. It is noteworthy that diseases and invasive species have not entered this discussion yet, but there are ongoing observations to monitor any environmental changes.

Further risky economic growth schemes involving oil and mineral extraction are still planned for the White Sea. Yet, there are few relevant climate change policies in place for the overall preservation of this unique ecoregion.

Falk Huettmann

Further Reading

Berger V. and S. Dahle, eds. *White Sea—Ecology and Environment.* St. Petersburg, Russia: Derzhavets Publishers, 2001.

Pertsova N. M. and K. N. Kosobokova. "Interannual and Seasonal Variation of the Population Structure, Abundance, and Biomass of the Arctic Copepod *Calanus Glacialis* in the White Sea." *Oceanology* 50, no. 4 (2010).

Stiansen, J. E. and A. A. Filin, eds. "Joint Polar Research Institute of Marine Fisheries and Oceanography (PINRO)/Institute of Marine Research (IMR) Report on the State of the Barents Sea Ecosystem 2006, With Expected Situation and Considerations for Management." *IMR/PINRO Joint Report Series* 2 (2007).

Solyanko, K., V. Spiridonov, and A. Naumov. "Biomass: Commonly Occurring and Dominant Species of Macrobenthos in Onega Bay (White Sea, Russia):

Data from Three Different Decades." *Marine Ecology* 32, no. 1 (2011).
Wilson Rowe, E., ed. *Russia and the North.* Ottawa, Canada: University of Ottawa Press, 2009.

Willamette Valley Forests

Category: Forest Biomes.
Geographic Location: North America.
Summary: The Willamette Valley has changed dramatically from its historical landscape, once dominated by oak woodlands and prairies with occasional coniferous stands, because of human disturbance.

The Willamette Valley is bounded on the west by the Oregon Coast Range and on the east by the Cascade Mountains. It spans a level alluvial plain with scattered groups of low basalt hills. The valley extends through northwest Oregon, and through a very small area in southern Washington state. The Willamette River, a major tributary of the Columbia River, meanders through this valley, flowing mainly south to north with historically great fluctuations in its flow, ranging from low summer flows to massive winter and spring floods. The river's bottomlands extend across low-lying floodplains, depositing rich sediment. The entire valley ecoregion stretches across approximately 5,800 square miles (14,900 square kilometers).

The climate of this region is relatively mild year-round, characterized by cool, wet winters and warm, dry summers. Temperatures seldom rise above 90 degrees F (32 degrees C), and even less frequently do temperatures drop below 0 degrees F (minus 18 degrees C). Most rainfall occurs in the late autumn, winter, and early spring, when temperatures are the coldest. The valley gets relatively little snow—5 to 10 inches (13 to 25 centimeters) per year.

This ecoregion was once dominated by a vast network of oak woodlands and grasslands interspersed with Douglas fir stands and wetlands, but human disturbance has dramatically changed the landscape over the past 150 years. Fire and seasonal flooding shaped the original characteristics of this landscape. The fire regimes, historically developed by seasonal wildfires and those set by Native Americans, were crucial for maintaining the grassland and savanna communities that dominated the region.

Oak prairies are able to endure on fire-prone landscapes where other forest types are unable to become established. The indigenous tribes of the Willamette Valley set fire to a large amount of land every year to regenerate the prairie plants on which they depended for food and medicine, and to generally clear the brush for hunting and other purposes. Some of the most important plants that were sought after were camas, tarweed, Oregon white oak acorns, and wapato. Other woodlands were purposely left unburned to provide areas where game would concentrate, so that animals could be hunted more successfully.

Biodiversity

Oregon white oak (*Quercus garryana*) occurs as scattered trees in savanna communities, and in pure or mixed-species closed-canopy woodlands. Pacific madrone (*Arbutus menziesii*), western serviceberry (*Amelanchier alnifolia*), ponderosa pine (*Pinus ponderosa*), and Douglas fir (*Pseudotsuga menziesii*) are among some of the most common tree species that co-occur with Oregon white oak. Poison oak (*Rhus diversiloba*) and common snowberry (*Symphoricarpos albus*) are some characteristic shrubs of Oregon white oak habitats.

In the absence of fire, the remaining prairie, oak woodland, and savanna have quickly filled with the fast-growing conifer Douglas fir, Oregon's state tree. Douglas fir trees are large, sun-loving conifers capable of living hundreds of years and reaching more than 200 feet (61 meters) tall and 10 feet (3 meters) in diameter. Douglas fir forests are the most extensive in Oregon and the most crucial for timber production. Douglas fir often forms immense, nearly pure stands. Western hemlock is the climax species for this conifer-dominated community in much of this region, while other associates can also be found, including western red cedar, noble fir, bigleaf maple, and red alder.

Douglas fir is used for a wide variety of building products, and millions of Douglas fir Christmas trees are exported from Oregon annually. Older Douglas firs provide an important habitat for an array of wildlife, including nesting birds and small mammals. Dying Douglas fir trees and snags are key sources of cavities for woodpeckers and other birds.

Despite all the changes to, and pressures on, the Willamette Valley's forest lands, they continue to play a vital role in wildlife health in the region. The now more common mixed stands of deciduous and conifer forest, and the watershed that sustains them, provide habitat for an array of fauna, including 18 species of amphibians, such as the Pacific giant salamander, and at least eight reptile species. Fish in the Willamette basin include 31 native species, among them cutthroat, bull, and rainbow trout; several species of salmon; and sturgeon.

Beaver and river otter are among 69 mammal species living in the watershed. Other mammal species include: elk, coyote, bobcat, mule deer, weasel, mink, muskrat, red fox, black bear; many small mammals such as voles, shrews, squirrels; and many species of bats. There are also 154 bird species that have been identified here, including the great horned owl, northern saw-whet owl, northern pygmy owl, various hawks, woodpeckers, grosbeaks, terns, quail, ducks, and hummingbirds. Many conservation efforts now focus on restoring, protecting, and maintaining the region's almost-lost oak woodland and grassland habitats.

The Taylor's checkerspot butterfly is among the species that have declined because of the loss of grassland in the Willamette Valley. The butterfly is now seen in only one local site, down from 70. (U.S. Fish and Wildlife Service)

Human Impact

The valley's fertile soils, mild climate, and gently sloping floodplains nestled within volcanic peaks and wet temperate forests made this area highly desirable to European settlers. The rich soils and plentiful rainfall made the Willamette region the most crucial agricultural area in the state. This is also the fastest-growing region to be developed, now housing the majority of Oregon's major cities and population.

Over the past 170 years, a significant change for the Willamette River Valley has been the loss of its floodplain forests, which in 1850 covered an estimated 89 percent of a 400-foot (120-meter) band along each river bank. By 1990, 63 percent of this forested area had been converted to farmland or to urban and suburban development.

The vast majority of land within the Willamette Valley is currently in private ownership. The landscape is a mosaic of farms, working forest lands, suburbs, and cities, with only small fragmented habitat of the types that were once present in continuous abundance. Less than 1 percent of the historically dominant oak savannas and prairies remain. Remnant pockets of historic oak woodlands and savannas are often invaded by conifers, Oregon ash, and an array of nonnative species.

The Willamette River itself has been disconnected from its floodplain. Dams now control the river's flow, and vast stretches of the riverside have been channelized. Invasive plant species are currently considered to be one of the major causes of species becoming threatened and endangered, second only to habitat loss. Population growth, land-use conversion, invasive-species infestations, pollution, and effects of global warming are all likely to continue to increase stress on all the habitats throughout the valley.

Mary Logalbo

Further Reading

Campbell, Bruce H. *Restoring Rare Native Habitats in the Willamette Valley.* West Linn, OR: Defenders of Wildlife, 2004.

McNab, W. Henry and Peter E. Avers. "Willamette Valley and Puget Trough." In U.S. Forest Service, eds., *Ecological Subregions of the United States.* Washington, DC: U.S. Department of Agriculture, 1996.

Vesely, David G. and Daniel K. Rosenberg. *Wildlife Conservation in the Willamette Valley's Remnant Prairies and Oak Habitats: A Research Synthesis.* Corvallis: Oregon Wildlife Institute, 2010.

Windward Islands Xeric Scrub

Category: Forest Biomes.
Geographic Location: Caribbean Sea.
Summary: This island group is characterized by wet forests and vital but small patches of xeric scrub vegetation.

The Windward Islands are the southernmost islands of the Lesser Antilles (in the West Indies) of the Caribbean Sea. Extending in a chain off the north coast of Venezuela from Grenada to Martinique, the Windward Island group typically includes Trinidad and Tobago and Barbados, which are east of the main archipelago. The extent of the island xeric scrub here is western Barbados and northwestern Trinidad and Tobago, as well as Grenada, the Grenadines, St. Vincent, St. Lucia, Martinique, and Dominica.

The rainy season in this tropical zone generally occurs in the summer and fall months, when hurricanes and tropical storms pass through the region. While annual rainfall in the interior highlands ranges from about 395 inches (1,000 centimeters) in Dominica to 150 inches (375 centimeters) for the lower-elevation mountains in Grenada, coastal areas that comprise much of the xeric scrub habitat of the islands receive lesser amounts, ranging from 40 inches (100 centimeters) on Dominica to 63 inches (160 centimeters) at the southern end of St. Vincent.

The Windward Islands are more geologically active than the Leeward Islands to the north, forming a volcanic arc between continental South America and the Leeward chain. Because of that, the topography is much more pronounced. The steep terrain has the effect of limiting the amount of low-elevation coastal habitat available, and thus further reduces the extent of this ecoregion type.

Flora and Fauna

The vegetation of the Windward Island Xeric Scrub biome is dry-adapted, often featuring stunted trees and shrubs found on poor soils. Typical vegetation consists of cactus and other succulent, spiny shrubs on the west coasts of the larger islands. Plant species include broadleaf lancepod (*Lonchocarpus pentaphyllus),* logwood *(Haematoxylon campechianum),* coralberry *(Myrsia atrifolia),* bastard redwood *(Chrysophyllum argenteum),* ratwood *(Erythroxylum ovatum),* and two species of cactus, *Opuntia dilenii* and *Pilosocereus royeni.*

Extensive coastal development in the area, as well as a long history of colonial occupation, have converted much of the native vegetation to agriculture or typical tropical, non-native resort gardens. An additional threat to this biome is the prevalence of numerous invasive species that either outcompete or feed upon the native flora.

Due to its larger size and isolation, Barbados has the higher degree of faunal endemism, that is, species found nowhere else. Notable among the Barbados endemic species is the tree lizard (*Anolis extremus*), which has adapted well to living in and around human infrastructure. Other notable lizards in the ecoregion are *Kentropyx borckiana* and *Phyllodactylus pulcher.* There is one endemic snake, the Barbados racer (*Liophis perfuscus*), which has a very restricted range and is considered to be endangered by the International Union for Conservation of Nature (IUCN) Red List. Many exotic species of reptiles have been introduced, including pit vipers (*Bothrops* spp.), which mongoose were subsequently introduced to control. Both have become invasive species in some areas.

The only endemic bird in this biome is the Grenada dove (*Leptotila wellsi*), which is found in remnant stands of native xeric vegetation. The Grenada dove is considered to be critically endangered (IUCN Red List) because of its extremely small and dispersed population, which has been declining because of habitat loss and fragmentation caused by fire, residential development, roads, and grazing and predation by invasive species.

No native terrestrial mammals are found in this biome; all native mammals are bats. However, a growing number of introduced mammals are becoming a threat to this and other islands in the region, especially rats, mongooses, goats, pigs, and opossums, all of which wreak havoc on increasingly rare native flora and fauna.

Human Impact

Unlike the Leeward Islands, which are predominantly a dry scrub habitat, the Windward Caribbean islands are typically host to wet forests, with small portions of dry scrub where climate is suitable. This means that this ecoregion type is naturally rare, but also that its small size and discrete locations make it very susceptible to disturbance, both human and natural. The dry portions of these islands are also targeted for human settlement and tourism development due to the more favorable climate—so, increasingly, this biome type is being converted by urbanization and resort development.

There is a clear need to take preventive measures to save the endangered native flora and fauna from continued habitat loss and introduction of aggressive exotic species. A long history of tobacco and sugarcane production, together with the more recent conversion of coastal habitats for tourism development, have taken a serious toll on biodiversity in the Caribbean, both generally, and more specifically in naturally rare ecoregions such as the Windward Islands xeric scrub. Due to climate change globally, threats of rising sea levels and associated impacts to the Caribbean islands will become more apparent in the future.

Jan Schipper

Further Reading

Carrington, S. *Wild Plants of the Eastern Caribbean.* London: Macmillan Education, 1998.

Censky, E. J. and H. Kaiser. "The Lesser Antillean Fauna." In *Caribbean Amphibians and Reptiles,* edited by Brian I. Crother. London: Academic Press, 1999.

Malhotra, A. and R. S. Thorpe. *Reptiles and Amphibians of the Eastern Caribbean.* London: Macmillan Education, 1999.

Pararas-Carayannis, George. "Tsunamis of Volcanic Origin in the Lesser Antilles Islands of the Caribbean." *Journal of Tsunami Hazards* 23 (2004).

Stoffers, A. L. "Dry Coastal Ecosystems of the West Indies." In E. Van der Maarel, ed. *Ecosystems of the World 2B: Dry Coastal Ecosystems Africa, America, Asia and Oceania.* Amsterdam, Netherlands: Elsevier Science Publishers, 1993.

Wingello State Forest

Category: Forest Biomes.
Geographic Location: Australia.
Summary: This temperate forest contains a mixture of non-native softwood pine plantations and native hardwood eucalypt forests.

Wingello State Forest is a mixed-use public forest located in Camden County in the Australian state of New South Wales, near the cities of Sydney and Goulburn; it is part of the Australian Southern Highlands region. Other forests located within New South Wales include the Penrose, Belanglo, Meryla, and Yalwal State Forests. The forests of New South Wales comprise approximately 44 percent of Australia's eucalypt open forests and approximately 19 percent of Australia's total plantation forest area.

Wingello features a temperate climate and is generally dry. Temperatures average from 53 degrees F (11 degrees C) in July to 82 degrees F (28 degrees C) in January. Some summer days can be much hotter. There are an average of 75 days of rain each year, providing an average of 25 inches (64 centimeters) of precipitation.

Wingello State Forest is home to such mammals as the southern brown bandicoot, eastern pygmy-possum, white-footed dunnart (a small marsupial), wallaby, grey-headed flying fox, a variety of bat species, wombat, and koala. This koala sits in a eucalyptus tree, which cover 80 percent of Australia's native forest area. (Thinkstock)

The soil is moderately fertile, and terrain consists of hills and valleys. The Wingello State Forest biome is comprised of a mixture of softwood pine plantations and native hardwood eucalypt forests. Radiata and other pine stands are grown for commercial purposes. Wingello's status as a state forest places it under the ownership of the New South Wales state government. Key environmental threats to the forest include logging operations, fires, and the introduction of non-native plants and animals.

Biodiversity

Much of Wingello's forest cover consists of non-native tree species planted for commercial purposes. The radiata pines (*Pinus radiata*), which are exotic to Australia and were first planted in the region in 1919 in nearby Belanglo State Forest, grow well in temperate climates with medium rainfall; these are the most common tree within Australia's softwood pine plantations. Wood produced from softwood conifers tends to be light in color as well as soft, and can be used for a variety of commercial and artisanal purposes.

Wingello State Forest also houses stands of native hardwood eucalypt forests, which are not accessible to logging operations. Eucalypt forests are among the most common and best-known of Australia's forest types, with most of the more than 700 species of eucalyptus trees being unique to the country. Eucalypts belong to the *Myrtaceae* family; in Australia, they appear in approximately 80 percent of the country's native forest area. Eucalypts grow in a variety of climates and rainfall levels, with species of eucalypts varying by forest and region. Native eucalyptus trees found within Wingello include peppermint, manna gum, and various stringybarks. These hardwoods have broad leaves, flowers, and dense wood, and are angiosperms.

Native forests generally feature greater density and wider variety of flora and fauna than their non-native counterparts do. They are also key to protecting soil and water quality. Many native forests in the region have been reduced to small remnant patches co-existing within other areas, such as within the pine plantations of Wingello.

However, Wingello State Forest supports a wide variety of wildlife. Mammals that are found here include southern brown bandicoot, koala, eastern pygmy-possum, white-footed dunnart (another small marsupial), wallaby, grey-headed flying fox, a variety of bat species, and, notably, the wombat. There are many bird species here, including parrot and owl species. There are at least six amphibian and three reptile species sustained by this forest.

Human Impact

Wingello State Forest's pine plantations support commercial logging, although the forest is not considered to be a large commercial forest operation. Overall, the Southern Highlands of Australia house approximately 13.5 square miles (35 square kilometers) of commercial pine plantations. Benefits of the forest plantation areas include windbreaks, erosion reduction, and a host of outdoors recreational activities. Camping is permitted with some environmental safety restrictions, such as limits to the times of the year when campfires may be set. The presence of commercial pine plantations has been controversial, however, because of their non-native nature and invasive threats to the ecosystem.

Direct threats to Wingello and other local forests have included natural and human-made fires, clearing for agricultural and urban development, heavy traffic from tourism, environmental degradation such as soil and water pollution, and the introduction of exotic and feral species of plants and animals.

New South Wales has implemented sustainable forest management practices and adapted its timber production regime to prevent environmental degradation and reduce threats to the forest's biodiversity. Codes and regulations include restrictions on road building, chemical and pesticide use, as well as fire protection requirements. The Australian national government's Department of Agriculture, Fisheries, and Forestry monitors forest issues and publishes an annual State of the Forests report.

Marcella Bush Trevino

Further Reading

Boland, D., M. Brooker, G. Chippendale, N. Hall, B. Hyland, R. Johnston, et al. *Forest Trees of Australia, 5th ed.* Melbourne, Australia: Commonwealth Scientific and Industrial Research Organisation, 2006.

Dargavel, John. *Fashioning Australia's Forests.* New York: Oxford University Press, 1995.

Davidson, J., S. Davey, S. Singh, M. Parsons, B. Stokes, and A. Gerrand. "The Changing Face of Australia's Forests." Canberra, Australia: Bureau of Rural Sciences, 2008.

Winnipeg, Lake

Category: Inland Aquatic Biomes.
Geographic Location: North America.
Summary: This relatively pristine freshwater habitat is under stress from anthropogenic sources of excess nutrients, invasive species, and climate change.

Lake Winnipeg is the largest lake in the Canadian province of Manitoba, and a valuable source of hydropower, commercial fishing, and recreation for the region. Its name is from the Cree term for *muddy waters.* The watershed is almost 40 times greater than the surface of the lake; it is shaped by environmental forces and human activities in an extensive area spanning Canada and the United States. The water quality and aquatic biodiversity of Lake Winnipeg are threatened by excess nutrients, invasive species, and climate change, while the large size and multiple jurisdictions of the watershed present further challenges for water management and policy decisions.

The 10th-largest freshwater lake in the world, Lake Winnipeg measures 270 miles (435 kilometers) north to south and spreads across approximately 9,460 square miles (24,500 square kilometers). It is the largest remnant of ancient Lake Agassiz, formed by glacial retreat in North America about 12,000 years ago. Lake Winnipeg consists of a large north basin and a smaller, comparatively shallow south basin; these are separated by a narrow channel that carries water from the south basin northward. Compared with other large lakes, Lake Winnipeg is relatively shallow, with an average depth of 39 feet (12 meters). This contributes to a fairly short water residence time of three to four years for the lake as whole, and further affects the water temperature, volume, and ecosystem functions.

The Lake Winnipeg watershed is nearly 386,000 square miles (1 million square kilometers), stretching from the Rocky Mountain foothills to within a few miles (kilometers) of Lake Superior and covering four provinces (Manitoba, Ontario, Saskatchewan, and Alberta) and four U.S. states (North Dakota, South Dakota, Minnesota, and Montana). River flow is the most significant source of water to the lake, which is fed by three major river systems: the Saskatchewan River from the west, the Red River from the south, and the Winnipeg River from the southeast. The Winnipeg River contributes nearly half of the total inflow to Lake Winnipeg, followed by the Saskatchewan River (25 percent), Red River (16 percent), other tributaries, and precipitation. The only outlet from the lake is the Nelson River, flowing northward to Hudson Bay, with evaporation also contributing to water loss.

Biodiversity

Walleye and carp are the most commonly fished species in the lake, accounting for over 90 percent of commercial fishing. Winnipeg was once the main source of goldeye. Other native fish include sauger, yellow perch, troutperch, berbot, freshwater drum, lake cisco, and emerald shiner. Fish designated as at risk (by the Committee on the Status of Endangered Wildlife in Canada) are silver chub, bigmouth buffalo, shortjaw cisco, and chestnut lamprey. The lake supports not only commercial and recreational fishing, but subsistence fishing for the local indigenous population.

Several species introduced into the lake are having considerable influence on ecological relationships, by competing with native species for food and other resources. Five aquatic invasive animal species are known to occur in Lake Winnipeg: the common carp, rainbow smelt, white bass, Asian tapeworm, and a zooplankton species (*Eusbomina coregoni*). Rainbow smelt are now the dominant prey fish species in the north basin of the lake, and scientists are investigating their effect on the harvest and production of walleye, an important commercial species.

Potential future invaders are zebra mussels, which have been found in the U.S. portion of the Red River watershed, and the spiny water flea, which was identified downstream from Pointe du Bois Dam on the Winnipeg River in 2010. Managers are concerned about these invasive species migrating up- or downstream, as they may affect ecosystem health and economic activities in Lake Winnipeg.

Human Impact

Lake Winnipeg provides important socioeconomic benefits to the region. It is the world's third-largest hydro-reservoir, and supplies Manitoba Hydro, the Crown Corporation hydroelectric utility, half of its storage and 75 percent of its generating capacity. Regulated since 1976, water flows around Lake Winnipeg now create hundreds of millions of dollars' worth of electrical power each year, with close to 75 percent of that revenue resulting from exports to other Canadian provinces and the United States.

This aquatic ecosystem also supports the most important commercial fishery in Manitoba; in 2010, the provincial government reported a value of $50 million per year for the economic activity related to commercial fishing on the lake. Recreation and tourism around Lake Winnipeg generate more than $100 million per year, as the area's beaches, parks, and cottages are popular destinations. Additionally, nearly 23,000 permanent residents along the shoreline in 30 communities, as well as populations downstream, rely on the water quality of the lake for drinking water.

Water quality in Lake Winnipeg has deteriorated during the past three decades, primarily from excess nutrients loaded into the lake through watershed sources. The catchment is dominated by agricultural use; it supports approximately 20 million livestock and 7 million people. Monitoring of watershed nutrient sources has revealed that cropland fertilizers carry the largest amounts of phosphorus to the lake through surface runoff and flooding, mainly through the Red River. Human and livestock wastes contribute additional loads.

Phosphorous and nitrogen are the two major nutrients contributing to the increased frequency and intensity of algal blooms that threaten the Lake Winnipeg biome. More than 80 percent of the algae that formed blooms in the lake during the ice-free seasons from 1999 to 2007 were blue-green algae or cyanobacteria, of particular concern because of their detrimental effect on the ecosystem here.

They are a less desirable food source than other algae types for organisms in the lake, leading to species accumulation and dominance. When these algae die and sink down to the bottom of the lake, their decomposition by bacteria consumes oxygen in the water; this reduces the available oxygen necessary for fish and other organisms in the food web. In addition, cyanobacteria have the potential to produce potent toxins that may pose health risks to humans and other animals. These water-quality changes can affect the lake's ecosystem sustainability, safety for the public, and aesthetic appeal.

Climate change has the potential to alter water temperatures and further affect ecosystem structure in the lake. From 1909 to the present, August water temperatures in the south and north basins of Lake Winnipeg have increased by 3.4 degrees F and 1.8 degrees F (1.9 degrees C and 1 degree C), respectively. Global climate models predict that the south basin's summer water temperatures could rise 9 degrees F (5 degrees C) by 2085. Research also suggests that the diversity of aquatic species in the lake will decline as water temperature rises.

Increased temperatures in central Canada will likely exacerbate the establishment of invasive species in Lake Winnipeg from southern regions, and favor species that thrive in higher water temperatures. From 1969 to 1990, the algal species composition during warm months shifted from a diverse assemblage to significant dominance of blue-green species with a tolerance for higher temperatures. Earlier spring melt and later freezes would lengthen the duration of the open water season and create warmer water temperatures, which could have detrimental effects for more than 12 coldwater fish species, including the commercial species of whitefish. Temperature-sensitive life cycles of aquatic insects may be sufficiently altered to cause shifts in abundance and species composition that disrupt food web functioning on a large scale.

In 2003, the Manitoba provincial government announced the Lake Winnipeg Action Plan, which aims to reduce phosphorous and nitrogen loading to Lake Winnipeg through several strategies. The plan strives to protect riparian growth, address fertilizer applications, introduce new sewage and septic regulations, reduce shoreline erosion, and engage other jurisdictions on nutrient management within the watershed. The Lake Winnipeg Stewardship Board was also initiated under this plan to identify necessary actions and make recommendations for reducing nutrient conditions by 10 percent. The board members represented a variety of interests, including fishing; agriculture; urban land use; First Nations; federal, provincial, and municipal government; and nongovernmental organizations.

Scientific research has also been advanced through intensive federal, provincial, collaborative, and independent research and monitoring programs. In 2008, Environment Canada and Manitoba Water Stewardship partnered in an effort to summarize accumulated scientific knowledge of the physical, chemical, and biological characteristics of Lake Winnipeg, focusing on 1999 through 2007. There are unique challenges to water resource management in the lake, given its variable water quality, socioeconomic importance, large size, and complex nature of the watershed ecology.

NICOLE MENARD

Further Reading

Lake Winnipeg Stewardship Board. "Reducing Nutrient Loading to Lake Winnipeg and its Watershed: Our Collective Responsibility and

Commitment to Action." Minister of Manitoba Water Stewardship. http://www.lakewinnipeg.org/web/content.shtml?pfl=public/downloads.param&page=000101&op9.rf1=000101.

North/South Consultants. "Literature Review Related to Setting Nutrient Objectives for Lake Winnipeg." Winnipeg, Manitoba: North/South Consultants, 2006.

Schindler, D. W. "The Cumulative Effects of Climate Warming and Other Human Stresses on Canadian Freshwaters in the New Millennium." *Canadian Journal of Fisheries and Aquatic Sciences* 58 (2001).

Wyoming Basin Shrub Steppe

Category: Grassland, Tundra, and Human Biomes.
Geographic Location: North America.
Summary: This rolling "sagebrush sea" extends over high plains in vast, cold desert basins, providing a transition from northern Great Plains grasslands into the semiarid steppe of the northern intermountain west.

The Wyoming Basin Shrub Steppe biome comprises some 50,000 square miles (129,500 square kilometers) of intermountain basins, high plains, and cold desert centered on the state of Wyoming. The area is a veritable sea of sagebrush and salt desert scrub, interspersed with rock outcroppings, badlands, and sand dunes. This classic western landscape known for its roaming antelope has set the scene for Plains Indian cultures and westward migrants, provided a rancher's or sportsman's paradise, witnessed the rise of railroads and coal mining, and most recently has transformed into a modern economy based mostly on energy development.

This region is defined by major watersheds of the Bighorn River and Upper Green River, the Red Desert, the Great Divide Basin, and other basins as far east as the Laramie Basin. Most of the area lies at 5,900–7,800 feet (1,800–2,380 meters) elevation. Mountains rising from the margins of each basin support woodlands and forests of Douglas fir, pine, and aspen.

Weather in the Wyoming basins is harsh. Climate here is defined as arid to semiarid. Although annual rainfall may reach 16 inches (40 centimeters) along the base of the mountains, the basins often fall within their rain shadow, resulting in precipitation from rain and snow of just 6–10 inches (15–25 centimeters) per year. Temperatures vary from bitter cold winters to hot summers, but freezing temperatures are possible in any month of the year. Snowfall is heavier in surrounding mountains than the basins, but across these semiarid basins, snowmelt enables soils to retain moisture longer than sudden rainfall events.

Flora and Fauna

Forming a transition from the northern Great Plains grasslands to the arid intermountain region, vegetation in the Wyoming Basins is mainly shrub-steppe. Shrub-steppe occurs in temperate latitudes, typically where semiarid climates support a mix of grass cover and scattered shrubs. Most of the Wyoming basins are sagebrush steppe, one of the most abundant and widespread types of vegetation in North America. Soils are typically deep, often with a thin, fragile crust of algae, lichen, and moss.

This shrub-steppe is dominated by perennial grasses, including western wheatgrass and bluebunch wheatgrass, with several distinctive subspecies of big sagebrush, silver sagebrush, and antelope brush. Areas with the deepest soils commonly support basin big sagebrush, but because these lands were among the only areas suitable for row-crop agriculture, they have mostly been plowed under. Drier soils tend to support Wyoming big sagebrush. In places of shallow soil and on windswept ridges, Wyoming big sagebrush may be replaced by black sagebrush or communities of cushion plants.

Historically, natural wildfire and grazing by bison probably maintained a patchy mosaic of open and closed shrubland among the grassland. Shrubs may increase in density following heavy grazing or with suppression of wildfire. In many areas where surface disturbance has occurred, the invasive annual cheatgrass or other annual brome grasses can be abundant. At somewhat higher elevations, mountain big sagebrush becomes dominant, extending up among woodlands and forests surrounding the basins. Other low sagebrush, snowberries, juneberries, and currants, along with a greater diversity of grass and forb plants, are common in these areas.

In the semiarid basin bottoms, water evaporation often leads to salt accumulation on the soil surface, limiting the plant species that can survive here. Throughout this region, saline basins often include open shrubland composed of one or more saltbush species, such as four-wing saltbush, shadscale, cattle saltbush, and spinescale. Other shrubs may include Wyoming big sagebrush, greasewood, rubber rabbitbrush, and winterfat. Some areas of the biome contain soils rich in clay. These areas expand and contract as they become wet and dry, preventing much plant growth. Derived from mudstones and shales, these zones are highly erodible, forming the characteristic buttes and cliffs called badlands.

Historically, this shrub-steppe was home to massive herds of bison. Today, it retains numerous birds such as greater sage-grouse, Brewer's sparrow, grasshopper sparrow, mountain plover, and prairie falcon. Increasingly intense cattle grazing in mixed-grass prairie leads to a shift in grass species toward needle-and-thread grass and, ultimately, lawns of blue grama grass.

These grasslands are also home to prairie dogs, whose range has been reduced to less than 2 percent of the area occupied just 150 years ago. Prairie dog colonies in turn are important to critically endangered black-footed ferrets, ferruginous hawks, swift foxes, mountain plovers, and burrowing owls. Elk, pronghorn, and coyote still inhabit the area, as do many smaller mammals, such as ground squirrel, mice, and shrew.

Human Impact

The cultural history of the Wyoming basins has been largely influenced by grazing animals. For the Plains Indians, these were the bison lands. Throughout the 1800s, massive migrations of European-American settlers passed through this region along the Oregon, Mormon, and Overland trails. Most of the area was homesteaded by the turn of the 20th century.

A greater sage-grouse, which is completely dependent on sagebrush habitat like that found in the Wyoming Basin Shrub Steppe. (U.S. Fish and Wildlife Service)

At first, large herds of cattle were brought into the area to fatten over summer and fall, trailed to markets in Denver and Cheyenne, and later shipped by rail to markets in the Midwest. With homesteads limited in size, grazing of cattle and sheep extended over the open range, leading in places to severe overgrazing and soil erosion. In most of the region, large roundup districts were created to move livestock throughout the landscape. Sheep were almost universally herded and moved to higher grasslands in the summer.

Traditional roundups and herding of livestock became more challenging during World War II, as men enlisted for the military. Fencing of common rangeland caused the landscape to be fragmented into smaller, intensively grazed units. It was not until the late 1940s and early 1950s that rotational grazing was first implemented among grazing allotments. Today, to remedy past environmental effects and to manage multiple values and uses, public land managers and private landowners work in an increasingly collaborative fashion here.

Another major influence shaping the face of the Wyoming basins has been industrial development, first via the railroads and then through mining and energy development. Nineteenth-century land ownership patterns were determined in part by the Union Pacific Railroad routing along the edge of southern Wyoming. This railroad route was influenced by the important coal seams at Hannah, Rawlins, and Rock Springs, Wyoming.

Most major basins in this region are important centers for oil, gas, coal, coal-bed methane, and soda-ash production. Additionally, recent years have seen the rapid expansion of renewable energy production. Extensive wind farms harness that abundant energy source along prominent low ridge lines across this windswept plain. Roads, power lines, and pipelines needed to support these industries continue to fragment and affect this regional landscape. Despite all these developments, this region remains one of the least affected and least populated in the United States.

Patrick J. Comer

Further Reading

Cutright, Paul R. *Lewis and Clark: Pioneering Naturalists.* Lincoln: University of Nebraska Press, 1969.

Dobkin, David S. and Joel D. Sauder. *Shrubsteppe Landscapes in Jeopardy: Distributions, Abundances, and the Uncertain Future of Birds and Small Mammals in the Intermountain West.* Bend, OR: High Desert Ecological Research Institute, 2004.

Knight, Dennis H. *Mountains and Plains: The Ecology of Wyoming Landscapes.* New Haven, CT: Yale University Press, 1994.

McPhee, John. *Basin and Range.* New York: Farrar, Straus and Giroux, 1982.

McPhee, John. *Rising From the Plains.* New York: Farrar, Straus and Giroux, 1986.

Waring, Gwendolyn L. *A Natural History of the Intermountain West: Its Ecological and Evolutionary Story.* Salt Lake City: University of Utah Press, 2011.

Xi River

Category: Inland Aquatic Biomes.
Geographic Location: Asia.
Summary: The major tributary of the Pearl River, vital to many habitats across southern and southwestern China, the Xi is confronted with pollution, overfishing, dams, sediment mining, desertification, and biodiversity loss.

Also called the West River, the Xi River is a major tributary of the Pearl River (Zhujiang), and a lifeline for agriculture, wildlife, and navigation in southern China. The name *Xi* means simply "west." The Xi River is known by other names as well. It has also been spelled Xijiang, Hsichiang, and Sikiang. Not to be confused with the other, smaller Xi River in Fujian Province, the Xi River flows from the Yunnan-Guizhou plateau over the South China Hills to the South China Sea. This takes it from high plateaus through mountainous and hilly regions, and then on to the Pearl River estuary and delta plain, with most of its length in rugged terrain.

The Xi River is sometimes referred to as the Pearl River, thus encompassing the Xi River and the Pearl River's other two tributaries. However, because it is the main tributary, referring to the Xi River covers the majority of the river system. The entire system, the third-longest and second-largest by volume in China, runs from west to east through Yunnan Province, Guizhou Province, Guangxi Zhuangzu Autonomous Region, and Guangdong Province.

The Xi River system consists of 1,378 miles (2,218 kilometers) of waterways. The main sections include Nanpanjiang, Hongshuihe, Qianjiang, Xunjiang, and Xijiang (another example of the name being used for a particular section of the river), along with the connecting tributaries of Beibanjiang, Liujiang, Yujiang, Guijiang, and Heijiang. The river drainage system, or catchment, is approximately 130,000 square miles (336,698 square kilometers). Annually, runoff consists of about 8 trillion cubic feet (230 billion cubic meters) of water. According to the Center for Columbia River History, this is close to the same tremendous volume as the Columbia River in the United States, a major river of the North American continent—but with around half the land base. The Xi River watershed area experiences high amounts of precipitation, particularly in the summer due to the East Asian monsoon.

Part of the Pearl River watershed, the Xi River acts as a tributary to the Pearl (Zhu) River, along with the Beijiang (North River) and Dongjiang

(East River). The Xi is the main tributary, with about three-fourths of the total water discharge. Its part of the total sediment load is even greater than that, around 90 percent. The Pearl River delta itself is highly populated and a major agricultural area for China, featuring tea and rice cultivation.

Flora and Fauna

The life along and in the Xi River is diverse and abundant. The river flows through several terrestrial ecoregions of the tropical and subtropical moist broadleaf forest biome, including Yunnan Plateau subtropical evergreen forest, Guizhou Plateau broadleaf and mixed forest, Jiannan subtropical evergreen forest, and South China-Vietnam subtropical evergreen forest. Such evergreen forests are not common at this latitude, making the area unique for that reason. In addition, the Xi River passes through the northern reaches of the Indo-Burma biodiversity hot spot, a region of high conservation priority because of the number of threats to native species.

The China wood oil tree, also called airy shaw or tung tree, is a hallmark deciduous species in the subtropical area of the Xi River watershed. Occurring both naturally and as a cultivated commodity it is prized for its fruit seeds that are rich in tung oil. Among the evergreens, pine, fir, camphor, and other laurels cover many hillsides here, giving way to the broadleaf woodlands, and then to larger and larger stands of bamboo and *Cycas,* or sago palm, as the Xi flows into its tropical zones. Rhododendron and lychee provide lush and fruitful cover in many areas.

Much of the downstream realm of the Xi has been long since taken over for agricultural purposes, with leading crops including maize, sorghum, hemp, rice, sugarcane, and various legumes. Fruits are also grown in abundance on the tropical slopes, particularly bananas, pineapples, oranges, and pomelos. Hillsides are often covered with tea plantations.

Mammals in the Xi River biome range from pandas, antelopes, and leopards, to ground squirrels, field and tree mice, and bamboo rats. Among amphibians, the Guangdong rice frog, or marbled pygmy frog, is a widespread species in this biome, breeding in vernal pools, paddy fields, or ponds, and tending to favor grasslands or other open areas. The Wanggao warty newt, a large species of salamander, prefers riverine habitats in the broadleaf forest areas here; it is considered endemic to the region (found nowhere else).

Considered a vulnerable population, the Wanggao warty newt has in recent years suffered from degradation and loss of habitat. Another denizen of the broadleaf zone, the Asiatic toad, is a characteristic reptile of the region and is most commonly found in and around the transition zones between cultivated fields and the forest fringe.

Among reptiles, the Chinese cobra enjoys broad distribution along the Xi River, from open forest in the mountains, to grassland areas along the open country, to the mangroves in the estuary and delta. At least 20 species of python are found along the river as well. The estuary and mouth of the river are frequented by red snapper, butterfish, Spanish mackerel, and grouper. Shellfish and fish are cultivated in small bamboo pens along the marshlands, as well as in very large commercial aquaculture establishments.

The Xi River biome attracts a great diversity of birds because of its plentiful habitat types and moderate climate. From 500 to 1,000 species are found here or pass through this region. Silver pheasant is a celebrated species that has been named the official provincial bird of Guangdong; it is afforded some protection in refuges such as Dinghu Mountain National Nature Reserve. Other protected species include Cabot's tragopan and Elliot's pheasant.

Human Impact

Significant locations around the delta and estuary include Guangzhou, Macau, and Hong Kong, all of which are economically important. Macau and Hong Kong both became Special Administration Regions in China in the late 1990s. A key internal migration site, Guangzhou has been an important hub for international trade for centuries, and was one of the first cities opened to the West during the economic reforms of the 1980s. In recent years, Guangzhou has seen a population and urbanization explosion as a center for rural-urban migration. Furthermore, it is a gateway for international migration, and many families receive remittances from relatives working in other countries.

The Xi River has historically been important for several reasons. Used for navigation, the river is vital for moving people and goods through the rugged terrain of south China, particularly to and from the coast. Over a sprawling network of waterways, the Xi connects the delta region with major cities such as Wuzhou, Jiangmen, Guilin, Nanning, Liuzhou, Lupanshui, and Kaiyuan. Although not always accessible the entire year, especially during times of drought, the Xi River's usefulness for navigation cannot be denied.

Another reason for the Xi's importance is its connection to agriculture. The area has historically been cultivated intensively, particularly in its eastern reaches. Recently, this has begun to shift to lucrative tropical crops, such as sugar cane and bananas. The river area also includes tourism sites, such as Yanshuo. Finally, the river itself has been an important source of fish. Despite the location's value, however, its resources have not been effectively protected. Many environmental laws in China are in place, but enforcement can be a challenge.

Although its water quality is better than that of some major northern rivers, such as the Yellow (Huang) River, the Xi River biome is not without problems, including pollution, overfishing, sediment loss, erosion, and even desertification. Humans have changed the flow of water in the Xi by adding bridges, irrigation systems, dams, and reservoirs for transportation, agriculture, flood control, and hydroelectric power, and also by mining and changing land use. In the 1990s, many dams were built on the Xi, and high rates of deforestation and erosion led to severe flooding.

To remedy this, national policies for reforestation and afforestation were implemented, which now affects the amount of water and sediment available as surface water. In addition, sediment mining takes materials out of the system that would normally flow down the river to the delta and the sea, affecting flow and erosion rates, as well as local nutrient availability.

Topography and land forms can interact with weather patterns and land use as well. The Xi River flows through south China, characterized by karst geology with limestone hills and tunnels. Karst areas are susceptible to rock desertification, an issue in the Xi River Basin, particularly in its western provinces. Drought and altered land use can exacerbate this condition. Human activity is also affecting the river in other ways.

A major issue for the river is acid rain pollution. Heavy industrialization and associated air pollution within China lead to serious problems such as acid rain, particularly in the south and east. Other sources of pollution are from sewage, both from city growth and agriculture. With high levels of fertilizer inputs, commonly a combination of artificial fertilizer and night soil from humans, too much nitrogen is being added to the system. The acid rain and excess nutrients affect water chemistry in a river whose fish are already depleted because of overfishing and the use of the river for

Flowers of the China wood oil tree, or tung tree (Vernicia fordii). The tree, which grows wild near the Xi River, is also cultivated for tung oil production. (Wikimedia/Malcolm Koo)

navigation. The Xi River is valuable for numerous reasons, such as agriculture, economics, and biodiversity, but the river and species within it face many challenges from changing ecosystem and land-use dynamics.

Sarah M. Wandersee

Further Reading

Gamer, Robert, ed. *Understanding Contemporary China*. Boulder, CO: Lynne Rienner Publishers, 2003.

Lu, X., et al. "Rapid Channel Incision of the Lower Pearl River (China) Since the 1990s as a Consequence of Sediment Depletion." *Hydrology and Earth System Sciences* 11 (2007).

Zhang, Shurong, Xi Xi Lua, David L. Higgitta, Chen-Tung Arthur Chenb, Jingtai Hanc, and Huiguo Sun. "Recent Changes of Water Discharge and Sediment Load in the Zhujiang (Pearl River) Basin, China." *Global and Planetary Change* 60, nos. 3–4 (2008).

Zhao, Songqiao. *Physical Geography of China*. New York: John Wiley & Sons, 1986.

Yala Swamp

Category: Inland Aquatic Biomes.
Geographic Location: Africa.
Summary: The Yala Swamp is one of Kenya's largest wetland areas, rich in flora and fauna, but it is threatened by large-scale development for commercial agriculture.

The Yala Swamp is a large wetland located in western Kenya; it lies to the northeast of Lake Victoria. The Yala Swamp forms part of the wetlands that contribute to the catchment for Lake Victoria, which is the second-largest freshwater lake in the world. Lake Victoria is an essential source of sustenance, both in terms of water and food, for tens of millions of people and uncounted wildlife in the Lake Victoria basin and beyond. The Yala Swamp itself is surrounded by a human population whose density exceeds 175 individuals per 0.4 square mile (1 square kilometer), one of the highest densities in eastern Africa.

The Yala Swamp is the third-largest wetland ecosystem in Kenya, and is a rich and diverse habitat for a wide range of flora and fauna. These include globally threatened species, as well as many that are typical of the Lake Victoria biome. The Yala Swamp has been isolated from Lake Victoria for some time, however; it is connected to a small lake, Lake Kanyaboli, that buffers and isolates it from the main body of Lake Victoria.

Several species that are extinct in the larger lake ecosystem are still found in the Yala Swamp, and some thrive in the adjacent Lake Kanyaboli, including various species of cichlid fishes. Among these are species that were extirpated by the 1950s introduction of the Nile perch into Lake Victoria.

The Yala Swamp biome consists of a complex of intact marshes and small lakes, and some disturbed areas in two zones of western Kenya: Siaya District in Nyanza Province and Busia District in Western Province. The Yala Swamp covers 19,768 acres (8,000 hectares) and lies 3,707–3,806 feet (1,130–1,160 meters) above sea level. Lake Kanyaboli covers about 2,471 acres (1,000 hectares), and is about 10 feet (three meters) deep. Studies have found the level of oxygenation in the waters of the Yala Swamp wetlands varies from highly oxygenated in its main channels and in Lake Kanyaboli, to very low in some of the stagnant areas of the swamp.

Biodiversity

The vegetation in the swamp is dominated by tall, dense, mono-dominant stands of papyrus (*Cyperus papyrus*). In some shallower areas, stands of tall reeds (*Phragmites* spp.) are common, along with a mixture of aquatic sedges and grasses. This dense vegetation, in particular the extensive network of papyrus, serves as a natural filter that holds silt and other sediments and pollutants back from surrounding habitats, releasing clean water to flow on into Lake Victoria.

Dragonflies and damselflies are abundant in the swamp. These insects have aquatic nymphs and are good indicators of water quality. One of the most common species along the edges of the swamp is the banded groundling (*Brachythemis leucosticta*), a dragonfly that can be found in such large numbers that swarms of them rise from the ground when an animal walks through the grass.

The Yala Swamp has been declared one of Kenya's Important Bird Areas; it is a site of both high diversity and high bird numbers. The avian life of the swamp includes notable species like the papyrus gonolek, papyrus yellow warbler, papyrus canary, great egret, and Bailon's crake, as well as numerous other ducks and migratory wetland birds such as the great snipe. The sitatunga, an antelope adapted to a semiaquatic lifestyle, is an emblematic mammal found within the swamp.

Fish species in Lake Kanyaboli include a tilapia (*Oreochromis esculentus*), part of the thriving local fishing industry. Many fish species breed in Lake Kanyaboli and the adjacent channels of the Yala Swamp. As the Nile perch is still absent from the swamp itself, the Yala Swamp's wetlands are a vital location for conservation of the fish species that are threatened, or likely even extinct, in Lake Victoria.

Human Activity

The Yala Swamp and its adjacent areas contain large amounts of nutrient-rich sediments, deposited by the Yala River over time. This has made the swamp a location for the development of commercial agriculture. The cash crops that are grown under intensive production include rice, cotton, and sugarcane. All of these crops are cultivated using irrigation with water drawn directly from the swamp or river channels. The Yala Swamp currently is the focus of large-scale agricultural development partnerships between the Kenyan government and multinational corporations. Some parts of the swamp have already been drained and are being cultivated. Local communities around the swamp also harvest and burn areas of papyrus periodically.

Global warming is expected to increase average temperatures in Kenya by several degrees over the next 65–85 years, continuing the current trend that is as much as double the global average rate of increase. Precipitation has declined in recent years; projections are variable, but tend not to be favorable. It is thought that drier conditions and drought may prevail. These trends are likely to affect the Yala Swamp biome by a combination of increasing evaporation rates, declining river flow levels, and destruction of wetland capacity to regenerate. All these factors will need to be taken into consideration if the area is to be sustainably developed and its vital habitats conserved.

Dino J. Martins

Further Reading

Abila, R., et al. "The Role of the Yala Swamp Lakes in the Conservation of Lake Victoria Region Haplochromine Cichlids: Evidence from Genetic and Trophic Ecology Studies." *Lakes & Reservoirs: Research and Management* 13 (2004).

Aloo, P. A. "Biological Diversity of the Yala Swamp Lakes, with Special Emphasis on Fish Species Composition, in Relation to Changes in the Lake Victoria Basin (Kenya): Threats and Conservation Measures." *Biodiversity and Conservation* 12 (2003).

Bennun, L. and P. Njoroge. *Important Bird Areas in Kenya.* Nairobi, Kenya: East Africa Natural History Society, 1999.

Shorrocks, B. *The Biology of African Savannahs.* Oxford, UK: Oxford University Press, 2007.

Yalu River

Category: Inland Aquatic Biomes.
Geographic Location: Asia.
Summary: The Yalu River is the border river between China and North Korea, an area rich

in biodiversity and vital to migrating raptors and waterfowl.

The Yalu River is located on the border between China and North Korea. It was known as the Ma Zi River during the Han Dynasty. Starting in the Tang Dynasty, it became known as the Yalu River. The word *yalu* in Chinese means "duck green," and the name is thought to originate from the fact that the clear waters of the river appear the same color as the feathers on a duck's head. Another possible source of the name is the combination of two river names, the Ya River and Lu River, which do come together to form the Yalu.

The Yalu River originates in the Changbai Mountains in northeastern China, then flows southwest through the cities of Changbai, Linjiang, Kuandian, and Dandong along the border between China and North Korea. As the various tributary branches of the Yalu combine, the river's southward flow takes it into the Yellow Sea. Approximately 493 miles (795 kilometers) long, the Yalu covers an area of over 39,767 square miles (64,000 square kilometers). The river ranges 459–1,279 feet (140–390 meters) wide in various sections; at its estuary, the Yalu is 3 miles (5 kilometers) wide.

Except for small areas of basalt in the upper portions, the entire riverbed of is mostly Precambrian rock formations; the river flows through deep river valleys in rugged mountainous terrain in most places. There are many waterfalls and hidden reefs. The lower part of the Yalu has slow-running water; it is often used for timber transportation. Alluvial sediment is plentiful in this segment, and the water is very shallow in winter. In some places, even a raft cannot float downriver easily.

South of Ji'an City, the river opens onto a coastal plain; the water runs deep and swiftly in this area. It is a spot conducive to hydroelectric plants, and several have been built there. The water then slows once more, depositing more alluvial sediment to form many large deltas and islands along the estuary to the Yellow Sea. The rate of sedimentation assures that the delta grows, and that the alluvial area of the river becomes more shallow, with mudflats and reed islands growing.

A view of the Sino-Korean Friendship Bridge that crosses the Yalu River, connecting China and North Korea. At least 90 species of fish can be found in the river. (Thinkstock)

The annual precipitation in the Yalu River area is 34–43 inches (870–1,100 millimeters), and is higher in the downstream region than upstream. Summer floods occur annually starting in mid-June, and the monthly flow reaches its highest point in August. The summer water flow makes up about 60 percent of the total water flow in an entire year, with the average annual flow rate at around 3,038 cubic feet (926 cubic meters) per second. The water volume decreases in the fall, and the river enters its dry season in October. A series of reservoirs have changed the character of these natural seasonal flows to a great extent, however.

Biodiversity

The Yalu River area is considered a humid, temperate climate zone. The winter is very cold and the summer is warm. There are vast differences in terms of natural conditions between the upper and lower parts of the Yalu River. The river surface is usually frozen from December to April.

The ample precipitation encourages the growth of conifer and deciduous trees, and vast forests cover much of the landscape. Forests provide ideal habitats for various wildlife; commonly found mammals along the river banks include the wild hog, wolves, tigers, leopards, bears, and foxes. Birds such as the thunderbird, pheasant, and many others are found here. In Liaoning Province, a major migration route and resting place for waterfowl and raptors, at least 163 species of water birds and 46 species of shorebirds have been recorded among the more than 400 total avian species cataloged here.

Some 90 fish species in the Yalu River give proof of a fairly robust riparian habitat. Most are native species, including types of carp and eel. They must contend, however, with a number of introduced fish species, such as silver carp, bighead carp, blue fish, grass carp, Changchun bream, and triangular bream. Each of these is considered a favorable type for subsistence or commercial fisheries.

Threats

Threats to the Yalu River include human-induced changes, such as the addition of hydroelectric plants that change the nature of the river and destroy habitat area, as well as the increased pressure upon the farmlands to produce more and more food. Climate change threats to the area include greater chances of floods. Combined atmospheric changes and direct human intervention, in the forms of logging and land-clearance for more cropland, are degrading the capacity of some ecosystems along the Yalu River to keep pace. Flora and fauna that cannot adapt must move to higher ground in some cases, or find shelter and sustenance in the expanding estuary area in others.

Zhiqiang Cheng

Further Reading

Alpine Birding. "Liaoning Province." http://www.alpinebirding.com/lview.asp?id=64.

Gao, Jian-hua, Shu Gao, Yan Cheng, Li-xian Dong and Jing Zhang. "Sediment Transport in Yalu River Estuary." *Chinese Geographical Science* 13, no. 2 (2002).

Yangtze River

Category: Inland Aquatic Biomes.
Geographic Location: Asia.
Summary: The longest river in Asia, the Yangtze supports a wide array of biodiversity and ecosystems but continues to be heavily reshaped by human activities.

The Yangtze River (also known as the *Cháng Jiāng*, or *Long River*) is the longest river in Asia and one of the world's major waterways, along with the Amazon in South America and the Nile in Africa. It flows through a wide array of ecosystems and environments, and supports rich biodiversity, although environmental degradation has worsened in recent decades. The river and its main tributaries have been immensely important in China's culture, economy, and history, providing water for a significant portion of the population.

Numerous historical sites along river banks reveal interconnectedness between the river and people. Nowadays, the Yangtze River is a major tourist destination, with some of China's most spectacular natural scenery. The lower reaches, especially the Yangtze River Delta, represent a major industrial and urban region. The ecosystems have not fared well, generally.

Scope of the River

The Yangtze River is 3,977 miles (6,400 kilometers) in length. Its source consists of glacial melt waters and tributaries in the northeast of the Tibet Autonomous Region. The upper course flows eastward through the Qinghai-Tibetan Plateau, a vast elevated area in central Asia covering most of Tibet and Qinghai Province, and then southeast through Sichuan Province in southwest China. The middle course, between Sichuan and Hubei Provinces in central China, flows through the Three Gorges (Qutang, Wuxia, and Xiling Gorges), a region of 75 miles (120 kilometers) in length, noted for its dramatic natural beauty, cultural importance, and the recently constructed mega-dam, Three Gorges Dam.

The lower course of the Yangtze traverses the lowland plains of central and eastern China, including the Yangtze River Delta. This delta covers 38,456 square miles (99,600 square kilometers) and is an economic and urban center, with major cities such as Hangzhou, Shanghai,

Suzhou, Ningbo, and Nanjing. Finally, the river widens into a large estuary and enters the East China Sea near Shanghai.

The Yangtze traverses several regions as it makes its way through China. Together, these areas are inhabited by 350 million people, or about one-third of the country's population. The human population is unevenly distributed along the river. The upper reaches are remote and mountainous, and sparsely populated by subsistence farmers who use natural resources for daily living. By contrast, the middle and lower reaches, especially the Yangtze River delta, represent major industrial, manufacturing, and urbanized areas with high population densities.

Environment

The Yangtze River flows through a wide array of ecosystems. Along its course, dense forests, deep valleys, grasslands, mountains, lakes, rivers, and wetlands support rich biodiversity. The river itself is an important freshwater biome, consisting of numerous other ecosystems, and is home to more than 600 species of aquatic plants and 400 species of fish.

Surrounding areas in the drainage basin are habitats for hundreds of terrestrial animal and plant species. There are endemic—found only here—and endangered species in the Yangtze, such as the Chinese alligator (*Alligator sinensis*), finless porpoise (*Neophocaena phocaenoides*), Chinese paddlefish (*Psephurus gladius*), and Yangtze sturgeon (*Acipenser dabryanus*). Some have become extinct, such as the Yangtze River dolphin (*Lipotes vexillifer*).

In recent decades, environmental degradation of the Yangtze River has intensified. Urban development and an increasing human population, especially in the delta, have converted large areas of natural habitat. Commercial uses of the river include agriculture, forestry, fishing, flood control, land reclamation, and tourism. Environmental problems include deforestation, desertification, soil erosion (especially in the upper and middle sections), increased runoff in floodplains, water pollution from sewage and industrial waste, reduction of lake and wetland areas, high sediment load and siltation, and exhaustion of fish resources.

Poverty in some areas, and dependence on livestock and fuelwood, with few viable economic alternatives, has contributed to unsustainable use of resources. Unsuccessful economic and social campaigns in China's past have also caused massive environmental damage, such as the Great Leap Forward (1958–60).

Conservation Efforts

Some areas of the Yangtze are protected as nature sanctuaries. An example is the Sanjiangyuan Nature Reserve, or Three Rivers Nature Reserve, at 2.3 million acres (939,441 hectares), surrounding the headwaters of the river. Conservation and habitat recovery initiatives being undertaken by both Chinese and international organizations include water conservation, environmental protection and sustainable development, assistance for local governments, environmental education and public awareness, scientific research and monitoring, restoration and sustainable use of wetlands, and initiatives to improve the livelihoods and economic opportunities of local inhabitants.

Cooperation among local and provincial officials, representatives of nongovernmental organizations, and representatives of international organizations have led to collective and integrated approaches to river management and environmental and social issues. However, many of these efforts are still in a preliminary stage. Several reserve areas have been awarded United Nations Educational, Scientific, and Cultural Organization (UNESCO) World Heritage Site status.

Floods and Droughts

Flooding is frequent along the Yangtze River, especially in the lower course, which is surrounded by low-lying terrain. Flooding is a natural event attenuated by high precipitation in the mountainous areas of the upper reaches, and also by seasonal heavy rains and monsoons. However, the frequency and magnitude of Yangtze floods have been altered by human activities, particularly deforestation, which increases river runoff and silt deposition on the riverbed, as well as the loss of lakes and wetlands, which reduces water storage capacity and tidal influence. Global warming effects in this region include heavier precipitation events, which also add to flooding woes and erosion.

Catastrophic floods have occurred periodically throughout history (most recently in 1911, 1931, 1954, 1991, 1998, and 2010), killing hundreds of thousands of people, destroying millions of homes and farms, and disrupting all types of habitats here. Flood control measures along the river include dams, dikes, embankments, levees, reservoirs, reforestation, and wetland restoration. Though most were created recently, sophisticated flood controls have existed in the Yangtze since ancient times, such as the Great Jinjiang Levee in 1548.

More recently, the Chinese government, with the guidance of international organizations, has developed flood control and prevention plans. A major objective of the Three Gorges Dam was to reduce flooding in the middle and lower stretches of the river, although its success is questionable. It should also be noted that floodwaters can benefit ecosystem functioning by cycling nutrients, fertilizing soil, and providing water resources to arid regions.

Irregular rainfall distribution in some areas of the Yangtze attenuates drought. A severe drought in 2011, which lasted several months, rendered hundreds of thousands of people without access to adequate drinking supplies. The drying of lakes, reservoirs, and wetlands threatens a large proportion of wildlife. Substantial agricultural and industrial losses resulted from power rationing in industrial districts, closure of parts of the river to shipping, reduced cargo loads, and stranding of boats.

To alleviate this drought and its consequences, the reservoir of the Three Gorges Dam was partially opened in 2011, and tens of thousands of fish were released in the river to replenish fish populations. Climate change is projected to deepen the region's latent drought conditions in coming years, however, and such countermeasures and preventive efforts will have to be redoubled.

Human Interaction

The Yangtze irrigates a large proportion of the country's total agricultural land and crop production; dominant crops are rice, barley, beans, cot-

The Three Gorges Dam

The Three Gorges Dam is the world's largest-capacity hydroelectric power station. It spans the Yangtze River in Yichang, Hubei Province, midway along the river's course. Construction of the dam began in 1993. The Three Gorges Dam aims to increase the availability of electricity and water in the region for industrial and domestic use, control flooding along the middle and lower reaches, and increase the volume of shipping. Although the Chinese government regards the dam as a success story, it is has been a very controversial project.

Opponents point out that the dam's benefits have been outweighed by environmental and social costs. Large areas of land were permanently flooded, including numerous archaeological and cultural sites, towns, and villages. More than 1 million people in the region had to be resettled. The dam's construction resulted in significant changes in the local ecology and large-scale environmental destruction. It is suspected that the dam has worsened the incidence and severity of droughts downstream. The final toll on species both upstream and downstream along the Yangtze will not be tallied for decades.

A view of the Three Gorges Dam, which has the greatest electrical generation capacity of any dam in the world. It may be contributing to worsening droughts in the region. (Wikimedia/Tomasz Dunn)

ton, corn, hemp, and wheat. Cultivation is intensive in the lower reaches and delta, where climatic conditions and fertile soils are highly favorable for agriculture. There are also extensive aquaculture and fishing industries. Most of these activities displace some proportion of native species; however, the human activity here has been going on for millennia, with consequent co-evolution of fauna and flora in many of the habitats.

Gareth Davey

Further Reading

Center for Global Environmental Education. "The Yangtze River." Hamline University. http://cgee.hamline.edu/rivers/Resources/river_profiles/Yangtze.html.

Hessler, Peter . *River Town: Two Years on the Yangtze.* New York: HarperCollins, 2006.

Wilkinson, P. *Yangtze.* London: BBC Books, 2006.

Winchester, Simon. *The River at the Center of the World: A Journey Up the Yangtze, and Back in Chinese Time.* New York: Henry Holt, 1996.

Yellow (Huang He) River

Category: Inland Aquatic Biomes.
Geographic Location: Asia.
Summary: This long river, dubbed Mother of China, has suffered extreme erosion, pollution, and riverbed degradation, but new approaches may help turn around its decline.

As the second-longest river in China after the Yangtze, and the sixth-longest in the world, the Yellow River, or *Huang He,* has attracted attention throughout history. Traditionally speaking, it is believed to be where Chinese civilization originated about 4,000 years ago. The huge drainage area of the river has nurtured thousands of generations and millions of people. Based on 2006 statistics, 107 million people had settled in the Yellow River basin, which was about 8.6 percent of the population in China then. Water from the Yellow River fosters an even greater number of people—140 million, which was 10 percent of the population in that year. Thus, the Yellow River is called Cradle of the Chinese People or Mother of China.

However, though it is the pride of the Chinese people, it also their sorrow. Flooding and droughts in the Yellow River basin have taken uncountable lives and caused unspeakable suffering to the people and the country. In recent years, pollution from domestic sewage, industrial wastewater, and other sources add additional layers of difficulty in managing the river. Progress has been made in preventing natural disasters and improving water quality, but there is still a long way to go to fully control the river and eliminate the problems that it has brought to the Chinese people.

Course and Character

The source water of the Yellow River has been a mystery and contentious issue for many years. Until recently, the twin lakes of Gyaring and Ngoring on the Tibetan plateau were thought to be the source of water for the river. Recently, Chinese scientists, irrigation experts, and flood control specialists traveled beyond the twin lakes and identified the true source of the river as a stream that seeps out of a gentle slope on the northern flank of the 17,854-foot-high (5,442-meter) mountain Yagradagze Shan.

The bubbling and crystal-clear water supplied by glaciers and underground springs eventually flows into the Bohai Gulf, joining the Yellow Sea northwest of the waters of the East China Sea. On the way to the gulf, a total distance of 3,395 miles (5,464 kilometers), the Yellow River wanders through the northern semiarid region, crosses the loess plateau (loess is the type of ochre-yellow calcareous rock that is found there), and passes through the eastern plain. Altogether, the twisting journey from the origin to the sea passes through nine provinces and autonomous regions. It passes over Qinghai, skirts Sichuan, cuts across Gansu, flows through the Ningxia Hui Autonomous Region and Inner Mongolia, forms the border between Shananxi and Shanxi, winds through Henan and Shandong, and enters the Gulf of Bohai.

Generally, it takes one month for water from the spring on Yagradagze Shan to reach the gulf. Along the way, it accumulates water from 286,659 square miles (742,443 square kilometers) of drainage area. This is 8.3 percent of the total area of China and is as big as Italy, Germany, and Great Britain combined. The river's course, if sketched roughly on a map, shows an outline of an angry and arch-backed dragon. Ironically, for many years, people who live in the Yellow River basin have believed that the river is occupied by a dragon—a fearsome, powerful, and unpredictable beast.

The Yellow River can be divided into three segments: the upper reach, which is between Qinghai Province and Hekouzhen in Inner Mongolia; the middle reach, which ends at Taohuayu in Zhengzhou City, Henan Province; and the lower reach, which ends in a delta on the Bohai Sea. The upper reach of the Yellow River, where the province of Qinghai is located, remains desolate, unpopulated, and remote. At this birthplace of the Yellow River, nature flourishes. This corner of the plateau is a paradise for animals and birds. Hares, rabbits, and marmots are common. Rare white-mouthed deer and elegant black-neck cranes are unique to this place.

The black-necked crane (Grus nigricollis) appears in the isolated upper reach of the Yellow River in China's Qinghai Province. (Wikimedia/Eric Kilby)

Sediment Load

At the middle reach, the Yellow River cuts the loess plateau in half like the blade of a sword, forming the longest continuous gorge in the whole drainage area of the river. The plateau, an area of about 247,105 square miles (640,000 square kilometers), is covered with a thick loess layer several hundred feet (meters) deep. During summer seasons, rain storms rip massive quantities of soil into the Yellow River. It is estimated that up to 11,023 tons (10,000 metric tons) of loess soil is washed away from each 0.4 square mile (1 square kilometer) of land annually. Like the Nile, which is full of silt, the Yellow River is rich in sediments.

The ochre mud washed down from the plateau gives the river a hue ranging from a golden glow under the sun to a sullen grayish tone under winter snow clouds. In recent decades, it is estimated that an average of 1.6 billion tons of sediment is carried by the Yellow River each year. For every 35 cubic feet (1 cubic meter) of water from the river, there are 84 pounds (38 kilograms) of sediment. Regarding soil content, the only major river in the world that comes close is the Colorado, with sediment content of 62 pounds (28 kilograms) per 35 cubic feet (1 cubic meter).

Of all the sediment in the water, only about 25 percent is emptied into the sea. The remainder is deposited in the riverbed and flood plains. As a consequence, the riverbed has risen 2–4 inches (50–100 millimeters) per year. Sediment deposition has raised flood control embankments and caused great difficulties in managing the river. More than 4,000 years ago, the legendary emperor Yu the Great, famous for his skills in hydraulic engineering, said, "Conquering the Yellow River is equal to controlling the whole of China."

Dams and Drought

Following Yu the Great, almost every emperor or king in the history of China tried to tame the Yellow River. Chairman Mao was no exception. As a result of his obsession with taming the river, the

350-foot-tall (107-meter-tall) Sanmenxia Dam was constructed in the 1950s. Though it provided electricity to the surrounding population, it forced people to leave homes where their families had settled for many generations. In addition, because of a shocking lack of foresight about how much sediment would end up in the river, the dam caused floods that ruined many lives.

A few years ago, the once largest hydroelectric plant on the Yellow River was superseded by another megastructure in Xiaolangdi. Its reservoir extends over 105 square miles (272 square kilometers). The Chinese government is planning to build 18 more dams by 2030, adding to the 20 major dams already interrupting the Yellow River.

At the lower reach, the Yellow River has lost its grandeur. Laden with sediments, the river looks exhausted and flows sluggishly. The delta area around where the river meets the Bohai Sea has been designated a nature reserve with an area of 378,070 acres (153,000 hectares). The humid climate has attracted more than 268 bird species, along with small mammals such as foxes.

For the lower reach, the problem is drought. Dry-up of the lower reach was observed from 1972 to 2000. The duration of dry periods increased rapidly in the 1990s. The worst year was 1997, when the main river close to the sea dried up for 226 days. The no-flow distance reached 437 miles (704 kilometers) from the river mouth.

This issue is caused by water overuse upstream. Approximately 80 percent of the Yellow River basin area is dry land. To use the land for agricultural purposes, irrigation was started in the basin more than 1,000 years ago. Vast irrigation projects were developed from the 1950s to the 1970s. In a decade after the 1970s, the irrigation areas were widely expanded to the outside of the basin. During the past 50 years, the irrigation area has increased by a factor of nearly 10. At the same time, over-irrigation, poor drainage systems, and increased demand have resulted in serious consequences of river dry-up.

Pollution and New Plans

In addition to flooding and drought, the water quality of the Yellow River is a serious concern. Annually, more than 4 billion tons (3.6 billion metric tons) of wastewater, which accounts for 10 percent of the river's volume, is dumped directly into the Yellow River. The untreated wastewater has led one-third of the river's native fish species to extinction and made long distances of the river unsuitable even for irrigation. It is estimated that 50 percent of the Yellow River is biologically dead. Toxic water has resulted in increased cases of cancer, birth defects, and waterborne diseases along the riverbanks.

The Chinese government has realized the sacrifice it made over the years to pursue fast economic growth. Many environmental laws, rules, and regulations have been written, but the true problem is that not many rules are put into full effect at the local level. The inability of local environmental protection agencies to prevent direct waste discharge has angered many environmental activists. As a result, they form groups and hold protests. To a certain degree, their involvement does make a difference. However, considering the scope of the problems that are facing the Yellow River, their efforts are still too small to have a huge effect on properly managing the river. Moreover, even if pollution could be controlled, the gap between water demand and water supply would still be too big to fill. This gap would have to be mitigated through different ways to reduce demand, which could be detrimental to economic growth.

During recent years, the Chinese government has been dedicating significant efforts to better managing the Yellow River. In light of the large drainage areas that are eroded—166,024 square miles (430,000 square kilometers) out of 286,659 square miles (742,443 square kilometers)—solutions have been implemented to reduce deforestation, overgrazing, and overworked hillsides by planting more hardy trees and shrubs and forming more terraces along the loess plateau. For one thing, this will reduce erosion of the plateau. For another, by preventing sediment deposition in the Yellow River, the flooding problems can be cut back due to the slower rise of the riverbed.

The drought issue is even harder to handle. In the future, water resource allocation will be a huge issue because of increasing population, rising living standards, increasing pressure of expanding agricultural areas, developing industries in the basin, and global climate change, which will tend to both cause more drought and precipitate more violent storm and flooding events. This issue may seem to belong to China only. But, in reality, its effect extends beyond this country, with international consequences in the trade of industrial products and food, as well as other activities.

Yanna Liang

Further Reading

Brook, Larmer. "Can China Save the Yellow—Its Mother River?" *National Geographic,* May 2008.

Pavan, Aldo. *Yellow River: The Spirit and Strength of China.* London: Thames & Hudson, 2007.

Sinclair, Kevin. *The Yellow River: A 5000 Year Journey Through China.* Los Angeles: Knapp Press, 1987.

Tetsuya, Kusuda. *The Yellow River: Water and Life.* Singapore: World Scientific Publishing, 2010.

Yellow Sea

Category: Marine and Oceanic Biomes.
Geographic Location: Asia.
Summary: This semi-enclosed sea is rich in marine life, but overfishing and pollution have caused serious concerns about the health of the sea and ensuring a sustainable future.

The Yellow Sea, located in the center of northeastern Asia, is one of the 25 major semi-enclosed seas in the world. It is surrounded by the East China Sea to the south, the Chinese landmass to the west and north, and the Korean peninsula to the east. It features two main bays, the Bohai Sea and Korea Bay.

Total surface area of the sea is 154,441 square miles (400,000 square kilometers), with a maximum length of 621 miles (1,000 kilometers) and maximum width of 435 miles (700 kilometers). The average depth of the Yellow Sea is 180–394 feet (55–120 meters), rather shallow compared to other seas in this part of the world.

The Yellow Sea contains more than 200 fish species, 140 phytoplankton species, a wide variety of zooplankton, some seabirds, and a few types of large marine mammals. In addition, aquaculture has been developed along the coasts of the sea. Yet, this unique marine system faces serious problems in maintaining sustainable development. Issues such as overfishing and pollution are continuously addressed by the three nations surrounding the sea.

Biodiversity

The Yellow Sea is rich in marine resources. As the home of more than 100 species possessing commercial value, the Yellow Sea has been intensively exploited over the years. Among 200 fish species that have been found here, 45 percent are warm-water species, 46 percent are warm-temperate forms, and 9 percent are cold-temperature species.

In addition, the long-tailed crustaceans comprise an amazing 54 species, which include 65 percent and 35 percent of warmwater and boreal forms, respectively. The cephalopod group contains 14 species, which consist of nine warm-water forms and five warm-temperate forms.

The habitats of marine populations in the Yellow Sea can be divided into two groups, near-shore and migratory. Near-shore species include those that are mainly found in bays, estuaries, and around islands, such as skate, greenline, black snapper, scaled sardine, and spotted sardine. During winter, with colder water temperatures, these species generally move to deeper waters.

The migratory species such as small yellow croaker, hairtail, and Pacific herring respond to water temperature very actively and have distinct seasonal movements. During winter, some species, such as chub mackerel, Spanish mackerel, and filefish, even migrate out of the Yellow Sea and enter the warmer East China Sea.

Reflecting the complicated oceanographic conditions of the Yellow Sea, microbial communities in the sea are fairly complex with regard to species composition, spatial distribution, and community structure. The total number of phytoplankton species is 140, including 91 diatoms, 37 dinoflagellates, seven euglenoids, four silicoflagellates, and one of cryptomonads. Composition of these species in the Yellow Sea has a distinct seasonal shift. In different seasons, different species dominate.

The overall biomass of zooplankton in the Yellow Sea is lower than that of adjacent seas; the species composition also varies with season and location. The most abundant group, copepods, accounts for 60 percent of the total species throughout the year. Zooplankton is a vital food for pelagic and demersal fish and invertebrates here.

Of marine mammals, the whales that have been observed in the Yellow Sea include the fin, humpback, and grey whale. They represent the remnant of far greater pods that used to migrate and breed here. Some marine mammals here are endangered species, including the black right whale, whitefin dolphin, Kurile harbor seal, and Japanese sea lion.

The Yellow Sea ecosystem also hosts seabirds. Two key types that breed off the eastern coast of China are the streaked shearwater (*Calonectris leucomelas*) and Bulwer's petrel (*Bulweria bulwerii*). Among the endangered seabirds of the China coast, the relict gull (*Larus relictus*) and the Chinese crested tern (*Sterna bernsteini*) live in the Yellow Sea region. On the Korea side, of 370 bird species identified in South Korea, 112 breed in the Yellow Sea region, and 17 locations have been designated as protected breeding grounds.

Effects of Human Activity

Fisheries in the Yellow Sea represent a multinational business involving China, the Koreas, and Japan. The total catch increased to approximately 5.5 million tons (5 million metric tons) in 1984 from about 3 million tons (2.7 million metric tons) in 1970. Of this amount, only one-third to one-half is considered sustainable. The number of species that are commercially harvested is about 100, including crustaceans and cephalopods.

Demersal, or deepwater, species are the major component of the fishery resources and account for 65 to 90 percent of the annual total catch. The two most commercially important demersal species are small yellow croaker and hairtail. Because of increases in fishing effort and intensive fishing of spawning stock and young fish, catches of both species have been declining.

Overfishing has also caused serious declines in stock abundance of Pacific cod, flatfish, sea robin, red seabream, and white croaker. Surprisingly, the abundance of species such as cephalpods, skates, and daggertooth pike-congers seems to be unaffected by fishing pressure. Two possible reasons could be their scattered distribution or their strong adaptive nature.

Besides abundant natural resources, the Yellow Sea has tremendous value in terms of beautiful coastlines, aquaculture, and beaches for recreation. All coastal provinces of China along the Yellow Sea have marine aquaculture, which uses seawater to culture different species. Shangdong and Liaoning are the two provinces where the aquaculture enterprises are most advanced. Similarly, South Korea has developed its own aquaculture. In 1977, the total invertebrate mariculture produced 52 percent of the fish yield, or 280,865 tons (254,796 metric tons).

Environmental Threats

Pollution of the Yellow Sea, especially near shore, has been a serious problem for both China and South Korea. Before strict environmental laws were put into effect, both countries dumped industrial and domestic wastes directly into the sea. In addition, oil discharged from vessels, ports, and oil exploration was released into the sea.

In the western part of the Yellow Sea, where China is, more than 100 million tons (91 million metric tons) of domestic sewage and about 500 million tons (454 million metric tons) of industrial wastewater are discharged into the sea each year. The major pollutants are organic chemicals that absorb quantities of oxygen; heavy metals such as mercury, lead, and cadmium; oils; and inorganic nitrogen. On the eastern side of the Yellow Sea, the pollution level has been similar. However, due to the dynamic nature of the sea, which leads to strong and rapid mixing and biodegradation, the marine environmental quality has been measured as roughly normal. Over the past several decades, climate change—particularly the slowly rising average seawater temperature—seems to have played a role in various algal and jellyfish blooms, as well as species shifts in the overall population.

The Yellow Sea is believed to have great potential for future oil and gas exploitation in the seabed; this will produce a new challenge in maintaining the biologic health of the ecosystem. Overfishing and pollution have been the two major issues concerning the sustainable use of this precious sea up until now. Numerous agreements between nations and a great number of environmental laws and rules have been put into action. Ideally, human beings will support efforts to enforce those mandates and will add new elements to the campaign to protect the natural habitats of the Yellow Sea.

Yanna Liang

Further Reading

Choon-ho, Park, Kim Dalchoong, and Lee Seo-Huang. *The Regime of the Yellow Sea: Issues and Policy Options for Cooperation in the Changing Environment.* Seoul, South Korea: Institute of East and West Studies, Yonsei University, 1990.

Park, Chul. *Yellow Sea and East China Sea Reported by PICES and Korean Monitoring Program.* Daejeon, South Korea: Chungnam National University, 2012.

Valencia, Mark J. *International Conference on the Yellow Sea, Transnational Ocean Resource Management Issues and Options for Cooperation.* Honolulu, HI: East-West Environment and Policy Institute, 1987.

Yenisei River

Category: Inland Aquatic Biomes.
Geographic Location: Asia.
Summary: Supporting rich fish communities amid its taiga and tundra environment, the Yenisei is threatened by industrialization and development.

As the largest river system flowing into the Arctic Ocean, the Yenisei River drains a large part of Asia and is considered to be one of the three great Siberian rivers. The Yenisei River ecosystem provides a rich habitat for a variety of flora and fauna.

As with many large rivers, however, the Yenisei has been heavily dammed so that a system of hydroelectric generating stations could be installed for human needs. This has affected the Yenisei River ecosystem and the species that rely on habitats throughout its watershed. Climate change effects in Siberia and the Arctic region, especially the thawing of permafrost lands here, are of mounting concern to many and portend further disruption of the river environment.

With a length of more than 3,400 miles (5,472 kilometers), the Yenisei is the world's fifth-largest river, exceeded only by the Nile, Amazon, Yangtze, and Mississippi. The Yenisei River's maximum depth is 80 feet (262 meters), and it has an average depth of 45 feet (148 meters). With more than 97 percent of its drainage basin located within Russia, the Yenisei also includes some areas of Mongolia in its drainage area.

The Yenisei River is fed by many tributaries, chief among them the Angara and Selenge Rivers. The Yenisei empties into the Kara Sea in the Arctic, which is icebound for more than half the year. Much of the river runs through sparsely populated areas, usually featuring taiga, also known as boreal forest. Taiga is the world's largest land biome and represents almost 30 percent of the planet's forest cover. After permanent ice caps and tundra biomes, taiga represents the ecosystem with the next lowest average temperatures, although during the depths of winter, taiga can be colder on a regular basis than tundra regions.

The short summer in the taiga ecosystem lasts one to three months on average, and winter extends five to seven months. Temperatures across the Siberian taiga vary greatly, from minus 65 to 86 degrees F (minus 54 to 30 degrees C). The Yenisei River watershed is not particularly moist; it experiences average annual precipitation of 7–30 inches (178–762 millimeters).

Soil supporting the taiga tends to be poor in nutrients. Lacking the deep, organically enriched

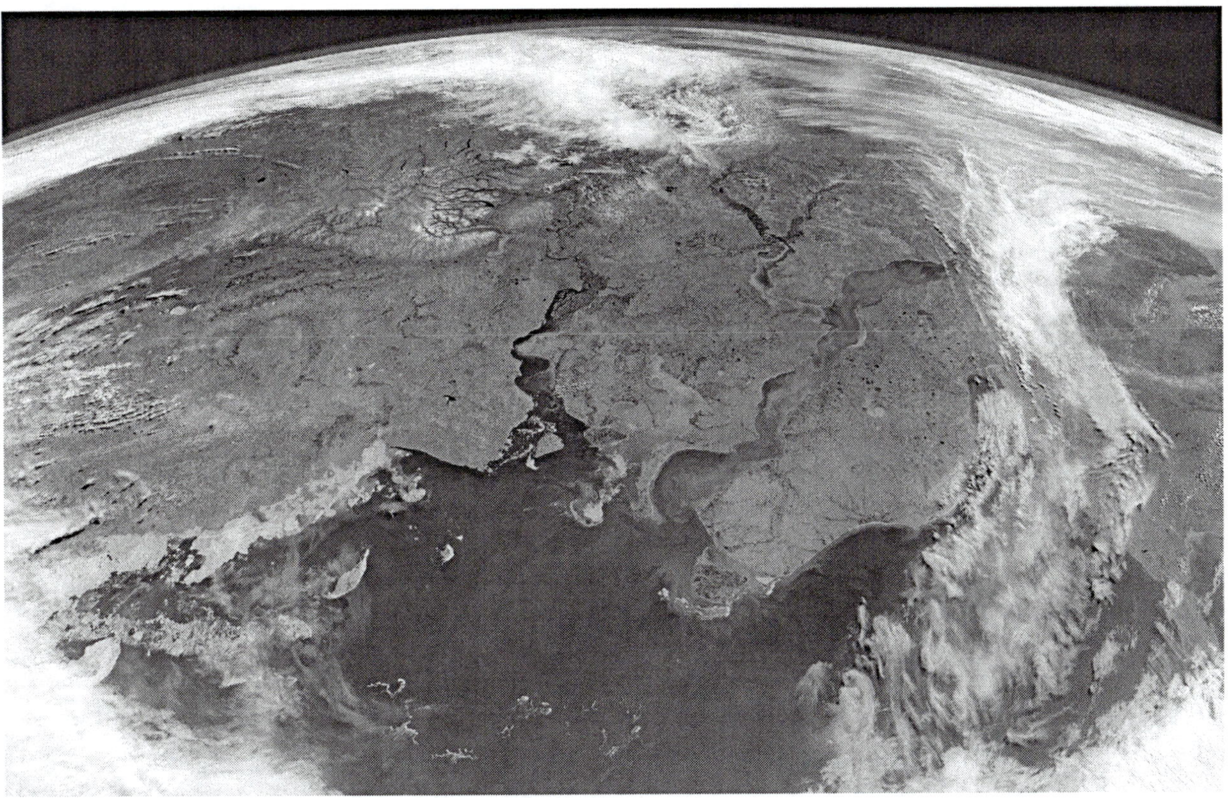

This June 29, 2012, satellite photo shows the Ob and Yenisei Rivers in north-central Russia at the point at which they meet the Kara Sea, which appears at bottom. Sediments and other dissolved organic material can be seen flowing from the rivers into the sea. After the Nile, Amazon, Yangtze, and Mississippi, the Yenisei ranks as the world's fifth-largest river. (NASA)

supplements provided by temperate deciduous forests, taiga soil is relatively thin and young. Cool temperatures discourage the contribution of fallen leaves to the soil, and evergreen needles make the soil acidic.

Flora and Fauna

Taiga forests are largely coniferous and are dominated by fir, larch, pine, and spruce. The southern parts of the taiga have some small-leaved deciduous trees, including alder, birch, poplar, and willow, although in the Yenisei River ecosystem, the most common species of tree is the larch, which can thrive in extremely low temperatures. Coniferous trees do better in this climate because the low sun in the region inhibits photosynthesis for much of the year, and evergreens are able to photosynthesize even in very low temperatures.

A large variety of wildlife thrives in the Yenisei River ecosystem, including many species that are endemic, or found only in this region, such as the Siberian thrush.

The region is especially varied in fish, with the ecosystem supporting more than 55 species in 14 families. Despite the huge geographic area covered by the Yenisei River, the distributional pattern of species is relatively the same along the river and among its tributaries. The fish found in the region can be categorized into four main ecological groups.

The first of these four groups includes cold-loving species that originated in the Arctic but now mostly inhabit lower reaches and adjacent sea areas; they tend to be anadromous. This group includes such fish as the Arctic cisco (*Coregonus autumnalis*), charr (*Salvelinus alpinus*), migratory tugun (*C. tugun*), muksun (*C. muksun*), inconnu (*Stenodus leucichthys*), peled (*C. peled*), and round whitefish (*Prosopium cylindraceum*).

The second group represents boreal river species, both resident and migratory, some anadromous, including the common dace (*Leuciscus leuciscus baikalensis*), European perch (*Perca fluviatilis*), ide (*Leuciscus idus*), lake minnow (*Phoxinus percnurus*), northern pike (*Esox lucius*), Prussian carp (*Carassius gibelio*), roach (*Rutilus rutilus lacustris*), Siberian sculpin (*Cottus sibiricus*), and Siberian sturgeon (*Acipenser baerii*).

The third group consists of a piedmont complex, such as Arctic grayling (*Thymallus arcticus*), East Siberian grayling (*T. palasii*), Eurasian minnow (*Phoxinus phoxinus*), lenok (*Brachymystax lenok*), and taimen (*Hucho taimen*).

The final group is comprised of originally marine species that have entered the lower reaches of the river, such as Arctic flounder (*Liopsetta glacialis*), Atlantic rainbow smelt (*Osmerus mordax*), fourhorn sculpin (*Triglopsis quadricornis*), and sardine cisco (*Coregonus sardinella*).

Lake Baikal, sometimes referred to as the Jewel of Siberia, is listed as a World Heritage Site by the United Nations Educational, Scientific, and Cultural Organization (UNESCO). The deepest and oldest lake in the world, Lake Baikal drains into the Angara tributary of the Yenisei River. The lake has a rich and diverse number of species, including more than 2,000 species of plants and animals, approximately 65 percent of which are endemic. These include the Baikal seal (*Pusa siberica*), also known as the nerpa, and the Baikal omul (*Coregonus migratorius*), a whitefish species of the salmon family. The richness of the Baikal system helps ensure that the Yenisei watershed retains the capacity to recover and regenerate species diversity following harsh climatic intervals.

Environmental Threats

For much of its history, the Yenisei River ecosystem was sparsely populated, save for intermittent visits by the nomadic indigenous peoples who lived throughout the central-southern area of Siberia near the river's banks. During the Soviet era, a tremendous amount of work was done, sometimes by gulag labor, to develop a series of dams in the middle section of the Yenisei River. These dams provide a tremendous amount of hydroelectric power—but also have produced industrial contamination, vast erosion and inundation, and other threats to the area's ecosystem.

Conservation efforts in the region seek to protect it from industrial pollution and other threats. Lake Baikal is home to the Baykalsk Pulp and Paper Mill (BPPM), which opened in 1966 and continues to operate, to the chagrin of many environmentalists and local residents. Despite protests, as recently as 2012, BPPM's continued operation was endorsed by the Russian government. Its toxic particulate output settles into the forests and grasslands that drain into the Yenisei system, while the tainted waters threaten the upper reaches of the watershed.

During the 1990s, the Russian national pipeline company Transneft announced plans to build the East Siberia-Pacific Ocean Oil Pipeline within 1 mile (1.6 kilometers) of Lake Baikal. This announcement caused a great deal of protest from environmental groups, which feared a pipeline spill would endanger the sensitive ecosystem in the area. After negotiation with the Russian government, resulting in an edict from president Vladimir Putin to move the pipeline at least 25 miles (40 kilometers) north of Lake Baikal, Transneft agreed to alter the proposed route for its construction.

In 2006, the Russian government made public its plans to build an International Uranium Enrichment Center at an existing nuclear facility in Angarsk, less than 60 miles (97 kilometers) from Lake Baikal, and to store in the region radioactive and other toxic materials generated as part of the process. Again, this announcement caused outcry from many quarters, although the Russian government has yet to retract its plans.

Environmental threats to the Yenisei River ecosystem are not new, and they are not likely to disappear. Because Siberia is in need of economic development, there is some hesitation on the part of Russian authorities to pursue policies that fully protect the Yenisei River ecosystem, despite the great benefits that the region presents to scientists, future generations, and, of course, the natural inhabitants.

Stephen T. Schroth
Jason A. Helfer

Further Reading

Goebel, Ted and I. Buvit. *From the Yenisei to the Yukon: Interpreting Lithic Assemblage Variability in Late Pleistocene/Early Holocene Beringia*. College Station: Texas A&M Press, 2011.
McCully, Patrick. *Silenced Rivers: The Ecology and Politics of Large Dams*. New York: Zed Books, 2001.
Woods Hole Research Center. "Investigating the Changing Arctic." http://www.whrc.org/global/arctic_system/index.html.

Yukon-Alaska Alpine Tundra, Interior

Category: Grassland, Tundra, and Human Biomes.
Geographic Location: North America.
Summary: This cold, harsh region has limited biodiversity, but important populations of certain wildlife species.

The south-central Yukon Territory in Canada and east-central Alaska in the United States form a contiguous ecoregion of alpine, subalpine, and boreal northern cordilleran regions. The prevailing climate here is one of short, cool summers averaging 50 degrees F (10 degrees C) and long, cold, dark winters with mean temperatures of minus 4 degrees F (minus 20 degrees C). Tree growth in this tundra biome is limited by the year-round low temperatures and short growing season.

In Russian, *tundra* means *treeless uplands,* though it is generally applied only when such areas have permafrost. The Yukon-Alaska Interior Alpine Tundra biome is the home of the coldest recorded temperature in North America of minus 81 degrees F (minus 63 degrees C), measured on the Kluane Plateau.

The ecoregion was heavily shaped by glacial movement and includes mountain ranges with peaks of 6,890–7,874 feet (2,100–2,400 meters), rolling hills, and incised plateaus separated by valleys and lowlands. Permafrost is common, especially in the north and at the highest elevations; on the coast, it is only sporadic. In the rain shadow of the coastal mountains, precipitation is less than 12 inches (0.3 meter) per year, but it is about 2 feet (0.6 meter) on the plateaus.

Vegetation

Alpine tundra vegetation has adapted to survive the low temperatures, limited water availability (though not as limited as in desert biomes), and increased exposure to ultraviolet radiation that are associated with higher elevations. The lower elevations of the Yukon-Alaska Interior Alpine Tundra region include forests as well, which thin out rapidly as elevations increase. The most common trees here are forests of black and white spruce (*Picea mariana* and *P. glauca*), in a matrix of quaking aspen (*Populus tremuloides*) or dwarf willow (*Salix*), as well as birch (*Betula*) and shrubs (*Ericaceae*).

In the lowlands, where drainage is poor, forests are predominantly black spruce, scrub willow, and birch. Floodplains are covered with balsam poplar (*P. balsamifea*). The permafrost slopes are home to forests of black spruce, willow, and paper birch (*Betula papyrifera*). At higher elevations, the vegetation is sparse and includes dwarf willow; birch; graminoid, or grass, species in the *Gramineae* and *Cyperaceae* genera; mosses; and mountain avens (*Dryas hookeriana*).

Other, less-common trees include the lodgepole pine (*Pinus contorta*) and subalpine fir (*Abies lasiocarpa*). Understory vegetation and vegetation at upper elevations use deep root systems to avoid water loss and reach maximum stomatal opening at midday, when the temperature is warmest. Ultraviolet radiation may be a contributing factor in stunting the growth of alpine tundra plant life; studies are inconclusive.

Many of the plant species in the tundra are food plants for *psyllids*, or jumping plant lice, of which

Ice crystals covering a willow branch in the Yukon Territory in Alaska. Willows provide plant lice with food and an escape from the cold. (U.S. Fish and Wildlife Service)

there are seven families: *Aphalaridae, Calophyidae, Carsidaridae, Homotomidae, Phacopteronidae, Psylllidae,* and *Triozidae*. Much of the region's species richness is in the *phyllidae* superfamily, in fact. Willows feed the greatest number of species; a willow stand may be home to more than one dozen species of plant lice. The lice feed on the living, nutrient-rich tissue, called pholem, of vascular plants. On a willow tree, this tissue is the innermost layer of the bark; the bark helps protect lice from the cold.

Many of the flowering plants of the region form their flowers early, in a process called preformation, so that they can flower immediately when the snow begins to thaw. Plants that undergo preformation protect their preformed inflorescence from frost damage in the winter by surrounding themselves with tightly packed leaves covered in hair, which serve as insulation. The advantage of flowering early is that any successfully germinated seeds have a longer period of time in which to mature before the next frost. Because few pollinators are available, though, their reproductive success is limited.

The opposite approach is attempted by late-flowering plants, which produce a large number of seeds toward the end of the brief spring. Most of the plants that grow from those seeds will die, but so many seeds are produced that some survivors will beat the poor odds. Plants using this adaptation tend to be self-pollinators.

At higher altitudes, flowering is less common; clonal propagation of grasses and reproduction by spores, as in the case of cryptogams (ferns, lichens, algae, and mosses), are more successful. Moss and lichens can flourish in the semiarid frigid biomes of the higher elevations, where wind, cold, snow, and soil quality make life difficult for other plants.

Wildlife

Wildlife includes grizzly and black bears (*Ursus arctos* and *U. americanus*), Dall's sheep (*Ovis dalli*), caribou (*Rangifer tarandus*), moose (*Alces alces*), red foxes (*Vulpes fulva*), wolves (*Canis lupus*), hares (*Lepus* spp.), ravens (*Corvus corax*), rock and willow ptarmigans (*Lagopus mutus* and *L. lagopus*), and golden eagles (*Aquila chrysaetos*).

The endemic (not found elsewhere) subspecies of caribou in the region is the porcupine caribou (*Rangifer tarandus granti*), named for the Porcupine River, which runs through its habitat. The porcupine caribou herd numbers about 125,000, with the caribou migrating 1,500 miles (2,414 kilometers) every year between their calving grounds in and near the Arctic National Wildlife Refuge and the Yukon-Alaska alpine tundra; this is the longest land migration route of any terrestrial mammal.

Threats

Most of the region remains intact. Road development and mining have contributed to degradation of about 15 percent of the area, mostly in valley bottoms that have been altered by mining activity. The Yukon contains no protected areas. Eastern Alaska includes four protected areas: the Yukon-Charley

Rivers National Preserve, Streese National Conservation Area, White Mountains National Recreation Area, and Arctic National Wildlife Refuge. These cover primarily the lowlands and do not include the alpine uplands surrounding them.

Climate change impacts in this biome may be seen in the permafrost, where scientists can measure warming trends. When permafrost melts, it can release greenhouses gases such as methane and carbon dioxide, contributing to a positive feedback loop that adds to the overall greenhouse effect, thereby accelerating further permafrost melting.

Bill Kte'pi

Further Reading

Hulten, Eric. *Flora of Alaska and Neighboring Territories.* Palo Alto, CA: Stanford University Press, 1968.

MacLean, S. F. and I. D. Hodkinson. "The Distribution of Psyllids (Homoptera: Psylloidea) in Arctic and Subarctic Alaska." *Arctic and Alpine Research* 12, no. 3 (1980).

Rand, A. L. "The Ice Age and Mammal Speciation in North America." *Arctic* 7, no. 1 (1954).

Scudder, G. G. E. "Present Patterns in the Fauna and Flora of Canada." *Memoirs of the Entomological Society of Canada* 111, no. 108 (1979).

Yukon Interior Dry Forests

Category: Forest Biomes.
Geographic Location: North America.
Summary: The Yukon interior dry forests host a diverse collection of plants; they also contain one of the northernmost grassland areas linked to the boreal forests.

The Yukon Interior Dry Forests ecosystem is located within the Yukon Plateau, corresponding in the north with the Yukon Plateau Central, and in the south with the Yukon-Southern-Lakes ecoregions of the Boreal Cordillera ecozone. Geographically, the Yukon interior dry forests are located predominantly in the southern portion of the Yukon Territory of Canada, with a small area extending into northwestern British Columbia.

The northern part of the ecosystem extends from Lake Laberge to the lower Stewart River, near Dawson. The southern part extends southward from Lake Laberge to the boundary of British Columbia, and includes the entire Teslin Plateau and parts of the Nisutlin and Lewes Plateaus. This southern region lies in the rain shadow of the St. Elias Mountains.

As a biome, the Yukon interior dry forests is classified as taiga, or boreal forest; it spans 24,100 square miles (62,419 square kilometers). The climate of the region is classified as dry sub-Arctic or boreal; this type of climate has long, cold winters and short, cool summers. The average annual temperature of the Yukon interior dry forests is around 27 degrees F (minus 3 degrees C). The winter temperature ranges from minus 2 to 2 degrees F (minus 17 to minus 19 degrees C), while the average summer temperature is 52 degrees F (11 degrees C). Annual precipitation is 9–16 inches (226–399 millimeters), with the higher elevations in the northeastern part receiving more precipitation.

The Yukon interior dry forests is composed of rolling uplands and nearly level plateaus that are separated by deeply and broadly cut U-shaped valleys. The biome makeup is 65 percent boreal-subalpine coniferous forest, around 25 percent alpine tundra, 5 percent lakes and wetlands, and 5 percent rocky ground. There are several small mountain groups scattered across the plateau. Most of the terrain of the biome lies at 1,969–4,921 feet (600–1,500 meters), with a few peaks reaching above 5,906 feet (1,800 meters). The Yukon Interior Dry Forests biome also features irregular, patchy, low-ice-content permafrost.

Eutric brunisols, an alkaline soil type, is the dominant soil of the ecoregion, and is found in the alpine zones and loose bodies of sediment, or colluviums, at the base of steep slopes. Cryosolic soils—which feature permafrost within 3.3–6.6 feet (1–2 meters) of the surface—are found beneath established forests on the base of northern-facing slopes and in some wetlands. Due to the ecoregion's low precipitation and elevated calcium carbonate deposits, weathering of rocks and soils is limited.

Flora

The Yukon Interior Dry Forests biome is in large part a subalpine coniferous forest. Southern-facing slopes in the lower elevations, however, contain a diverse collection of grassland communities that include juniper, sagewort, rose, and Kinnikinnick *(Arctostaphylos uva-ursi).* Willows, shrub birch, sedge tussocks, shrubby cinquefoil, cowberry, Canada buffaloberry, and various peat mosses, or sphagnum, dominate the wetlands areas.

Trees of the subalpine areas above 3,937 feet (1,200 meters) are white and black spruce, lodgepole pine, and alpine fir, intermingled with quaking aspen and balsam poplar. White and black spruce are the most common trees, with black spruce more common in the wetter areas. White spruce typically forms the treeline; alpine fir or paper birch occasionally form the treeline.

Lodgepole pines are typically found in recently burned or very dry areas. Recurring natural fires have resulted in developing many seral communities, an intermediate stage prior to full climax forest status. In the upper treeline and subalpine area, shrub birch and willows are common. Trees in the subalpine areas—such as fir, mountain hemlock, whitebark pine, white spruce, and Englemann spruce—are stunted, crooked, bent, and twisted (*krummholz*).

Moving above the treeline, the weather is characteristic of the alpine climate. This alpine area lacks vegetation and is covered by ice, snow, and rocks, while the alpine areas with vegetation consist of lichens, herbs, moss, forbs, mountain blueberry and crowberry, grasses, dwarf shrubs, and mountain avens.

Fauna

The Yukon Interior Dry Forests biome supports an abundant number of mammalian species of the Yukon. Common mammals here include woodland caribou from the Ibex, Atlin, and Carcross-Squanga herds; moose; mountain goats; Dall's and stone sheep; mule deer and elks; wood bison; Arctic ground squirrels and red squirrels; beavers; and snowshoe hares. There are also predators such as grizzly bears, black bears, gray wolves, cougars, coyotes, wolverines, foxes, lynxes, and martens.

Avian species, too, are abundant. Examples of waterfowl are the northern pintail; Canada goose; Pacific and common loon; and horned, pied-billed, and red-necked grebes. Frequently seen shorebirds include the common snipe, sora, American golden plover, semipalmated plover, lesser yellowlegs, semipalmated and pectoral sandpipers, and American coot. Game birds include willow ptarmigan and the spruce, sharp-tailed, and ruffed grouse. Other common birds are black-billed magpies, gray jays, boreal chickadees, common ravens, rusty blackbirds, three-toed woodpeckers, red-breasted nuthatches, Lapland longspurs, and Wilson's warblers. The abundance of waterfowl, songbirds, shorebirds, and game birds attracts numerous winged predators, such as great horned owls, bald and golden eagles, Merlin's and peregrine falcons, and gyrfalcons.

The only amphibians in the Yukon interior dry forests are western toads and wood and spotted frogs. The ecoregion has no reptiles.

Land Use and Conservation

Approximately 31,000 people of the Yukon Territory live within these forests. Land uses in the alpine and subalpine area include hunting, trapping, tourism, and recreation. Lower valleys below 2,789 feet (850 meters) are used for forestry and forage-based agriculture. Mining operations are scattered throughout the ecoregion. It is estimated that 75 percent of the biome remains intact, with the upland areas being better preserved than the valley bottoms. The Yukon Interior Dry Forests biome is considered be ecologically vulnerable. Threats to biodiversity are forestry and mining activities, major transportation routes, and urban sprawl around Whitehorse. The Yukon government has discontinued its wolf-control program.

Climate change is likely to stress these forests by reducing the amount of available water and undermining soil structure by the accelerating melt of permafrost. This in turn, will add to concentrations of global greenhouse gasses.

Currently, there are no large protected areas in the Yukon interior dry forests. There are small areas, such as the Charlie Cole Creek Ecological Reserve, established in 1981 and located in British Columbia. This reserve is 400 acres (162 hectares). Another such area is the Nisutlin River Delta National Wildlife Area, established in 1998 in accordance with the Teslin Tlingit Council Final Agreement, and located in southern Yukon. This wildlife area is 13,541 acres (5,480 hectares) and is only partly protected. The proposed Kusawa Lake Territorial Park is currently a special management area, and negotiations are continuing between Canadian and tribal governments regarding its final status.

Andrew Hund

Further Reading

Ecological Stratification Working Group. *A National Ecological Framework for Canada.* Ottowa, Ontario: Centre for Land and Biological Resources Research, 1996.

McKenzie, Don, David L. Peterson, and Jeremy Littell. "Global Warming and Stress Complexes in the Forests of Western North America." Seattle, WA: USDA Forestry Service, 2007.

Ricketts, T., E. Dinerstein, D. Olson, C. Loucks, W. Eichbaum, D. DellaSalla, et al. *Terrestrial Ecoregions of North America: A Conservation Assessment.* Washington, DC: Island Press, 1999.

Yukon River

Category: Inland Aquatic Biomes.
Geographic Location: North America.
Summary: The fifth-longest river in North America is being threatened by oil exploration.

The Yukon River is the fifth-longest river in North America, and is considered to be the longest free-flowing river on the continent. Only one dam has been constructed on the river. *Yukon,* in the Gwich'in language, means great river. The Yukon River begins near Whitehorse, British Columbia, and flows for about 1,988 miles (3,200 kilometers) to the Yukon delta and the Bering Sea in Alaska. The Yukon River lies in the largest freshwater ecoregion in North America, which includes all but the southern portion of Alaska and several other rivers draining into the Bering and Beaufort seas.

Major rivers draining into the Yukon include the Teslin, Big Salmon, Nordenskiold, Pelly, White, Stewart, and Klondike. The Yukon basin is formed as the river flows through Alaska; the population of the entire Yukon River basin is about 100,000, mostly concentrated in Whitehorse on the upper Yukon River and Fairbanks, Alaska, on the lower stretch.

The Yukon basin lies in the continental climate zone, characterized by cool summers, cold winters, and low annual precipitation. The basin includes several terrestrial ecoregions: the Yukon Interior Dry Forests in the upper part, the Interior Yukon-Alaska Alpine Tundra in the middle basin, the Interior Alaska-Yukon Lowland Taiga in the northern part, and the Beringia Lowland Tundra in the lower part. The landscape of the basin is natural tundra and boreal forest.

The headwaters of the Yukon collect in Teslin, Atlin, Tagish, and Bennett Lakes in the northwestern corner of British Columbia. From Whitehorse to Dawson City, the Yukon receives water from large tributaries, and then flows into the Yukon Flats area, which is currently protected as the Yukon Flats National Wildlife Refuge Area. Flowing out of the flats, the river crosses the Trans-Alaskan Pipeline and flows west through the Nulato Hills along the west coast of Alaska. There, the Tanana, Innoko, and Koyukuk Rivers join the Yukon, forming the Yukon Delta. This delta is similar in topography to the Mississippi River Delta—but is twice as large.

Like that of many northern rivers, the hydrology of the Yukon River is influenced by a combination of seasonal and annual variations in temperature and precipitation, and their effect on the supply of glacial meltwater and the distribution of permafrost. The mean annual discharge of the Yukon is 223,895 cubic feet (6,340 cubic meters) per second, which is the sixth-largest in North America. Because of the cold environment, evapotranspiration is relatively low, and annual runoff is a high fraction of precipitation. The strong influence of temperature on runoff indicates that climate change will influence the overall function and seasonal dynamics in the future.

Flora and Fauna

Common trees in the Yukon River biome include the white and black spruce, alpine fir, lodgepole pine, balsam poplar, white birch, and green alder in the boreal forests; the higher-elevation areas feature shrubs and sedges. The area holds rich floral diversity, with many species still being discovered. Wetlands and grasslands spread across great expanses between hilly zones.

Because of the extreme climate, the overall diversity of invertebrates is low in the Yukon basin. Few species of mollusks, clams, or snails have been reported. *Diptera* (true flies), mayflies, and stoneflies dominate the insect community in the basin.

Similarly, only 30 fish species have been reported in the Yukon River. Two fish, the Arctic grayling and the Alaska blackfish, are considered to be the icons of the Yukon River. The Arctic grayling is found throughout the river, and the Alaska blackfish is common in weedy areas in the lower part of the river.

In addition, several ecologically and culturally important salmonids—lake trout, dolly varden, and chinook salmon—are found in the Yukon River. Other notable fish species are the endemic (found only here) Arctic lamprey, as well as such regionally-distributed types as wintering spawning burbot, Bering cisco, whitefish, chum salmon, coho salmon, northern pike, and rainbow trout.

Some of the keystone terrestrial fauna in the Yukon are bald eagle, beaver, muskrat, brown and black bear, and moose.

Effects of Human Activity

Recently in its evolutionary history, the Yukon River basin has been a part of a transportation route to the south for oil and gas. Hydrocarbon fuel exploration in Alaska began in the mid-1950s, before statehood in 1958. In the mid-1960s, a major oil discovery on the north slope of Alaska led to the construction in 1977 of the Trans-Alaskan Pipeline, which runs through the Yukon River basin. Additional oil exploration has been conducted on the northern slope of Alaska and in and around the Arctic National Wildlife Refuge. The ongoing gold mining and oil exploration activities of the region have directly affected the Yukon River basin.

The Yukon Flats National Wildlife Refuge in fall. The Yukon River flows through the refuge before moving west through the Nulato Hills along the west coast of Alaska. (U.S. Fish and Wildlife Service/Ted Heuer)

Major effects include direct pollution of lower streams from mining activities, loss and fragmentation of wildlife habitat, disruption of water flow, and secondary ecological degradation from oil exploration and infrastructure. However, the effects can worsen still from mining, oil exploration, forestry, dams, and climate change. Climate change will have significant effects on the hydrology and ecosystems of the river, as these are largely controlled by seasonal temperature regimes. Warmer temperatures can change the hydrological patterns regulated by autumn freezes and spring melts.

Even though the effect from the oil pipeline has been modest so far, further oil exploration and development have the potential for devastating consequences. A major oil leak from the pipeline could have catastrophic effects, similar to the Gulf of Mexico oil disaster in 2010, where long-term damage remains to be assessed.

KRISHNA ROKA

Further Reading

Benke, Arthur C. and Colbert E. Cushing. *Rivers of North America.* New York: Elsevier Academic Press, 2005.

Environment Yukon. "Yukon Plants." Government of Yukon. http://www.env.gov.yk.ca/wildlife biodiversity/plants.php.

Roland, Carl, Mary Beth Cook, Amy Larsen, and Claudia Rector. *Results of an Inventory of Vascular Plants of Yukon-Charley Rivers National Preserve.* Fairbanks: Central Alaska Network, National Park Service, 2004.

Zacatonal

Category: Grassland, Tundra, and Human Biomes.
Geographic Location: North America.
Summary: Found high above the tree line on Mexico's tallest peaks, this ecoregion is dominated by the alpine bunch grasslands—*zacates amacollados* in Spanish—that give the region its name.

The Zacatonal ecoregion is made up of small, isolated patches of high alpine grasslands and shrublands occurring above the natural treeline on mountains in Mexico. This name of this ecoregion is a derivation of the Spanish word for grass: *zacate.* Grasses, such as plants from the genera *Fetusca, Calamogrostis, Muhlenbergia,* and *Stipa,* are the dominant vegetation of the ecoregion. Although these grassy patches are neither extensive in size nor widespread, the Zacatonal biome is a hot spot for biodiversity, home to many endemic (not found elsewhere) species of plants, birds, and mammals.

Zacatonal occurs in enclaves among pine-oak forests of the Trans-Mexican Volcanic Belt, at elevations of 3,600–8,200 feet (1,100–2,500 meters), and in even higher elevations above the natural treeline. The Trans-Mexican Volcanic Belt stretches across south-central Mexico for about 560 miles (900 kilometers), from Jalisco on the west coast to Veracruz on the east coast. It is a band of volcanic peaks, some of which are the highest in Mexico. Some of the volcanoes are active and some are dormant. Zacatonal can be found on volcanic mountains here, such as Pico de Orizaba, Iztaccihuatl, Popocatépetl, Nevado de Toluca, Nevado de Colima, Malinche, and Ajusco.

At the elevations where zacatonal is found, the climate is subhumid continental; however, there are climatic changes at higher altitudes. Temperatures usually range from 36 to 41 degrees F (2 to 5 degrees C). On average, there are 24 to 31 inches (60 to 80 centimeters) of precipitation per year, much of which is snow; the snow usually does not persist for long periods. The soils on which zacatonal occurs are volcanic in origin, can be quite deep, and tend to have high organic content.

Flora and Fauna

Biological evidence indicates that zacatonal represents relict flora from a time during the Pleistocene, when climatic conditions were such that bunch grasslands covered much of Mexico. The volcanic activity that shaped these mountains, along with subsequent changes in the climate, made the area favorable to alpine bunch grasslands and shrublands. At present, zacatonal occurs mainly as isolated patches within the temperate pine-oak forest.

This habitat fragmentation led to the radiation of new species. Approximately 75 percent of the plants found in zacatonal are endemic to this ecoregion. Zacatonal is characterized by grasses and sedges such as *Agrostis perennans, Calamagrostis* spp., *Bouteloua* spp., *Deschampsia elongate, Festuca* spp., *Fimbristylis mexicana, Hilaria cenchroides, Muhlenbergia* spp., *Poa* spp., *Sporobolus* spp., and *Stipa* spp., as well as other herbaceous plants in the *Arenaria, Draba, Lupinus,* and *Potentilla* genera.

Creeping junipers (*Juniperus monticola* f. *compacta*) can be found growing on rocky outcrops and along the banks of streams.

This ecoregion also has high levels of vertebrate diversity. More than 200 species of birds have been observed in the Zacatonal biome, more than 20 of which are endemic, including the striped sparrow (*Oriturus superciliosus*), several species of sedge wrens (*Cistothorus platensis potosinus, C.p. potosinus,* and *C.p. jalapensis*), the Sierra Madre sparrow (*Xenospiza baileyi*), the long-tailed wood-partridge (*Dendrortyx macroura*), and montezuma quail (*Cyrtonyx montezumae*). Several of these species are threatened with extinction. A total of 85 species of migratory birds make use of zacatonal.

Zacatonal is also home to 10 species of amphibians, 42 species of reptiles, and 48 species of mammals, including endemic mammals such as Saussure's shrew (*Sorex saussurei*), the black-eared mouse (*Peromyscus melanotis*), and the volcano rabbit (*Romerolagus diazi*). Other mammals in the ecoregion include skunk, opossum, grey fox, bobcat, and notably the puma.

Human Impact

Threats to the Zacatonal ecoregion include hunting, land clearing for agriculture, overgrazing, fires, and climate change. Several of the mountains on which zacatonal occurs are protected as national parks. Some of the threatened or endangered species occurring in this ecoregion have also been given protected status.

Given that preservation of a small area of zacatonal could ensure the survival of hundreds of species of plants and animals, this ecoregion is important for conservation. Zacatonal also provides ecosystem services such as climate regulation and the maintenance of hydrological cycles. This is a fragile ecosystem that is unable to tolerate disturbances.

Human activity threatens this ecoregion with destruction. Habitat loss is a significant issue. Overgrazing by cattle, goats, and sheep depletes grasslands. Erosion from grazing on steep hillsides is also a significant problem. Plants are often cleared for the cultivation of food crops. Fires resulting both from negligence and arson are becoming more common. Hunting and the capture of wild birds for the exotic bird trade are further endangering some species. The effects of climate change on zacatonal could be severe because of the high sensitivity of the grasses to temperature and moisture.

Five peaks with zacatonal—Izta-Popo Zoquiapan, Malinche, Nevado de Colima, Nevado de Toluca, and Pico de Orizaba—are protected by the Mexican national park system. Endangered and threatened species, such as the volcano rabbit and Saussure's shrew, are protected under Mexico's National Commission of Protected Natural Areas.

Melanie Bateman

Further Reading

Bobbink, R., G. W. Heil, and Nuri Trigo Boix, eds. *Ecology and Man in Mexico's Central Volcanoes Area.* Dordrecht, Netherlands: Kluwer Academic Publishers, 2003.

Dávila, Aranda P., R. L. Saade, and J. B. Reyna. "Endemic Species of Grasses in Mexico: A Phytogeographic Approach." *Biodiversity and Conservation* 13 (2004).

Delgadillo, Claudio, Jose L. V. Rios, and Patricia Dávila Aranda. "Endemism in the Mexican Flora: A Comparative Study in Three Plant Groups." *Annals of the Missouri Botanical Garden* 90, no. 1 (2003).

Ramamoorthy, T. P., A. Lot, and J. Fa, eds. *Biological Diversity of Mexico: Origins and Distribution.* New York: Oxford University Press, 1993.

Zagros Mountains Forest Steppe

Category: Forest Biomes.
Geographic Location: Middle East.
Summary: An ancient source of many modern foods, which is now an ecosystem rich in faunal diversity, this forest steppe region is at risk from human action and global warming.

The Zagros Mountains is the largest mountain range in Iran, extending roughly 900 miles (1,500 kilometers) from the northwest to the southeast and paralleling the border with Iraq to its west. It ends in the south, in rocky cliffs by the coasts of the Persian Gulf and the Gulf of Oman. Comprised mainly of shale and limestone, the range has numerous folds and ridges created by the pressure between the Arabian and Eurasian tectonic plates.

This ecoregion is marked by many peaks, some of which reach elevations of more than 10,000 feet (3,000 meters). Iran's main oil fields are located in the southwestern foothills of the Zagros. Agricultural development and livestock overgrazing pose ongoing threats to this ecoregion, as does petroleum extraction.

Deciduous broadleaf forests, dominated by Persian oak, and endangered animals such as the wild goat and the Persian leopard are characteristic of the forest steppe biome here. The mountain forest steppe ecoregion features numerous deep and narrow valleys eroded by small rivers that separate a series of parallel mountain ridges, primarily in the northern and central portions of the range. Its southern face is marked by a steep descent, leading into Khuzestan Province in southwestern Iran near the Tigris-Euphrates river delta. The eastern face is marked by a smoother transition into the Iranic Plateau. The three highest peaks of the Zagros—Dena, Oshtoran Kuh, and Zard Kuh—reach elevations around 14,000 feet (4,000 meters) and are permanently covered in snow.

This ecoregion features numerous waterfalls, pools, and lakes. Many large rivers, including the Karun, Dez, and Kharkeh, originate here, draining either southward into the Persian Gulf or north to the Caspian Sea. Snowmelt feeds the twin Gahar Lakes, located more than 8,300 feet (2,500 meters) high in the mountains near several deep canyons, including the Bactiara River Canyon, the Sezar River Gorges, and the Karun River Canyon. There are also numerous caves, notably Ali Sadr, a calcite crystal water cave near the city of Hamadan, which contains a clear lake that extends for 9 miles (14 kilometers).

The Zagros Mountains Forest Steppe ecoregion experiences a semiarid temperate climate. Annual precipitation ranges from 15 to 30 inches (40 to 80 centimeters), with rain falling primarily in the winter and spring months. The summer and fall months are quite dry. The winters are extremely cold, with temperatures dipping below minus 13 degrees F (minus 25 degrees C).

Flora and Fauna

The boundaries of this biodiverse ecoregion correspond with two biologically mature steppe forest communities: a xerophilous, or low-moisture, deciduous highland forest covering the western slopes, dominated by Persian oak (*Quercus brantii*); and a lowland forest dominated by cashew (*Pistacia*) and almond (*Prunus amygdalus*) scrub. Other trees and shrubs include Syrian pear (*Pyrus syriaca*), Persian walnut (*Juglans regia*), hornbeam (*Carpinus*), hawthorn (*Crataegus*), hackberry

*A six-week-old Persian leopard cub (*Panthera pardus*), which is endangered but can still be found in the Zagros ecoregion. (Thinkstock)*

(*Celtis*), olive (*Olea*), and plane (*Plantanus*). The genetic predecessors of many familiar foods—including wheat, barley, lentils, almonds, apricots, plums, pomegranates, and grapes—can be found growing wild throughout this region.

The ecoregion supports a wide variety of invertebrate species. Mammals recorded in this area include the brown bear (*Ursus arctos*), Asiatic black bear (*U. thibetanus)*, Syrian jackal (*Canis aureus syriacus*), fox (*Vulpes vulpes*), marten (*Martes foina*), sheep (*Ovis orientalis*), wolf (*Canis lupus*), mouselike hamster (*Calomyscus bailwardi*), mongoose (*Herpestes ichneumon*), jungle cat (*Felis chaus*), and wild pig (*Sus scrofa*). The striped hyena (*Hyaena hyaena*), also found here, is listed as threatened.

Bird species include the lesser spotted eagle (*Aquila pomarina*), golden eagle (*Aquila chrysaetos*), rock partridge (*Alectoris chukar* and *A. graeca*), ase-see partridge (*Ammoperdix griseogularis*), little bustard (*Tetrax tetrax*), houbara bustard (*Chlamydotis undulata*), black-bellied sandgrouse (*Pterocles orientalis*), and black vulture (*Aegypius monachus*). There are also five taxa of endemic lizards.

The wild goat (*Capra aegagrus*), known for its large, curved horns, is classified as a vulnerable species on the International Union for Conservation of Nature (IUCN) Red List, because of hunting and habitat loss caused by timber production. The Persian leopard (*Panthera pardus*) is listed as endangered by IUCN, because of over-hunting for supplying the fur trade and in defense of livestock. One of the world's rarest foxes, Blandford's fox (*Vulpes cana*), also lives in the Zagros Mountains Forest Steppe ecoregion.

Believed to have gone extinct, the Persian fallow-deer (*Dama mesopotamica*) was rediscovered here in 1956. An IUCN assessment in 2011 found that the total population is approximately 250 or more mature individuals. The species is currently listed as endangered.

Human Impact

The Zagros mountains forest steppe is threatened primarily by human activity. The spread of agriculture and overgrazing livestock have destroyed much of the region's natural flora. Oil development in the southwest remains a cause of concern for conservationists. In many areas, populations of native tree species such as oak and hornbeam have been decimated. Climate change has already been cited for increased frequency of droughts and dust storms in the region, exacerbating the trend of desertification.

The establishment of several protected areas supports a wide range of vertebral species in various habitats. The Arjan Protected Area and Biosphere Reserve spreads for more than 250 square miles (650 square kilometers) across the southwestern face of the Zagros range. Lake Parishan and Dasht-e-Arjan are Wetlands of International Importance (Ramsar) and provide habitat for many bird species. Bamou National Park provides habitat for a variety of mammals, some vulnerable, including the Bezoar goat, caracal, Blanford's fox, Indian crested porcupine, and several bat species.

Reynard Loki

Further Reading

Firouz, Eskandar. *The Complete Fauna of Iran*. New York: I. B. Tauris, 2005.

Frey, W. and W. Probst. "A Synopsis of the Vegetation in Iran." In *Contribution to the Vegetation of Southwest Asia*, edited by H. Kurschner. Weisbaden, Germany: Ludwig Reichert Verlag, 1986.

Heshmati, G. A. "Vegetation Characteristics of Four Ecological Zones of Iran." *International Journal of Plant Production* 1 (2007).

Masih, I., S. Uhlenbrook, S. Maskey and V. Smakhtin. "Streamflow Trends and Climate Linkages in the Zagros Mountains, Iran." *Climatic Change* 104, no. 2 (2011).

National Science Foundation. "Ancient Alteration of Seawater Chemistry Linked With Past Climate Change." http://www.nsf.gov/news/news_summ.jsp?cntn_id=124844.

Olszewski, Deborah I. *The Paleolithic Prehistory of the Zagros-Taurus*. Philadelphia: University of Pennsylvania Museum of Archaeology and Anthropology, 1993.

Zambezian Cryptosepalum Dry Forests

Category: Forest Biomes.
Geographic Location: Africa.
Summary: On deep and infertile Kalahari soils and with a dry season comparable to savanna regions, the Zambezian Cryptosepalum Dry Forests form the largest area of tropical evergreen forest outside the equatorial zone.

This unique forest, largely in the western parts of Zambia, eastern Angola, and southern Democratic Republic of the Congo, is dominated by mukwe trees, or *Cryptosepalum exfoliatum pseudotaxus.* This evergreen tree, actually a tree legume, is usually less than 98 feet (30 meters) in height, with a dense and flat crown. The lower areas of vegetation characteristic of this forest, consisting of a very dense evergreen shrub or thicket matrix, remain about 13 feet (4 meters) in height. This tangled understory contains liana species including *Combretum microphyllum, Uvaria angolensis, Artabotrys monteiroae,* and *Landolphia.* These dry forests are found on hills of sandy soil drained by the Kabompo River.

Within Zambia, this forest is called the Mavunda and includes parts of West Lunga National Park. Although not densely settled by people, it has some valuable resources. The fragrant white flowers of the mukwe tree are a valued source for honey, and the bark is used for constructing beehives. Other trees in the Zambezian *Cryptosepalum* Dry Forests biome include rosewood and *Guibourtia coleosperma.*

Further south, the character of the forests gradually changes, and they become dominated by Zambezi teak trees (*Baikiaea plurijuga*). Logging is potentially a serious threat to these species. However, the lack of surface water and relatively infertile Kalahari soils mean that this ecoregion is less attractive for human settlement and agrarian activities. Disturbance by natural fire is not significant, but small-scale farming on the edges can change *Cryptosepalum* forest to a more open and fire-prone vegetation type known as chipya.

The Kalahari soils are deep, infertile, and well drained. The apparent lack of surface water should be a limitation on the evergreen trees. The mean annual rainfall of 31–47 inches (80–120 centimeters) becomes higher to the north away from the Kalahari Desert; nevertheless, there is a distinct dry season lasting up to eight months. The evidence gathered suggests that trees have access to groundwater during the dry season, thus providing the natural conditions for a higher biomass.

There is a question as to how the evergreen trees have access to this subsurface water. The water table in this part of Africa is more than 328 feet (100 meters) below the surface; therefore, the suggestion is that there must be "perched" water up to and around 66 feet (20 meters) below the surface. This also helps to explain the mosaic quality of this vegetation.

Biodiversity

This biome is distinctive in terms of its biota, but is not especially species-rich, nor does it support many endemic (found only here) or near-endemic populations. The only species that is virtually endemic to the ecoregion is a striped African grass mouse (*Lemniscomys roseveari*).

Among the mammals found are the elephant (*Loxodonta africana*), bush pig (*Potamochoerus porcus*), warthog (*Phacochoerus africanus*), blue and yellow-backed duiker (*Cephalophus monticola* and *C. sylvicultor*), kudu (*Tragelaphus strepsiceros*), and buffalo (*Syncerus caffer*).

Cryptosepalum forests are exceptionally rich in bird fauna, with 381 known species. Some of these include the *Guttera edouardi kathleenae* subspecies of the crested guinea fowl, the olive long-tailed cuckoo, Cabanis's greenbul, purple-throated cuckoo-shrike, African crested flycatcher, olive sunbird, forest weaver, and black-tailed waxbill. Interestingly, the highest levels of species richness are found where local habitat disturbance has resulted in a patchwork of tree savanna, thicket, savanna woodland, and forest habitats.

The tree *Cryptosepalum exfoliatum* is not an economically important timber species, and the arable potential of the ecoregion is very low. The soils are nutrient-poor Kalahari sands, cultivation of rain-fed crops is limited to one or two seasons, and the dense vegetation is difficult to clear. These factors, combined with this ecoregion's remoteness from roads and modern urban settlement, and consequent low population density, have prevented its destruction or stark transformation.

West Lunga National Park falls within this ecoregion. Hunting restrictions form the main conservation management strategies of the area. This park covers an area of 650 square miles (1,700 square kilometers) in Mwinilunga District. The park occupies an area between two major rivers: West Lunga to the west and Kabompo to the east and south. It was declared a national park in 1972. In the following years, it fell into some decline; however, in 2002, several local stakeholders formed the West Lunga Development Trust. Their aim is to conserve the secluded nature of the area and the interesting features of the vegetation in this part of Zambia.

Local chiefs have helped mobilize communities into village action groups, helping patrol and monitor ecological conditions of the area. Community resource boards control the natural resources in the surrounding areas and derive a financial benefit from any operations there. Community-based natural resource management has been adopted as a strategy for sustainable wildlife management in many places. It aims to contribute to poverty alleviation and help improve household food security.

Small-scale farming, logging, and poaching are challenges this ecoregion still faces. However, impact continues to be low because of the inaccessibility of the region. Another concern is how climate change will ultimately affect this area in terms of water resources, because the arable potential of the ecoregion is already minimal and projected rainfall will fall, while temperatures are slated to rise.

Nicholas James

Further Reading

Aregheore, Martin Eroarome. "Country Pasture/Forage Resource Profiles: Zambia." United Nations Food and Agriculture Organization. http://www.fao.org/ag/AGP/AGPC/doc/Counprof/zambia/zambia.htm.

Coates Palgrave, Keith. *Trees of Southern Africa.* Cape Town, South Africa: Struik Publishers (Pty.), 1996.

Scholes, Richard J., P. R. Dowty, K. Caylor, D. A. B. Parsons, P. G. H. Frost, and H. H. Shugart. "Trends in Savanna Structure and Composition Along an Aridity Gradient in the Kalahari." *Journal of Vegetation Science* 13, no. 3 (2002).

Zambezian Flooded Grasslands

Category: Grassland, Tundra and Human Biomes.
Geographic Location: Africa.
Summary: An ecoregion of extraordinary productivity and species abundance, this biome is at risk from accelerating deforestation and related threats.

The Zambezian Flooded Grasslands ecoregion is a land-based archipelago of grasslands spread across south-central and central-east Africa. It consists of low-lying areas inundated seasonally or year-round, either in the immediate watershed of the vast Zambezi River or in smaller portions of other neighboring river basins. The ecoregion is located in pockets throughout a band stretching diagonally from Tanzania, in the miombo and mopane woodlands south of Lake Victoria in its northeast range, through Zambia and the Central African Plateau, to Namibia and Botswana at the verge of the Kalahari Desert at its southwest extreme.

This scattered biome is characterized by unusual vegetation productivity among grasses, sedges, reeds, and other swamp types, as well as faunal abundance. Both stem from the availability of water and food for most of the year. This abundance of both animal and plant species is in stark contrast with the drier, nutrient-poor expanses of woodlands surrounding much of the ecoregion's land islands, which can only support animals in relatively low densities.

Climate

The region is characterized by a seasonal tropical climate with two distinct seasons: hot and dry from April to October, and a hot and wet summer between November and March. Temperatures vary due to the broad spread across the continent, with high temperatures ranging at 64–81 degrees F (18–27 degrees C) and low temperatures ranging at 48–64 degrees F (9–18 degrees C). Altitude of a given locality here is a strong factor in the temperature regime.

Precipitation amounts vary similarly, from a lower range of 18–24 inches (450–600 millimeters) annually, to a maximum of up to 55 inches (1,400 millimeters). There are pulses of stream inflow from associated highlands, and when these hit the generally clay-bed soils of these flatlands, a waterlogged situation easily develops and often persists throughout the year. This depends on evaporation and precipitation rates, as well as particular drainage outlets and wind conditions.

An elephant charging through flooded grassland in the Zambezi River region. Productive grasses and other plants in the biome support a high density of wildlife. (Thinkstock)

Flora and Fauna

Floristically, the Zambezian Flooded Grasslands biome is situated at the heart of some of the largest centers of endemism (species found nowhere else) within Africa. Given the spread of these fertile and fecund marshes and grasslands across a broad belt of the continent, it is in many ways ideal as a corridor system for an incredible variety and richness of species. Papyrus is a familiar and dominant reed form here, and is joined by grasses of genera *Phragmites, Echinochloa, Oryza, Acrocera,* and others. Mudflats, thickets, and floating vegetation join with these grasses to provide a range of habitat. These are supplemented by such accents as never-flooded dry islands built up atop generational termite mounds, and some stands of trees upon the better drained fringes.

Wildlife species here range from crocodiles to herd antelopes, as well as charismatic mega-fauna and carnivores, to an incredible variety of avian species. The ungulate fauna feature very large herds of Burchell's zebra, eland, lechwe, puku, wildebeest, African buffalo, tsessebe, waterbuck, reedbuck, and kudu. Elephants and hippopotamuses are in great supply—as are many large carnivores such as lion, leopard, cheetah, and, not least, the crocodile. Endemic and globally rare avian species include several species of weaver, Chaplini's barbet, egrets, blue crane, guineafowl, and black korhaan.

Human Interaction

The Zambezian Flooded Grasslands biome has a built-in attraction for farmers, herders, hunters and poachers—it also is a vital water supply for people struggling to get though dry seasons. Given the attractiveness to tourists, growing indigenous populations, and climate patterns tending toward hotter and drier, this ecoregion is increasingly being threatened by biodiversity- and habitat-loss, despite the numerous protected areas and conservation measures already in place to protect these delicate ecosystems.

The biome is also increasingly at risk from deforestation of its surrounding woodlands, diminution of water flow from human diversion, and pollution of the water quality by ill-controlled land-use practices upstream.

Protected areas include some areas set aside in a sprawling range of ecozones, including Barotse Floodplain and Kafue Flats in Zambia, Lake Chilwa in Malawi, Okavango Delta in Botswana, and some Tanzanian floodplains.

NARCISA G. PRICOPE

Further Reading

Chabwela, H. N. and W. Mumba. "Integrating Water Conservation and Population Strategies on the Kafue Flats." In A. de Sherbinin and V. Dompka, eds., *Water and Population Dynamics: Case Studies and Policy Implications. Zambia*. Washington, DC: American Association for the Advancement of Science, 1998.

Zambezi River

Category: Inland Aquatic Biomes.
Geographic Location: Africa.
Summary: The fourth-longest river in Africa, the Zambezi is the site of the world-famous Victoria Falls, one of the largest waterfalls in the world.

The source of the Zambezi River is in the furthest-northwest corner of Zambia, where it borders Angola to the west and the Democratic Republic of the Congo to the north. The elevation here is about 4,900 feet (1,500 meters). The river flows south through Angola for about 145 miles (230 kilometers), before crossing the border back into Zambia. After traversing Zambian territory in a southward direction, it then turns east and forms the international boundaries between both Zambia and Namibia and Zambia and Zimbabwe. Entering Mozambique at Luangwa, it flows southeast across Mozambican territory into the Indian Ocean. The Zambezi, at 2,200 miles (3,540 kilometers) in length, drains the entire region of south-central Africa, an area of about 540,000 square miles (1.4 million square kilometers).

Because of their vast ecological differences, the Zambezi River system can be divided into three separate sections: the Upper, Middle, and the Lower Zambezi. Each possesses different geographical features, landscapes, history, and biome characteristics.

Major Segments

The Upper Zambezi contains the source of the Zambezi River, which originates from a marshy wetland region. Near the source is the divide between the Zambezi and Congo River drainages; the two river systems do not connect. After exiting Angola and re-entering Zambia, the river is approximately 1,300 feet (400 meters) wide, and fast-flowing during the rainy season. The beginning of floodplains is marked by the Zambezi's first major tributaries, the Kabompo River and the Lungue River. As the river travels the remaining stretch through Zambia, it flows over a thick mantle of sand that contributed to the formation of the Barotse Floodplain.

After the floodplain, the Zambezi goes through the Caprivi Swamps along the Botswana border, where it briefly forms the border between four countries (Botswana, Zimbabwe, Namibia, and Zambia). After emerging from the swampy areas, the Chobe River joins the Zambezi and the river goes through the Katambora rapids. As it travels toward the Middle Zambezi, the river is set just over 3,000 feet (900 meters) above sea level on the basalt sheet of the south-central African plateau. The separation of the Upper and Middle Zambezi sections is marked by an area that cuts across the river bed; this forms Victoria Falls.

The Middle Zambezi is commonly defined as the area between Victoria Falls and the Lupata Gorge. This section of the river forms the border between Zambia and Zimbabwe. It is also very popular for tourism, as it contains many sandy beaches and clear waters, and several sections that are international attractions for whitewater

The Upper Zambezi, which flows on the basalt sheet of the south-central African plateau at an elevation of over 3,000 feet (900 meters) above sea level, becomes the Middle Zambezi at Victoria Falls, shown here. (Thinkstock)

rafting and kayaking. On the Zimbabwe side, the area is also home to the Zambezi National Park. The Middle Zambezi has seen the most human modification out of the Zambezi's three sections. After passing through Victoria Falls, the river cuts through gorges of basalt rock between 650 to 800 feet (200 to 250 meters) high.

The river then emerges from the basaltic plateau and turns eastward to enter one of the largest human-made lakes, Kariba Lake. In 1959, the Kariba Dam was constructed and enabled hydroelectric power to be generated. Today, the dam provides electricity for much of Zambia and Zimbabwe. The river is again joined by a major tributary, the Kafue River, and then by the Luangwa River. It then enters Lake Cahora Bassa; this water body was created in 1974 when the Cahora Bassa Dam was constructed. Downstream from here, the Zambezi is fairly shallow in many places in the dry season. The river stretches to a wide range, except in one area that is confined between high hills; this is known as the Lupata Gorge, which marks the division between the Middle and Lower Zambezi.

The Lower Zambezi is the area from the Lupata Gorge through the remainder of the river. Just 200 miles (320 kilometers) from the mouth of the Zambezi, the Lupata Gorge stretches only 650 feet (200 meters) wide and flows on top of a sandy river bed. During the rainy seasons, streams come together and form fast-flowing waters here. Aside from the Lupata Gorge, the Lower Zambezi mainly spans a range of 3–5 miles (5–8 kilometers) in width. As the Zambezi River approaches the Indian Ocean, it splits into several branches to form a wide delta. Before entering the ocean, the Zambezi waters flow through four principal mouths, each obstructed by sandbars: the Milambe, Kongone, Luabo and Timbwe Deltas. Prior to the construction of the Kariba and Cahora Bassa dams, this delta was about half as wide as it is today.

Biome Divisions

Just as the Zambezi River is commonly considered in sections, its basin can also be divided into four main biomes. The river basin flows through the Congolian biome, Zambezian biome, a montane biome, and a coastal biome. Starting at the headwaters, the Congolian biome is associated with northwestern Zambia and northeastern Angola. The Congolian biome is moister and warmer in climate than the rest of the basin. This biome is considered tropical, is high in rainfall, and is without a marked dry season. It is also linked to the Congo Basin.

The Zambezian biome is subtropical and covers the majority of the basin, up to 95 percent. This biome contains woodland, grassland, swamp, and lakes. The climate is marked by a dry season and is primarily based on seasonal changes. Within the Zambezian biome, there can also be a subdivision made between the drier and moister areas.

The montane biome is temperate, with some dry-season precipitation; it is cooler and wetter overall. It is part of the Eastern Arc Mountains; the species in this biome are similar to those in nearby mountains.

Finally, the coastal biome is tropical with a mild, stable climate. The coastal biome encompasses the small part of the Zambezi basin affected by coastal climate influences—mainly the delta area. It is marked by a dry season, but temperatures do not greatly fluctuate. The habitats here include dry forest, woodlands, and grasslands; most species found here are widespread along the east African coast.

Flora and Fauna

The Zambezi hosts a diverse range of plant and animal species. There are estimates of between 6,000 and 7,000 species of flowering plants, more than 200 species of mammals, 700 species of birds, 290 species of reptiles and amphibians, at least 190 species of freshwater fish, 210 species of dragonflies, 1,100 species of butterflies, and 100 species of freshwater mollusks.

The river supports large populations of many animals. Hippopotamus and crocodile are abundant along most of the calm stretches of the river. Monitor lizards are found in many places. There are four Important Bird Areas in the basin, including the Lower Zambezi National Park. Species present in large numbers include heron, pelican, egret, trumpeter hornbill, Meyer's parrot, narina trogon, and the African fish eagle. Cormorant, openbill stork, spur-winged goose, common pratincole, caspian plover, whiskered tern and African skimmer are found in the wetlands.

Woodlands in the basin also support many large mammals, such as spot-necked otter, buffalo, reedbuck, eland, zebra, giraffe, and elephant. These animal population numbers have declined, however, since the 1970s.

The Zambezi also supports diverse species of fish, some of which are endemic to the river (found nowhere else). Catfish, carp, and eel species are common. The bull shark, sometimes known as the Zambezi Shark, is found around the world. It normally inhabits coastal waters, but has been found far inland in many large rivers including the Zambezi.

Human Impact

Within such a widespread ecoregion, there are several possible threats to the Zambezi River's ecosystems. The Zambezi provides water for more than 32 million people who live in its watershed. Much of the water used is for agriculture. Poor land management practices, practiced mainly by indigenous farmers of the basin, have contributed excessive sedimentation through extreme soil erosion throughout much of the area. In urban areas, sewage effluent is a major cause of water pollution, as inadequate water treatment facilities allow for the release of untreated sewage into the river. This has resulted in eutrophication of the river water, and has facilitated the spread of diseases such as cholera, typhoid, and dysentery.

In recent years, groups and organizations have been recruited or formed to protect the fairly intact Zambezi River, such as the Nature Conservancy, Ramsar Convention on Wetlands, and the transfrontier Okavango-Zambezi Conservation Park. Land- and water-use policy, industrial development, maintaining migratory routes, and preparing for climate change are all issues being addressed relative to the Zambezi and its watershed.

William Forbes
Lori Beshears

Further Reading

Milich, Lenard and Robert G. Varady. "Openness, Sustainability, and Public Participation in Transboundary River-Basin Institutions: Part I: The Scientific-Technical Paradigm of River Basin Management." *Aridlands Newsletter* 44 (1998).

Munjoma, Leonissah. "Benefit Sharing in Integrated Water Resources Management." *Zambezi* 7, no. 2 (2006).

Purchase, G. K., C. Mateke, and D. Purchase. *A Review of the Status and Distribution of Carnivores, and Levels of Human Carnivore Conflict, in the Protected Areas and Surrounds of the Zambezi Basin.* Bulawayo: Zambezi Society, 2007.

Timberlake, J. *Biodiversity of the Zambezi Basin Wetlands.* Harare, Zimbabwe: International Union for Conservation of Nature, Regional Office for Southern Africa, 1998.

Tweddle, Denis. "Overview of the Zambezi River System: Its History, Fish Fauna, Fisheries, and Conservation." *Aquatic Ecosystem Health & Management* 13, no. 3 (2010).

White, F. *The Vegetation of Africa, A Descriptive Memoir to Accompany the UNESCO/AETFAT/UNSO Vegetation Map of Africa*. Paris, France: United Nations Educational, Scientific, and Cultural Organization (UNESCO), 1983.

Zapata Swamp

Category: Marine and Oceanic Biomes.
Geographic Location: North America.
Summary: The coastal Zapata Swamp is the largest wetland in Cuba.

Zapata Swamp is widely considered to be the best-preserved wetland throughout the entire Caribbean. Located on the southwestern coast of Cuba about 124 miles (200 kilometers) away from the capital city Havana and adjacent to the infamous Bay of Pigs, the Zapata Swamp Biosphere Reserve is Cuba's largest protected area. To date, more than 900 species of plants, 175 species of birds, 31 species of reptiles, and more than 1,000 unique invertebrates have been cataloged within the swamp. Among these are several species that are endemic (found only here), including at least five endemic plants, three endemic birds—the Zapata wren, the Zapata rail, and the Zapata swallow—an endemic rodent, the Zapata hutia, and an endemic fish, the Cuban gar.

The Zapata Swamp, locally called *Ciénaga de Zapata,* is also home to the West Indian manatee and the Cuban crocodile, both of which are considered to be highly endangered. More than 75 percent of the 2,300-square-mile (6,000-square-kilometer) reserve is marsh and brackish lagoon, making the Zapata Swamp the largest wetland in Cuba.

Except in the mountains, the climate of Cuba is semitropical or temperate. The average minimum temperature is 70 degrees F (21 degrees C); the average maximum is 81 degrees F (27 degrees C). The mean temperature at Havana, which is just north of the swamp, is about 77 degrees F (25 degrees C). The trade winds and sea breezes make coastal areas more habitable than temperature alone would indicate. Cuba has a rainy season from May to October; it averages about 49 inches (125 centimeters) of precipitation per year.

Biodiversity

Within the Zapata Swamp Biosphere Reserve are white-sand beaches, mangrove swamps, fresh- and saltwater marshes, semi-dry deciduous forests, and even some evergreen forests. This multitude of habitat types enables a wide variety of organisms to use the different areas of the swamp, and leads to higher biodiversity. Of particular importance are the marshes, which are used as nursery areas by juvenile fish, which in turn become food for overwintering migratory birds.

One of the great wonders of the Zapata Swamp is the multitude of birds that call the swamp home, either year round or as a stopover. More than 170 bird species have been identified here. The island of Cuba hosts 25 endemic bird species, 22 of which can be found nesting within the swamp reserve's boundaries. Zapata Swamp is also an important stopover point for migratory birds; at least 65 species of migratory birds use the swamp during migrations from North to South America.

The swamp is a popular birdwatching destination because of the great diversity of bird species offered. A birdwatcher can potentially see everything from the common black hawk, peregrine falcon, and the greater flamingo to the world's smallest bird, the bee hummingbird, all in one visit to the reserve. Without the presence of the intact marshes and wetlands that the Zapata Swamp offers, migrating birds would miss a vital resting point during their migrations, and native birds would lack important nesting grounds.

Human Impact

The first efforts to protect the Zapata Swamp happened in 1936, when the area was declared a National Refuge for Fishing and Hunting. However, the declaration was never enforced, and modest development in the area continued unabated.

Cuban Crocodile

The most noteworthy resident of the Zapata Swamp is the Cuban crocodile (*Crocodylus rhombifer*). One of only four species of crocodile found in the Americas, the Cuban crocodile is considered to be highly endangered. At one time, the Cuban crocodile ranged throughout the Caribbean, and fossilized remains have been found on Grand Cayman Island and in the Bahamas. However, hunting pressure and habitat degradation have nearly wiped out the species.

Zapata Swamp is home to the last remaining wild population of Cuban crocodiles, although recent efforts have been made to establish a population at the nearby Isla de la Juventud. Occasionally, the Cuban and American crocodiles, both of which live in the swamp, will mate, creating a hybrid crocodile.

*The Cuban crocodile (*Crocodylus rhombifer*) is highly endangered; the last few found in the wild live in the Zapata Swamp. (Wikimedia/Ltshears)*

In 1961, Ciénaga de Zapata National Park was established, but again, this designation was seen as largely symbolic. Luckily, and despite the swamp's proximity to the capital city, deforestation and development (plagues that destroyed other Cuban wetlands) never took hold in Zapata.

In 1971, international attention to the swamp increased when the Ramsar Convention, an international treaty for the conservation and sustainable utilization of wetlands, added the Zapata Swamp to its list of Wetlands of International Importance. Wetlands selected for this list are recognized as having international significance in terms of their ecology, botany, zoology, or hydrology. At the time of the designation, however, Cuba was not part of the Ramsar Convention, so the Zapata Swamp was not included as a Ramsar site; it would be formally added to the convention in 2001.

In 2000, the Zapata Swamp was designated a United Nations Educational, Scientific, and Cultural Organization (UNESCO) Biosphere Reserve, in recognition of efforts to reconcile conservation and development by emphasizing the involvement of the local community. Zapata Swamp National Park remains one of the most important tourist destinations in Cuba, attracting birdwatchers from around the world. However, development pressure in the forms of additional tourist infrastructure and the broadening of local communities continue to threaten the swamp. Protection of swamp habitat is vital, not only for the myriad of organisms that call Zapata Swamp home, but also for the local people who depend on the swamp for their livelihoods.

Global warming is bringing new stresses to the Zapata Swamp habitat, in the forms of harsher heat waves that accelerate evaporation and alter sprouting, hatching, and mating cycles, and bunched rainfall cycles leading to severe flooding. Cuba has already recorded a decrease in annual rainfall in the range of 15 percent since the 1960s. Now, rising sea level threatens the swamp with the impact of greater storm damage and erosion, as well as inundation of root systems and disruption of the intertidal habitats here.

Robert D. Ellis

Further Reading

Batze, Darol P. and Andrew H. Baldwin, eds. *Wetland Habitats of North America: Ecology and Conservation Concerns.* Berkeley: University of California Press, 2012.

Echenique, Lazaro Miguel. "Zapata Swamp: Cuba's Largest, Wildest Wetland." *International Journal of the Wilderness* 4, no. 2 (1998).

Wildlife Conservation Society. "Zapata Swamp, Cuba." http://www.wcs.org/saving-wild-places/latin-america-and-the-caribbean/zapata-swamp-cuba.aspx.

Chronology

4.5–3.8 billion years ago: The Hadean period, during which the solar system was formed.

3.8–2.5 billion years ago: The Archaean period, which includes the first appearance of life on Earth.

2.5 billion–543 million years ago: The Proterozoic Era, which includes the formation of stable continents on Earth and the appearance of eukaryotic cells (ca. 1.8 billion years ago).

543–248 million years ago: The Paleozoic Era, including the Cambrian "explosion" (great increase in diversity) of animal life, and mass extinction of about 90 percent of marine animal species at the end of the Era.

248–65 million years ago: The Mesozoic Era, including the appearance and extinction of dinosaurs, and the appearance of gymnosperms (e.g., conifers) and angiosperms (flowering plants).

6.5–1.8 million years ago: The Tertiary Period of the Cenozoic Era, a period that indicates through fossils the dawning of modern orders of mammals (in the Eocene Epoch, 54.8–33.7 million years ago), and the appearance of many grasses (in the Oligocene Epoch, 33.7–23.8 million years ago).

1.8 million years ago: Beginning of the Quaternary Period of the Cenozoic Era, which includes the present day.

1.8 million–ca. 10,000 years ago: The Pleistocene Period of the Quaternary Period of the Cenozoic Era; large mammals such as mastodons roamed North America during this period but went extinct by the end of the period; *Homo sapiens* (modern humans) spread throughout much of the world.

ca. 10,000 years ago: Beginning of the Holocene Period, which starts at the end of the last global ice age, and runs to the present day.

ca. 500 B.C.E.: Large-scale irrigation begins in Sri Lanka.

206 B.C.E.–222 C.E.: In China, records from the Han Dynasty indicate the use of forestry management techniques.

ca. 1200 C.E.: Cahokian Indian culture in Illinois (U.S.) reaches its peak.

1492: The arrival of Christopher Columbus in North America begins sustained contact between Europeans and North American native tribes.

1498: The Portuguese explorer Vasco da Gama lands in India and departs with a load of spices, beginning the European competition to access the riches of India.

1537: In what is now the southern United States, the Indian population is swept by the first major epidemic of disease brought by European colonists.

1556: An earthquake with an estimated magnitude of 8 in Shaanxi (Shensi), China, causes over 830,000 deaths, and damage is recorded over 270 miles from the epicenter.

1601–1603: A major famine in Russia is estimated to have killed one-third of the population.

1607: English settlers establish a successful colony at Jamestown, Virginia.

1626:The Plymouth Colony in Massachusetts passes an ordinance regulating timber harvesting and sales.

1680s: In South Carolina, rice cultivation begins in North America.

1681: On the island of Mauritius, the dodo bird becomes extinct.

1669: In France, a Forest Code is established to ensure sufficient wood supply for the navy.

1731–1742: Mark Catesby, an English naturalist, publishes his *Natural History of Carolina, Florida, and the Bahama Islands.*

1755: An earthquake on All Saint's Day in Lisbon, Portugal, kills an estimated one-quarter of the city's population; the earthquake's magnitude is estimated at 8.7 and waves of 20 feet (6 meters) in height (from the associated tsunami) are observed in Morocco, Spain, and Portugal.

1786: There is localized extinction of sea otter populations from trapping in the U.S. northwest.

1791: William Bartram publishes *Travels Through North & South Carolina, Georgia, East & West Florida.*

1793: The cotton gin is perfected by Eli Whitney; this invention makes cotton production highly profitable and leads to cotton becoming the most important crop in the American south.

1804: The estimated world population reaches 1 billion.

1807: In the United States, the National Ocean Service, which later becomes the National Oceanic and Atmospheric Administration (NOAA), is created.

1811–1812: A series of earthquakes centered on New Madrid, Missouri, are among the most violent ever recorded; contemporary reports state that the effects were felt as far away as Cleveland, Ohio, and Washington, D.C.

1813: The Charter Act (Great Britain) allows missionary activity in the subcontinent of India, and limits the monopolies of the East India Company.

1817: A cholera epidemic in Calcutta, India, kills over 10,000 British troops, as well as unknown numbers of Indians; the epidemic spreads to other countries and arrives in North America in 1832.

1820s: Overtrapping nearly eliminates the sea otter population in North America.

1843: In India, construction begins on the Ganges Canal, the first major colonial irrigation scheme in India.

1849: In the United States, the Department of the Interior is established.

1852: Dr. John Snow theorizes that cholera is transmitted through water, and convinces London authorities to shut down a water pump in an area heavily affected by a cholera epidemic.

1854: Henry David Thoreau advocates for a simple life lived in harmony with nature in *Walden*.

1859: In Australia, 12 pairs of rabbits are released from a farm; they multiply rapidly and cause extreme environmental damage, leading to repeated attempts to exterminate them.

1859: In the Cape Colony, South Africa, passage of the Forest Protection Act occurs.

1859: Charles Darwin publishes *On the Origin of Species*; together with his 1871 book *The Descent of Man*, it explains his theory of natural selection.

1862: In the United States, the Homestead Act spurs European settlement of the Great Plains states.

1864: In the United States, Yosemite becomes the first state park.

1864: The Forest Service is created in India.

1869: Ernst Haeckel coins the term *ecology* from the Greek words *logos* (meaning study of) and *oikos* (meaning dwelling place) to denote the study of how plants, animals, and the environment interact with each other.

1869: The Suez Canal connects the Red Sea to the Mediterranean Sea, shortening the sea route from Bombay, India, to London, England, by over 4,500 nautical miles in the process.

1869: The coffee crop in Kandy, Sri Lanka, is devastated by a fungus, which leads to the end of plantation cultivation of coffee in Sri Lanka.

1872: English chemist R. A. Smith coins the term *acid rain*.

1872: In the United States, Yellowstone becomes the first national park.

1875: The American Forestry Association is founded.

1878: John Wesley Powell, in his *Report on the Lands of the Arid Regions of the United States*, suggests that irrigation will be necessary to make much land in the western United States suitable for agriculture.

1879: The first national park in Australia, Royal National Park, is established in New South Wales.

1879: Creation of the U.S. Geological Survey.

1880s: Chicago, Illinois, and Cincinnati, Ohio, become the first U.S. cities to pass ordinances limiting smoke emissions.

1883: In Indonesia, eruption of the Krakatau volcano causes a tsunami that kills an estimated 36,000 people.

1885: Canada's first national park, Banff National Park, is established.

1891: In the United States, the Forest Reserve Act gives the president the authority to create forest reserves from lands in the public domain.

1892: German scientist Robert Koch identifies the bacteria that causes cholera.

1892: The boll weevil enters the United States; within 30 years, this pest will end cotton monoculture in most of the American south.

1893: Algonquin Provincial Park becomes Canada's first provincial park.

1894: Tongariro National Park becomes the first national park in New Zealand.

1896: The Sierra Club is founded by John Muir.

1896: Japan is struck by the Sanriku tsunami, which has waves estimated at 75 feet (23 meters) tall, and kills over 26,000 people.

1896: Swedish chemist Svante Arrehenius notes that burning coal, while essential to the Industrial Revolution, could also increase atmospheric carbon dioxide and thus induce global warming.

1897: Thailand creates the Royal Forest Department; in 1899, all forests are declared property of the government.

1900: The United States passes the Lacey Act, which regulates interstate shipment of wild animals.

1903: President Theodore Roosevelt establishes the first wildlife refuge in the United States, Pelican Island, Florida.

1905: Formation of the National Audubon Society in the United States.

1905: British scientist H. A. Des Voeux coins the term *smog* to describe the mix of smoke and fog that regularly settles over London.

1906: The United States and Mexico sign the Convention on the Equitable Distribution of the Waters of the Rio Grande, regulating the use of Rio Grande waters for irrigation.

1906: San Francisco, California, is devastated by an earthquake and subsequent fire.

1908: A magnitude 7.2 earthquake in Messina, Italy, kills an estimated 72,000 people, including over 40 percent of the city's population; tsunamis are observed on the Sicilian coast and aftershocks from the earthquake are felt until 1913.

1910: The U.S. Bureau of Mines is created to regulate mine safety.

1914: The passenger pigeon, once numerous in North America, becomes extinct when the last known passenger pigeon dies in an Ohio zoo.

1916: In the United States, the National Park Service is established as an agency of the Department of the Interior.

1916: Frederick Clements publishes *Plant Succession*, which argues that plant communities undergo a predictable series of stages until reaching the steady-state phase of climax.

1917: Joseph Grinnell uses the term *ecological niche* to describe the way the California thrasher (a bird) differentiated its use of habitat from that of similar species.

1918: An influenza pandemic (the Spanish flu) infects about a quarter of the world's population and kills an estimated 3 to 6 percent; in contrast to the usual pattern in epidemics, many of the flu's victims are healthy young adults.

1920: A magnitude 7.8 earthquake in Haiyuan, Ningxia (China), kills an estimated 200,000 people; this earthquake is sometimes called the Kansu earthquake.

1920: In the United States, the Mineral Leasing Act is passed to govern the exploitation of fuel and fertilizer minerals on public lands.

1921: Thomas Midgley and colleagues discover that adding tetraethyl lead to gasoline increases the power and efficiency of automobile engines.

1922: The Izaak Walton League, a conservation organization originally focused on water pollution and threats to sport fishing, is founded.

1923: A magnitude 7.9 earthquake in Kanto, Japan, causes extreme destruction in the Tokyo-Yokohama region, killing over 140,000 people.

1926: South Africa passes the National Parks Act.

1927: The estimated world population reaches 2 billion.

1927: The U.S. Army Corps of Engineers begins planning an extensive system of levees to control the Mississippi River after a major flood inundates much of the lower Mississippi region.

1927: British ecologist Charles Elton publishes *Animal Ecology*, introducing many common concepts in ecology, including food chains, food webs, and trophic layers.

1928: CFCs (chlorofluorocarbons) are discovered by Thomas Midgely.

1929: Serengeti National Park is established in northern Tanzania in Africa; the park is expanded in 1940.

1930: Construction begins on the last major colonial irrigation project in India, the Sarda Canal.

1933: The Tennessee Valley Authority (TVA) is created by the U.S. Congress; the TVA will radically change the landscape of the upper American south through the construction of dams and reservoirs, tree planting, and rural electrification.

1934: In the United States, the Taylor Grazing Act regulates grazing on public lands.

1935: Aldo Leopold founds the Wilderness Society.

1935: In *Deserts on the March*, Paul Sears argues that the Dust Bowl was caused by human misuse of the land.

1935: In the United States, the Soil Conservation Service is established to combat the Dust Bowl.

1943: The Alaska Highway, which links Alaska with the lower 48 U.S. states, is completed.

1943: Norman Borlaug begins work in Mexico that leads to the "Green Revolution" and marked increases in food production in developing countries.

1945: In New Mexico, the United States conducts the first nuclear weapons test.

1946: The U.S. Bureau of Land Management is established.

1946: An earthquake in Alaska (U.S.) generates a tsunami which kills 159 people in Hilo, Hawaii, and causes millions of dollars in damage.

1947: Marjory Stoneman Douglas publishes *The Everglades: River of Grass* to call attention to the poor conditions in the Everglades in Florida.

1948: Swiss scientist Paul Hermann Müller receives the Nobel Prize for his discovery of the effectiveness of DDT (dichlorodiphenyltrichloroethane) as an insecticide.

1948: A magnitude 7.3 earthquake in Ashgabat, Turkmenistan, destroys most brick buildings in the area and kills an estimated 110,000 people.

1948: In the U.S. industrial town of Donora, Pennsylvania, a temperature inversion traps air pollution that darkens the skies and leads to 20 deaths; this incident points out the importance of nonsmoke elements (e.g. nitrogen sulphide, sulphur dioxide, and particulate metals) in air pollution and spurs efforts to create effective clean air legislation.

1948: In Switzerland, the International Union for the Conservation of Nature and Natural Resources, later the World Conservation Union, is formed to protect wildlife and the environment; membership is composed of governmental and nongovernmental organizations.

1949: The British Nature Conservancy Council becomes the world's first nature conservancy; it is replaced by three separate councils (in Scotland, Wales, and England) in 1991.

1949: Creation of the International Whaling Commission (IWC), following the 1946 International Convention for the Regulation of Whaling; the IWC sets quotas for whaling that represent a compromise between allowing the whaling industry to continue and the need to protect whales from extinction.

1949: Aldo Leopold introduces the idea of a land ethic in *A Sand County Almanac*.

1951: Murray Bookchin publishes *The Problem of Chemicals*, drawing attention to the potentially harmful effects of chemicals on the environment and human health.

1952: The Population Council is established in New York City to address the world's population growth; the council's activities include conducting research into birth control methods and examining the effects of family size on health and well-being in different countries.

1952: In the United Kingdom, a temperature inversion leads to a thick fog with high levels of smoke and sulphur dioxide blanketing the London area; over 2,000 people die as a result, and the incident leads to passage of the Clean Air Act of 1956.

1952: The United States tests nuclear fusion devices on the Marshall Islands in the South Pacific.

1953: Construction on the Kalabagh Dam Project begins in Pakistan.

1954: In the United States, the Atomic Energy Act gives the federal government the authority to regulate nuclear power facilities.

1956: The first victims of Minamata disease, caused by mercury poisoning, are reported in and near Minamata, Japan.

1957: In Nepal, the Private Forests Naturalization Act is introduced.

1957: In the United States, the Price-Anderson Act limits the liability of civilian producers of nuclear power in the case of a nuclear accident.

1958: The United Nations Conference on the Law of the Sea codifies the right to fish freely on the high seas (outside of territorial seas).

1959: The Great Lakes are linked to the Atlantic Ocean through completion of the St. Lawrence Seaway.

1959: The Cuyahoga River, a polluted tributary of Lake Erie near Cleveland, Ohio, catches fire for the first time; the river will again catch fire in 1969.

1960–1969: Construction of the Aswan High Dam on the Nile River in Egypt.

1960: The estimated world population reaches 3 billion.

1960: India and Pakistan sign the Indus Waters Treaty.

1960: In the United States, President Dwight D. Eisenhower establishes the Arctic National Wildlife Refuge in Alaska.

1960: Iran, Iraq, Kuwait, and Venezuela found the Organization of Petroleum Exporting Countries (OPEC); other countries have since become members, and OPEC members are believed to control about two-thirds of the world's oil reserves.

1961: The World Wildlife Fund is founded to help create nature reserves and preserve the habitats of endangered species.

1962: Rachel Carson publishes *Silent Spring*, creating greater awareness of the environmental effects of pesticide use.

1962: Oregon becomes the first U.S. state to establish a comprehensive program to regulate air pollution.

1963: In the United States, the first Clean Air Act is passed.

1963: The United States and the Soviet Union sign the Nuclear Test Ban Treaty, agreeing to end the atmospheric testing of nuclear weapons.

1964: In the United States, the Wilderness Act creates legal protection for wilderness areas.

1964: In India, foundation of the Dasholi Gram Swarjya Mandal (DGSM), a movement by peasants to limit environmental damage and create a self-reliant village society.

1965: In the United States, the Water Quality Act creates federal water quality standards.

1965: The Australian Conservation Foundation is founded.

1966: In the United States, the Wild and Scenic Rivers Act provides protection of rivers classified as either wild or scenic.

1966: In Canada, Dr. Harold Harvey identifies the effects of acid rain on fish populations.

1967: Off the coast of England, the supertanker *Torrey Canyon* sinks and causes a major oil spill.

1967: In the United States, author Cleveland Amory founds the Fund for Animals.

1968: The Club of Rome is founded to study the relationships among human activity, the natural environment, and population growth.

1968: Garrett Harding publishes "The Tragedy of the Commons" in *Science*, arguing that private ownership is necessary in order to protect the environment; this theory has been challenged by the existence of numerous societies in which other systems of ownership have produced sustainable management of natural resources.

1968: American scientist Paul Ehrlich publishes *The Population Bomb* and founds Zero Population Growth, an organization dedicated to fostering a sustainable relationship between natural resources and the human population.

1969: The U.S. National Environmental Policy Act requires all federal agencies to consider the environmental consequences of their actions.

1969: Robert Paine coins the term *keystone species*, referring to species whose loss would have drastic consequences on a biological community (e.g., predators that control the density of other species).

1969: The Environmental Law Institute is founded to promote research on environmental law.

1970: A magnitude 7.9 earthquake in Chimbote, Peru, kills an estimated 70,000 people, including about 20,000 residents of the town of Yungay, which is buried in an avalanche of mud, rock, and ice.

1970: Norman Borlaug is awarded the Nobel Peace Prize for his work toward combating hunger and malnutrition in developing countries.

1970: The first Earth Day is celebrated on April 22, pioneered by John McConnell at a 1969 United Nations Educational, Scientific and Cultural Organization (UNESCO) Conference in San Francisco.

1971: Michael Jacobson founds the Center for Science in the Public Interest.

1971: The Ramscar Convention on Wetlands of International Importance, held in Iran, adopts an agreement to protect migratory wildfowl.

1971: In Canada, the Department of the Environment (now Environment Canada) is created.

1971: The international environmental organization Greenpeace is founded.

1972: The United States passes the Clean Water Act, which includes the goal of making all U.S. surface waters swimmable and fishable.

1972: The U.S. Congress passes the Marine Mammal Protection Act, a comprehensive federal law that prohibits the taking (including hunting) of marine mammals on U.S. vessels; limited exceptions include animals captured for educational purposes or by Native Americans for subsistence or the production of traditional crafts.

1972: The United States phases out use of the pesticide DDT.

1972: The world's first national green party, the Values Party, is established in New Zealand; it is later renamed the Green Party of New Zealand.

1972: *The Limits to Growth* is published by Donella H. Meadows and others, articulating the view of the Club of Rome that sustainability, rather than unlimited growth, should become the global ideal.

1972: In the United States, the Ocean Dumping Act (or the Marine Protection, Research and Sanctuaries Act) regulates ocean dumping, establishes a process for creating marine sanctuaries, and establishes marine research programs.

1972: In the United States, the Ports and Waterways Safety Act regulates oil handling facilities and the transport of oil.

1973: In the United Kingdom, the first European green party is founded; first called People, it later changes its name to the Ecology Party, and finally the Green Party.

1973: The Cousteau Society to promote marine research is founded.

1973: An embargo by the Arab members of OPEC (the Organization of Petroleum Exporting Countries) causes sharp increases in the price of oil, and fosters interest in developing alternate sources of energy.

1973: The Norwegian philosopher Arne Naess coins the term *deep ecology*, challenging *shallow ecology*, which assumes that humans are the central species on Earth and that the desires of humans should guide environmental efforts.

1973: In *Small is Beautiful: Economics as if People Mattered*, E. F. Schumacher advocates for self-reliance and a simple life lived in harmony with nature.

1973: The United Nations Environment Programme (UNEP) is founded to coordinate environmental policies among U.N. agencies, national governments, and nongovernmental organizations (NGOs).

1973: In the United States, the Endangered Species Act authorizes the Secretary of the Interior to designate plant and animal species as "endangered" or "threatened," and prohibits destruction of habitat critical to endangered species, as well as hunting of endangered species.

1974: The estimated world population reaches 4 billion.

1974: In Australia, the Environmental Protection (Impact of Proposals) Act requires that major resource projects include a consideration of the environmental impact of the project.

1974: In the United States, the Safe Drinking Water Act establishes minimum safety standards for municipal water supplies.

1974: Economist Lester R. Brown founds the environmental thinktank WorldWatch Institute with funding from the Rockefeller Brothers' Fund.

1974–1975: In the Sahel (Africa), a severe drought displaces an estimated 10 million farmers.

1975: Annie Dillard is awarded the Pulitzer Prize for *A Pilgrim at Tinker Creek* (published in 1974), a meditation on the natural world of Roanoke Valley, Virginia.

1975: In *The Monkey Wrench Gang*, E. F. Abbey advocates for radical actions in defense of the environment, including "ecotage" (sabotage in the service of environmental protection).

1975: Peter Singer publishes *Animal Liberation*, proposing that animals have rights and deserve moral consideration from human beings.

1976: A magnitude 7.5 earthquake in Tangshan, China, kills over 240,000 people (other estimates run as high as 655,000), with almost 800,000 injured.

1976: In the United States, the Resource Conservation and Recovery Act gives the EPA authority to regulate municipal waste.

1976: The Mediterranean Action Plan, coordinated by the United Nations Environment Programme, sets forth legislation to combat pollution in the Mediterranean Sea.

1977: In the United States, the Surface Mining Control and Reclamation Act empowers the Department of the Interior to develop environmental standards for surface mining, to limit environmental problems such as erosion and water pollution.

1977: The United Nations Conference on Desertification meets in Kenya.

1978: In northern New York State, residents of Love Canal are evacuated because of contamination of the area by chemical wastes.

1978: In the United States, the Energy Tax Act (also known as "the gas guzzler tax") includes a tax on passenger cars whose fuel efficiency falls below a set standard.

1979: The Convention on the Conservation of Migratory Species of Wild Animals meets in Bonn, Germany, and drafts an agreement on international management of wild animals that regularly cross national borders.

1979: In Canada, the Department of Fisheries and Oceans is established.

1979: In Pennsylvania, the cooling systems fail in the Three Mile Island nuclear reactor and the plant comes close to reaching the point of a nuclear meltdown.

1980: In Washington state, the Mt. St. Helens volcano erupts.

1980: PETA (People for the Ethical Treatment of Animals) is founded.

1980: In the United States, a "Superfund" is established through the Comprehensive Environmental Response, Compensation, and Liability Act to clean up environmental damage caused by the improper disposal of hazardous waste.

1982: The United Nations convenes a Conference on the Law of the Sea; the resulting "constitution for the world's oceans" increases the territorial sea zone to 22 nautical miles, and establishes an Exclusive Economic Zone of 200 nautical miles.

1983: Dian Fossey publishes *Gorillas in the Mist*, drawing attention to the mountain gorillas of Zaire and the threat they face from poaching.

1984: In Bhopal, India, chemical leakage from a Union Carbide plant kills thousands of people.

1985: The Rainforest Action Network is founded.

1985: A hole in the ozone layer is observed over Antarctica.

1985: In Canada, the Minister of Agriculture cancels the registration of alachlor, a herbicide in use since 1969, due to evidence that it increases human risk for cancer.

1986: In Missouri, the town of Times Beach is evacuated due to contamination from dioxin.

1986: In Ukraine (which was then part of the Soviet Union), the Chernobyl Nuclear Power Station undergoes a meltdown and spreads radioactive waste over much of northern Europe and the Soviet Union.

1987: The U.S. Department of Energy designates Yucca Mountain as the country's first permanent repository for radioactive waste.

1987: *Our Common Future*, commonly referred to as "the Brundtland Report," is published at the request of the Secretary-General of the United Nations. The report identifies global concerns for the environment and addresses issues of sustainable development and cooperation between industrialized and developing countries.

1987: Twenty-four countries sign the Montreal Protocol on Substances that Deplete the Ozone Layer; the Montreal Protocol is the first international agreement to set target dates for the reduction or elimination of the use of chemicals thought to harm the Earth's ozone layer.

1987: The United States passes the Marine Plastic Pollution Research and Control Act, prohibiting the dumping of plastics at sea and restricting the dumping of other types of ship-generated garbage in the ocean.

1987: Dave Foreman publishes *Ecodefense: A Field Guide to Monkeywrenching*, which continues with E. F. Abbey's concept of "ecotage" (sabotage in the service of environmental causes) and describes techniques such as spiking trees to hamper forestry.

1987: Foundation of the American Cetacean Society, an organization dedicated to the protection of whales, dolphins, and porpoises.

1987: In the United States, the National Appliance Energy Conservation Act sets minimum efficiency standards for both home appliances and home heating and cooling systems.

1988: The Canadian Environmental Protection Act becomes law, expanding federal regulatory powers and establishing a comprehensive plan for the control of toxic substances.

1988: In the United States, the Alternative Motor Fuels Act encourages development of automobiles that can run on fuels other than gasoline (e.g., ethanol).

1988: The first genetically modified plants are grown.

1989: The Basel Convention on the Control of Transboundary Movements of Hazardous Wastes and their Disposal, an international treaty, is adopted to regulate the global shipment of wastes; it becomes active in 1992.

1989: In Prince William Sound, Alaska, the oil tanker *Exxon Valdez* runs aground and spills an estimated 11 million gallons of oil.

1989: At the First World Congress of Herpetology, many scientists report the global decline and occasional disappearance of amphibian populations and species.

1990: Publication of the *Canadian Long-Range Transport of Air Pollutants and Acid Deposition Assessment*, which draws attention to the damage caused by acid rain and the effects of cross-border pollution on forests.

1990: A magnitude 7.4 earthquake in western Iran kills an estimated 50,000 people, with an additional 400,000 made homeless.

1991: Mt. Pinatubo in the Philippines erupts.

1992: The United Nations Earth Summit (the United Nations Conference on Environment and Development, or UNCED) is held in Rio de Janeiro, Brazil; the resulting Rio Declaration on Environment and Development sets forth 27 principles for guiding environmentalism, balancing the need for development with environmental protection.

1994: NAFTA (the North American Free Trade Agreement), an international agreement among the United States, Canada, and Mexico, includes environmental protections as well as the planned phasing out of tariffs between the countries.

1994: In China, construction begins on the Three Gorges Dam, which is planned to include the world's largest power station.

1996: The International Organization for Standardization adapts ISO 14000, a series of standards for environmental management.

1998: An offshore earthquake causes a tsunami which kills over 2,000 people living near the northern coast of Papua New Guinea.

1999: The estimated world population reaches 6 billion.

2001: The United States withdraws from the Kyoto Protocol.

2002: The Larsen B ice shelf in Antarctica breaks apart, a result generally attributed to global warming.

2003: The Philippines establishes national emissions standards for motorcycles.

2004: A magnitude 9.1 earthquake in Sumatra, Indonesia, kills over 227,000 people and displaces 1.7 million; the earthquake causes a tsunami affecting east Africa and south Asia.

2004: The *Arctic Climate Impact Assessment*, published by the Arctic Council, predicts that continued global warming will devastate life in the Arctic.

2005: In the Arctic, the Ayles Ice Shelf breaks off the northern coast of Ellesmere Island.

2005: Hurricane Katrina sweeps through the U.S. Gulf Coast region, causing more than 1,800 deaths, extensive property damage, and the flooding of New Orleans, bringing attention to the environmental conditions (including loss of wetlands) that exacerbated the flooding.

2005: A magnitude 7.6 earthquake in northern Pakistan destroys entire villages and kills at least 86,000 people.

2006: In a Canadian national election, the Green Party wins 4.5 percent of the popular vote.

2006: Scientists in China declare the Yangtze dolphin extinct, marking the first extinction of a large vertebrate in over 50 years.

2007: Costa Rica announces plans to become the world's first carbon-neutral country (by 2021).

2007: The International Biofuels Forum (IBF) is launched by the United States, Brazil, China, India, South Africa, and the European Commission.

2008: A magnitude 7.9 earthquake in eastern Sichuan Province, China, kills over 87,000 people and affects over 45.5 million; over 5 million buildings collapse due to the earthquake, and the total economic losses are estimated at $86 million.

2008: World oil prices rise to over $100 per barrel, creating renewed interest in conservation and alternative fuels.

2009: An influenza pandemic (H1N1 or "swine flu") begins in Mexico and sweeps the globe, infecting an estimated 10–20 percent of the world's population.

2009: Computer hackers breach a server at the University of East Anglia and steal thousands of emails and other documents related to climate change; dubbed "climategate," some argue that the stolen documents show an attempt to manipulate data related to climate change, while others dispute that interpretation.

2010: A magnitude 7.0 earthquake in Haiti kills over 300,000 people, injures another 300,000, and displaces 1.3 million.

2010: Eruption of the Eyjafjallajökull volcano in Iceland interrupts air traffic on some airline routes for weeks.

2010: The Deepwater Horizon oil spill in the Gulf of Mexico, caused by an explosion on an offshore oil rig, spills almost 200 million gallons of crude oil into the gulf; in response, U.S. President Obama's administration bans new offshore drilling (until 2017) in the Gulf of Mexico and off the Atlantic and Pacific coasts, but not in Arctic waters.

2010: The United Nations declares 2010 to be the International Year of Biodiversity.

2011: The estimated world population reaches 7 billion.

2011: The Horn of Africa experiences its worst drought in 60 years, with the lack of rainfall attributed to the cyclical *La Niña* phenomenon.

2011: Josh Fox's documentary *Gasland* becomes a finalist for an Academy Award; the resulting publicity popularizes the term *fracking* (hydraulic fracturing) and highlights the dangers of this method of drilling for natural gas.

2011: In Japan, an earthquake and resulting tsunami kills over 15,000 people and causes meltdowns at the Fukushima Daiichi power plant.

2011: Canada withdraws from the Kyoto Protocol on climate change.

2011: The United States experiences major flooding from the Mississippi River system, exacerbated in part by rainfall accompanying one of the largest tornado storm systems observed in the last 100 years.

2011: In Germany, Chancellor Angela Merkel announces that Germany will phase out the use of nuclear power by 2022.

2012: In January, a magnitude 6.3 earthquake strikes coastal Peru.

2012: In July, a brief period of extreme rainfall (over 10 inches in two days) causes severe flooding in Krasnodar Krai, Russia.

2012: In September, 52 percent of the continental United States is estimated to be suffering from moderate to extreme drought, and 37 percent from severe to extreme drought, according to the Palmer Drought Index.

2012: In October, a research study by Katharina Fabricius and colleagues reports that half of Australia's Great Barrier Reef has disappeared in the past 27 years.

2012: In October, Hurricane Sandy, the second-largest and second-costliest Atlantic hurricane on record, sweeps across the Caribbean before making landfall in New Jersey and causing flooding along much of the east coast of the United States.

SARAH BOSLAUGH

Glossary

Abiotic: Not living.

Acid Rain: Precipitation (rain, snow, dew, etc.) with a pH more acidic than normal precipitation (~5.7); acid rain is formed by the combination of oxygen and water vapor with industrial pollutants, including nitrogen oxides and sulfur dioxide.

Activated carbon: A highly adsorbent form of charcoal used to remove toxic substances and odors from gases and liquids.

Adsorption: A process in which a solid surface collects gases or vapors; the adsorbed molecules attach in a thin layer to the solid surface.

Afforestation: Creating forests by seeding or planting trees in a land that was previously not forested.

Agenda 21: An agenda for sustainable development passed in 1992 at the United Nations Conference on Environment and Development in Rio de Janeiro, Brazil.

Agroecology: The study of relationships between the environment and agricultural practice.

Agroforestry: An approach to rural land use that integrates cultivation of trees and shrubs with traditional agriculture (crop production) and animal husbandry, with the goal of providing households with fuel wood and other tree products while also controlling erosion and enhancing soil fertility.

Alar: Brand name for a chemical (daminozide) used on fruit, including apples, to enhance their color and regulate their growth; Alar was used in the United States from 1963 to 1989, but is now prohibited from use on food crops due to concerns that it is a carcinogen.

Algal bloom: Rapid increase in algae (rootless plants growing in sunlit waters), caused by an increase in nutrients in the water.

Alpine: The region in mountains that is below permanent snow but above the tree line.

Ambient air: Outdoor air in the lower atmosphere; measures of air pollution usually refer to the ambient air in a particular area.

Anaerobic: Characterized by an absence of oxygen.

Anthropogenic: Caused by humans; the term is often applied to changes in the natural world due to human activities.

AQCR: Air Quality Control Region, a concept defined by the U.S. Clean Air Act; an AQCR is a contiguous area within which air quality is relatively uniform.

Aquaculture: Raising fish or shellfish in an intense production environment (as opposed to fishing in the wild) similar to that employed in agriculture or livestock production; aquaculture implies individual or corporate ownership of the fish stocks and intervention in the production process (e.g., feeding, stocking).

Alien species: A species, usually a plant, which is not native to the area in which it lives; also called an exotic species or an introduced species.

Aquifer: An underground geologic formation containing water.

Arable land: Land suitable for cultivation of crops.

Arid zone: An area that receives less than 250mm annually of rain.

ASEAN haze: A visible aerosol haze common in the ASEAN (Association of Southeast Asian Nations) region, which occasionally (e.g., in August 1990 and September–October 1991) becomes so severe as to limit plant growth and require airports to shut down.

Assimilative capacity: The capacity of the natural environment to carry waste materials (e.g., pollutants) without adverse effect.

Atmosphere: The layer of gas surrounding the Earth; it consists primarily of nitrogen (78 percent) and oxygen (21 percent), and is primarily bound to the Earth by gravity.

Avoidance costs: Costs incurred in an effort to protect the environment from deterioration; these can be opportunity costs (from failing to engage in economic activities) or costs due to alternate production and consumption activities.

Backcasting: A method of working backward from a desired outcome to determine which policy measures would be required in the present to reach that future state.

Background level: The amount of a chemical (e.g., lead or arsenic) or of radiation normally present in a given environment.

Badlands: An area, usually without vegetation and with intricate erosion, in a semiarid or arid environment.

Bajada: Debris spread around mountain ranges by streams, generally in arid climates; often a bajada is formed by the coalescence of several alluvial fans near the mouth of a river.

Barrier island: An island parallel to a coastline, but mostly separate from it, and generally formed by sand deposited by wind and water actions.

BDR: Bycatch Reduction Device, a device used in fishing to prevent the capture of fish or other marine life that are not the focus of a fishing expedition; also called an "excluder device," e.g., a turtle excluder device (TED).

Benthic fish: Fish that live at the lowest depths (benthic zone) of a body of water; also called groundfish.

Bioavailability: The ease or difficulty with which a chemical is taken up by an organism; higher bioavailability means a substance is more likely to enter the food chain.

Biodegradable: Materials that can be broken down and absorbed into the environment; for instance, wood is biodegradable while most plastics are not.

Biodepletion: The mass extinction of a species.

Biodiversity: The condition of having a large number of different life forms in an environment, including plants, animals, and microorganisms.

Biological accumulation: The accumulation of harmful substances (e.g. pesticides, heavy metals) in a living organism.

Biocontrol: Biological pest control, the use of organisms rather than chemicals to control unwanted plants or animals.

Bioindicators: Plants or animals that are sensitive to, or particularly tolerant of, specific types of pollution; also called indicator organisms.

Biomass: The total weight of living organisms in an area or habitat; usually measured as dry weight and sometimes expressed per unit area.

Biome: A large ecological unit or region that is defined by its distinct environmental conditions, animal life, and vegetation.

Biosphere: All the living organisms and their environment on Earth, and/or the parts of the atmosphere and Earth that can support life.

Biosphere Reserve: A designation by the United Nations Educational, Scientific and Cultural Organization granted to specific areas (over 300 as of 2011) that are noted for their biodiversity, and are centers of research and environmental monitoring.

BOD: Biological Oxygen Demand, a measurement of the amount of oxygen in water consumed by microorganisms in the decomposition of organic matter. It is used as a gauge of pollution.

Brackish water: Water containing salts, but at a lower concentration than found in sea water; often defined as water having 0.035 ounces to 0.35 ounces (1,000–10,000 milligrams) of dissolved salts per 33.8 ounces (1 liter).

Broad-leaved: Plants that have thin flat leaves (e.g., many deciduous tress), as opposed to needles.

Brownfield: An abandoned site formerly used for industry or urban uses; often the term also implies contamination or perceived contamination resulting from the former use.

Bycatch: Marine life captured during fishing operations that is not the desired catch, and which is often discarded.

CAFE: Corporate Average Fuel Economy, a set of standards adopted in the United States in 1975 in response to the gas crisis caused by the oil embargo; CAFE requires that the fleet of vehicles sold by any manufacturer meet minimum fuel economy levels.

Cairo Plan: Recommendations passed in 1994 at the United Nations International Conference on Population and Development in Cairo, Egypt, for stabilizing the world's population through the provision of family planning services and improved educational opportunities for girls.

Carbon cycle: The natural process by which carbon is circulated among the land, ocean, atmosphere, and the biosphere.

Carbon dioxide: CO_2, a greenhouse gas that occurs naturally in the Earth's atmosphere, and whose concentration has increased substantially since the Industrial Revolution.

Carbon monoxide: CO, a gas produced by incomplete combustion of fossil fuels; CO is harmful to humans because it combines with hemoglobin in the bloodstream and reduces the blood's capacity to carry oxygen.

Carcinogen: A substance that causes cancer.

Carbon tax: A tax applied to hydrocarbon fuels (e.g., coal, petroleum) that release carbon dioxide in combustion.

Carrying capacity: The maximum number of a species, or of multiple species, that an area can support in the least favorable season.

CEE: The Consortium for Energy Efficiency, an organization of U.S. and Canadian energy efficiency program administrators.

CFCs: Chlorofluorocarbons, chemical compounds containing chlorine, fluorine, carbon, and possibly hydrogen; CFCs are human-made and used primarily as refrigerants, although their use has been restricted due to evidence that they damage the Earth's protective ozone layer.

CGIAR: The Consultative Group on International Agricultural Research, an organization founded to improve food production in developing countries.

Channelization: Deliberate manipulation of a river by, for instance, widening, deepening, straightening, or the construction of dikes and embankments.

Chemiepolitik: German for "chemical policy," a term coined in 1984 to refer to regulating the chemical industry to reduce environmental harm.

Chemosterilant: A pesticide that destroys the ability of pests to reproduce.

Chlorine: A chemical element (Cl) that is often used to disinfect water (e.g. for drinking and in swimming pools); it is a highly reactive halogen gas and is found in many common compounds, including table salt (sodium chloride, NaCl).

Chlorophyll: A green pigment found in plants that is essential for photosynthesis.

CITES: The Convention on International Trade in Endangered Species of Wild Fauna and Flora, an international agreement adopted in the 1970s that controls trade in endangered and threatened plant and animal species.

Clearcutting: A type of logging in which all the trees in an area are cut down at once.

Climate: Weather patterns occurring over long periods of time (e.g., decades or centuries) that encompass temperature, precipitation, atmospheric pressure, and other meteorological elements; climate is contrasted with weather, which refers to local atmospheric conditions at a given point in time.

Cloud forest: A forest growing in a mountainous area where cloud cover and fog occur regularly.

Cogeneration: The simultaneous production of electrical power and heat where excess heat from steam turbines is used for purposes such as heating.

Compound leaf: A leaf that is fully subdivided, with multiple leaflets that are distinct and are separated along a vein (primary or secondary).

Community: A group of organisms sharing an environment.

Contact pesticide: A pesticide that kills pests upon contact, rather than requiring ingestion.

Critical habitat: Habitat essential to the survival of a species; as used in the 1973 U.S. Endangered Species Act, critical habitat includes migration routes, breeding grounds, feeding grounds, and the area where the species lives.

Criteria pollutant: An air pollutant for which governmental standards have been set.

Critical load: In studies of pollution, an estimate of the safe level of exposure to a pollutant.

Crop rotation: The agricultural practice of growing different crops in succession on the same land, often to allow the soil to replenish.

Cyanobacteria: Protistans, also referred to as blue-green algae, that are found in water and soil, can fix nitrogen, and can photosynthesize.

DDT: Dichlorodiphenyltrichloroethane, a synthetic insecticide that accumulates in the food chain and was banned for agricultural use in 2001 by the Stockholm Convention.

Debt for nature swap: An arrangement in which national debt is reduced or canceled if a country agrees to enact specified environmental protections.

Deciduous: Plants that shed their leaves in non-growing seasons; the term is usually used in reference to shrubs and trees.

Defensive environmental costs: Costs incurred in protecting the environment, compensating for or repairing environmental deterioration, and preventing or neutralizing decreases in environmental quality.

Demand-Side Management: A method of meeting energy needs that focuses on reducing consumption levels (e.g., through economic incentives) rather than increasing supply.

Desertification: Degradation of dry, arid, or semi-arid land; contributors to desertification include natural environmental fluctuation and human activities.

DGSM: Dasholi Gram Swarjya Mandal, a movement founded in India in the 1960s by Chandi Prasad Bhatt; goals include resisting environmental degradation and exploitation, and forming a self-reliant rural society.

Dioxin: Any of a number of chemical compounds formed during manufacturing and incineration; dioxins are environmental pollutants and some are highly carcinogenic.

Dissolved oxygen: DO, the amount of gaseous oxygen present in water.

Drift net: A net used in commercial fishing that is suspended from surface floats and may stretch for miles across the ocean; drift nets are prone to bycatch and were banned in 1992 by the United Nations General Assembly.

Dryland farming: Farming practices developed to grow crops in low-rainfall areas without irrigation.

Duff: Partially and entirely decomposed vegetable matter composed of litter or humus, typically found on the forest floor.

Ecology: The study of the relationships among humans, other organisms and their environment.

Ecosystem: A community of organisms and their environment, including both living and nonliving elements, that function together, recycle nutrients, and maintain a flow of energy.

Ecotage: Sabotage in the service of an environmental cause, a term popularized by E. F. Abbey in *The Monkey Wrench Gang* (1975).

Ecotourism: Tourism based on access to natural resources such as coral reefs, caves, or scenic areas.

EIS: Environmental Impact Statement, a report examining the environmental consequences of planned activities; since the 1969 passage of the National Environmental Policy Act, U.S. federal agencies have been required to file an EIS for any major project.

El Niño: a climate event that often occurs in December (hence the name, which means "Christ Child" in Spanish); caused by shifts in Pacific Ocean currents, El Niño increases rain in some areas and decreases it in others, causes shifts in climate, and may impact human activity.

Endemic species: Species of plants and animals that occur exclusively in a single region (from a small area or island to an entire continent).

Ephemeral species: Species of plants and animals that live only a few months or weeks.

Epiphyte: Also known as an air plant, an epiphyte grows on another plant or physical structure non-parasitically using the host only for support.

EPR: Extended Producer Responsibility, a policy that requires the producers of products to also be responsible for postconsumer waste, thus creating incentives for recycling, use of environmentally friendly materials, etc.

Estuary: An inlet or bay in which seawater and freshwater mix; estuaries are often located at river mouths and may be important breeding or nursery grounds for marine life.

Exotic: An introduced or nonnative species.

Factory farming: Agriculture carried out on the model of a factory, at a large scale and with a high degree of mechanization.

Fauna: All the animal life of a particular time or geographic location.

Feral: An adjective applies to animals from a domesticated species that are living and breeding in the wild.

Floodplain: An area built up by river or stream deposition; floodplains are generally flat and subject to periodic flooding.

Flora: All the plant life existing at a particular time or in a particular geographic location.

Flyway: The paths taken by waterfowl and migratory birds when travelling between seasonal homes and their breeding grounds.

Food chain: A term describing the feeding sequence of a biological community, through which energy flows; food chains are described in terms of trophic levels, beginning with autotrophs (green plants), through primary consumers (herbivores, i.e., plant eaters), secondary consumers (carnivores that eat herbivores), tertiary consumers (carnivores that eat other carnivores), and detrivores (animals and microbes that eat dead organisms).

Fossil fuel: Any fuel produced by the natural decomposition of buried dead animal and plant materials; examples include coal, natural gas, and petroleum.

Fungicide: A product intended to kill fungi; effective fungicides include mercury and cadium, although because of their toxicity, they are less used today.

Gaia Hypothesis: A hypothesis by James Lovelock, first articulated in the 1970s, that all of Earth can be considered as a single living organism.

GEMS: The Global Environmental Monitoring System, a data collection project by the United Nations Environment Program begun in 1975; GEMS monitors changes in weather and climate, the health of plant and animal species, changes in soils, and the environmental impact of human activities on a global basis.

Genetic diversity: A component of biodiversity, genetic diversity refers to the amount of genetic variation in a species or population.

Geophyte: One of the life forms classified by Christen Raynkiaer; a plant that is perennial and has its bulb or corn in a protected location below the surface.

Geothermal energy: Energy fueled by naturally occurring heat sources located below the Earth's surface; geothermal energy is used for both direct heating and electricity generation.

Glacial till: Sediment deposited by a glacier, generally heterogenous and coarsely graded.

Grazing: The practice of allowing herbivores (e.g., sheep, cattle) to freely feed on grasses and plants, rather than being fed in feedlots.

GMO: Genetically Modified Organism, typically applied to plants that have been genetically engineered to be resistant to pests.

Green Revolution: A period beginning in the late 1950s in which crop yields in the Third World were dramatically improved through improvements in seed varieties, use of irrigation, and application of chemical pesticides.

Greenwash: A term combining "green" and "whitewash," used to suggest that an organization promotes an environmentally responsible image without making significant changes in business and industrial practices.

Groundwater: Water that exists below the surface of the Earth, and that feeds springs and wells.

Hectare: A unit of area, equivalent to 2.5 acres, used in the metric system.

Herbicide: A chemical used to kill unwanted plants, often weeds.

High seas: The parts of the ocean that are not under the jurisdiction of any country.

Homeostasis: A concept developed by Walter Bradford Cannon, explaining how an organism can maintain a fairly constant internal environment despite changes in the external environment.

Humus: Partially decayed plant matter.

Hydrofluorocarbons: Organic compounds used as refrigerants (replacing chlorofluorocarbons to some extent), cleaners, and solvents; they are believed to be less harmful to the ozone layer.

Hydroponics: Growing plants in water enriched by nutrients, rather than in soil.

Hydropower: Power (generally electricity) created by harnessing the energy of moving water.

Hypoxia: Lower than normal levels of oxygen; in water, hypoxia is associated with massive fishkills (e.g., as seen in the 1980s in the United States) as pollutants increase the nutrients available in the water, which increases algae growth, thus depleting the oxygen supply in the water and killing the fish.

ICPD: The International Conference on Population and Development, held in Cairo, Egypt, in 1994, where the steering document for the United Nations Population Fund was adopted.

ICS: The International Commission on Stratigraphy, a scientific body within the International Union of Geosciences.

Industrial species: Crops or domestic livestock bred for maximum productivity, with high inputs (e.g., fertilizer, water, feed) and genetic uniformity.

Introduced species: A species transported outside its natural area by humans; the transportation may be accidental or deliberate.

Intercropping: The practice of planting two or more crops in the same field, either in alternating rows or intermingled.

Intertropical Convergence Zone (ITC): The area on the globe where the Northern Hemisphere and Southern Hemisphere trade winds meet; the ITC moves with the annual migrations of direct solar radiation, and creates the typical alternation, in the topics, of dry and rainy seasons.

IPCC: Intergovernmental Panel on Climate Change, a joint program of the World Metereological Organization and the United Nations Environment Programme, founded in 1988 to assess and disseminate scientific information about climate change.

Keystone species: A species whose removal, it is believed, would cause drastic change in the environment, including the abundance of other species of plants and animals.

Landfill: A method of garbage disposal in which garbage is deposited, and ultimately covered with dirt.

Landings: In fishing, the catch that is brought to shore and to market.

Landrace: A domesticated species (plant or animal) that has developed largely through the influence of natural processes, rather than human intervention; members of landraces generally are more variable than members of breeds.

Leaching: The dissolution and transportation of chemicals by water.

LEED: Leadership in Energy and Environmental Design, a certification program to recognize buildings created using principles of sustainable design.

Liana: A vine, also called a climber, that is rooted in the ground but uses supports to get sunlight.

Liebling's Law of the Minimum: A principle developed by Justus Liebig in the mid-19th century, which states that the growth of a plant is limited by the relationship between the nutrients available and its needs.

Limnology: The branch of environmentalism concerned with studying inland waters, including lakes, ponds, rivers, and wetlands.

Longline fishing: A method of fishing using lines with hundreds of hooks on main and branch lines, and that may stretch for miles.

Loess: Sediment formed by the deposition of fine, wind-blown particles.

MAB: Man and the Biosphere, a global system of biosphere reserves organized by UNESCO (the United Nations Educational, Scientific and Cultural Organization).

Mass extinction: The large-scale disappearance of most or all living creatures; Earth has experienced several mass extinctions, the most severe during the Permian period (ca. 250 million years ago).

Minimata Disease: a condition caused by mercury poisoning from industrial pollution, identified in Minamata, Japan; symptoms include neurological disorders, retardation, and sensory disorders.

Mixing zone: The area in which wastewater is discharged into surface water.

Monkey-wrenching: Ecological sabotage (ecotage) to prevent actions the activist believes will cause environmental destruction; the term comes from Edward Abbey's 1975 novel *The Monkey Wrench Gang*.

Monoculture: The practice of cultivating only a single crop, often on large plantations.

Malthusian: Ideas based on the theories of Thomas Robert Malthus, a British economist working in the early 19th century, who argued that population growth would inevitably tend to outstrip the food supply if not checked by disease, famine, or war.

Methyl bromide: CH_3Br, a chemical compound used as an insecticide, but which is harmful to the ozone layer.

Multicropping: The agricultural practice of producing more than one crop in the same plot, whether in sequence or simultaneously.

Multiple use forestry: A method of managing forested areas to serve a variety of purposes, including wildlife habitat, timber production, recreational use, and soil protection.

Mycorrhizal association: A mutually beneficial relationship between the roots of a vascular plant and a fungus.

NAFTA: The North American Free Trade Agreement, an international accord enacted in 1994 between the United States, Canada, and Mexico; NAFTA includes a number of environmental agreements as well as its main concern, which is the elimination of tariffs.

Needle-leaved: Plants such as pines and spruce that have long, pointed leaves.

Neurotoxin: Metabolic poisons that disrupt or attack the nervous system; many pesticides, such as Malathion, DDT, and Parathion, are neurotoxins.

Nitrogen Oxide: NO_x, a term encompassing nitric oxide (NO) and nitrogen dioxide (NO_2), both of which are created by the combustion of fossil fuels, and both of which contribute to global warming.

Noncriteria pollutant: A pollutant for which no standards have been set.

Nonpoint source: Where the origin of pollution is diffuse and has no fixed location; examples include agricultural runoff and emissions from forest fires.

Nonrenewable resources: Natural resources, such as minerals and fossil fuels, which can be replaced only on a geological time scale.

Oil shale: Sedimentary rock containing kerogen, a solid organic substance that is a form of oil.

Open canopy: An arrangement in which the crowns of adjacent plants do not touch, allowing sunlight through the upper layer of a forest.

Over-grazing: The harmful practice of allowing livestock to graze on land too long and/or too intensively, so that the land cannot recover; over-grazing can lead to a number of ill effects, including eroding or desertification of the landscape.

Ozone: O_3, a molecule formed of three oxygen atoms; it is protective in the upper atmosphere, shielding the Earth from ultraviolet radiation, but a pollutant in the lower atmosphere.

Paleoecology: The study of ancient environments, including the interrelationships of plants, animals, and the environment.

PBDE: Polybrominated Diphenyl Ether, a type of chemical used as a flame retardant in consumer goods; in the 1990s PBDEs were discovered to be present in human breast milk.

Pelagic fish: Fish that live near or on the surface of bodies of water.

Permafrost: Soil, subsoil, or bedrock that remains continuously frozen for at least two years.

Persistent compound: A chemical compound that is slow to degrade in the environment and hence may tend to accumulate, causing environmental and health hazards; examples include lead, PCBs (polychlorinated biphenyls) and PAHs (polycyclic aromatic hydrocarbons).

Pesticides: Chemicals used to kill organisms, such as insects or weeds, in order to protect human life, agricultural products, etc.

PETA: People for the Ethical Treatment of Animals, an animal rights organization founded in 1980.

pH: A scientific measure of the alkalinity or acidity of a substance; pH is measured on a logarhythmic scale in which 7 is neutral, 0 is extremely acidic, and 14 is extremely alkaline.

Photosynthesis: A natural process in which plants, algae and bacteria use energy from the sun to convert oxygen and carbon dioxide to chemical energy (carbon compounds, particularly sugar).

Plate tectonics: A theory describing movement of the Earth's crust, including continental drift, spreading of the sea floor, and mountain upthrust.

PM10: Particulate matter smaller than 10 microns (10 micrometers, 10μ) in diameter.

ppb: Parts per billion.

ppm: Parts per million.

ppt: Parts per trillion.

Point source: A discrete and fixed origin of pollution emissions, such as a pipe or smokestack.

Positional good: A good whose value depends on scarcity or exclusivity; examples include wilderness, fame, and luxury goods.

Prescribed burning: Burning used to prevent uncontrolled wildfires or to maintain desirable environmental conditions; prescribed burning is often practiced in forests, prairies, and savannas.

PSI: The Pollutant Standard Index, a scale ranging from 0 to 500 that classifies air quality based on the concentration of various air pollutants.

Radon: Rn, a radioactive, water-soluble gas formed by the disintegration of radium; radon can leach into a watershed and can contaminate homes.

Reef: A feature below sea level; coral reefs are the best known, but the term also applies to rocks, sandbars, and even human constructions such as sunken ships.

Reef fish: Fish that live in or near coral reefs, as opposed to open waters.

Raunkiaer plant life form: A system of classification developed by the Danish naturalist Christen C. Raunkiaer for plants.

RDF: Refuse-Derived Fuel, for example, fuel derived from discarded materials, such as municipal solid waste.

Restoration ecology: The practice of returning a damaged ecosystem to its prior state; for instance, to return abandoned midwestern farmland to its prairie state.

Rhizobia: Paraphyletic bacteria that establish themselves within nodules on the roots of some plants and fix nitrogen to the plant.

Rhizome: A type of root structure that is primarily horizontal.

Riparian zone: The area between a river or stream and the land.

Sagebrush Rebellion: A movement in the western United States, from the 1970s to the present day, to resist federal attempts to manage federal lands; disputes include matters such as grazing fees and restrictions on mineral extraction.

Savanna: An ecosystem characterized by unbroken grassland and an open canopy of trees that allows sunlight to reach the ground.

Second growth forest: A forest that has been harvested once and has grown back.

Selection cutting: A method of forestry in which mature trees are harvested in small groups or individually, leaving gaps in the canopy that allows understory trees to develop; selection cutting protects against soil erosion and offers seedlings protection against the wind and the sun.

Shifting cultivation: A method of farming in which a tract of land alternates between cultivation and disuse; when not cultivated, the land is regenerated through the growth of native plants.

Smog: Fog containing pollutants such as ash, soot, carbon dioxide and sulfur dioxide; smog is irritating to human lungs and is implicated in some respiratory diseases.

Soot: An airborne contaminant created during combustion of fossil fuels.

Stockholm Conference: The United Nations Conference on the Human Environment, held in 1972 in Stockholm, Sweden; it was the first major conference held by the U.N. on international environmental issues.

Straddling stock: Fish that move between the high seas (international waters) and territorial waters (waters controlled by a nation).

Subsoil: Soil lying beneath the surface soil; subsoils are generally not as productive as surface soils, because they tend to contain more clay and less organic matter.

Succession: A theoretical description of how a plant community develops over time; the process begins with a barren site, then a series of communities, and finally ending with a stable, persistent community (known as the climax).

Succulent: A type of plant typical in deserts and saline habitations that has the ability to store water in its leaves, stem, roots, etc.

Sustainable yield: The amount of a resource that can be harvested without harming the capacity of the ecosystem to produce an equal amount in the future.

Taxon: A single level of a taxonomy; the plural is taxa.

Teratogen: A substance that can cause abnormalities in a developing organism; examples include the drug thalidomide, ionizing radiation, the rubella virus, and alcohol.

Tipping point: The point at which steady change in an ecosystem begins to accelerate, and may become uncontrollable.

Toxicology: The study of the adverse effects of chemicals on humans and other living organisms.

Tragedy of the Commons: A phrase popularized by Garrett Hardin, referring to the tendency of individuals to exploit common holdings (e.g., pasture) for their own benefit, without regard for the common interest; Hardin's theory has been challenged by scholars who point to instances where voluntary agreements have allowed numerous individuals to share common holdings without conflict or environmental degradation.

Transboundary pollution: Pollution that crosses national borders; one example is the radioactive contamination from the Chernobyl nuclear plant.

Trawls: Fishing nets that have a wide mouth tapering to a pointed end, and are towed behind a vessel.

UNCLOS: United Nations Conferences on the Law of the Sea, several international meetings (first held in 1958) that addressed economic and environmental issues regarding fishing and other maritime activities.

UNEP: The United Nations Environment Programme, created by the United Nations General Assembly in 1972.

UNFPA: The United Nations Fund for Population Activities, founded in 1969.

Urban heat island: An urban area that is substantially warmer than surrounding areas.

Vadose zone: In geology, an unsaturated zone between the water table and the land surface.

Virgin forest: A forest that has never been logged.

VOC: Volatile Organic Compounds, carbon-based compounds that generally exist only in the gaseous state in the atmosphere due to vapor pressures above 10^{-2} kiloPascals.

Weathering: Part of the process of soil creation, in which rocks are broken down into particles by natural forces such as wind and water.

Wise use: A term used by Gifford Pinchot in 1910 to describe scientific management of natural resources. Wise use has become associated in the United States with the desire to exploit natural resources (forests, mines, etc.) without the oversight of governmental environmental regulation.

Sarah Boslaugh

Resource Guide

Books

Adams, W. M., A. S. Goudie, and A. R. Orme. *The Physical Geography of Africa*. New York: Oxford University Press, 1996.

Alongi, Daniel M. *The Energetics of Mangrove Forests*. Dordrecht, Netherlands: Springer, 2009.

Andel, Jelte van and James Aronson, eds. *Restoration Ecology: The New Frontier*. Oxford, UK: Blackwell Publishing, 2006.

Armstrong, Patrick. *Darwin's Other Islands*. New York: Continuum, 2004.

Bennett, Michael and David W. Teague, eds. *The Nature of Cities: Ecocriticism and Urban Environments*. Tucson: University of Arizona Press, 1999.

Black, K. S., D. M. Paterson, and A. Cramp, eds. *Sedimentary Processes in the Intertidal Zone*. London: The Geological Society, 1998.

Boomgaard, P. *Southeast Asia: An Environmental History*. Santa Barbara, CA: ABC-CLIO, 2007.

Breckle, Siegmar-Walter. *Walter's Vegetation of the Earth: The Ecological Systems of the Geo-Biosphere*, transl. and rev. by Gudrun Lawlor and David Lawlor. 4th ed. New York: Springer, 2002.

Buckles, Mary Parker. *Margins: A Naturalist Meets Long Island Sound*. New York: North Point Press, 1997.

Catron, Jean-Luc E., Gerardo Ceballos, and Richard Stephen Feiger, eds. *Biodiversity, Ecosystems, and Conservation in Northern Mexico*. New York: Oxford University Press, 2004.

Clewell, Andrew F. and James Aronson. *Ecological Restoration: Principles, Values, and Structure of an Emerging Profession*. Washington, DC: Island Press, 2007.

Corlett, Richard T. and Richard B. Primack. *Tropical Rain Forests: An Ecological and Biogeographical Comparison*. Hoboken, NJ: Wiley-Blackwell, 2011.

Crul, R. C. M. *Management and Conservation of the African Great Lakes: Lakes Victoria, Tanganyika, and Malawi: Comparative and Comprehensive Study of Great Lakes*. Paris: UNESCO Publishing, 1999.

Cumbler, John T. *Northeast and Midwest United States: An Environmental History*. Santa Barbara, CA: ABC-CLIO, 2005.

Davis, Donald E., et al. *Southern United States: An Environmental History*. Santa Barbara, CA: ABC-CLIO, 2006.

Dovers, Stephen, Ruth Edgecombe, and Bill Guest, eds. *South Africa's Environmental History: Cases and Comparisons.* Athens: Ohio University Press, 2003.

Dubinsky, Zvy and Noga Stambler, eds. *Coral Reefs: An Ecosystem in Transition.* New York: Springer, 2011.

Dumont, Henri J., ed. *The Nile: Origin, Environments, Limnology, and Human Use.* Dordrecht, Netherlands: Springer, 2009.

Eisma, Doeke. *Intertidal Deposits: River Mouths, Tidal Flats, and Coastal Lagoons.* Boca Raton, FL: CRC Press, 1998.

Elkind, Sara S. Bay *Cities and Water Politics: The Battle for Resources in Boston and Oakland.* Lawrence: University Press of Kansas, 1998.

Farndon, John. *Atlas of Oceans: An Ecological Survey of Underwater Life.* New Haven, CT: Yale University Press, 2011.

Fennell, David A. *Ecotourism.* 3rd ed. New York: Routledge, 2008.

Fritsch, Albert J. *Ecotourism in Appalachia: Marketing the Mountains.* Lexington: Unversity Press of Kentucky, 2004.

Garden, Donald S. *Australia, New Zealand, and the Pacific: An Environmental History.* Santa Barbara, CA: ABC-CLIO, 2005.

Goldsmith, F. B., ed. *Tropical Rain Forest: A Wider Perspective.* New York: Chapman & Hall, 1998.

Goldschmidt, Tijs. *Darwin's Dreampond: Drama in Lake Victoria*, trans. Sherry Marx-Macdonald. Cambridge, MA: MIT Press, 1996.

Golley, Frank B. *A Primer for Environmental Literacy.* New Haven, CT: Yale University Press, 1998.

Goodman, Steven M. and Jonathan P. Benstead, eds. *The Natural History of Madagascar.* Chicago: University of Chicago Press, 2003.

Goodwin, Edward J. *International Environmental Law and the Conservation of Coral Reefs.* New York: Routledge, 2011.

Grayson, Donald K. *The Great Basin: A Natural Prehistory.* Berkeley: University of California Press, 2011.

Haslam, Sylvia Mary. *The Riverscape and the River.* Cambridge, UK: Cambridge University Press, 2008.

Hawksworth, David L. and Alan T. Bull, eds. *Forest Diversity and Management.* Dordrecht, Netherlands: Springer, 2006.

Hill, Christopher V. *South Asia: An Environmental History.* Santa Barbara, CA: ABC-CLIO, 2008.

Hill, Michael J. and Niall P. Hanan. *Ecosystem Function in Savannas: Measurement and Modeling at Landscape to Global Scales.* Boca Raton, FL: CRC Press, 2011.

Holst-Warhaft, Gail and Tammo Steenhis, eds. *Losing Paradise: The Water Crisis in the Mediterranean.* Burlington, VT: Ashgate, 2010.

Honey, Martha. *Ecotourism and Sustainable Development: Who Owns Paradise?* Washington, DC: Island Press, 2008.

Hughest, J. Doald. *The Mediterannean: An Environmental History.* Santa Barbara, CA: ABC-CLIO, 2005.

Kappelle, M., ed. *Ecology and Conservation of Neotropical Montane Oak Forests.* New York: Springer, 2006.

Kole, Chittranjan. *Forest Trees.* Berlin, Germany: Springer-Verlag GmbH, 2007.

Koster, Eduard A., ed. *The Physical Geography of Western Europe.* New York: Oxford University Press, 2005.

Lal, Rattan and B. A. Stewart, eds. *World Soil Resources and Food Security.* Boca Raton, FL: CRC Press, 2012.

Larkum, Anthony W. D., Robert J. Orth, and Carlos M. Duarte, eds. *Seagrasses: Biology, Ecology, and Conservation.* Dordrecht, Netherlands: Springer, 2006.

Laurence, William F. and Carlos A. Peres, eds. *Emerging Threats to Tropical Forests.* Chicago: University of Chicago Press, 2006.

Maddox, Gregory. *Sub-Saharan Africa: An Environmental History.* Santa Barbara, CA: ABC-CLIO, 2006.

Manning, Richard. *Grassland: The History, Biology, Politics, and Promise of the American Prairie.* New York: Viking, 1995.

Marzluff, John, Reed Bowman, and Roarke Donnelly, eds. *Avian Ecology and Conservation in an Urbanizing World.* Boston: Kluwer Academic Publishers, 2001.

McIntyre, Alasdair D., ed. *Life in the World's Oceans: Diversity, Distribution, and Abundance*. Chichester, UK: Blackwell Publishing, 2010.

Mistry, Jayalaxshmi and Andrea Berardi, eds. *Savannas and Dry Forests: Linking People With Nature*. Burlington, VT: Ashgate, 2006.

Mortimer, Clifford Hiley. *Lake Michigan in Motion: Responses of an Inland Sea to Weather, Earth-Spin, and Human Activities*. Madison: University of Wisconsin Press, 2004.

Müller, Norbert, Peter Were, and John G. Kelcey, eds. *Urban Biodiversity and Design*. Hoboken, NJ: Wiley-Blackwell, 2010.

Nelson, Darby. *For Love of Lakes*. East Lansing: Michigan State University Press, 2012.

Ogden, Laura. *Swamplife: People, Gators, and Mangroves Entangled in the Everglades*. Minneapolis: University of Minnesota Press, 2011.

Oliveira, Paulo S. and Robert J. Marquis, eds. *The Cerrados of Brazil: Ecology and Natural History of a Neotropical Savanna*. New York: Columbia University Press, 2002.

Orme, Antony R., ed. *The Physical Geography of North America*. New York: Oxford University Press, 2002.

Osaki, Mitsuru, Ademola K. Braimoh, and Ken'ichi Nakagami, eds. *Designing our Future: Local Perspectives on Bioproduction, Ecosystems, and Humanity*. New York: United Nations University Press, 2011.

Purser, Bruce H., and Dan W.J. Bosence. *Sedimentation and Tectonics in Rift Basins: Red Sea—Gulf of Aden*. New York: Chapman & Hall, 1998.

Ranganathan, Janet, Mohan Munasinghe, and Frances Irwin, eds. *Policies for Sustainable Governance of Global Ecosystem Services*. Northampton, MA: Edward Elgar, 2008.

Richardson, Curtis J. *The Everglades Experiments: Lessons for Ecosystem Restoration*. New York: Springer, 2008.

Schumm, Stanley Alfred. *River Variability and Complexity*. New York: Cambridge University Press, 2005.

Serreze, Mark C. and Roger G. Barry. *The Arctic Climate System*. New York: Cambridge University Press, 2005.

Shahgedanova, Maria, ed. *The Physical Geography of Northern Eurasia*. New York: Oxford University Press, 2003.

Sherow, James E. *The Grasslands of the United States: An Environmental History*. Santa Barbara, CA: ABC-CLIO, 2007.

Sowards, Adam M. *United States West Coast: An Environmental History*. Santa Barbara, CA: ABC-CLIO, 2007.

Spalding, Mark. *A Guide to the Coral Reefs of the Caribbean*. Berkeley: University of California Press, 2004.

Talling, John Francis. *Ecological Dynamics of Tropical Inland Waters*. New York: Cambridge Press, 1998.

Veblen, Thomas T., Robert S. Hill, and Jennifer Read, eds. *The Ecology and Biogeography of Nothofagus Forests*. New Haven, CT: Yale University Press, 1996.

Webersik, Christian. *Climate Change and Security: A Gathering Storm of Global Challenges*. Westport, CT: Praeger, 2010.

West, Philip W. *Growing Plantation Forests*. Berlin, Germany: Springer-Verlag, 2006.

Wetzel, Robert G. *Limnology: Lake and River Ecosystems*. 3rd ed. San Diego, CA: Academic Press, 2001.

Whited, Tamara L. *Northern Europe: An Environmental History*. Santa Barbara, CA: ABC-CLIO, 2005.

Whitmore, Timothy Charles. *An Introduction to Tropical Rain Forests*. New York: Oxford University Press, 1998.

Wickens, Gerald E. *The Baobabs: Pachycauls of Africa, Madagascar, and Australia*. Dordrecht, Netherlands: Springer Science + Business Media, 2007.

Wieser, Gerhard and Michael Tausz, eds. *Trees at their Upper Limit: Treelife Limitation at the Alpine Timberline*. Dordrecht, Netherlands: Springer, 2007.

Wohl, Ellen E. *A World of Rivers: Environmental Change on Ten of the World's Great Rivers*. Chicago: University of Chicago Press, 2010.

Wright, L. Donelson. *Morphodynamics of Inner Continental Shelves.* Boca Raton, FL: CRC Press, 1995.

Wyn, Graeme. *Canada and Arctic North America: An Environmental History.* Santa Barbara, CA: ABC-CLIO, 2007.

Journals

Advances in Ecological Research
African Journal of Ecology
African Journal of Environmental Science and Technology
African Journal of Plant Science
Agricultural and Forest Entomology
Agriculture, Ecosystems and Environment
American Fern Journal
American Journal of Botany
American Midland Naturalist
American Naturalist
Animal Behaviour
Animal Biodiversity and Conservation
Animal Conservation
Annals of the Missouri Botanical Garden
Annual Review of Entomology
Annual Review of Environment and Resources
Annual Review of Marine Science
Annual Review of Plant Biology
Antarctic Science
Applied Ecology and Environmental Research
Aquatic Ecology
Aquatic Ecosystem Health and Management
Aquatic Insects
Aquatic Living Resources
Aquatic Microbial Ecology
Arctic, Antarctic, and Alpine Research
Arctic Journal
Arid Land Research and Management
Asia Pacific Journal of Climate Change
Asian Journal of Plant Sciences
Australian Journal of Botany
Australian Journal of Soil Research
Australian Journal of Zoology
Australian Mammalogy
Avian Conservation and Ecology
Behavioral Ecology and Sociobiology
Biochemical Systematics and Ecology
Biodiversity and Conservation
Biodiversity Informatics
Biodiversity Science
Biological Conservation
Biological Diversity and Conservation
Biology Bulletin
BIOTROPICA
Bird Conservation International
Botanical Journal of Scotland
Botanical Review
Botany
Botany Research Journal
Bulletin of Entomological Research
Bulletin of the Peabody Museum of Natural History
Canadian Entomologist
Canadian Journal of Fisheries and Aquatic Sciences
Canadian Journal of Forest Research
Canadian Journal of Zoology
Community Ecology
Condor: A Magazine of Western Ecology
Conservation
Conservation and Society
Contemporary Problems of Ecology
Copeia
Coral Reefs
Ecological Abstracts
Ecological and Environmental Anthropology
Ecological Restoration
Ecology
Ecology and Society
Ecology Letters
Ecology of Freshwater Fish
Écoscience
Ecosystem Health
Ecosystems
Edinburgh Journal of Botany
Endangered Species Research
Entomological Research
Environment, Development and Sustainability
Environment International
Environmental and Ecological Statistics
Environmental Chemistry
Environmental Conservation
Environmental Management
Environmental Modelling and Assessment
Environmental Monitoring and Assessment

Environmental Practice
Environmental Research
Environmental Research Journal
Environmentalist
Estuarine, Coastal and Shelf Science
European Journal of Forest Research
Evolution
Evolutionary Ecology Research
Forest Ecology and Management
Forest Science
Forestry
Freshwater Biology
Geobiology
Global Environmental Change
Global Journal of Environmental Research
Harvard Papers in Botany
Human Ecology
Ibis
ICES Journal of Marine Science
Inland Water Biology
Insect Conservation and Diversity
Integrated Environmental Assessment and Management
International Environmental Agreements: Politics, Law and Economics
International Forestry Review
International Journal of Biodiversity and Conservation
International Journal of Biodiversity Science, Ecosystem Services and Management
International Journal of Ecology
International Journal of Ecology and Development
International Journal of Ecology and Environmental Sciences
International Journal of Forestry Research
International Journal of Limnology
International Journal of Plant Sciences
International Journal of Soil Science
International Journal of Sustainable Development & World Ecology
Invasive Plant Science and Management
Journal of Applied Ecology
Journal of Arachnology
Journal of Arid Environments
Journal of Avian Biology
Journal of Biodiversity and Ecological Sciences
Journal of Botany
Journal of East African Natural History
Journal of Ecology
Journal of Ecology and the Natural Environment
Journal of Environment and Development
Journal of Environmental Biology
Journal of Environmental Monitoring
Journal of Environmental Sciences
Journal of Ethnobiology
Journal of Evolutionary Biology
Journal of Forest Research
Journal of Great Lakes Research
Journal of Horticulture and Forestry
Journal of Human Ecology
Journal of Mammalogy
Journal of Marine Biology
Journal of Mediterranean Ecology
Journal of Natural and Environmental Sciences
Journal of Natural History
Journal of Nature Conservation
Journal of Oceanography
Journal of Ornithology
Journal of Paleolimnology
Journal of Plankton Research
Journal of Plant Ecology
Journal of Rural Studies
Journal of Sea Research
Journal of Shellfish Research
Journal of the American Water Resources Association
Journal of the Limnological Society of Southern Africa
Journal of the Marine Biological Association of the United Kingdom
Journal of the Torrey Botanical Society
Journal of Tropical Ecology
Journal of Wildlife Management
Journal of Zoological Systematics and Evolutionary Research
Journal of Zoology
Limnology
Madroño
Mammal Review
Mammalian Biology
Mammalian Species
Marine and Freshwater Research
Marine Biology
Microbial Ecology

Molecular Ecology
Molluscan Research
Mountain Research and Development
Mycology
Natural Resources Forum
Nature
Nature Climate Change
New Forests
New Zealand Journal of Botany
New Zealand Journal of Ecology
Northeastern Naturalist
Northwest Science
Open Ecology Journal
Organisms Diversity and Evolution
Ornithological Science
Oryx
Pacific Science
Pan-Pacific Entomologist
Plant and Soil
Plant Biology
Plant Biosystems
Plant, Cell and Environment
Plant Ecology
Plant Ecology and Diversity
Plant Ecology and Evolution
PLoS Biology
Polar Biology
Politics and the Life Sciences
Primates
Proceedings of the National Academy of Sciences
Radiation and Environmental Biophysics
Rangeland Ecology and Management
Rangeland Journal
RAP Bulletin of Biological Assessment
Renewable Agriculture and Food Systems
Restoration Ecology
Scandinavian Journal of Forest Research
Science
Science of the Total Environment
Scientific American
Small-Scale Forestry
South African Forestry Journal
Southeastern Naturalist
Southern Forests
Southern Hemisphere Forestry Journal
Southwestern Naturalist
Studies on Neotropical Fauna and Environment
Sustainability Science
Tree-Ring Research
Tropical Ecology
Urban Ecosystems
Vadose Zone Journal
Water and Environment Journal
Water Research
Water Resources Management
Water Science and Technology
Waterbirds
Weed Science
Western North American Naturalist
Wetlands
Wetlands Ecology and Management
Wildlife Research
Wilson Journal of Ornithology
World Journal of Fish and Marine Sciences
Zoology

Internet

American Fisheries Society
http://www.fisheries.org/afs
The Atlas of Canada (Natural Resources Canada)
http://atlas.nrcan.gc.ca/auth/english/learning resources/facts/lakes.html
Australian Government. Environment and Natural Resources
http://australia.gov.au/topics/environment-and-natural-resources
Bay of Bengal Large Marine Ecosystem Project
http://www.boblme.org
Coral Reef Alliance
http://www.coral.org
The Encyclopedia of Earth
http://www.eoearth.org
Environmental Literacy Council
http://www.enviroliteracy.org
European Atlas of the Seas
http://ec.europa.eu/maritimeaffairs/atlas/index_en.htm
European Commission. Fisheries.
http://ec.europa.eu/fisheries/index_en.htm
European Commission Joint Research Centre, Institute for Environment and Sustainability
http://efdac.jrc.ec.europa.eu

European Environment Agency.
European Waters—Overview
http://www.eea.europa.eu/themes/water/european-waters
European Forest Institute
http://www.efi.int/portal
Fisheries and Oceans Canada
http://www.dfo-mpo.gc.ca/index-eng.htm
Forest Europe: Ministerial Conference on the Protection of Forests in Europe
http://www.foresteurope.org
Great Lakes Information Network
http://www.great-lakes.net
International Boreal Conservation Campaign
http://www.interboreal.org
International Ecotourism Society
http://www.ecotourism.org/site/
International Society of Limnology
http://www.limnology.org
National Aeronautics and Space Administration.
NASA Science: Earth
http://science.nasa.gov/earth-science
National Geographic. Environment.
http://environment.nationalgeographic.com/environment
National Oceanic and Atmospheric Administration: NOAA Coral Reef Conservation Program
http://coralreef.noaa.gov
Natural Geography in Shore Areas Project
http://www.nagisa.coml.org
Natural Resources Canada
http://www.nrcan.gc.ca/home
The Oceanography Society
http://tos.org/oceanography
Rainforest Alliance
http://www.rainforest-alliance.org
Seagrass Watch
http://www.seagrasswatch.org/home.html
Tropical Forest Foundation
http://www.tropicalforestfoundation.org
U.S. Environmental Protection Agency.
Our Waters
http://water.epa.gov/type
U.S. Fish and Wildlife Survey:
National Wetlands Inventory
http://www.fws.gov/wetlands
U.S. Geological Survey
http://www.usgs.gov
UNESCO World Heritage List
http://www.seagrasswatch.org/home.html
United Nations Environment Programme
http://www.unep.org
University of California Museum of Paleontology.
The World's Biomes
http://www.ucmp.berkeley.edu/exhibits/biomes
Washington State Department of Ecology.
Saving Puget Sound
http://www.ecy.wa.gov/puget_sound/index.html
World Biomes
http://worldbiomes.com
World Lakes
http://www.worldlakes.org/index.asp
World Wildlife Foundation. Science
http://www.worldwildlife.org/science/index.html

Sarah Boslaugh

Index

Index note: Text and page numbers in **boldface** refer to main topics

B

D

F

Q

R

X

Y

Z